RECENT ADVANCES
IN
PRIMATOLOGY

Volume Three
EVOLUTION

RECENT ADVANCES IN PRIMATOLOGY

RECENT ADVANCES
IN
PRIMATOLOGY

Volume Three
EVOLUTION

Edited by

D. J. CHIVERS

*Sub-Department of
Veterinary Anatomy,
University of Cambridge,
Cambridge, U.K.*

K. A. JOYSEY

*Department of Zoology,
University of Cambridge,
Downing Street,
Cambridge, U.K.*

1978

ACADEMIC PRESS
London . New York . San Francisco

A Subsidiary of Harcourt Brace Jovanovich, Publishers

ACADEMIC PRESS INC. (LONDON) LTD.
24/28 Oval Road,
London NW1

United States Edition published by
ACADEMIC PRESS INC.
111 Fifth Avenue
New York, New York 10003

Library of Congress Catalog Card Number: 77-75366
ISBN: 0-12-173303-3

Printed in Great Britain by
Galliard (Printers) Ltd, Great Yarmouth

LIST OF CONTRIBUTORS

Andrews, P.J., British Museum (Natural History), Cromwell Road, South
  Kensington, London SW7 5BD, UK.
Basmajian, J.V., McMaster University, Hamilton, Ontario, Canada.
Beard, J.M., Department of Anthropology, University College London, Gower
  Street, London WC1E 6BT, UK.
Bilsborough, A., Department of Physical Anthropology, University of
  Cambridge, Pembroke Street, Cambridge, UK.
Brandon-Jones, D., Department of Dental Anatomy, Guy's Hospital Medical
  School, London Bridge, London SE1 9RT, UK.
Brockelman, W.Y., Department of Biology, Faculty of Science, Mahidol
  University, Rama VI Road, Bangkok, Thailand.
Brush, A.H., Biological Sciences Group, University of Connecticut, Storrs,
  Connecticut 06268, USA.
Butler, H., Department of Anatomy, University of Saskatchewan, Saskatoon,
  Saskatchewan, Canada.
Cartmill, M., Department of Anatomy, Duke University Medical Center, Durham,
  N.C. 27710, USA.
Charles-Dominique, P., Laboratoire d'Ecologie Generale, Museum National
  d'Histoire Naturelle, 4 Avenue du Petit Chateau, Brunoy 91800, France.
Chiarelli, B., Instituto di Antropologia, via Accademia Albertina 17, 10123
  Torino, Italy.
Clark, Anne B., Primate Behavior Research Group, Department of Psychology,
  University of Witwatersrand, Johannesburg, South Africa.
Conroy, G.C., Department of Cell Biology, School of Medicine, New York
  University, 550 First Avenue, New York, N.Y. 10016, USA.
Cook, C.N., Department of Anthropology, University College London, Gower
  Street, London WC1E 6BT, UK.
Cramer, D.L., Zoo, W. 79th Street, New York, N.Y. 10024, USA.
Cronin, J.E., Department of Anthropology, New York University, New York,
  N.Y., USA.
Day, M., Department of Anatomy, St. Thomas's Hospital Medical School,
  London SE1 7EH, UK.
Doyle, G.A., Department of Psychology, Witwatersrand University, Johannesburg,
  South Africa.
Etter, H.-U. F., Anthropologisches Institut, Universität Zürich, Künstlergasse
  15, 8001, Zürich, Switzerland.
Fink, B.R., Anesthesiology Research Center, School of Medicine, University of
  Washington, Seattle, Washington 98195, USA.
Frederickson, E.L., Woodruff Medical Center, Emory University, Atlanta,

Georgia 30322, USA.

Gingerich, P.D., Museum of Paleontology, The University of Michigan, Ann Arbor, Michigan 48109, USA.

Gittins, S.P., Sub-Department of Veterinary Anatomy, University of Cambridge, Tennis Court Road, Cambridge CB2 1QS, UK.

Goffart, M., Institute Léon Fredericq, University of Liège, 17 Place Delcour, B.4000 Liège, Belgium.

Goodman, M., Department of Anatomy, Wayne State University, 540 E. Canfield Avenue, Detroit, Michigan 48201, USA.

Hewett-Emmett, D., Department of Biochemistry, University of Bristol, Medical School, University Walk, Bristol BS8 1TD, UK.

Ishida, H., Faculty of Human Sciences, Osaka University, 286-1 Ogawa, Yamada, Suita, Osaka, Japan (new address).

Jelinek, J., Moravske Museum, Nam. 25, Unora 7, Brno, Czechoslovakia.

Jerison, H.J., Neuropsychiatric Institute, The Center for the Health Sciences, University of California, 760 Westwood Plaza, Los Angeles, California 90024, USA.

Joysey, K.A., University Museum of Zoology, Downing Street, Cambridge, UK.

Kay, R.F., Department of Anatomy, Duke University Medical Center, Durham, N.C. 27710, USA.

Khajuria, H., Eastern Regional Station, Zoological Survey of India, Risa Colony, Shillong 793003, Meghalaya, India.

Kimura, T., Department of Legal Medicine, Teikyo University School of Medicine, 11-1 Kaga, 2 Chome, Itabashi-Ku, Tokyo 173, Japan.

Kortlandt, A., Psychologie en Ethologie der Dieren, Universiteit van Amsterdam, Nieuwe Prinsengracht 126, Amsterdam 1004, The Netherlands.

Luckett, W.P., Department of Anatomy, School of Medicine, Creighton University, Omaha, Nebraska 68178, USA.

McArdle, J., Department of Anatomy, University of Chicago, 1025 E. 57th Street, Chicago, Illinois 60637, USA.

McGeorge, L.W., Duke University, Department of Zoology, Durham, N.C. 27710, USA.

MacKinnon, J., Sub-Department of Veterinary Anatomy, Tennis Court Road, Cambridge CB2 1QS, UK.

Manfredi-Romanini, M.G., Istituto di Istologia, Embriologia e Antropologia, Pavia, Italy.

Malmgren, Linda A., University of Connecticut, College of Liberal Arts & Sciences, The Biological Sciences Group, Storrs, Connecticut 06268, USA.

Martin, R.D., Wellcome Institute of Comparative Physiology, The Zoological Society of London, Regent's Park, London NW1 4RY, UK.

Moor-Jankowski, J., New York University Medical Center, 550 First Avenue, New York, N.Y. 10016, USA.

Oxnard, C.E., Department of Anatomy, The University of Chicago, 1025 E. 57th Street, Chicago, Illinois 60637, USA.

Pages, Elisabeth, Laboratoire d'Ecologie, MNHN, 4 Avenue du Petit Chateau, Brunoy 91800, France.

Passingham, R.E., Department of Psychology, University of Oxford, South Parks Road, Oxford, UK.

Prasad, K.N., Mukarramjahi Road, Manoranjan Building, 5-5-449 Hyderabad, 500001 AP, India.

Preuschoft, H., Ruhr-Universität Bochum, Institut für Anatomie, Arbeitsgruppe für Funktionelle Morphologie, 463 Bochum, Universitätsstrasse 150, Gebäude Ma. 01/551, Postfach 2148, West Germany.

Rijksen, H.D., Zoological Laboratory, State University, Gröningen, Netherlands.

List of Contributors

Romero-Herrera, A.E., Department of Anatomy, School of Medicine, Gordon H. Scott Hall of Basic Medical Sciences, 540 E. Canfield Avenue, Wayne State University, Detroit, Michigan 48201, USA.

Sakka, M., 31 rue Erlanger, 75016 Paris, France.

Sarich, V.M., Department of Biochemistry, University of California, Berkeley, California 94720, USA.

Schwartz, J.H., Department of Anthropology, University of Pittsburgh, Pittsburgh, Pa. 15260, USA.

Scheffrahn, W., Anthropologisches Institut der Universität, Künstlergasse 15, 8001 Zürich, Switzerland.

Schwegler, Thea, Anthropologisches Institut der Universität, Künstlergasse 15, 8001 Zürich, Switzerland.

Siegel, M.I., Department of Anthropology, University of Pittsburgh, Pittsburgh, Pa. 15260, USA.

Sirianni, Joyce E., Department of Anthropology, State University of New York at Buffalo, 4242 Ridge Lea Road, Buffalo, N.Y. 14226, USA.

Socha, W.W., New York University Medical Center, 550 First Avenue, New York, N.Y. 10016, USA.

De Stefano, G., Istituto di Istologia, Embriologia e Antropologia, Piazza Botto 10, Pavia, Italy.

Stringer, C.B., Sub-Department of Anthropology, British Museum (Natural History), Cromwell Road, London SW7 5BD, UK.

Susman, R.L., Department of Anatomical Sciences, Health Sciences Center, State University of New York at Stony Brook, Stony Brook, New York 11794, USA.

Sussman, R.W., Department of Anthropology, Washington University, St. Louis, Mo. 63130, USA.

Szalay, F.S., Department of Anthropology, Hunter College, CUNY, New York, N.Y. 10021, USA.

Tattersall, I., Department of Anthropology, American Museum of Natural History, Central Park West at 79th Street, New York, N.Y. 10024, USA.

Tobias, P.V., Medical School, Hospital Street, Johannesburg 2001, South Africa.

Trinkaus, E., Department of Anthropology, Peabody Museum, Harvard University, Cambridge, Mass. 02138, USA.

Tuttle, R.H., Department of Anthropology, 1126 East 59th Street, The University of Chicago, Chicago, Illinois 60637, USA.

Vincent, F., Laboratoire de Psychophysiologie, Université de Paris X, 2 rue de Rouen, 92001 Nanterre, France.

Wiener, A.S., New York University Medical Center, 550 First Avenue, New York, N.Y. 10016, USA.

Wilson, W.L., Regional Primate Research Center, University of Washington, Seattle, Wa. 98195, USA.

Wind, J., Departments of Otorhinolaryngology and Human Genetics, Free University, Amsterdam, Netherlands.

Wood, B.A., Department of Anatomy, Middlesex Hospital Medical School, Cleveland Street, London W1P 6DB, UK.

Yamazaki, N., Keio University, Kyoto Japan.

Zihlman, Adrienne L., University of California, Santa Cruz, USA.

PREFACE

The Congresses of the International Primatological Society, held bien-
nially since 1966, are noted for their social success - for the enjoyment and
mutual benefits of mixing scientists from diverse disciplines, who share an
interest in the one order of animals that is rather special for man. It is
these very differences in training and specific interests in primates, however, which hinder the real academic success of these Congresses.

    When starting to plan the Sixth Congress three years ago, we wondered
whether to continue the tradition of a programme of very varied topics, with
the Congress inevitably fragmenting into specialist groups, or to attempt to
focus the multi-disciplinary interests on a particular theme, since primatology had expanded so much. The latter might still be the more constructive
challenge for the future, but we opted for the traditional format (with
modifications), because we felt unable to attract the whole range of primatologists for any of the themes proposed.

    Nevertheless, we resolved, in the sessions we would convene on specific
topics in primatology, to ensure (1) a real, broad synthesis of recent
advances, with a careful selection of papers, (2) the inclusion of review
papers of general interest near the start of each session, and (3) the exploration of each topic with increasing informality (and specialization) - from
symposia, to short-paper sessions, to round-table discussions - with ample
time for discussion.

    While soliciting topics and papers, care was taken to avoid an exces-
sive overlap of sessions. The resulting five-day programme was derived
almost entirely from the interests expressed by prospective delegates,
with some bias towards local interests.

    Nearly 600 delegates attended the Sixth Congress of the International
Primatological Society between August 22nd and 28th 1976 on the Sidgwick
Avenue Site of the University of Cambridge, England. Nearly half the delegates presented papers, and an extensive selection of these is published
here. In editing the four volumes which comprise the proceedings of the
Congress it has been more necessary than in compiling the programme to select
really relevant material. The aim has been to produce thorough syntheses
of recent developments in the fourteen or so areas discussed, rather than
just another collection of congress papers.

    Thus, the bulk of the four volumes is composed of the fourteen symposia -
6 on Behaviour, 3 on Conservation, 2 on Evolution and 3 on Medicine - with
important supplementary material from 7 short-paper sessions (in the Behaviour and Evolution volumes) and from 6 (out of the 8 convened) round-
table discussions.

Volume 1 contains papers relating mainly to Behaviour - inter-individual relations and group structure, early social behaviour (mother-infant relations and play), demography and social organization, behavioural terminology, feeding behaviour in relation to food availability and composition, sexual and aggressive behaviour (including sexuality in apes), motor coordination, hearing and acoustic communication, language and its origins, cognition and learning, and aspects of the visual system.

Volume 2 contains papers relating mainly to Conservation - current problems in primate conservation in Africa, Asia and the Americas, the trade and supply of primates, and breeding primates in captivity.

Volume 3 contains papers relating mainly to Evolution - on primate evolution in general, and molecular and chromosomal evolution in particular, on behavioural factors in prosimian evolution, the phylogeny of tarsiers, S.E. Asian primates, methods of phylogenetic inference, and hominid evolution.

Volume 4 contains papers relating mainly to Medicine - infectious diseases including zoonoses, research on transmissible cancer, and the use of primates in research on human reproduction including fertility control.

Special thanks are due to those who contributed to the financial success of the Congress: - All-type Tools (Woolwich), Boots Company Ltd., Barclays International, British Council, Cambridge City, Cambridge University, Commonwealth Foundation, Hope Farms of Holland, Huntingdon Research Centre, I.C.I., I.U.B.S., Labshure Animal Foods, Shamrock Farms (GB) Ltd., Roche Products, Ltd., The Royal Society, Sandoz Ltd. (Basel), Wellcome Foundation, Wenner-Gren Foundation, W.H.O., and the delegates themselves.

We owe a great debt of gratitude to those who acted as session chairmen and section editors (listed in each volume), as well as to all the speakers. It was the responsibility of the chairmen to plan and conduct each session and to edit the proceedings, and it is they who have helped to ensure that publication can be rapid. I am especially grateful to Bill Lane-Petter, Edward Ford, Joe Herbert and Ken Joysey for the work they did in editing and collating the four volumes.

It was made clear during the Congress that events of the next few years will be critical for the survival of many primate species. And yet there are increasing signs of disunity among those most able to help them survive, with a corresponding delay in implementing the necessary education, habitat protection and research. Any tendency to polarise into "preservationists" and biomedical researchers must be resisted, so that there is united effort to help achieve the conservation of various ecosystems to the advantage of their occupants and the countries concerned. This is where "primatologists", with their unique spectrum of talents, should be able to play a leading role, and why regular meetings of an international society are important.

January 1977   D.J.C.

PREFACE TO VOLUME 3

When the Organising Committee of the Sixth Congress began to plan the sessions dealing with evolutionary topics they were faced with various practical constraints which have come to be reflected in the layout of the Evolution volume. They first subdivided the field between 'Hominid evolution', under the chairmanship of Michael Day and Phillip Tobias, and 'Primate evolution' (by inference excluding hominids) under the chairmanship of Bob Martin and myself. Thereafter the organisation of these two Symposia evolved along divergent lines.

The contributions delivered during the sessions on Hominid evolution have been welded together in a single section of this volume. The papers range through palaeontology, functional morphology, comparative anatomy, the problems of phylogeny and classification and, as they should, they trespass slightly into the fields of ethology and the ecosystem. These topics are brought together because of their relevance to the unifying theme of hominid evolution, but diverse specialists, who may have little interest in hominids as such, will find much to interest them in this section of the volume.

The Symposium on Primate evolution (excluding hominids) started with a general session in which the first four papers were intended to survey the field and introduce major themes which would be developed in subsequent short-paper sessions. It is no secret that the chosen themes tended to reflect the special interests of the organisers. Bob Martin's introductory paper on prosimians was followed up by short-paper sessions on behavioural factors in prosimian evolution, on the phylogeny of *Tarsius* and on other aspects of prosimian biology. My own introductory paper on myoglobin was followed up by a short-paper session on molecular and chromosomal evolution which, together with that on the phylogeny of *Tarsius*, naturally gave rise to a round-table discussion on methods of phylogenetic inference. It all seemed so logical in the planning stage with one session leading on to the next but, as a result, papers on molecular evolution are now to be found under three different headings in this volume!

Short papers which fell outside the chosen themes were retained in the first general Symposium and these appropriately follow up the introductory papers by Glen Conroy on anthropoid ancestry and by Peter Andrews on fossil apes. David Chivers persuaded us to include the ecology and distribution of his beloved South-East Asian primates under the umbrella of evolution. His proposal for a round-table discussion seemed harmless enough but somehow it matured into a more formal short-paper session and it is published as such.

The resulting volume provides a broad survey of the present state of knowledge of Primate Evolution (including hominids) but, by developing several major themes, some emphasis is given to those areas which the organisers regarded as being of particular interest in current research.

I am indebted to the chairmen of the various sessions for their help in the planning stages, in the conduct of the meetings and in the subsequent editorial work; they are not individually acknowledged here because their responsibilities are listed elsewhere. I wish to thank David Chivers and Bob Martin for their understanding when my editorial work was delayed by illness and bereavement within the family. I am especially grateful to Edward Ford for his help in reading some of the proofs and to the staff of Academic Press for all the hard work they have put into the production of this volume.

K.A.J.

CONTENTS

Contents

Contents

Contents

Contents

*Wishing that I had found more time for her while
it was still possible, this volume is dedicated to
the memory of my mother, Queenie Florence Joysey,
who died in 1977.*

K.A.J.

SECTION I

PRIMATE EVOLUTION
*Chairmen and Section Editors: R.D. Martin (London)
and K.A. Joysey (Cambridge)*

MAJOR FEATURES OF PROSIMIAN EVOLUTION: A DISCUSSION
IN THE LIGHT OF CHROMOSOMAL EVIDENCE

R.D. MARTIN

*Wellcome Laboratories of Comparative Physiology, The Zoological Society
of London, Regent's Park, London, NW1 4RY, UK.*

INTRODUCTION

The last 15 years have seen a considerable increase in the volume of research into prosimian biology in general and prosimian evolution in particular (e.g. see Martin, Doyle and Walker, 1974; Tattersall and Sussman, 1975).
This intensified research effort has vastly increased our knowledge of the
behaviour, morphology, functional anatomy and physiology of these so-called
"lower primates". Chromosomal and biochemical data have further enhanced
hypothetical reconstruction of phylogenetic relationships. Yet we are still
far from reaching a real consensus of opinion about some of the major features
of prosimian phylogeny, which may be held (following Simpson, 1945) to encompass the early radiation of the Primates and the subsequent diversification of
the more primitive members of the order (viz. the lemurs, lorises and tarsiers).
A recent conference devoted to major issues in reconstruction of primate phylogenetic history (Luckett and Szalay, 1975) highlighted the continuing problems of interpretation of evolutionary relationships.
The main areas which have recently attracted attention in research on prosimian evolution are as follows:
1. The relationship between the tree-shrews and the primates.
2. The details of the early (Cretaceous-Eocene) radiation of the primates.
3. The details of the relationships between the lemurs of Madagascar and
    the members of the Afro-Asian loris group.
4. The relationships of the tarsiers.
Some degree of consensus has been reached on a number of the points raised
under these headings. For instance, it is now generally accepted that the relationship between the tree-shrews and the primates is far more tenuous than
was originally suggested by Le Gros Clark's pioneering work on the Tupaiidae
(see Simpson, 1945; Le Gros Clark, 1971; and contrast with Campbell, 1974;
Romer, 1968; Martin, 1973; and Simons, 1972). Further, it is now widely
agreed that the lemurs and lorises can all be traced back to a distinct
common ancestry within the primates (Cartmill, 1975; Charles-Dominique and
Martin, 1972; Dene et al., 1976; Martin, 1972; Szalay and Katz, 1973;
Tattersall and Schwartz, 1974), though the details of the subsequent radiation
of this primate sub-group remain a matter for controversy. There is, however,
no commonly accepted view concerning the early radiation of the primates and
the relationships of the tarsiers. Indeed, the special session held at this
conference to discuss the phylogenetic affinities of the tarsiers showed that
the conclusions drawn are still as varied as was the case half a century ago.
Thus, although relevant information continues to accumulate rapidly, we are

still some way from reconstructing a reliable picture of primate evolutionary history. One may ask why this is so and whether we will ever be able to reconstruct a generally acceptable picture of primate evolution.

It is probably true to say that there are practical limits to the accuracy with which we can reconstruct evolutionary history, and that for this reason alone divergence of opinion is likely to persist. However, there are at least three factors which are at present hindering progress in the field of phylogenetic reconstruction and exaggerating the disparities between the conclusions drawn by different workers. These are:
1. Confusion between the two methodologically distinct operations of classification and phylogenetic reconstruction.
2. The lack of consensus with respect to the methods used for reconstructing phylogenetic relationships.
3. Increasing specialization of research interests, which is gradually precluding broad coverage of characters in phylogenetic reconstruction.
In all three areas, there is considerable room for clarification which would enhance the value of discussions about primate evolution.

The question of the relationship between classification and phylogenetic reconstruction has now become the focus of a conceptual clash between two main schools of thought. On the one hand, there is the "classical" school represented by Mayr (1974), Simons (1972; 1974) and Simpson (1945; 1961; 1975), among others. The essential tenet of this school is that classification should be based on the concept of *morphological divergence* between species, involving a combined assessment of (1) recency of common ancestry, and (2) rates of evolution subsequent to divergence from a given common ancestor. Since it appears that in any group of animals, such as the primates, some living descendants are more primitive than others (i.e. exhibit lesser morphological divergence from the common ancestor of the group as a whole), the concept of the *grade* of evolution (Huxley, 1958; Simpson, 1961) has become closely linked with that of morphological divergence. In Simpson's classification of the Primates (1945), the suborder Prosimii includes all of those living forms (lemurs, lorises and tarsiers) which are estimated to have diverged the least from the ancestral primate condition. The monkeys, apes and man — classified in the suborder Anthropoidea — are commonly regarded as showing a greater degree of morphological divergence from the common primate ancestor, and these "simian" primates are collectively viewed as representatives of a higher grade of primate evolution. It follows from the logic of such a classification that the early fossil representatives of the primate radiation are classified in the lower grade Prosimii.

In contrast to this approach, there is another which can be traced back to the writings of Hennig (1966) and is widely referred to as the "cladistic" approach (see Mayr, 1974). The basic tenet of this cladistic school is that classification should more accurately reflect the fine details of phylogenetic history and should therefore be based exclusively on the branching-points of the phylogenetic tree. Rates of evolution subsequent to divergence from branching-points are not taken into account in "cladistic" classifications. It is therefore possible to argue (e.g. see Mayr, 1974) that such classifications omit vital information on differential rates of evolution; but conversely it can be argued that cladistic classifications more accurately reflect apparent ancestral relationships. However, there is a more fundamental point which was overlooked in Mayr's otherwise excellent review and has been similarly passed over by those favouring the cladistic approach. Phylogenetic trees represent hypotheses about evolutionary relationships, and as such they tend to vary — especially in terms of the arrangement of nodes (branching-points) — from one author to another. The implacable logic of the cladistic school dictates that every distinct hypothetical branching pattern (or "clado-gram", if one has a penchant for jargon) should be accompanied by a different classification. As it happens, classificatory terms represent a large compo-nent of the language with which primatologists communicate about the animals they study, and a prime requirement of any successful language is *maximal sta-bility*. For this reason alone, the classical approach to primate classifica-tion is preferable, in that a classification such as Simpson's (1945) is com-patible with a large number of different phylogenetic trees. These points are illustrated in Figure 1. It is, of course, arguable that if everyone were to

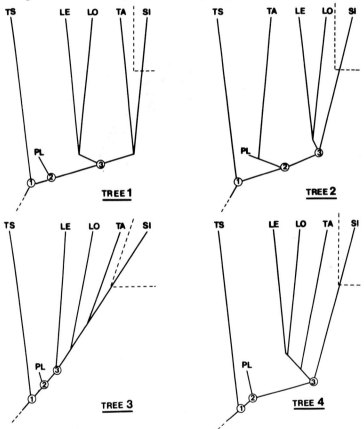

Figure 1. *4 different hypothetical branching diagrams relating 6 major groups.*
   TS = tree-shrews   LO = loris group
   PL = plesiadapoid fossils TA = tarsiers
   LE = Madagascar lemurs   SI = simians (monkeys, apes and man)
   Although each "tree" would require a distinct classification from the cladistic school, all are compatible with Simpson's classifi-cation (1945). The simians can in all cases be classified in the higher grade Anthropoidea (dotted line), leaving the other pri-mates in the lower grade Prosimii. For each tree, any one of three branching points (1, 2 or 3) could be arbitrarily defined as the ancestral primate stock.

agree on a single reliable interpretation of the details of primate evolutionary history, a cladistic classification would be as stable as a classical
classification (provided that no further fossil taxa were discovered). However, it is quite clear from the current literature that we are not yet in
sight of such an objectively established consensus, and in any case stability
is not the only criterion for preferring the classical approach to classification (see Mayr, 1974 and Simpson, 1975).

A more important point is that both schools of thought have given rise to
actual confusion between the operations of classification and phylogenetic reconstruction, the first unwittingly and the second deliberately. The question
"Is the tree-shrew a primate?" contains two logically distinct elements, which
are obvious from Figure 1. One element involves definition of the likely evolutionary relationships between the tree-shrews and the other groups, fossil
and living, included in the order Primates by Simpson (1945). This could
yield any one of the hypothetical trees shown in Figure 1, or even some other
pattern. The second element involves a decision on a formal classification
which is compatible with the hypothetical tree which emerges and also serves
a number of practical functions. If Tree 1 in Figure 1 were selected as the
preferred hypothesis, any one of the nodes (branching-points) labelled 1, 2
and 3 could be selected as the "ancestral primate stock" (see Martin, 1973, for
a definition of ancestral stocks). Accordingly, it would be purely a matter
of definition whether the tree-shrews were included in the order Primates,
and the same could be said of the plesiadapoid fossils. As one illustration
of this, one can take Simons' (1974) statement that the "plesiadapids are of
course primates". Does this mean (1) that Tree 4 (which seems to most closely
reflect Simons' views) is "of course" the correct hypothesis concerning the
branching sequence involved in primate evolution, or merely (2) that one should
"of course" define the order Primates such that the plesiadapoids are included
at any cost? There is unfortunately no objective certainty in either respect.
It is also worth noting that Hill's (1953) classification of the Primates,
which excluded the tree-shrews, differed from those of Simpson (1945) and Le
Gros Clark (1971) not on the basis of some fundamental disagreement about the
evolutionary relationships involved but largely on the basis of formal definition of the order Primates. In simple terms, Hill preferred node 2 of the
trees in Figure 1 as the ancestral primate stock, whereas Simpson and Le Gros
Clark preferred node 1. By contrast, it is the present author's view that
node 1 of the trees in Figure 1 is practically indistinguishable from the ancestral placental mammal stock, that node 2 in Trees 1, 3 and 4 is at present
indistinguishable from node 1 and that Tree 1 represents the most likely hypothesis concerning primate evolutionary relationships. All of these latter
points are exclusively concerned with phylogenetic history and not with formal classification. With respect to the latter, whether one follows classical
or cladistic principles of classification, arbitrary definition must enter
into the names selected for taxonomic categories. Hence, the statement "The
tree-shrew is a primate" conveys little information about primate evolutionary
history unless the underlying hypothesis concerning evolutionary relationships
is explicitly stated.

That said, it should be emphasised that the rest of this paper concerns
the reconstruction of evolutionary relationships and not the procedure of
classification. For this reason, the concept of the grade — together with
the closely allied concept of the *Scala naturae* (Martin, 1973) — will play
no part. This particular concept, which has been invaluable in the establishment of relatively stable classifications of the classical type, has been a

major factor underlying past confusion between the operations of classification and evolutionary reconstruction. In the latter operation, every species
and every character should be treated on its own merits and not be loaded with
some *a priori* qualitative assumption about grade of evolution. Much of the
argument in the past about the relationships of the tree-shrews has centred
around the partially subjective decision as to whether or not the tree-shrews
have attained the "primate grade" of evolution. This has added little but
confusion to the issue and has obscured the fundamental question about phylogenetic relationships: "Did the tree-shrews derive from a recognisable common ancestral stock which also gave rise to the lemurs, lorises, tarsiers and
simians and which was demonstrably distinct from (and later than) the ancestral placental mammal stock?" The author's conclusion is that the answer to
that question is negative (Martin, 1968a, 1968b, 1973, 1975).

This leads on to the second source of confusion arising from the lack of a
defined methodology for reconstructing evolutionary relationships. The truth
of the matter is that there is still no objectively defined, universally acceptable set of methods for reconstructing phylogenetic relationships. Until
we can agree on the *methods* we use, we can hardly agree about the conclusions
reached. There are, understandably, those who believe that they have infallible
methods for phylogenetic reconstruction. They are responsible for much of the
confusion which reigns in this field, especially when they immediately trans-
late their phylogenetic hypotheses into idiosyncratic classifications.

Confusion caused by the lack of a common methodology is compounded by increasing specialization in the research conducted on primates. Without a
common basis for interpretation, any specialist is limited to drawing phylogenetic conclusions from his own data and either accepting on trust the conclusions from workers in other fields, or rejecting such conclusions without
adequately understanding them. It is therefore vital that those concerned
with primate evolution should work actively towards a common methodology for
reconstructing phylogenetic relationships. This is a subject which deserves
a symposium in its own right.

In the absence of a widely accepted, objectively defined set of methods,
the best that any worker can do is to set out his own methods in detail. In
the morphological field, an excellent example of this is provided by Cartmill
(1975), and in the biochemical field such a procedure is standard practice
(see Joysey, this volume). However, there is a universal problem which has
not yet proved amenable to direct solution by objectively defined procedures.
This problem resides in the fact that if there are differential rates of evolution of given characters within a group of animal species, the degree of
phylogenetic relationship between any two species is *not* proportional to the
number of shared homologous characters (Martin, 1968a, 1973, 1975). In other
words, the fact that two species share large numbers of homologous characters
does not necessarily indicate recency of common ancestry; those characters
may have been retained from a very early stock. For example, it can be argued
that tree-shrews resemble primates in many ways because both have retained
characters from the ancestral placental mammal stock, not because they have
retained characters from a later stock definable as the ancestral primate
stock (Martin, 1973). Comparisons of living primate species involve consideration of degrees of morphological divergence — a combined function of time
elapsed since separation from a common ancestor and rates of subsequent evolution. To obtain a scientifically valid result, one would need to be able to
evaluate two of these components (e.g. morphological divergence and rates of
evolution) to determine the third (time elapsed since separation); but exist-

ing techniques unfortunately provide direct information only about morphological divergence. In all justice, one must acknowledge the contribution made by the cladistic school in recognition of this problem, and the establishment of explicit guidelines for reconstruction of phylogenetic relationships. Whatever one may feel about Hennig's (1966) views on classification, he and his followers have certainly helped to clarify our approach to phylogenetic reconstruction. The problem has therefore been recognised, but not solved. If one can measure only one component in an equation containing three, circularity is inevitable and one must be content with an approximation procedure. This applies to all techniques currently used for reconstructing phylogenetic relationships, though some are more rigorous than others. There are, however, a number of explicit principles which can be applied to optimise such approximation:

1. It is essential to consider as many different characters as possible.
2. Comparisons should be as broad as possible and should extend outside the group of primary interest (e.g. broad consideration of placental mammal evolution is an indispensable adjunct to reconstruction of primate evolution).
3. It must be recognised that many alternative hypotheses may be erected with respect to the phylogenetic relationships between the species in a particular group.
4. At present, the only objective basis that we have for selecting one hypothesis and rejecting all others is the *parsimony principle* (see Cartmill, 1975).
5. Any phylogenetic tree which is proposed must be internally consistent (e.g. if a character of a given species seems to have evolved relatively slowly with respect to that of another species, the same perspective must emerge in comparison with other species).

    These principles will now be illustrated with respect to the chromosomal evidence for prosimian evolution, and then an attempt will be made to integrate the conclusions with some other evidence.

CHROMOSOMES AND PRIMATE EVOLUTION

    Only two decades have elapsed since Darlington and Haque (1955) reported diploid chromosomes numbers for man and three Old World monkey species, emphasising the need for study of a wide range of primate species. Progress has since been so rapid that it is now possible to summarize the chromosome complements (karyotypes) of a minimum of 81 different primate species (Egozcue, 1969; Chiarelli, 1974; Buettner-Janusch, 1973; Hsu and Benirschke, 1967-75; Rumpler, 1975; de Boer, 1973a, 1973b). A stage has thus been reached where a preliminary overall study of chromosomal evolution within the order Primates is possible. However, recent detailed analyses of the available data have been generally limited to discussion of chromosomal evolution within relatively small taxonomic groups of the Primates, such as the Galaginae (de Boer, 1973a), the Lorisinae (de Boer, 1973b), members of the genus *Lemur* (Egozcue, 1972), and the Ceboidea (Staton, 1976a; de Boer, 1974). In addition, these studied have been hampered by a common tendency to take the karyotype of one living species of a group as primitive and to derive the karyotypes of other species from this hypothetical "archetype". Here, as in classical comparative anatomy (Martin, 1973), it is highly improbable that any living species may be taken as completely equivalent to an ancestral form. A process of inferential reconstruction is required. With chromosomal features, as with any other characters, the essential problem in reconstruction of evolutionary

history is that there is no sound *a priori* basis for determining the primitive
condition for any group. For this reason, it seemed more appropriate to apply
to the data on primate chromosomes a broad comparative approach which has
been previously applied in evolutionary analysis of anatomical and behavioural
characters (Martin, 1968, 1973, 1975).

Within certain very narrow limits, the numbers and shapes of the chromosomes
in somatic cells of a given species are constant, and such species-specific
characters may reasonably be expected to yield an accurate reflection of evolutionary history.

MECHANISMS OF CHROMOSOMAL CHANGE

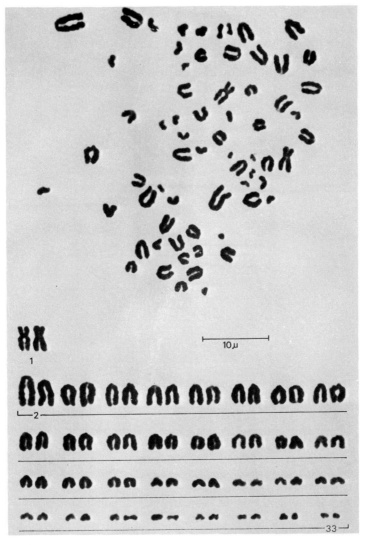

Figure 2.   Metaphase chromosomes of a female Microcebus murinus (2N = 66).
64 chromosomes are acrocentric; only the 2 X-chromosomes are
metacentric.

Chromosomes in the karyotype of any given species are now widely classified according to the position of the centromere (the point of attachment to the spindle during cell-division). Chromosomes with the centromere close to one end can be classified as "acrocentrics" and those with the centromere close to the centre are termed "metacentrics" (Fig. 2). If the centromere were actually in a terminal position, the term "telocentric" could be applied, but there is considerable debate about whether such chromosomes occur naturally (Matthey, 1949; White, 1954; Todd, 1967). There is also the problem that a range of intermediates between acrocentrics and metacentrics could poten- tially be defined, and some authors use additional intervening categories (e.g. "submetacentrics"). However, there seems to be general agreement in the lit- erature about a primary distinction between acrocentric chromosomes and those with more central centromeres. Hence, a simple division into acrocentrics and metacentrics is adequate for purposes of broad analysis. This distinction is crucial because it is now fairly well established (Matthey, 1949; White, 1954; Bender and Metler, 1958; Chu and Bender, 1961; Bender and Chu, 1963) that one of the main processes in chromosomal evolution in animals generally must be interconversion of acrocentric and metacentric chromosomes (Fig. 3).

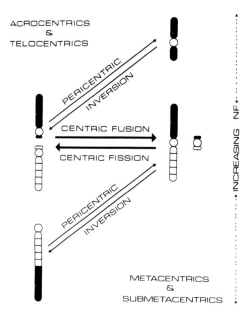

Figure 3. Diagram of the major pathways of chromosomal evolution (White, 1954). Centric fusion and/or fission probably account for most chromosomal evolution (NF maintained constant), while subsid- iary processes such as pericentric inversion alter the NF.

This may involve either the conversion of two acrocentrics into one metacentric ("centric fusion"), or a conversion in the opposite direction ("centric fis- sion"), or possibly a combination of events in both directions. Centric fusion is probably a somewhat misleading term, since the actual process involved seems likely to be *reciprocal translocation* (White, 1954), which would convert two acrocentric chromosomes to give a larger metacentric chromosome and a minute chromosome (consisting of a centromere and two very short arms) which could be

lost at some subsequent time. If this is the case, then the process of centric
fission (an equally misleading term) is likely to be relatively rare compared
to centric fusion. Definitive resolution of this question awaits more detailed
description of the nature and function of the heterochromatic region of the
chromosome referred to as the centromere (White, 1954). However, the terms
"centric fusion" and "centric fission" are so well established in the literature that they will be used here for convenience.

If centric fusion, also known as Robertsonian fusion or translocation after
its discoverer (Robertson, 1916), is assumed to be the major process involved
in chromosomal evolution, one can extrapolate back to a hypothetical situation
in which all chromosomes would have been acrocentrics. This extrapolation,
equivalent to counting the total number of arms on the interphase chromosomes,
gives the fundamental number (a term usually attributed to Matthey, 1949, but
actually first used by Robertson). It is an empirical fact (e.g. see Matthey,
1945) that the fundamental number (NF) often provides a more stable and useful
guide to evolutionary relationships within animal groups than does the diploid
number (2N), which can vary widely for a given fundamental number even between
closely-related species. For example, the two species *Galago senegalensis* and
G. crassicaudatus may differ greatly in their diploid numbers (e.g. 38 vs 62),
but their fundamental numbers are the same (66).

With respect to the Primates, some authors have postulated that centric
fusion is the major or exclusive process involved in chromosomal evolution
(Matthey, 1945, 1949; Bender and Chu, 1963; Hamerton, 1963) while others
have claimed that centric fission is the primary process (Staton, 1967; Todd,
1967). On the present evidence, no clear-cut choice between these two alternative postulates can be made. For certain purposes of analysis, this choice
is not necessary since the total number of chromosome arms (NF) would remain
constant in either case. As Robertson himself suggested (1916) it is possible
that *both* processes occur (as indicated in Fig. 3), and if they do there can
be no simple unidirectional guideline to chromosomal evolution. It is preferable, therefore, to refer to "centric rearrangement" (centric fusion and/or
centric fission) in the initial analysis of fundamental numbers. One can
accept that centric rearrangement is the primary process in chromosomal evolution without necessarily postulating that it is unidirectional.

Whatever the actual balance between centric fusion and centric fission may
be, it is obvious that centric rearrangement does not account for all chromosomal evolution in vertebrates. If it did, all living mammals (for example)
would have the same NF, which is far from true (see Hsu and Benirschke, 1967-
1975). It is apparent that accessory processes, such as pericentric inversion
and perhaps chromosome fragmentation (Tobias, 1953) have acted to increase or
reduce fundamental numbers in the course of chromosomal evolution through rearrangement in the vertebrates. Nevertheless, numerous examples indicate that
centric rearrangement is the most common pathway of chromosomal evolution. This
is further borne out by the frequent cases of natural polymorphism involving
centric rearrangement. Often, within a single species or subspecies three
different karyotypes can be found: in one form (1st homozygous condition),
two specific chromosome pairs occur as acrocentrics; in the intermediate form
(heterozygote), one of each pair of acrocentrics occurs combined in a metacentric; in the third form (2nd homozygous condition), there is a pair of
such metacentrics. This situation was originally suspected for the locust
Jamaicana subguttata by Robertson (1916) and was subsequently demonstrated in
the iguanid lizard *Gerrhonotus scincicauda* (Matthey, 1949). Among mammals,
such polymorphisms have been detected in feral mice *Mus musculus* (Gropp et al.,

1972), in the shrew *Sorex araneus* (Ford et al., 1957), in the gerbil *Gerbillus pyramidum* (Wahrman and Zahavi, 1955), in domestic cattle (Gustavsson and Rockborn, 1964), sheep (Bruére, 1974) and goats (Riek, 1974), in the bushbaby *Galago senegalensis* (de Boer, 1973a), and in the owl monkey *Aotus trivirgatus* (de Boer, 1974; Brumback et al., 1971 - Fig. 4). Recently, it has also been shown that a polymorphism involving centric rearrangement occurs in natural populations of impala (Wallace and Fairall, 1967) in the Kruger National Park. Many more cases will undoubtedly emerge as future studies embrace larger numbers of each species. Finally, in man three types of polymorphism involving centric rearrangement have been reported (Hamerton, 1968). All three types involve chromosomes of the D and G groups, which typically occur as acrocentrics. Although the incidence of D/D centric fusion may be as high as 1:1,000 in the human population, the relative scarcity of metacentric chromosomes in this category clearly suggests centric fusion rather than fission. Studies of human chromosomal polymorphisms of this kind have confirmed a theoretical

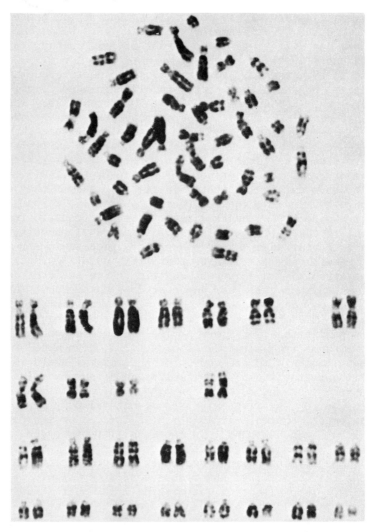

Figure 4a. Metaphase chromosomes of female Aotus trivirgatus griseimembra. A: homozygous condition (2N = 54).

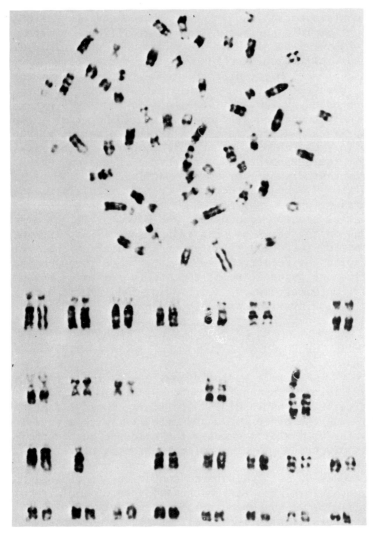

*Figure 4b. Metaphase chromosomes of female Aotus trivirgatus griseimembra.
B: heterozygous condition (2N = 53). Banding techniques, as
shown here, permit reliable matching of chromosome pairs.*

expectation that centric rearrangement may, in the heterozygous condition,
give rise to unbalanced segregation of chromosomes at meiosis. Both of the
relatively uncommon human centric fusions (D/G and G/G) are associated with
an increased incidence of Down's syndrome. Significantly, however, the far
more common D/D centric fusions have not been found to have such an association (Hamerton, 1968). Since chromosomal rearrangements of this kind clearly
do arise and spread through animal populations, either certain centric con-
versions do not give rise to major problems at meiosis, or there must be strong
selection pressures favouring such rearrangements and outweighing the eventual
disadvantages of disruption at meiosis.
    The overall conclusion that chromosomal evolution in mammals takes place
predominantly through structural rearrangement (cf. Fig. 3) is further sup-

ported by data indicating that somatic cell nuclei of different placental mammal species contain approximately equivalent quantities of DNA, approaching 7×10^{-9} mg DNA per diploid nucleus (Atkin et al., 1965; Manfredi-Romanini, 1972; Manfredi-Romanini et al., 1972). These data virtually rule out the possibility of multiplication of chromosome numbers through polyploidy, which would in any case present major difficulties with the X/Y sex chromosome system in mammals. However, they do not rule out evolution by chromosome fragmentation, as suggested by Tobias (1953), nor do they rule out relatively minor changes in DNA content through gene duplication (Ohno et al., 1968; Ohno, 1970) or small losses. These latter processes, if they occur at all, probably play a very minor role in gross chromosomal change compared to the established mechanisms of centric rearrangement and pericentric inversion, which have been confirmed by recent studies of chromosome banding.

There is additional evidence that some overall regularity is involved in chromosomal rearrangement. Analysis shows that in the karyotypes of individual placental mammal species acrocentric and metacentric chromosomes tend to cluster in groups on the basis of size (Bengtsson, 1975). In more than 10% of the mammal species surveyed, this tendency is statistically significant ($p < 0.025$), despite the relatively small numbers of chromosomes involved. Thus, several lines of evidence point to centric rearrangement as the major process in evolution of chromosome size, shape and number among placental mammals. Pericentric inversion probably accounts for much of the remaining variation in shape, and accessory processes may contribute to some extent.

ANALYSIS OF PRIMATE CHROMOSOMES

Much recent work on primate evolution has tended to confirm Hill's proposition (1953) that living primates fall into two distinct groups: "strepsirhines" (lemurs and lorises) and "haplorhines" (tarsiers, monkeys, apes and man). Chromosomal analysis was therefore conducted primarily with a view to testing this proposition, taking as a starting point the evidence that centric rearrangement is the main process of chromosomal evolution in placental mammals generally.

As might be expected, straightforward comparison of diploid numbers of strepsirhines and haplorhines provides no evidence of a clear-cut dichotomy. The histograms of diploid numbers overlap extensively (Fig. 5, A & B), and there is no statistically significant difference between the two sets of data ($p > 0.075$, Mann-Whitney U test, 1-tailed, $U = 657$). Grouping primates on the basis of their diploid numbers alone would hence be of little value.

In line with the concept that chromosomal evolution occurs mainly through centric rearrangement, the next step is to compute fundamental numbers (NF) for all 81 species involved in the comparison. Histograms of NF (Fig. 6, A & B) clearly show a tendency for strepsirhines to have fundamental numbers lower than those of haplorhines. The difference between the two primate groups is highly significant ($p > 0.0003$, Mann-Whitney U test, 1-tailed, $U = 314.5$). In each group, the distribution of fundamental numbers is unimodal. The modal value for strepsirhines is NF = 66 (mean = 69); that for haplorhines is NF = 84 (mean = 86). On the other hand, strepsirhines tend to have a greater percentage of their smaller number of chromosome arms present in the form of acrocentrics (Fig. 7, A & B). Again, this difference from the haplorhines as a group is highly significant ($p < 0.0003$; Mann-Whitney U test, 1-tailed, $U = 319.5$). In all cases, there is some overlap between strepsirhines and haplorhines, so individual species cannot be allocated confidently to either

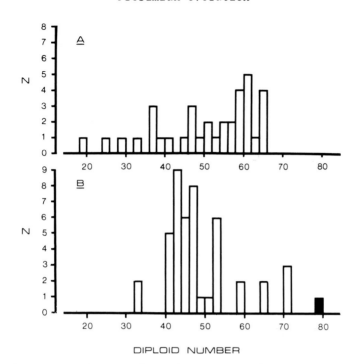

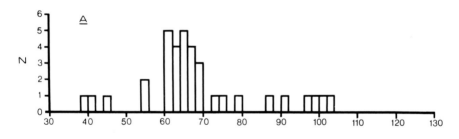

Figure 5. *Histograms showing distributions of diploid numbers for strep-*
sirhine (A: N = 36) and haplorhine (B: N = 45) primates.
Tarsius indicated in black.

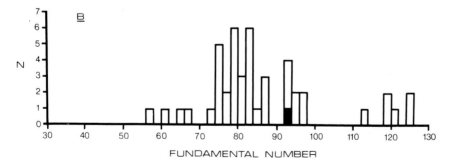

Figure 6. *Histograms showing distributions of fundamental numbers for*
strepsirhines (A: N = 36; modal NF = 66) and haplorhines
(B: N = 45; modal NF = 84). Tarsius indicated in black.

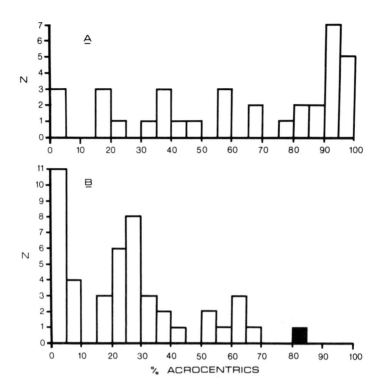

*Figure 7. Histograms showing % acrocentric chromosomes for strepsirhines
(A: N = 36) and haplorhines (B: N = 45). Tarsius indicated in
black.*

group on this basis alone. However, it is noteworthy that *Tarsius* is closest
to the simian mode in terms of its fundamental number, whilst it resembles
the strepsirhine primates in exhibiting a large percentage of acrocentric
chromosomes. It is also striking that the considerable overlap between
diploid numbers of strepsirhines and haplorhines is a product of two opposing
tendencies; the lower fundamental numbers of strepsirhines are balanced by a
higher percentage of acrocentric chromosomes, and *vice versa* for haplorhines.
Hence, two statistically significant differences are cancelled out in terms
of overall chromosome numbers. There is a possibility that this convergence
of diploid numbers is strepsirhines and haplorhines is not a chance effect,
but a reflection of a trend towards an optimal diploid number (see later).
  The analysis can be taken further by examining statistically the relationship between the numbers of metacentric chromosomes (M) and the numbers of
acrocentric chromosomes (A). If all species in a group had the same NF, with
differences in diploid numbers arising solely from centric rearrangement, the
data would show a perfect fit to a line of formula: M = 0.5 (NF-A). As already
shown (Fig. 6, A & B), there is some variation in NF values within each of the
two groups of primates (strepsirhines and haplorhines), representing departures from this perfect fit. Positive or negative departures from the perfect
fit may be attributed to relatively unusual processes - such as pericentric
inversion or chromosome fragmentation - which modify the fundamental number.
If numbers of metacentrics are plotted against numbers of acrocentrics
for the various primates (Fig. 8), the distribution of points can be analysed in more detail, with the aim of identifying "representative NF

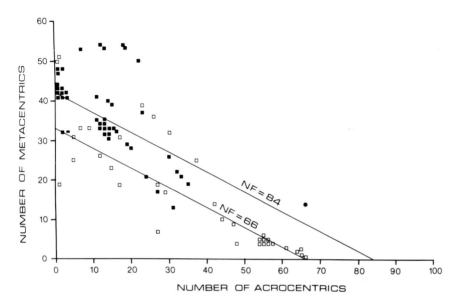

Figure 8. *Graphical analysis of numbers of metacentrics against acro-*
centrics for strepsirhines (open squares; N = 36) and haplor-
hines (black squares + circle for Tarsius; N = 45). Sloping
lines indicate ideal relationships involving only centric re-
arrangement for modal NF values indicated in Fig. 6.

values" for groups of species. Provided that departures from the perfect fit
(e.g. through pericentric inversion) are relatively restricted, and provided
that such departures involve both increases and decreases in NF values without a marked directional bias, representative NF values should emerge as statistical probabilities. One approach is to calculate simple linear regressions. For strepsirhines, the regression formula is M = 37.0 - 0.57A (correlation coefficient, r = 0.86); for haplorhines, it is M = 44.1 - 0.56A
(r = 0.66). In both cases, the slope is close to the expected value of -0.50,
but slightly steeper. For the strepsirhines, the regression indicates a representative NF value between 65 (A-intercept) and 74 (2xM-intercept), whereas
for haplorhines an NF value between 77 (A-intercept and 88 (2xM-intercept) is
indicated. The similarity in slope between the two regressions itself indicates a consistent relationship in the scatter of points around the representative NF values. A second approach is to follow the theoretical formula more
closely and assume that the slope *must* be - 0.50, using this value to find
the best fit line giving the least squared deviation of points. The M-intercept
for this best fit line can be simply calculated from the formula:

$$\text{M-intercept} = \frac{NF}{2} = \frac{2\ \Sigma M_i +\ \Sigma A_i}{2n}\ ,$$

where M_i, A_i are paired values of M and A for n species. For strepsirhines,
the representative NF value thus obtained is 69; for haplorhines, the value
is 87. Hence, all three possible indicators of the representative fundamental
number (modal value; linear regression; best-fit line of fixed slope) are in
good agreement for strepsirhines and haplorhines treated separately, and all
indicate a clear distinction between these two primate groups. Strepsirhines

are demarcated from the haplorhines by a statistically significant difference
of 12-18 between their representative NF values.

In fact, the minor differences between the three alternative indicators
of representative NF values are explicable in terms of the distribution pattern in Fig. 8. Species with a relatively large proportion of acrocentric
chromosomes generally lie quite close to the "ideal" line, while species with
a large proportion of metacentrics show a greater scatter - with a slight bias
in an upward direction. This could be interpreted as confirmation that centric
fusion is the predominant process in chromosomal evolution, with the resulting evolutionary trend towards increased numbers of metacentric chromosomes
reinforced by pericentric inversions predominantly converting acrocentrics to
metacentrics. Circumstantial evidence agrees with this. The lesser mouse
lemur (*Microcebus murinus* – Fig. 2) has 64 acrocentric somatic chromosomes,
and only the 2 X-chromosomes are metacentric. Demidoff's bushbaby (*Galago
demidovii*) has 50 acrocentric somatic chromosomes, 6 metacentric chromosomes
and 2 metacentric X-chromosomes (de Boer, 1973). *Tarsius spectrum* has 66
acrocentric somatic chromosomes and 14 metacentric chromosomes (probably including the sex chromosomes). These are all primate species which have been
judged on morphological and behavioural grounds to be relatively primitive
members of their respective groups and hence closer to the ancestral primate
condition (Charles-Dominique and Martin, 1970; Martin, 1972), and they all
have a large proportion of acrocentric chromosomes.

INTERPRETATION OF PRIMATE EVOLUTION FROM CHROMOSOMAL EVIDENCE

It has been shown that strepsirhines as a group are statistically different
from haplorhines in terms of NF value distribution. However, this would only
provide evidence of a specific common ancestor for tarsiers and simians if the
higher NF values of haplorhines could be reliably interpreted as a later specialization away from the ancestral primate condition. One must also consider
an alternative possibility that the ancestral primates had a high NF value
(in the region of 84) which was reduced (to about 66) as a specialization in
the evolution of the strepsirhines. If that were the case, tarsiers could
have diverged separately from the ancestral primate stock; the high NF value
would be no more than a primitive primate feature (Fig. 9). This problem can
be tackled by analysis of chromosomes from a wide range of placental mammal
species. In addition to the 81 primate species already discussed, chromosomal
data for 385 placentals are readily available for analysis (Hsu and Benirschke,
1967-1975). Treatment of the overall sample of 466 placental data points,
with the same techniques as for the primate data alone, gives directly comparable results. Again, diploid numbers provide little direct indication of
evolutionary relationships (Fig. 10), but it is important to note that the
spread of 2N values is limited and that there is a modal value of 2N = 48
for the placentals as a group. As far as NF values are concerned (Fig. 11),
the modal value (NF = 68), the range of values indicated by linear regression
(NF = 63-72) and the value derived from the best fit line of slope - 0.50
(Fig. 12; NF = 67) are all in good agreement. These values are virtually
indistinguishable from those obtained for strepsirhine primates alone, whereas
the haplorhine values are considerably in excess of the range of representative
NF values for the placentals overall (77-88 vs. 63-72). Hence, it may be
reliably concluded that the high NF value of *Tarsius* does indeed represent a
specialized feature shared with simian primates, possibly indicating close
phylogenetic relationship. However, it can be seen From Fig. 6 that some

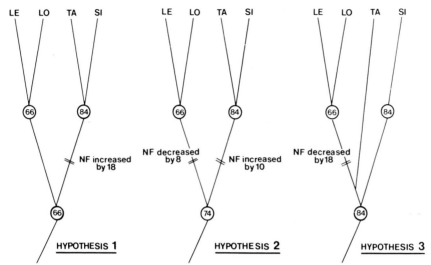

Figure 9. *Three alternative hypotheses for the evolution of modal NF values in strepsirhine and haplorhine primates. Interpretations may vary according to the NF value accepted as ancestral for all primates.*

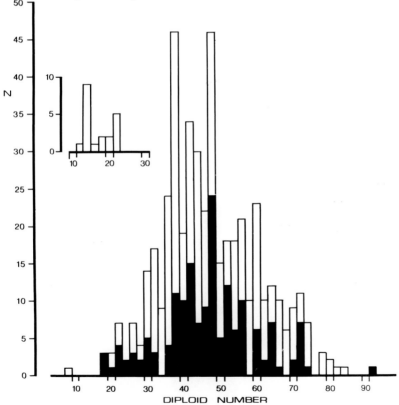

Figure 10. *Histograms showing distributions of diploid numbers for placental mammals (N = 466) and marsupials (inset; N = 20). Rodents, indicated by black shading, show a similar distribution to placental mammals generally.*

strepsirhine species have unusually high NF values in the range 77-104, representing a similar specialization. But the karyotypes of all of these species
contain a large proportion of metacentric chromosomes and could relatively
easily be derived from an ancestral strepsirhine type with approximately NF =
66 by a number of pericentric inversions (see Fig. 7). Since the karyotype of
Tarsius contains 14 metacentrics *in addition* to 66 acrocentrics, derivation
from such an ancestral condition would require a more complex sequence of
changes (e.g. a number of pericentric inversions in large acrocentrics to
produce metacentrics followed by centric fissions to produce small acrocentrics, thus increasing both the fundamental number and the proportion of
acrocentrics). Hence, strepsirhine species with high NF values may be reasonably regarded as secondarily specialized from a hypothetical ancestral strepsirhine condition of NF = 66, whereas *Tarsius* has a karyotype which is far more
distinct from those of all the strepsirhine species and is at the same time
very close to the hypothetical ancestral haplorhine condition.

Other features of the analysis of placental mammals as a group (Figs. 10-
12) are identical to those found in analysis of primates alone (Figs. 5-8).
The distributions of NF values are unimodal in both cases, and there is consistency in the greater scatter of NF values for species with a larger proportion of metacentric chromosomes (Figs. 8 and 12). This latter feature
is further reflected in the similarly skewed distributions of NF values for
primates and for placentals in general (Figs. 6C and 11). In fact, since
rodents comprised a disproportionately large part of the placental mammal
sample (163 out of 466 species), they have been displayed separately as a
further check on the consistency of the relationships between NF values and
numbers of metacentrics/acrocentrics. Figs. 10, 11 and 12 show that rodents

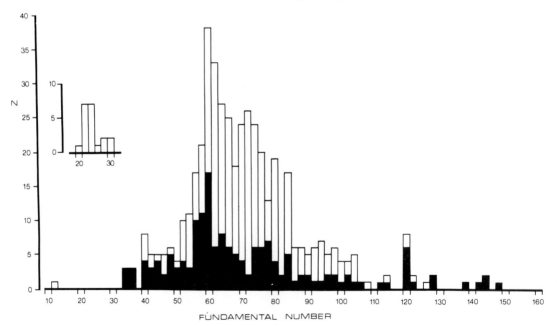

*Figure 11. Histograms showing distributions of fundamental numbers
for placentals (N = 466) and marsupials (inset; N = 20).
Rodents again show a similar distribution to placentals g
generally.*

do indeed conform to the basic pattern. Finally, the clear distinction between
the small sample of marsupials and the placentals, in terms of 2N and NF values (Figs. 10 and 11), provides additional evidence of the clear reflection
of phylogenetic relationships in chromosomal characters.

Overall, it can be concluded that the placental mammals generally follow
the rule that centric rearrangement (probably dominated by centric fusion) is
the major process of chromosomal evolution, with the trend to increasing proportions of metacentrics reinforced by pericentric inversion. Since there is also a
general tendency towards a standard diploid number of about 48 (Fig. 10) regardless of evolutionary relationships, it can further be postulated that there
is some advantage in maintaining the total number of chromosomes at about this

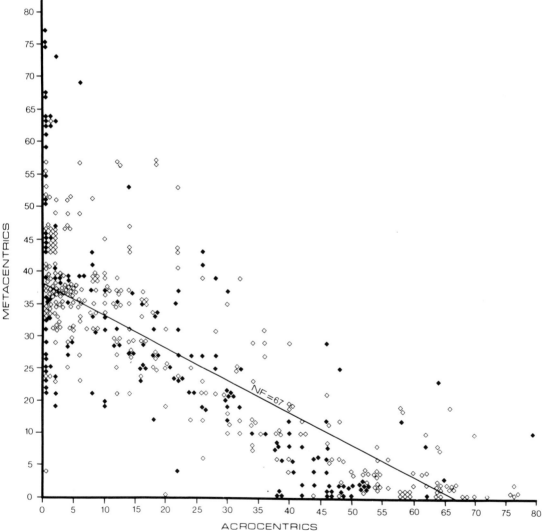

Figure 12.   *Graphical analysis of numbers of metacentrics against acro-*
centrics for 466 placentals (rodents indicated in black). The
sloping line indicates the ideal relationship involving only
centric rearrangement for an NF value of 67.

level. The most parsimonious hypothesis to explain all of the data would be
that the placental mammals are derived from a common ancestor with a fundamental number of approximately 66 and a diploid number of approximately 48
(i.e. 30 acrocentrics and 18 metacentrics). Evolution from that hypothetical
ancestral stage would have occurred primarily by centric fusion, with pericentric inversion further increasing the proportion of metacentrics in many
species. However, there has undoubtedly been some evolution in the other
direction as well, increasing the proportion of acrocentrics in some species
through centric fission and pericentric inversion. Viewed in this light,
the strepsirhine primates would seem to have remained primitive in terms of
retaining NF values close to 66, but variations in chromosomes brought about
primarily by centric rearrangement and pericentric inversion have produced
a wide scatter of diploid numbers. Species such as *Microcebus murinus* and
Galago demidovii have apparently remained relatively primitive in retaining
NF values close to 66 and in possessing a majority of acrocentric chromosomes,
but there is a definite possibility that they are somewhat specialized in
having more than 65% of their chromosomes acrocentric. This is reflected in
the strepsirhines generally by the high frequency of diploid numbers exceeding 48 (cf. Figs. 5 and 6). Such a specialization may already have been present in the ancestral primates, however.

CONCLUSIONS

It is clear that the chromosomal evidence can be interpreted in a parsimonious way to support the hypothesis that the living tarsiers branched off
the line leading to the living simians. However, the chromosomal evidence
alone does not provide such firm support for other details of primate evolutionary relationships, and one must of course rely on different lines of evidence to infer the relationships of fossil forms such as the plesiadapoids.
The first point to emphasize in conclusion, then, is the necessity for effective integration of many different kinds of evidence. On the one hand, this
requires a readiness on the part of primatologists to cross interdisciplinary
boundaries; on the other, it is completely dependent upon acceptance of a
common methodology for reconstructing evolutionary relationships.

With respect to the specific question of the evolutionary relationships of
the tarsiers, the chromosomal evidence accords well with a number of other
lines of evidence. It has long been suggested on the grounds of reproductive characters that the tarsiers shared a distinct common ancestry with the
simians. Hubrecht (1908) noted that the tarsiers have in common with the
simians discoidal, haemochorial placentation quite distinct from the diffuse,
epitheliochorial placentation of the lemurs and lorises. This dichotomy was
further emphasized by Hill (1932), and Luckett (1974, 1975) has recently
underlined the fact that the tarsiers also share with the simians other characters such as loss of the choriovitelline component of the placenta and
considerable reduction of the allantois. The reproductive similarities between tarsiers and simians are further reinforced by the observation that
within this sub-group of primates the weight of the neonate is uniformly related to the mother's weight by a simple allometric formula (Leutenegger, 1973;
Martin, 1975). In lemurs and lorises, by contrast, there is a significantly
different allometric relationship between neonatal and maternal weight. In
simple terms, it emerges that at any given maternal weight a tarsier or a
simian produces a neonate approximately three times heavier than that produced
by a lemur or loris of comparable size.

The division of the Primates into Strepsirhini (lemurs and lorises) and Haplorhini (tarsiers and simians), used by Hill (1953), was originally based on a difference in the morphology of the nasal area in living forms (Pocock, 1918). In the living strepsirhine primates, a naked rhinarium is present surrounding the nostrils and joining the upper lip. This is widely presumed to be a retained primitive mammalian character. In the living haplorhines, this feature has been lost. In fact, presence of the rhinarium in strepsirhines is typically correlated with a median gap between the upper incisors and (usually) the possession of an oral opening to the Jacobson's organ. These features may also be regarded as retentions from the primitive mammalian condition. Haplorhines typically lack both the median gap between the upper incisors and the oral opening to the Jacobson's organ. Once again, tarsiers share with simians what would appear to be specialized features. In this case, the median gap between the upper incisors can be traced back to the Eocene Adapidae among the strepsirhine primates, whereas even in Eocene tarsioids and Oligocene simians the gap is already absent. These apparent specializations among haplorhines have been interpreted (Martin, 1973) as a result of the early adoption of diurnal habits by the ancestral haplorhines, with consequent reduction of olfactory mechanisms and additional emphasis on vision (and hence brain expansion). The reflecting tapetum found in the eye of living strepsirhines (even diurnal forms) is lacking in haplorhines (even nocturnal forms), indicating that the latter were subjected to lesser selection for improved nocturnal vision in the course of their evolution. Noback (1975) has also shown that strepsirhines differ from haplorhines in the laminary patterns of the lateral geniculate body. There is, in sum, a great deal of morphological evidence to suggest that there has been an essential dichotomy in primate evolution between the lemurs and lorises, on the one hand, and tarsiers and simians, on the other (see also Hoffstetter, 1974). Certain serological studies (Goodman, 1973; Goodman et al., 1974; Dene et al., 1976) have independently indicated an early phylogenetic link between tarsiers and simians, and amino acid sequence studies of the haemoglobin of *Tarsius* (Beard et al., 1976) also suggest such a link. The consensus of present evidence therefore indicates a considerable number of characters shared by tarsiers and simians which do not seem to be primitive retentions from the ancestral primates (or ancestral placental mammals), and which cannot easily be explained away as the products of convergence. Thus, although some dental and craniological evidence has been adduced to suggest that the primates instead underwent an early dichotomy into plesiadapoids + tarsioids and lemurs + lorises + simians (Gingerich, 1976; Schwartz, this volume), the bulk of the evidence does not support such a sharp division between tarsiers and simians. It is certainly premature to propose a new classification on the basis of this as yet fragile alternative hypothesis.

In considering the question of primate classification, as a final note, it must be remembered that there are currently two alternative hypotheses about the initial division into lemurs, lorises, tarsiers and simians. It may be some time before one hypothetical tree can be reliably accepted as the most accurate reflection of primate phylogeny. We must also bear in mind Simons' point (1974) that a classificatory division based primarily on differences between living representatives does not help us in the classification of early fossil forms. Therefore, although it may be reasonably concluded from comparisons of living primates that there was an early dichotomy into "strepsirhines" (lemurs and lorises) and "haplorhines" (tarsiers and simians) in primate evolution, this dichotomy might not provide the best basis for a classification which must also cover the early primate fossil evidence. It remains to be seen whether a well-constructed classification based on a division into Strepsirhini and Haplorhini would prove to be more widely understandable, more enduring, more informative and more logical than Simpson's original classification of 1945.

ACKNOWLEDGEMENTS

Preparations of karyotypes were kindly conducted with blood samples from *Aotus trivirgatus griseimembra* by Dr. Jennifer Parrington and Mrs. Lynne West (Galton Laboratory, University College, London) and from *Microcebus murinus* by Dr. L.E.M. de Boer (Institute of Genetics, University of Utrecht). Advice and help with statistical techniques were provided by Dr. B. Seaton, Mr. C. Owen and Dr. K.M. Homewood. Thanks also go to my wife, Anne-Elise, for assistance in the preparation of illustrations, and to Mr. T.B. Dennett for photographic help. Dr. A.F. Dixson kindly provided advice for the revision of the manuscript, and comments from Prof. P.V. Tobias were invaluable.

REFERENCES

Atkin, M.B., Mattinson, G., Beçak, W. and Ohne, S. (1965). *Chromosoma*, 17, 1-10.
Beard, J.M., Barnicot, N.A. and Hewett-Emmett, D. (1976). *Nature*, 259, 338-340
Bender, M.A. and Chu, E.H.Y. (1963). *In* "Evolutionary and Genetic Biology of Primates", (J. Buettner-Janusch, ed), vol. II, pp. 261-310, Academic Press, New York and London.
Bender, M.A. and Metler, L.E. (1958). *Science*, 128, 186-190.
Bengtsson, B.O. (1975). *Hereditas*, 79, 287-292.
Bruere, A.N. (1974). *Proc. 1st World Congr. appl. Livest. Prod., Madrid*, 1, 151-175.
Brumback, R.A., Staton, R.D., Benjamin, S.A. and Lang, C.M. (1971). *Folia primat.*, 15, 264-273.
Buettner-Janusch, J. (1973). "Physical Anthropology: a Perspective", John Wiley, New York.
Campbell, C.B.G. (1974). *Mammal Rev.*, 4, 125-143.
Cartmill, M. (1975). *In* "Phylogeny of the Primates: a Multidisciplinary Approach", (W.P. Luckett and F.S. Szalay, eds), pp. 313-354, Plenum Press, New York.
Charles-Dominique, P. and Martin, R.D. (1970). *Nature*, 227, 257-260.
Chiarelli, B. (1974). *In* "Prosimian Biology", (R.D. Martin, G.A. Doyle and A.C. Walker, eds), pp. 871-880, Duckworth, London.
Chu, E.H.Y. and Bender, M.A. (1961). *Science*, 133, 1399-1405.
Darlington, C.D. and Haque, A. (1955). *Nature*, 175, 32.
de Boer, L.E.M. (1973a). *Genetica*, 44, 155-193.
de Boer, L.E.M. (1973b). *Genetica*, 44, 330-367.
de Boer, L.E.M. (1974). *Genen Phaenen*, 17, 1-115.
Dene, H., Goodman, M., Prychodko, W. and Moore, G.W. (1976). *Folia primat.*, 25, 35-61.
Egozcue, J. (1969). *In* "Comparative mammalian Cytogenetics", (K. Benirschke, ed.), pp. 353-389, Springer-Verlag, New York.
Egozcue, J. (1972). *Folia primat.*, 17, 171-176.
Ford, C.E., Hamerton, J.L. and Sharman, G.B. (1957). *Nature*, 180, 392.
Gingerich, P.D. (1976). *Mus. Pal. Univ. Michigan, Pap. Pal.*, 15, 1-140.

Goodman, M. (1973). *Symp. zool. Soc. Lond.*, 33, 339-375.
Goodman, M., Farris, W. (Jr.), Moore, W., Prychodko, W., Poulik, E. and
    Sorenson, M. (1974). *In* "Prosimian Biology", (R.D. Martin, G.A. Doyle,
    and A.C. Walker, eds), pp. 881-890, Duckworth, London.
Gropp, A., Winkling, H., Zech, L. and Müller, H. (1972). *Chromosoma*, 39
    265-288.
Gustavsson, I. and Rockborn, G. (1964). *Nature*, 203, 990.
Hamerton, J.L. (1963). *Symp. zool. Soc. Lond.*, 10, 211-219.
Hamerton, J.L. (1968). *Cytogenetics*, 7, 260-276.
Hennig, W. (1966). "Phylogenetic Systematics", University of Illinois Press,
    Urbana, Illinois.
Hill, J.P. (1932). *Phil. Trans. Roy. Soc. Lond* ., (B) 221, 45-178.
Hill, W.C.O. (1953). "Primates", vol. I ("Strepsirhini"), Edinburgh Uni-
    versity Press, London.
Hoffstetter, R. (1974). *J. Hum. Evol.*, 3, 327-350.
Hsu, T.C. and Benirschke, K. (1967-1974). "An Atlas of Mammalian Chromo-
    somes", 8 vols. Springer-Verlag, New York.
Hubrecht, A.A.W. (1908). *Quart. J. micr. Sci.*, 53, 1-181.
Huxley, J. (1958). *Uppsala Univ. Arsskr.*, 6, 21-39.
Le Gros Clark, W.E. (1971). "The Antecedents of Man", Edinburgh University
    Press, London.
Leutenegger, W. (1973). *Folia primat.*, 20, 280-293.
Luckett, W.P. (1974). *In* "Prosimian Biology", (R.D. Martin, G.A. Doyle and
    A.C. Walker, eds), pp. 475-488, Duckworth, London.
Luckett, W.P. (1975). *In* "Phylogeny of the Primates: a Multidisciplinary
    Approach", (W.P. Luckett and F.S. Szalay, eds), pp. 157-182, Plenum
    Press, New York.
Luckett, W.P. and Szalay, F.S. (eds) (1975). "Phylogeny of the Primates: a
    Multidisciplinary Approach", Plenum Press, New York.
Manfredi-Romanini, M.G., de Boer, L.E.M., Chiarelli, B. and Tinozzi-Massari, S.
    (1972). *J. Hum. Evol.*, 1, 473-476.
Martin, R.D. (1968a). *Man,* 3, 377-401.
Martin, R.D. (1968b). *Z. Tierpsychol.*, 25, 409-532.
Martin, R.D. (1972). *Phil. Trans. Roy. Soc. Lond.*, (B) 26, 295-352.
Martin, R.D. (1973). *Symp. zool. Soc. Lond.*, 33, 301-337.
Martin, R.D. (1975). *In* "Phylogeny of the Primates: a Multidisciplinary
    Approach", (W.P. Luckett and F.S. Szalay, eds), pp. 265-297, Plenum Press,
    New York.
Martin, R.D., Doyle, G.A. and Walker, A.C. (1974). "Prosimian Biology",
    Duckworth, London.
Matthey, R. (1945). *Experientia*, 1, 50-56 and 78-86.
Matthey, R. (1949). "Les Chromosomes des Vertébrés", F. Rouge, Lausanne.
Mayr, E. (1974). *Z. zool. Syst. Evol.-forsch* ., 12, 94-128.
Nobach, C.R. (1975). *In* "Phylogeny of the Primates: a Multidisciplinary
    Approach", (W.P. Luckett and F.S. Szalay, eds), pp. 199-218, Plenum Press,
    New York.
Ohno, S. (1970). "Evolution by Gene Duplication", Springer-Verlag, Berlin.
Ohno, S., Wol, U. and Atkin, N.B. (1968). *Hereditas*, 59, 169-187.
Pocock, R.I. (1918). *Proc. zool. Soc. London*, 1918, 19-53.
Rieck, G.W. (1974). *Proc. 1st World Congr. appl. Livest. Prod.*, Madrid, 1,
    177-190.
Robertson, W.R.B. (1916). *J. Morph.*, 27, 129-331.

Romer, A.S. (1968). "Notes and Comments on Vertebrate Palaeontology", University of Chicago Press, Chicago.

Rumpler, Y. (1975). *In* "Lemur Biology", (I. Tattersall and R.W. Sussman, eds), pp. 25-40, Plenum Press, New York.

Simons, E.L. (1972). "Primate Evolution", Macmillan Co., New York.

Simons, E.L. (1974). *In* "Prosimian Biology", (R.D. Martin, G.A. Doyle and A.C. Walker, eds), pp. 415-433, Duckworth, London.

Simpson, G.G. (1945). *Bull. Amer. Mus. Nat. Hist.*, 85, 1-350.

Simpson, G.G. (1961). "Principles of Animal Taxonomy", Columbia University Press, New York.

Simpson, G.G. (1975). *In* "Phylogeny of the Primates: a Multidisciplinary Approach", (W.P. Luckett and F.S. Szalay, eds), pp. 3-19, Plenum Press, New York.

Staton, D. (1967). *Mammal. Chromosomes Newsl.*, 8, 203-219.

Szalay, F.S. and Katz, C.C. (1973). *Folia primat.*, 19, 88-103.

Tattersall, I. and Schwartz, J.H. (1974). *Anthrop. Pap. Amer. Mus. Nat. Hist.*, 52, 139-192.

Tattersall, I. and Sussman, R.W. (eds) (1975). "Lemur Biology", Plenum Press, New York.

Tobias, P.V. (1953). *S. Afr. J. Sci.*, 50, 134-140.

Todd, N.B. (1967). *Mammal. Chromosomes Newsl.*, 8, 268-279.

Wahrman, J. and Zahavi, A. (1955). *Nature*, 175, 600-602.

Wallace, C. and Fairall, N. (1967). *S. Afr. J. Sci.*, 63, 482-486.

White, M.J.D. (1954). "Animal Cytology and Evolution", Cambridge University Press, London.

CANDIDATES FOR ANTHROPOID ANCESTRY: SOME MORPHOLOGICAL AND PALAEOZOOGEOGRAPHICAL CONSIDERATIONS

G.C. CONROY

Department of Cell Biology, New York University Medical Center, New York, New York 10016, USA and Department of Anthropology, New York University, New York, New York 10003, USA.

INTRODUCTION

During the 20th century, several little-known (and perhaps best forgotten?) candidates have been put forward as representing the earliest known anthropoid-like primates in the fossil record. All of the specimens have been described previously (some more than once) and opinions vary considerably as to their proper taxonomic placement. It is often difficult to separate what we *really* know in palaeoprimatology from what we *think* we know and it is clear (to me, at least) that the specimens reviewed here are no basis for some of the sweeping generalizations deduced from them elsewhere. At the risk of alienating some readers, I would suggest that any dogmatic interpretation of the prosimian-anthropoid transition goes far beyond what the facts warrant.

This somewhat pessimistic assessment stems from two basic concerns. Firstly, I have yet to be convinced that fossil primate taxonomy, the way it has been practiced in the past, has anything necessarily to do with biological reality. Appropriate biological controls are often lacking. For example, there has never been an adequate study of dental variability *within* a natural population of extant non-human primates (J. Phillips, pers. comm.). This type of study can be the only appropriate standard by which to judge dental variability in fossil primates. The second concern involves the preoccupation of most paleoprimatologists with determining *direct* fossil lineages. If one recalls that all the known primates from the Oligocene and Miocene of Africa occur in a total geographic area sampling only 1/3,000,000th of the present day land surface of that continent (Simons, 1975), it stretches the imagination to believe that we have found any *direct* ancestors of later apes and ultimately man. I believe it more realistic to speak in terms of 'morphological grades' of evolution. Thus, we can speak of the Fayum primates, for example, as having attained an 'anthropoid grade' of evolution that entitles them to be considered broadly ancestral to later un-doubted hominoids (and cercopithecoids) without necessarily being directly an-cestral. By 'broadly ancestral', I mean that the palaeontological record samples a minute part of a 'population pool' whose members shared a basically similar morphological grade and from which later forms evolved. The fossils themselves present an image of what this 'pool' must have looked like morpho-logically, but they need not represent the actual direct ancestor.

In terms of the specimens reviewed here, some features of morphology and palaeozoogeography will be considered. It seems apparent that by Oligocene times an 'anthropoid grade' of evolution had been reached which in strict morphological terms can be said to be more 'platyrrhine-like' than anything else.

ANTHROPOID CANDIDATES

Pondaungia cotteri

This genus of fossil primate, discovered near Pangan, Pakokku district, Burma, was among Dr. G. Cotter's collection of fossil vertebrates from the Pondaung Sandstones of Upper Eocene age. Although the holotype was discovered before 1916, it was not described by Pilgrim until 1927 (Pilgrim, 1927). The holotype, GSI 201-203, consists of a left maxillary fragment with M^{1-2}, a left posterior horizontal ramus with M_{2-3}, and a right posterior horizontal ramus with M_3. Although there seems to be a 'general consensus' that these remains are of a fossil anthropoid, this 'consensus' is hardly unanimous. It has been considered both a lemuroid (Madden, 1975) and a condylarth (Von Koenigswald, 1965).

The main 'lemuroid' similarities are found in the morphology of the upper molars. They possess a pseudohypocone which results from cleavage of the proto-cone, a condition typical of Eocene notharctines (e.g., *Notharctus*). This would seem to be a clearcut similarity between *Pondaungia* and Eocene lemuroids if it were not for the fact that certain South American primates exhibit a rather similar morphology. Thomas (1913; cited in Rosenberger, 1976) in his study of ceboid maxillary molars, noted that the size of the hypocone and its connection to the trigon could be arranged in a graded series from a virtual tri-cuspid tooth to a quadricuspid one with no abrupt transitions. At one extreme hypocones are virtually absent in callithricids, but become progressively lar-ger in a graded series through the cebids, with *Saimiri* and *Callimico* occupying the 'middle zone' (Rosenberger, 1976). The interpretation of this morphocline illustrates one of the difficulties in determining just what the ancestral morphotype in ceboids was. The determination of the ancestral morphotype is crucial, of course, in cladistic analysis. Gregory (1920) considered a morpho-logy like that seen in *Callicebus* (and *Alouatta*) as ancestral, whereas others (Rosenberger, 1976; Orlosky and Swindler, 1975) have considered a *Saimiri* model as primitive and all other morphologies as derived. In any event, it has yet to be demonstrated that the presence of a pseudohypocone has any zoo-logical significance in modern populations for taxonomic purposes, particularly at the sub-ordinal level. As an example of how variable cusp formation can be, a single modern population of *Papio anubis* has disto-lingual cusps which appear to be formed either by cleavage of existing cusps (analagous to the mode of development of pseudohypocones in notharctines?) or as fully separate cusps (analagous to hypocone development in adapines?).

In the upper dentition, *Pondaungia* exhibits broad, 3-rooted molars having low cusps with a tritubercular pattern and a pseudohypocone. In the lower molars, the M_3 is long and narrow with a median hypoconulid, the six main cusps enclosing a large, but shallow, basin. The trigonid is slightly higher than the talonid and a small paraconid is present. Pilgrim (1927) clearly pointed out the overall similarity of the lower molars to *Pelycodus* but noted that the crests on the trigonid and talonid characteristic of that genus were absent in *Pondaungia*. In spite of these general resemblances to *Pelycodus*, Pilgrim con-sidered *Pondaungia* to be anthropoid due mainly to the brachyodont cusps and the large basin-shaped area of M_3.

Von Koenigswald (1965), in his 'critical observations' of the Burmese fossils, considered *Pondaungia* a condylarth for some of the same reasons that Pilgrim (1927) had considered it an anthropoid, namely, the presence of low, rounded cusps. He also considered that the lack of an 'outer wall' in the upper molars

formed by the paracone and metacone and the presence of external cingulum in
the upper molars without styles aligned *Pondaungia* with condylarths rather than
prosimian or anthropoid primates. These features of the external part of the
upper molars in *Pondaungia* are difficult to assess, however, since this surface is extremely weathered. The lack of ectocingular styles is no reason (a
priori) to exclude *Pondaungia* from Anthropoidea, however. In platyrrhines
there is a positive correlation between tooth size and ectocingular expressions such as styles. For example, in the smaller subspecies of the howler
monkey, *Alouatta palliata mexicana*, the ectocingular styles are reduced so
much that the mesostyle is absent on M^2, and M^3 is completely devoid of styles.
Other platyrrhines such as *Ateles* and *Lagothrix* also lack most ectocingular
expression even though they are relatively large monkeys (Zingeser, 1973).
Certain fossil adapids also lack significant ectocingular development (Wilson
and Szalay, 1976).

Von Koenigswald's (1965) point that other condylarths are known from the
early Tertiary of Asia (*Promioclaenus* from Pakistan and *Phenacolophus* from
Mongolia) is well taken, but the morphology of *Pondaungia* does not necessarily
make it one.

Both Madden (1975) and Von Koenigswald (1965) have questioned the anthropoid
affinities of *Pondaungia* (and *Amphipithecus*) on palaeozoogeographic grounds,
arguing that these fossils are representative of an endemic Eurasian fauna which
could not have given rise to undoubted anthropoids found in the Early Oligocene
of North Africa (Fayum, Egypt). This question will be discussed in more detail
below, but suffice it to say that certain Eurasian mammals did reach Africa by
Early Oligocene times, notable creodonts and anthracotheres (Simons and Wood,
1968; Colbert, 1938; Simons, 1971a, b; Simons and Gingerich, 1974).

Amphipithecus mogaungensis

This specimen, AMNH 32520, was discovered by Barnum Brown in the Upper
Eocene Pondaung Sandstones near Mogaung, Burma, in 1923. It was subsequently
overlooked until Colbert's description in 1937 (Colbert, 1937). The fossil
consists of a left mandibular ramus with P_4-M_1 preserved. It is distinguished
by the great depth of the mandibular ramus and the short, vertical symphyseal
region. Colbert (1937) inferred the dental formula to be ?133, and if correct,
there is no diastema between the canine and premolar.

As is the case with *Pondaungia,* opinions have differed widely as to *Amphipithecus*' affinities. Thus, Von Koenigswald (1965) and later Szalay (1970,
1972) considered it a lemuroid whereas its original describer, Colbert (1937),
and Simons (1971b, 1972) among others, considered it to be anthropoid. Colbert
(1937) and Simons (1971b) convincingly argued against its placement in either
condylarths (e.g., *Hyopsodus*) or the bunodont artiodactyls (e.g. *Wasatchia*),
by demonstrating the distinctiveness of *Amphipithecus* in premolar and mandibular morphology.

In his comparison of *Amphipithecus* with various anthropoids, Colbert (1937)
noticed certain similarities with New World primates. 'The presence of three
premolars in *Amphipithecus* at once suggests the possibility of a relationship
with the South American Cebidae. Not only the dental formula but also the deep
mandibular ramus and the abbreviated symphysis are characters by which it
resembles after a fashion New World monkeys.' He remarked specifically on the
strong morphological resemblance to *Alouatta*. Colbert's major reason for dis-
missing the similarities with the New World monkeys was his contention that
the latter had very large second premolars whereas this tooth in *Amphipithecus*

was obviously small. He thus considered the small premolar in *Amphipithecus*
as a retention of a lemuroid or tarsioid characteristic in this primitive
anthropoid. It remains to be pointed out, however, that not all platyrrhines
have enlarged second premolars. The size of the second premolar in the platyrrhines is functionally related to the size of the paracone blade of the upper
canine with which it forms a 'honing notch' (Zingeser, 1973). Thus, P_2 is not
a relatively enlarged tooth in *Brachyteles*, for example.

     After pointing out various morphological similarities to certain platyrrhines,
Colbert (1937) concludes that the 'resemblance is not close enough to indicate
any true affinity' and that 'it may be rather a parallelism in the development of these teeth'. Resorting to parallelism to explain morphological similarity should be used judiciously, otherwise the palaeontological tenet that
genetic similarity underlies morphological similarity becomes practically in-
operable. Recourse to parallel evolution should be invoked ideally when one
has evidence from sources beyond the realm of gross morphology alone which
makes parallelism the only viable explanation (e.g., evidence from geophysics
or biochemistry). The phenomenon of 'disjunct endemism' is a case in point.
Three Late Cretaceous dinosaurs, *Laplatosaurus*, *Titanosaurus*, and *Antarcto-
saurus* are present in South America and India. *Titanosaurus* is also present
in both Europe and Africa whereas *Laplatosaurus* is also found in Madagascar.
By late Cretaceous, most geophysicists agree that India had been moving away
from both South America and Africa for some 100 million years. By Late
Cretaceous India should have been isolated in the Indian Ocean and yet it
shares similar dinosaurs with South America. There are three possibilities
for solving the dilemma: the fossils have been misidentified, the land connection between India and South America could not have been severed so early,
or remarkable degrees of parallel evolution must have taken place among the
three genera (Hallum, 1972).

     Colbert (1937) concluded his description of *Amphipithecus* by provisionally
placing the specimen in ?Simiidae mainly due to its deep mandibular ramus, well
developed lingual torus, possession of a deep genioglossus pit, the abbreviated
symphysis, and the crowding of the canine and premolar teeth. This interpretation was accepted and expanded upon by Simons (1971b). For an opposing view,
see Szalay (1970, 1972).

Cercamonius brachyrhynchus

     This specimen, first described by Stehlin in 1912 as *Protoadapis brachy-
rhynchus*, is from the Quercy phosphorites of southern France. Thus, it is
presumably of Late Eocene age (Bartonian). The fossil has been recently re-
described by Gingerich (1975a) as a new genus, *Cercamonius*, and subfamily,
Cercamoninae, of fossil primate. Gingerich (1975a) believes the specimen represents a genus closely related to the origin of Old World anthropoid primates.

     In molar structure, *Cercamonius* resembles other adapid primates such as
Protoadapis and *Notharctus*, but differs from them in having a shorter, relatively deeper mandible. The major reasons for considering this primate as
close to the origin of Higher Primates (besides the relatively deep mandible)
are that the lower canine is large and vertically implanted, the P_3 is large
and set obliquely in the jaw and functioned with P_2 as a hone for the larger
upper canine (Gingerich, 1975a). These, indeed, would be interesting similarities to Higher Primates; unfortunately, the upper canine, lower canine,
P_2 and P_3 do not exist. The honing function of the P_3 is inferred by the
position of the alveoli for its roots, which are set somewhat obliquely in the

jaw. However, such an oblique orientation of P_3 does not necessarily imply anthropoid-type honing, as reference to Gregory's (1920, p. 189) illustration of *Protoadapis recticuspidens* clearly shows. As far as large, vertically implanted canines go, they seem to be typical of many adapids.

One difference between this specimen and many other adapines is the apparent absence of P_1 (however, there is slight damage between the alveolar margins of C and P_2 in the type and only specimen of this genus). *Cantius* also shows this apparent reduction in the number of premolar teeth (Simons, 1972).

Gingerich (1975a) compares *Cercamonius* most favourably with *Aegyptopithecus* of the known Oligocene anthropoids. This comparison seems to be based mainly on similarity in length of teeth and the fact that they both share large, vertically oriented canines. However, besides lacking the anteroposteriorly crowded and relatively broad cheek teeth of both *Aegyptopithecus* and *Amphipithecus*, the preserved cheek teeth in *Cercamonius* are morphologically quite different from those of *Aegyptopithecus*.

"Kansupithecus"

The previous three candidates for the ultimate anthropoid ancestor, *Pondaungia*, *Amphipithecus*, and *Cercamonius*, all come from Late Eocene sites in Eurasia. In spite of this supposed wide distribution of Late Eocene primitive anthropoids in Eurasia, virtually no primates (prosimian or anthropoid) have been discovered in subsequent Oligocene deposits in Europe in spite of numerous sites of this age (e.g., Ronzon and part of the Quercy phosphorites in France, the Mainz Basin in Germany). In Asia, however, a poorly known fossil, "Kansupithecus", was discovered from what was thought to be Oligocene age deposits and was considered an anthropoid by its discoverer (Bohlin, 1946). The specimen, a symphyseal fragment, was recovered from the Tabun-Bulek region of China and, if correctly identified, would be the only occurrence of an anthropoid primate in the Old World outside Africa in the Oligocene. It would leave open to question the current opinion that anthropoids differentiated exclusively in Africa during the Oligocene.

The presence of an anthropoid primate in Oligocene age deposits in China would be remarkable for at least one reason: the Late Eocene Tientong Sandstone fauna of Kwansi Province and the Oligocene faunas of Mongolia (Ulan Gochu and Hsanda Gol Formations) and southern China (Tsaichiachung marls) have not produced any known specimens of anthropoid primates (Chow, 1957; Young and Bien, 1939). This, admittedly negative, evidence suggests that the relative dating of "Kansupithecus" may be incorrect (Conroy and Bown, 1974).

There are several mammals found in the Tabun-Bulek fauna which suggest that a fauna younger than the Late Oligocene might be present in the deposits. Of particular significance is the presence of the ctenodactylid rodent, *Sayimys*. *Sayimys* is known from Middle Miocene deposits in the Siwaliks and in North Africa (Wood, 1937; Thenius, 1959). Significantly, *Tataromys plicidens*, which is thought to be ancestral to *Sayimys*, is associated with Late Oligocene faunas and is not present in those sites yielding *Sayimys* in the Tabun-Bulek region (Bohlin, 1946). Remains of both proboscideans and bovids also suggest a younger age for the "Kansupithecus" specimen. Proboscideans did not reach Eurasia much before Burdigalian times (circa 19-16.5 m.y.) (Van Couvering, 1972). The earliest known bovids are boselaphines (*Eotragus*) from Burdigalian and Vindobonian deposits of Europe (Gentry, 1970). Bovids from the Chinji Formation share at least one genus, *Protragoceros*, with Ft. Ternan and Ngorora in Kenya, whereas the Nagri and lower Dhok Pathan bovids are more Eurasian in

character. A new wave of African bovids appears in the Upper Dhok Pathan (H.
Thomas, pers. comm.). Thus it seems most reasonable, pending further discoveries in China, to view the provenance of "Kansupithecus" as Middle Miocene, a
time when anthropoids had spread widely throughout Eurasia (Conroy and Bown,
1974).

FAYUM PRIMATES

The primate fauna from the Jebel el Qatrani Formation of Egypt has been
extensively reviewed in recent years (Simons, 1972; Delson and Andrews, 1975;
Cachel, 1975; Conroy, 1976a). Thus, I will restrict my remarks to some more
specific developments in the study of these early anthropoids.

Of particular importance is the recent description concerning the anatomy
of the temporal bone in *Apidium phiomense* (Gingerich, 1973). One petrosal
fragment (YPM 23968) preserves the cochlea, semicircular canals, round and
oval windows, portions of the facial canal and the canal for the internal carotid artery. The most interesting aspect of the morphology is the apparent
lack of a stapedial branch of the internal carotid artery. This is a resemblance to Anthropoidea and differs from prosimians. It should be remembered,
however, that anthropoids do have a stapedial artery in fetal life. The
stapes in *Homo* is formed mainly from mesenchyme of the second branchial arch
(Reichert's cartilage) and both the mesenchymal and cartilagenous stages of
the stapes are perforated by the stapedial artery up until the third month of
intrauterine life at which time the artery normally disappears (Arey, 1974).
The second fragment, YPM 23968, is a right squamosal of *Apidium phiomense*.
Gingerich (1973) concluded that the tympanic ring was quite near the lateral
edge of the skull, which precluded any extension forming an osseous auditory
tube. He sees the greatest overall similarity in morphology with small platyrrhines and with lemuroids. The configuration of the ectotympanic is different between lemurids and platyrrhines, however, being intrabullar in the
former and forming part of the lateral bullar wall in the latter. He suggests
that *Apidium* was "lemur-like" in having a free intrabullar ectotympanic with
an unfused articulation between the anterior crus and the squamosal. This
morphology is not necessarily "lemur-like". The unfused squamosal-tympanic
articulation characterizes early postnatal development in *Homo* as well as
specimens of *Tarsius* and various platyrrhines (Cartmill, 1975; Hershkovitz,
1974). If the morphological details of the two specimens are considered together (and if they truly represent one species of fossil primate) the overall
similarities are clearly with platyrrhines.

Recent investigations of the appendicular skeleton suggest that postcranial
morphology of the Fayum primates also most closely resembles that of platyrrhines among extant primates (Conroy, 1976a). Postcranial remains attributed
to *Apidium* (and possibly *Parapithecus*) closely resemble small, arboreal quadrupeds such as *Cebus* and *Saimiri* in overall morphology and presumed locomotor
habits. A large primate ulna has been referred to the hominoid *Aegyptopithecus
zeuxis*. It is virtually identical in size and morphology to extant *Alouatta*
(Conroy, 1976a; Fleagle, Simons, Conroy, 1975; Preuschoft, 1974). This
specimen (YPM 23940) comes from Yale Quarry M, the same quarry in which the
skull of *A. zeuxis* was discovered.

Until recently, this ulna was the only known postcranial specimen of an
Oligocene hominoid. However, an hallucial metatarsal has been found which
also has been referred to as *A. zeuxis* (Conroy, 1976b). This specimen, then,
represents our only knowledge of hallucial morphology in the earliest anthro-

poids. Several hallucial metatarsals of Eocene primates are known, e.g., *Notharctus* and *Hemiacodon* (Gregory, 1920; Simpson, 1940; Szalay, 1976). The Fayum specimen (YPM 25806) consists of the proximal two-thirds of a right hallucial metatarsal. Its overall size suggests an animal whose body weight might be approximated by *Ateles belzebuth*. Of particular interest is a well defined, raised, oval facet on the medioplantar aspect at the base of the metatarsal. This undoubtedly reflects the fact that a large prehallux bone participated in the hallucial tarsometatarsal joint. Thus, this joint consisted of three bones in primitive anthropoids, the hallucial metatarsal, medial cuneiform, and prehallux bone (Conroy, 1976b). This condition is found only in platyrrhines and hylobatids among extant primates. The prehallux is absent in cercopithecoids and pongids (Lewis, 1972). In those primates possessing a prehallux, it is found in the tendon of tibialis anterior muscle and forms part of a synovial joint with the first metatarsal and medial cuneiform (Lewis, 1972).

The basal surface for articulation with the medial cuneiform is saddle-shaped, an obvious adaptation for conjunct rotation during opposition of the hallux (see MacConaill, 1953). Another adaptation to opposition is the degree of torsion seen in the metatarsal shaft (Conroy, 1976b).

The peroneal tubercle is a large, blunt projection for the insertion of the peroneus longus tendon. This muscle, an everter of the tarsus and adductor of the hallux, was obviously well developed in this primitive, arboreal, fossil ape. By way of contrast, *Simopithecus*, a large, terrestrially adapted Pleistocene monkey, had a rather small peroneus muscle. This is suggested by the relatively small peroneal tubercle and the small facet for the peroneus longus sesamoid on the cuboid (Jolly, 1972). It is interesting to note that *Dryopithecus africanus*, a putative descendant of *A zeuxis* , also apparently had a prehallux incorporated into its hallucial tarsometatarsal joint (Lewis, 1972). Thus, there is nothing in the anatomy of the hallucial tarsometatarsal joint which would preclude *Aegyptopithecus* from being broadly ancestral to later dryopithecines and hylobatids.

It is evident from reference to Simpson's (1940) and Szalay's (1976) illustrations that the hallucial metatarsal of omomyids (*Hemiacodon*) was quite different from that of *Aegyptopithecus*, particularly in the morphology of the peroneal tubercle.

The small tarsal bones referred to *Apidium* and/or *Parapithecus* are distinct in several ways from those of paromomyiforms or omomyids (Conroy, 1976a). Unlike omomyids, the calcaneal body is not elongated anteriorly and there is a peroneal tubercle. Calcaneal elongation characterizes *Hemiacodon, Teilhardina, Tetonius, Necrolemur,* and *Nannopithex* (Szalay, 1975, 1976) among Eocene primates. In this aspect of its morphology, *Apidium* cannot be considered primarily a leaping form, but rather, a more generalized arboreal quadruped. The morphology of the calcaneocuboid facet also differs. The articular surface for the cuboid in the Fayum primates is a crescent-shaped, gently concave facet. The articular surface is more extensive on the lateral side than on the medial side. This is opposed to the condition seen in the condylarth *Protungulatum* (Szalay and Decker, 1974). Due to the "screw-like" nature of the transverse tarsal joint, this is an adaptation for inversion of the tarsus (Conroy, 1976a). In omomyids, this articular surface is a more rounded, pivotal surface.

The long axis of the posterior calcaneal facet is approximately 15° from the long axis of the calcaneum in the Fayum specimens. This is in contrast to both the 34-40° regarded by Szalay and Decker (1974) as the primitive eutherian condition and the large value for this angle often seen in certain extant jumping mammals such as *Zapus* (Stains, 1959).

Szalay and Decker (1974) considered the morphology of the astragalocalcaneal facet and calcaneoastragalar facets in omomyids, adapids, and paromomyiforms as indicative of "screw-like" rotation at the lower ankle joint. Such rotation is considered by them as a primitive character for the Order Primates as a whole. However, such a mechanism was noted long ago in the highly specialized foot of *Homo* (Manter, 1941). This "screw-like" rotation is also apparent in the Fayum tarsal specimens (Conroy, 1976a).

The talus of *Apidium* also shows certain distinctions from Eocene primates. For example, there is no posterior extension of the trochlear body into a "posterior shelf" as there is in adapids and some omomyids (e.g., *Teilhardina*) (Decker and Szalay, 1974). The astragalar canal, present in some paromomyiforms such as *Plesiadapis tricuspidens* and some lemuroids, like *Archaeolemur* and *Megaladapis,* is likewise absent.

The articular surface on the talar head is more extensive on the lateral side than the medial side, reflecting a disposition towards tarsal eversion. However, this talar head morphology does not necessarily mean that such animals were adapted primarily to terrestrial, plantigrade quadrupedalism, *contra* Szalay and Decker (1974). For example, the normal human foot cannot achieve as highly an everted position as that shown to be the common terrestrial position of the chimpanzee. In fact, the human foot develops an anterolateral process on the calcaneum which prevents excessive eversion of the foot (Elftman and Manter, 1935). It is clear from numerous other characteristics of the Fayum tarsals that *Apidium* was not a terrestrially adapted primate (Conroy, 1976a).

Much controversy has recently surrounded the taxonomic placement of *Apidium* and *Parapithecus* (Simons, 1974; Cachel, 1975; Delson and Andrews, 1975; Delson, 1975 among others). Earlier views on their taxonomy has been reviewed by Simons (1972) and are really only of historical interest due to the inadequate nature of the fossil material before the 1960s.

Simons (1974) has argued that *Parapithecus* and the closely related *Apidium* have reached an anthropoid grade of evolution and should be in their own family, Parapithecidae, within the superfamily Cercopithecoidea. This view is based on the acquisition of a suite of Higher Primate characters found in these two genera: fused mandibular symphysis, lack of metopic suture, postorbital closure, dental eruption patterns, and resemblances in the cheek teeth between *Parapithecus* and the extant cercopithecine *Cercopithecus talapoin* (Simons, 1972; Conroy, Schwartz and Simons, 1975). Delson (1975) and Delson and Andrews (1975), arguing from cladistic principles, have minimized these resemblances by stressing that *C. talapoin* is a highly derived modern species and is not a proper model for the ancestral morphotype of cercopithecoids. One example of a perceived derived feature in *Cercopithecus* is the loss of M_3 hypoconulids (Delson, 1975). However, a perusal of Delson's papers gives the clear impression that the major (perceived) derived feature in *Parapithecus* which excludes it from ancestry of later cercopithecoids is the P_2 honing present in these three premolared catarrhines. He believes that such a specialization is unlikely to have been lost and then redeveloped again on P_3 as it is in later cercopithecoids. Gingerich (1975a), however, has nicely demonstrated (although in a different context) how this transformation of honing from one premolar to the next might have taken place through time. In a specimen of *Leptadapis magnus,* the upper canine honed against the small P_1 and the larger P_2. As the P_1 atrophies, the honing between the upper canine and P_2 continues without interruption. Thus, premolars can be reduced through time

without disturbing the honing function (Gingerich, 1975a). A similar mechanism
may be seen in the fossil adapid, *Margarita stevensi* (Wilson and Szalay, 1976).

Based on his deductions of what the ancestral cercopithecoid morphotype
should be, Delson (1975) considers other features found in *Parapithecus* to be
unsuitable for its inclusion in Cercopithecoidea: these include small P_4 metaconids, short M_3's with little or no hypoconulid development, and the P_2 honing
mentioned above. However, none of these features is uncommon in modern cercopithecoids (except P_2 honing of course). Small P_4 metaconids are found in
colobines, short M_3's in many smaller species, and lack of hypoconulids on M_3
in *Cercopithecus* and *Erythrocebus*. It is my feeling that cladistics provides
a necessary rigor in paleontological systematics, but only the fossil record
in conjunction with a good biostratigraphical sequence can be the final arbiter
as to whether or not inferred primitive morphotypes are correct. As Simpson
(1975) points out, primitiveness and ancientness are not necessarily related
but they usually are.

Because of the seemingly minor morphological details cited above, some
would remove *Parapithecus* and *Apidium* from Cercopithecoidea and create instead
a new superfamily, Parapithecoidea, for them. The problem of taxonomic placement is difficult for many reasons, not least of which is the lack of taxonomic
standards in palaeontology that are biologically relevant and universally
applied. It is sometimes difficult to distinguish between modern primate
species even when ecology, behaviour, dentition, blood, saliva, dermato-
glyphics, hair samples and whole skeletons are known. One cannot help but
wonder, then, whether obsession with taxonomic minutiae in palaeontology is
a biologically viable endeavour or not. Some palaeontologists have even gone
so far as to suggest that evidence from "soft anatomy" (presumably including
placentation and biochemistry) should be ignored in working out phylogeny since
it cannot be ascertained in fossils (Gingerich, 1973). One may legitimately
ask, then, whether such phylogeny has any relation to biological reality?

At the other extreme, we have the situation where most authorities would
include in one Eocene prosimian family, Omomyidae, not less than 14 genera
which differ from one another in many significant ways (Szalay, 1975, 1976).
For example, in the basicranium the petromastoid is inflated in some (*Necrolemur* and *Tetonius*) but not in others (*Rooneyia*); the number of teeth differs
widely (*Rooneyia* is reconstructed with 8 upper teeth, while many other omomyids
have 9 and *Teilhardina* is reconstructed with 10; the number of lower teeth
varies from 7-9). The dentition of members of the two subfamilies, omomyines
and anaptomorphines, differs morphologically, particularly in details of the
third molars. The inferred dietary habits of omomyids also differ more than
those known in a family of modern monkeys - *Macrotarsius* is considered a herbi-
vore and *Omomys* an insectivore. Accordingly, functional areas of cheek teeth
between these two genera differ markedly.

Thus given the variability seen in various Eocene primate *families*, it
seems to me unwarranted to place *Apidium* and *Parapithecus* in a different *superfamily* from Cercopithecoidea. This does not necessarily mean that parapithecids
are directly ancestral to later cercopithecoids; the evidence is not clear
enough on this point. It is not necessary to include only direct ancestors
of later cercopithecoids in that superfamily. Too much emphasis is placed
on the discovery of direct ancestors. I believe the fossil record is more
manageable if one considers "grades" of evolution rather than postulating
direct lineages (unless the fossil record is exceedingly good, which it is not
for primates generally). The Fayum primates are a diverse lot, yet they have
all reached an anthropoid "grade" of evolution (or "monkey-like" grade, if you

will). This morphological "grade" is clearly similar to that seen in living
platyrrhines. All this implies is that living South American primates have
retained the "grade" of evolution that was typical of the earliest anthropoids
from the Fayum. In a strict morphological sense, there is little more reason
for considering *Aegyptopithecus* a hominoid than there is for considering
Alouatta one! Our conception of primate fossil taxonomy is rooted (subconsciously perhaps) as much in palaeogeography as it is in morphology. The
important fact of primate evolution (to me at least) is that sometime in the
Late Eocene-Early Oligocene an anthropoid "grade" of evolution was reached in
both the Old and New Worlds presumably via an Eocene family which had a holarctic distribution. A "hominoid" grade of evolution was not reached until Early
Miocene times (typified by *Dryopithecus africanus*) and was restricted to the
Old World (leaving Dr. Napier's treatise on Bigfoot aside!).

PALAEOZOOGEOGRAPHIC CONSIDERATIONS

    The numerous similarities between the Oligocene anthropoids and extant
platyrrhines have prompted several workers to suggest an African origin for
the New World monkeys by rafting across the Atlantic some time in the early
Tertiary (Lavocat, 1974; Hoffstetter, 1974). Others have the rafting going
in the other direction, from South America to Africa (Szalay, 1975, 1976).
Various estimates have been calculated for the initial break-up of western
Gondwanaland. Given the overall estimate for sea-floor spreading rates of
2cm/yr, South America and Africa began their separation some 130 million years
ago and by early Tertiary times are calculated to have been some 3,000 km
apart. Estimates for sea-floor spreading rates in the northern South Atlantic
are a bit slower, some 1.6-2.0 cm/yr (Maxwell et al., 1970; Tarling and Tarling,
1971; Keast, 1972). Final separation between Africa and South America was
most probably a mid-Cretaceous event. Evidence for this comes from various
sources. Upper Jurassic beds in Senegal contain marine deposits and marine
evaporite-type beds are found in the Lower Cretaceous in Senegal, Congo, and
Gabon. Also, freshwater ostracods of Upper Jurassic to Lower Cretaceous
deposits of Congo, Angola, and Gabon are similar to those in Brazil (eighteen
species are common to the two areas). The same pattern also holds for Lower
Cretaceous marine trigoniid molluscs. A boa (*Madtsoia*) from the Upper Cretaceous
of Madagascar is similar to that found in the Paleocene of Patagonia (Del
Corro, 1968; cited in Keast, 1972). As far as plants go, Cretaceous spore-
pollen floras from Brazil and the Congo-Gabon area share 34 or 39 taxa (Keast,
1972).
    If the Atlantic was approximately 3,000 km wide by the end of the Eocene,
could primates possibly have crossed it by rafting? It seems highly unlikely
for various reasons. The presence of the fossil monkey, *Xenothrix*, on Jamaica,
some 600 km from South America, is the longest (inferred) rafting journey of
a primate. Also, neither rodents nor primates have been rafted great distances
to isolated oceanic islands. In spite of the fact that the Hawaiian Islands
have been in existence for some 70 million years (and are also some 3,000 km
from the nearest mainland), no placental mammals have ever reached them (see
review in Simons, 1975), nor have they reached Australia.
    The relative position of North and South America to one another is a little
more problematical. It seems reasonable to conclude that some sort of land
bridge or island chain existed in the early Tertiary between the two continents.
Late Cretaceous dinosaurs of South America are more similar to North American
ones (approximately 60% of the genera are in common) than to those in Africa

and Eurasia. However, certain characteristic North American dinosaurs are
absent from South America, namely ceratopsids and hadrosaurs. This suggests
a filter effect of some sort across Central America (Keast, 1972). Interesting in this regard is the presence of *Arctostylops*, a notoungulate, in Late
Paleocene deposits in North America and some enigmatic fused cervical vertebrae
similar to those of edentates from the Middle Eocene of Wyoming (McKenna, 1975).

It is unlikely that the connection between the Americas ever included freshwater streams during the Tertiary. If one compares the freshwater fish between
North and South America, it seems that no North American family of fish can be
ancestral to South American characoids. Characoids are entirely freshwater
forms, having osmoregulatory limitations which prevent them from entering sea-
water. Connections between the two continents in early Tertiary (and late
Mesozoic?) were probably via island arcs (Keast, 1972).

It appears that primates did not reach South America much before early
Oligocene times (Deseadan), when *Branisella* appears in Bolivia (Hoffstetter,
1969). Their absence from the Paleogene is not an artifact of a poor palaeontological record as it is in Africa. South America has produced a more complete
palaeontological sample of Tertiary deposits than has any other southern continent. Unlike the situation in Africa, palaeontological sites yielding
mammals are known for every epoch of the Cenozoic (Patterson and Pasqual, 1972).

The fossil mammals of South America are the result of three major "coloniza-
tions". Mammals first appear in the Cretaceous and consist mainly of condylarths and marsupials. From Late Paleocene to Middle Eocene (Riochican to
Mustersan) the fauna consisted of a diverse group of marsupials, ungulates,
and Edentates (Xenarthra), which presumably arrived as "waif immigrants" from
the north. Of the seven orders present in latest Paleocene, four are regarded
as having come from the north; however, the absence of insectivores and placental carnivores suggests an obvious filtering effect of this presumed land
route (Patterson and Pasqual, 1972). Similarities with African or Eurasian
mammals in the early Tertiary are minimal. The second colonization occurred
in the Late Eocene and Early Oligocene and consisted of primates and caviomorph
rodents (Hoffstatter, 1974; Lavocat, 1974). Haffer (1970, cited in Patterson
and Pasqual, 1972) has found geological evidence suggestive of a discontinuous
land bridge in the Middle Eocene between northern Central America and the
slowly rising Andes of Columbia. No other major group of mammals seems to
have entered South America before the Panamanian isthmus arose some time in
the later Pliocene (except for some procyonids in the Early Pliocene).

The presence of caviomorph rodents in the Early Oligocene of South America
(along with primates) does not necessarily imply a migration from Africa by
rafting. It was thought that the ancestors of the caviomorph rodents were to
be found among the hystricognathous Phiomorpha of Africa (Lavocat, 1974; Hoffstetter, 1974). One support for such a thesis was the apparent lack of ancestral caviomorph-like rodents from the Eocene of North America or South America.
With the recent discoveries of hystricognathous rodents in the Eocene of
Texas and the early Tertiary of Utah this thesis loses much support (Wood,
1972, 1973; Patterson, pers. comm.). Besides the certain faunal resemblances
in the early Tertiary between North and South America discussed above, geological evidence would also suggest a land connection of some sort up until approximately 50 million years ago. An uplift which closed the Central American "flood-
gate" was caused by a northward moving section of seafloor of the Galapagos
Rift Zone descending into an extension of the Middle American Trench along
the coasts of Costa Rica and Panama sometime prior to the end of the Cretaceous
(Sullivan, 1974).

Thus, platyrrhine monkeys (and caviomorph rodents) are best interpreted as arriving from North America sometime in the Late Eocene or Early Oligocene. It is well known that climatic cooling in the northern latitudes was commencing about this time, perhaps causing this southern migration of primates in both the Old and New Worlds (Frakes and Kemp, 1972; Conroy, 1976a).

As mentioned earlier, palaeozoogeographic considerations have weighed heavily in discussions of the systematics of *Amphipithecus* and *Pondaungia* (Von Koenigswald, 1965). The clear implication in Von Koenigswald's paper is that the Tethys Sea constituted an impassable water barrier which prohibited mammal migration from Eurasia to Africa in Late Eocene-Early Oligocene times. There are many lines of evidence that cast doubt on this hypothesis. One of the most telling is the fact that the mid-Eocene fauna from the northern part of the Indian subcontinent (16 species) contains only holarctic elements. This implies that by this time India had migrated sufficiently far northwards that some intermittent land connection with the Eurasion plate was possible. Indeed, some paleomagnatic data would indicate that India might have stayed fairly close to Africa during part of its migration northwards, suggesting the interesting possibility of faunal interchange (limited) between India, Eurasia, and Africa in the early Tertiary (see review in Keast, 1972). It has even been suggested that lemurs migrated in one fashion or another to Madagascar from India (Gingerich, 1975b).

Unfortunately, our knowledge of early Tertiary mammals in Africa is extremely limited. Only one specimen of a Jurassic mammal is known in Africa - from Tendaguru in southern Tanzania. It is an edentulous jaw considered by Simpson to be a pantothere (*Branctherulum*). As yet, no Cretaceous or Paleocene mammals are known from the continent. The mammal record picks up in the Upper Eocene in scattered localities across northern Africa. These include the Qasr el Sagha Formation in Egypt; M'Bodione Dadere in Senegal; Tafidet and Tamaquelel in Mali; and Gebel Coquin in Libya (Simons and Wood, 1968; Vondra, 1974; Cooke, 1972; Lavocat, 1953). The preserved faunas appear basically similar, including the primitive proboscidean *Moeritherium*, hyracoids, creodonts etc. No undoubted primates are known from these Eocene localities. The recently discovered Eocene *Azibius*, described as a possible primate by Sudre (1975), might possibly be a hyopsodontid-like condylarth (Szalay, 1975).

The earliest undoubted primate is from the Oligocene Jebel el Qatrani Formation, Fayum, Egypt. This specimen, *Oligopithecus savagei*, is from Yale Quarry E (Simons, 1962). Other primates higher up in the section include *Apidium*, *Parapithecus*, *Aeolopithecus*, *Propliopithecus*, *Aegyptopithecus*. These are well known dentally (Simons, 1972; Conroy, Schwartz, and Simons, 1975) and fairly well known cranially and postcranially (Conroy, 1976a).

In North Africa, the Lower Oligocene seems to correspond with a reduction of the Tethys Sea, associated with mountain building activity suggestive of some sort of land connection with Eurasia. The lack of marine Oligocene beds in Northern Pakistan would reinforce this suggestion of such a land connection (Cooke, 1972). It should be pointed out as well that some geophysical evidence suggests that Burma and the Malay peninsula might have been part of Gondwanaland and not Eurasia and thus more closely linked to India than previously thought (Ridd, 1971; Tarling, 1972). Leaving the primates aside, certain faunal elements in the Fayum are closely related to earlier or contemporary Eurasian forms. Thus, the creodonts are similar to earlier European genera (Simons and Gingerich, 1974), artiodactyls include cebochoerids referred to European Eocene *Mixotherium*; anthracotheres include *Rhagotherium* and *Brachyodus* which are also found in Europe (and Pondaung Sandstone fauna of Burma) (Cooke,

1972; Colbert, 1938). Other Oligocene localities in North Africa, such as Zella in Libya and Gebel Bon Gobrine in Tunisia seem to sample a similar fauna (without primates however) (Cooke, 1972).

Thus, palaeozoogeographic evidence cannot be used to exclude the Burmese fossils from anthropoid ancestry.

CONCLUSIONS

Four fossil primate species, considered to be transitional forms between prosimians and anthropoids, are reviewed. Theories of primate evolution based upon them, either in terms of morphology or paleozoogeography, are not well grounded in factual data. By Oligocene times, an "anthropoid grade" of evolution had been reached in both the New and Old Worlds. Descendants of this grade were subsequently restricted (with a few minor exceptions in North America) to the southern continents until Middle Miocene times when hominoids radiated into Eurasia.

REFERENCES

Arey, L. (1974). "Developmental Anatomy", W.B. Saunders Co., Philadelphia.

Bohlin, B. (1946). *Pal. Sinica (N.S.) C,* 8.

Cachel, S. (1975). *In* "Primate Functional Morphology and Evolution", (R. Tuttle, ed.), pp. 23-36, Mouton, The Hague.

Cartmill, M. (1975). *In* "Phylogeny of the Primates",(W. Luckett and P.S. Szalay, eds), pp. 313-354, Plenum Press, N.Y.

Chow, M. (1957). *Vert. Palasiatica* 1 (3), 201-214.

Colbert, E. (1937). *Am. Mus. Novit.,* 951, 1-18.

Colbert, E. (1938). *Bull. Amer. Mus. Nat. Hist.,* 74, 255-436.

Conroy, G. (1976a). "Primate postcranial remains from the Oligocene of Egypt", *Contrib. Primatol.,* 8, 1-134.

Conroy, G. (1976b). *Nature* 262, 684-686.

Conroy, G. and Bown, T. (1974). *Yb. Phys. Anthrop.,* 18, 1-6.

Conroy, G., Schwartz, J. and Simons, E. (1975). *Folia Primatol.,* 24, 275-281.

Cooke, H. (1972). *In* "Evolution, Mammals and Southern Continents", (A. Keast, F. Erk and B. Glass, eds), pp. 89-139, State Univ. of New York Press, Albany.

Decker, R. and Szalay, F. (1974). *In* "Primate Locomotion", (F. Jenkins, ed.), pp. 261-291, Academic Press, New York and London.

Del Corro, G. (1968). *Com. Museo Argentino Cien. Nat., "Bernardino Rivadavia",* Paleo 1, 21-26.

Delson, E. (1975). *Contrib. Primatol.,* 5, 167-217.

Delson, E. and Andrews, P. (1975). *In* "Phylogeny of the Primates" (W. Luckett and F. Szalay, eds), pp. 405-446, Plenum Pub. Co., New York and London.

Elftman, H. and Manter, J. (1935). *J. Anat.,* 70, 56-67.

Fleagle, J., Simons, E. and Conroy, G. (1975). *Science,* 189, 135-137.

Frakes, L. and Kemp, E. (1972). *Nature,* 240, 97-100.

Gentry, A. (1970). *In* "Fossil Vertebrates of Africa", vol. II, (L. Leakey and R. Savage, eds), pp. 243-323, Academic Press, New York and London.

Gingerich, P. (1975a). *Contrib. Mus. Paleo., Univ. of Michigan,* 24, 163-170.

Gingerich, P. (1975b). *In* "Lemur Biology", (I. Tattersall and R. Sussman, eds), pp. 65-80, Plenum Pub. Co., New York and London.

Gregory, W. (1920). *Mem. Am. Mus. nat. Hist.,* 3, 49-243.

Haffer, J. (1970). *Caldasia,* 10, 603-636.

Hallum, A. (1972). *Scientific Amer.*, 227, 56-66.
Hershkovitz, P. (1974). *Folia Primatol.*, 22, 237-242.
Hoffstetter, R. (1969). *C.R. Acad. Sci. Paris*, 269, 434-437.
Hoffstetter, R. (1974). *J. Human Evol.*, 3, 327-350.
Jolly, C. (1972). *Bull. Brit. Mus. nat. Hist. Geol.*, 22, 1-122.
Keast, A. (1972). *In* "Evolution, Mammals and Southern Continents", (A. Keast, F. Erk, B. Glass, eds), pp. 23-87, State Univ. of N.Y. Press, Albany.
Lavocat, R. (1953). *C.R. somm. Seanc. Soc. geol. Fr.*, 7, 109-110.
Lavocat, R. (1974). *J. Human Evol.*, 3, 323-326.
Lewis, O. (1972). *Am. J. Phys. Anthrop.*, 37, 13-34.
MacConaill, M. (1953). *J. Bone Jt. Surg.*, 35, 290-297.
Madden, C. (1975). "*Pondaungia* needs restudy", unpublished ms.
Manter, J. (1941). *Anat. Rec.*, 80, 397-410.
Maxwell, P. and Von Herzen, R., Hsu, K., Andrews, J., Saito, T., Percival, S., Milow, E. and Boyce, R. (1970). *Science*, 168, 1047-1059.
McKenna, M. (1975). *In* "Phylogeny of the Primates", (W.P. Luckett and F.S. Szalay, eds), pp. 21-46, Plenum Pub. Co., New York and London.
Orlosky, F. and Swindler, D. (1975). *J. Human Evol.*, 4, 77-83.
Patterson, B. and Pasqual, R. (1972). *In* "Evolution, Mammals and Southern Continents", (A. Keast, F. Erk, and B. Glass, eds), pp. 247-309, State Univ. N.Y. Press, Albany.
Pilgrim, G. (1927). *Pal. Ind.* (N.S.), 14, 1-26.
Preuschoft, H. (1974). *Symp. 5th Cong. Int'l. Primat. Soc.*, 345-359.
Ridd, M. (1971). *Nature*, 234, 531-533.
Rosenbarger, A. (1976). "*Xenothrix* and ceboid phylogeny", unpublished ms.
Simons, E. (1962). *Postilla*, 64, 1-12.
Simons, E. (1971a). *In* "Dental Morphology and Evolution", (A. Dahlburgh, ed.), pp. 193-208, Univ. of Chicago Press, Chicago.
Simons, E. (1971b). *Nature*, 232, 489-491.
Simons, E. (1972). "Primate Evolution: An Introduction to Man's Place in Nature", Macmillan, New York.
Simons, E. (1974). *Postilla*, 66, 1-12.
Simons, E. (1975). *Burg Wartenstein Symp.*, 65 (in press).
Simons, E. and Gingerich, P. (1974). *Ann. Geol. Surv. Egypt.*, 4, 157-166.
Simons, E. and Wood, A. (1968). *Bull. Peabody Mus. Nat. Hist.*, 28, 1-105.
Simpson, G. (1940). *Am. Mus. Nat. Hist. Bull.*, 77, 185-212.
Simpson, G. (1975). *In* "Phylogeny of the Primates", (W.P. Luckett and F.S. Szalay, eds), pp. 3-19, Plenum Pub. Co., New York and London.
Stains, H. (1959). *J. Mammal.*, 40, 392-401.
Sudre, M. (1975). *C.R. Acad. Sci. Paris ser. D.*, 280, 1539-1542.
Sullivan, W. (1974). "Continents in Motion", McGraw Hill, New York.
Szalay, F. (1970). *Nature*, 227, 355-357.
Szalay, F. (1972). *Nature* 236, 179-180.
Szalay F. (1975). *In* "Phylogeny of the Primates", (W.F. Luckett and F.S. Szalay, eds), pp. 357-404, Plenum Pub. Co., New York.
Szalay, F. (1976). *Bull. Amer. Mus. Nat. Hist.*, 156, 159-449.
Szalay, F. and Decker, R. (1974). *In* "Primate Locomotion", (F. Jenkins, ed.), pp. 223-259, Academic Press, New York and London.
Tarling, D. (1972). *Nature*, 238, 92-93.
Tarling, D. and Tarling, M. (1971). "Continental Drift", Doubleday, New York.
Thenius, E. (1959). *In* "Handbuch der stratigraphischen Geologie", Tartiar II, (F. Lotze, ed.), pp. 1-328, Enke, Stuttgart.
Thomas, O. (1913). *Ann. Mag. Nat. Hist.*, 11, 130-136.

Van Couvering, J. (1972). *In* "Calibration of Hominoid Evolution", (W. Bishop and J. Miller, eds), pp. 247-271, Scottish Academic Press.
Vondra, C. (1974). *Ann. Geol. Surv. Egypt*, 4, 79-94.
Von Koenigswald, G. (1965). *Kon. Neder. Akad. Wetensch. ser. B*, 68, 165-167.
Wilson, J. and Szalay, F. (1976). *Folia primatol*., 25, 294-312.
Wood, A. (1937). *Am. J. Sci.*, 34, 64-76.
Wood, A. (1972). *Science*, 175, 1250-1251.
Wood, A. (1973). *The Pearce-Sellards Series, Tex. Mem. Mus.*, 20, 1-41.
Young, C. and Bien, M. (1939). *Proc. 6th Pacific Sci. Cong.*, 531-534.

DISCUSSION

Kay: I have recently completed a study of molar occlusion among Fayum
Oligocene catarrhines. Most Fayum primates share with Miocene to Recent
catarrhines the presence of a unique derived wear facet on the posterior
aspect of the molar trigonids. This supports the notion that they have
special affinities with Old rather than New World Anthropoidea. *Oligopithecus* on the contrary shares no known derived molar features with Fayum
or later catarrhines. Thus the upper canine-lower third premolar hone was
most probably achieved independently in several catarrhine lineages.

TAXONOMY AND RELATIONSHIPS OF FOSSIL APES

PETER ANDREWS

Department of Palaeontology, British Museum (Natural History), London, UK.

INTRODUCTION

A great number of new specimens of fossil hominoid primates have been discovered in recent years. With the new discoveries have come many new names,
some of them justified and some of them not. These new specimens will be
briefly reviewed in the context of the last major revision of the dryopithecines (Simons and Pilbeam, 1965), and a revised taxonomy will be put forward
to clarify the status of the specimens. Basically the taxonomy will follow
that of Simons and Pilbeam.

The first dryopithecine specimen, *Dryopithecus fontani* Lartet 1856, was
found 120 years ago in France. *D. fontani* is known from several good specimens from the type locality, but little additional material has come to light
since except for some doubtfully assigned isolated teeth and postcranial bones.
No maxilla or definitely assigned upper teeth are known, and owing to the incomplete nature of the mandibular specimens it is not known for sure what the
shape of the symphysis was.

These shortcomings in the material of *D. fontani* lead to difficulties in
distinguishing other groups of dryopithecines. For instance, the African
Proconsul group is well represented by cranial, dental and postcranial material, but the only good basis for comparison with the type species of *Dryopithecus* is the lower dentition and parts of the mandible, and it so happens
that in these parts they are quite similar, whereas indications are that they
may be quite dissimilar in other parts. The distinctions between *Proconsul*,
Rangwapithecus (Andrews, 1974), and *Sivapithecus* (Pilbeam, in prep.) are best
seen in the upper dentition and maxilla, and the lower dentitions differ only
slightly within the three groups and between them and *Dryopithecus fontani*.
It is difficult to know, therefore, at what level the various groups of dryopithecine primates should be distinguished. At the moment, following Simons
and Pilbeam (1965), they are regarded as subgenera of *Dryopithecus*, but it is
becoming increasingly apparent that they constitute much more distinctive
categories than was thought ten years ago, a distinctiveness best recognised
by reinstating *Sivapithecus* and *Proconsul* as full genera in the Dryopithecinae.

Whatever systematic scheme is followed, it is becoming clear that there are
several subdivisions discernible in the Miocene apes. The African apes of
the early and middle Miocene, *Proconsul*, *Limnopithecus* and *Rangwapithecus*, all
clearly belong in one group though exhibiting varying degrees of difference,
and all are quite different from the single remaining African hominoid species,
Ramapithecus wickeri. The Asian apes of the middle Miocene can be similarly
grouped, and include *Sivapithecus*, *Ramapithecus* and *Gigantopithecus*. Some

of these forms are now also known from Europe: Hungary, Austria, Greece, and
perhaps even Germany. The other European species, the type species of *Dryo-*
pithecus from France and Germany and a smaller species from Spain, are dubiously linked by Simons and Pilbeam (1965) into subgenus *Dryopithecus*, but it is
quite possible that the Spanish form, *D. laietanus*, should go with the
Asian group, to which it is very similar, rather than with *D. fontani*. Once
again the incompleteness of the type material makes it difficult to be sure

NEW HOMINOID FOSSILS FROM AFRICA

In East Africa 271 new specimens have been described recently from the
Early Miocene (Pilbeam, 1969; Andrews, 1970, 1973, 1974; Fleagle, 1975). These
are divided between *Proconsul* (137), *Rangwapithecus* (60) and *Limnopithecus*
(74). In addition 9 *Proconsul* and 14 *Limnopithecus* specimens have been described from middle Miocene deposits at Fort Ternan (Andrews and Walker,
1976), and 4 specimens of *Ramapithecus wickeri* from the same site (Andrews,
1971; Walker and Andrews, 1973; Andrews and Walker, 1976).
Limnopithecus is regarded as a separate genus of the Dryopithecinae. It is
very distinctive because of its diminutive size, and it is at present known
from at least one species, *L. legetet;* but it is very probable that at least
two other species exist. One of these would be based on the middle Miocene
material from Fort Ternan which, although basically very similar to *L. legetet*,
differs in being slightly larger, having molars with reduced cingula, and having M_3 reduced in size.
Another new species of uncertain generic affinities is known from early
Miocene deposits at Napak, Uganda (Bishop, 1964). The most complete specimen,
a palate with molars and premolars present on both sides, has recently been
described (Fleagle, 1975). Although he did not name it at this time, Fleagle
claimed that this specimen indicated hylobatid affinities for the taxon, and
that it differed from *L. legetet* in being considerably smaller. Some of the
other features given by Fleagle as diagnostic include the broad arcuate palate,
the great orbital breadth, the broad nose, and the shallow and short face, and
these were combined in a multivariate analysis which placed the Napak palate
closest to the gibbons. If, however, log regressions of the dimensions used by
Fleagle to make up his indices are plotted, it can be shown that, at least in
the case of nasal breadth, palatal shape, and face length, the Napak palate
falls on the same line as the other dryopithecines. In other words the change
in shape, very evident between the Napak palate and the *Proconsul major* palate
from Moroto, is allometrically controlled so that the short face, rounded
dental arcade and broad nose in the former are direct consequences of its
smaller size. Even if size has been eliminated from a multivariate analysis,
changes in proportion due to size are not eliminated unless the allometry is
allowed for, so that proximity of two animals of different size in such an
analysis indicates convergence rather than homology between them. The new
palate from Napak does not, therefore, provide evidence of either relationship to gibbons or difference from a similarity to dryopithecines. Further
work on the Napak collection as a whole may change this, but at present there
is no clear indication of the relationships of these specimens.
Proconsul was originally described as a genus of fossil ape by Hopwood
(1933) and Le Gros Clark and Leakey (1951). It was referred to as a subgenus
of *Dryopithecus* by Simons and Pilbeam (1965) and this is its presently recognised status. More recently, however, the matter has been complicated by the
addition of a new subgenus of *Dryopithecus* named *Rangwapithecus* (Andrews, 1974).

At the time this was the simplest way of fitting *Rangwapithecus* into the systematics proposed by Simons and Pilbeam, but it is becoming clear that their scheme is oversimplified and that a greater level of difference needs to be recognised for *Proconsul*, and with this comes the problem of whether *Rangwapithecus* should be left as a subgenus in *Dryopithecus*, transferred to *Proconsul*, or raised to generic station itself. Since it is closer morphologically to *Proconsul* than to *Dryopithecus*, and as it is not as distinct from *Proconsul* as the latter is from *Dryopithecus*, the most reasonable course is to transfer it as a subgenus to *Proconsul*.

Proconsul contains three species which differ mainly in size (Le Gros Clark and Leakey, 1951; Andrews, 1973). They range from the size of a siamang to the size of female gorillas. Although they are characteristic of the early Miocene sites of East Africa two of the species are also known from middle Miocene deposits at Fort Ternan. Seven specimens are referred to *P. nyanzae* and two to *P. africanus*, and as far as can be seen these nine specimens are indistinguishable from the early Miocene material (Andrews and Walker, 1976).

Rangwapithecus contains two species, both of them newly described: *P. (R.) gordoni* (Andrews, 1974) and *P. (R.) vancouveringi* (Andrews, 1974). These also differ mainly by size, and they overlap the bottom half of the *Proconsul* range. They differ from *Proconsul* in having elongated upper molars which are longer than broad, elongated premolars, posterior tooth size increase with P^4 larger than P^3 and M^3 larger than M^2 and M^1, massive cingulum development on molars and premolars, gracile maxillary bodies with extensive maxillary sinus development, and gracile mandibles, very deep relative to thickness. In most of these characters the two species of *Rangwapithecus* are even further from other dryopithecine species than they are from *Proconsul* species.

NEW DRYOPITHECINE FOSSILS FROM EUROPE AND ASIA

Quite a number of new specimens have been found recently in India and Pakistan, but most of these are undescribed and cannot be mentioned except in passing. Tattersall and Simons (1969) and Prasad (1969) have described some new specimens that apparently are assignable to existing species. *Adaetontherium icognitum*, referred to by Tattersall and Simons, is now known not to be primate, while Prasad advocates, probably rightly, the retention of generic status for *Sivapithecus* and *Proconsul*. A new species of *Sivapithecus*, *S. lewisi* Pandey and Sastri (1968) was based on a large mandibular fragment with the crowns of P_4 and M_1. In most dimensions it is bigger than any previously described speciment of *Dryopithecus* or *Sivapithecus indicus*, but on its own it is insufficiently diagnostic to justify a new species name. There are several specimens in undescribed collections of Siwalik primates at present held in Europe which could belong with *S. lewisi*, but nothing similar has been found in the recent important collections from Pakistan by D.R. Pilbeam of Yale University.

The presence of a new large species of dryopithecine is, however, supported by specimens found recently in Greece and Turkey. One Turkish specimen was described in 1957 by Ozansoy as *Ankarapithecus meteai*, and it was included in *Dryopithecus (Sivapithecus) indicus* by Simons and Pilbeam (1965). It is very large, however, and the anterior end of the mandible and symphysis, the only parts preserved, are very deep, corresponding with the great body depth of *S. lewisi*. An undescribed complete palate from the same deposits in Turkey, and new material from Greece, described as *Dryopithecus macedoniensis* de Bonis et al., 1974, 1975) establish convincingly the presence of a new species of large fossil ape. The Greek material from Macedonia consists of three mandibles so far described. They show great variation in size and morphology, as great as is known for *S. indicus* with its much larger sample size, but it is unlikely that more than one species is represented. In most dimensions these new specimens do not overlap with even the largest specimens of *S. indicus*, although overlap does occur for some of the more variable teeth. *A. meteai* falls into the bottom half of the range of the Greek mandibles and *S. lewisi* into the top half (see Figs 1 and 2).

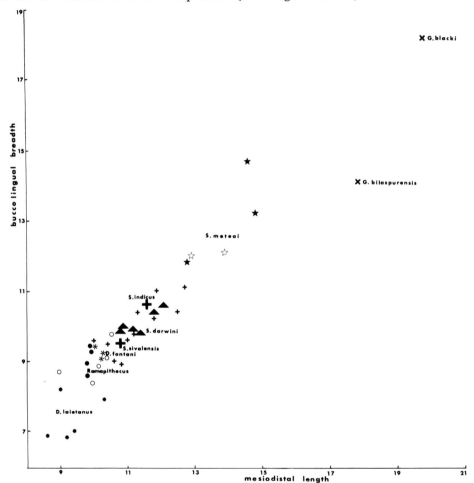

Figure 1. *Bivariate plot of breadth against length for* M_1 *of European and Asian hominoids. Circles = Ramapithecus (closed circles: R. wickeri; open circles: R. punjabicus). Crosses = Siwalik primates, with the sample means for Sivapithecus sivalensis and S. indicus taken from Simons and Pilbeam (1965). Inverted triangles = the new Turkish specimens assigned here to S. darwini. Stars = the new specimens of sivapithecus (open stars: A. meteai from Turkey and the Indian form "S. lewisi"; closed stars: the 3 specimens originally described as "D. macedoniensis").*

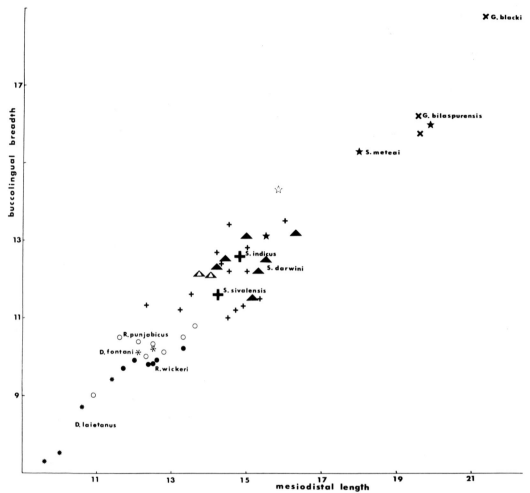

Figure 2. Bivariate plot for M₃. Symbols as for Fig. 1. The inverted triangles representing S. darwini are divided into the new Turkish specimens (solid symbols) and the type specimens described by Abel from the Vienna basin (white centres).

One other specimen is the crushed mandible from Pyrgos in Greece named *Graecopithecus freybergi* (von Koenigswald, 1972). This specimen is so poorly preserved it is difficult to assign it to any species with confidence, but it was evidently a large hominoid species with large molars and small canines and incisors. Because of the latter it has been thought that it might represent *Ramapithecus*, presumably a new and much larger species because of its large molars. However, the first described of the Macedonian mandibles (de Bonis et al., 1974) has this same combination of large molars and small canines and incisors and it now seems more reasonable to group these two specimens. The overall dimensions of M_2, the only measurable tooth on the Pyrgos mandible, are similar to those of the Macedonian specimens.

The ranges of variation for the new specimens from Greece, Turkey and India are shown in Figs 1 and 2. These bivariate plots for M_1 and M_3 show the amounts of variation in the least variable and most variable molar respectively, and it can be seen both that the sample range of the new specimens is very considerable, more actually than the presently recognised range for *S. indicus*, and that there is very little overlap between this sample and that of *S. indicus*. It seems quite clear, therefore, that these specimens should be assigned to at least one new species of *Sivapithecus* and probably two. Exactly how these should be divided is not yet certain, but it could be that both *meteai* and *macedoniensis* might be valid species of *Sivapithecus*. Particularly important in this respect are three as yet undescribed palates of large dryopithecine from Pakistan, Turkey and Greece, as these suggest the existence of at least three distinct but closely related species.

Several new specimens have been added recently to the extant species of *Sivapithecus*, *S. indicus* and *S. sivalensis*. One specimen of the latter is being described by Dr. E.L. Simons of Yale University. A good collection of Siwalik primates is being obtained at present from late Miocene deposits in Pakistan by David Pilbeam, also of Yale. These have not been described yet, but it is already clear that many specimens of *S. indicus*, as well as *Ramapithecus*, have been recovered. Finally there are three specimens from Hungary that are referred to a new genus and species, *Bodvapithecus altipalatus* (Kretzoi, 1975), but these are indistinguishable from *S. indicus* and must be assigned to that species.

It seems likely that the relationship between *S. indicus* and *S. sivalensis* will have to be revised. Many of the larger specimens at present attributed to the latter species would seem to belong to the former, and *S. sivalensis* should be reserved for smaller specimens with relatively lightly built maxillary and mandibular bodies, for example specimens YPM 13811 and AMNH 19411, probably representing female and male respecively. At present it is very difficult to distinguish the two species, and they each have restricted ranges of variation compared, for instance, with those known to be present in dryopithecines of equivalent size from East Africa. Further study of the original material is necessary for this matter to be resolved.

The three species of *Sivapithecus* just discussed are all similar morphologically. They all have molars with relatively bunodont cusps crowded together so that the occlusal fovea are restricted; thick enamel on the molars; relatively simple occlusal patterns; stoutly built central incisors and canines which are low crowned and robust; relatively deep mandibular bodies and symphysis; and greatly expanded zygomatic regions of the maxilla. They range in size from animals about the size of siamangs to female gorilla sized animals, which is a similar size range to the three *Proconsul* species. The distinctiveness of this group suggests that it should be recognised by a higher category than a subgenus, and the position adopted here is that *Sivapithecus* should be reinstated as a valid genus of the Dryopithecinae.

In addition to these three species of *Sivapithecus*, there is some evidence for the presence of a fourth species in Turkey and Czechoslovakia. Unfortunately much of this material from Turkey is not yet described, but it consists of about 45 isolated teeth that are intermediate in morphology between *P. major* and *S. indicus* (Becker-Platen et al., 1975). In occlusal morphology the crowns of the upper molars are essentially similar to those of *S. indicus*, except for the retention of molar cingula, and the lower molars show greatest resemblance to *P. major*. These specimens from Pasalar in Turkey are associated with a Vindobonian fauna which indicates an earlier date than for the majority of the collection from the Siwaliks, although they would have been contemporary with the Chinji dryopithecines of reliable provenance. They were

also contemporary with, and identical in size and morphology to, the specimens
from the Vienna basin described by Abel (1902) as *Dryopithecus darwini*, which
were later referred by Lewis (1937) to *Sivapithecus*. The specimens from the
Vienna basin were referred to *Dryopithecus fontani* by Simons and Pilbeam (1965),
but the M_3 differs in being relatively broader, considerably bigger, and in
having much more prominent buccal cingula than specimens of *D. fontani*. In my
opinion these two sets of specimens should be assigned provisionally to a separate and relatively primitive species of *Sivapithecus*, *S. darwini* (Abel,
1902), which could well represent the ancestral condition from which the other
three *Sivapithecus* species were derived.

These specimens are particularly important because they are intermediate
both in age and morphology between the early Miocene *Proconsul* species of
Africa and the late Miocene *Sivapithecus* species of Europe and Asia. They
are slightly larger than the equivalent teeth of *Proconsul major* and the same
size as those of *S. indicus*. Whether they are regarded as advanced *P. major*
or primitive *S. indicus*, or (as suggested here) as a third intermediate species, is unimportant, for whatever the taxonomy they do seem to represent an
emigrant population from Africa linking the African species with the later
Eurasian ones. Moreover, associated with these specimens in Turkey are a number of smaller teeth that are indistinguishable from *Ramapithecus wickeri*, and
it will be suggested in the next section that these specimens represent an
early stage of the *Ramapithecus* lineage soon after it divided from a parallel
Sivapithecus lineage, both being derived from species of *Proconsul*. It is not
clear yet whether they both originated from a single *Proconsul* species, but the
indications are that more than one species was involved.

Gigantopithecus is represented by one important new discovery. A nearly
complete mandible from Dhok Palton deposits in India was described as a new
species, *G. bilaspurensis* (Simons and Chopra, 1969). This is evidently the
same species as the isolated M_3 described by Pilgrim (1915) as *D. giganteus*,
and it is shown in Figs 1 and 2 that these specimens are only slightly larger
than *Sivapithecus meteai*. They are known from deposits rather later in time
than those characteristic of *S. meteai*, and it seems very probable that *S.
meteai* is ancestral to *G. bilaspurensis*. Since *G. bilaspurensis* is probably
ancestral to the Chinese Pleistocene form *G. blacki* (von Koenigswald, 1935),
there is now a lineage definable from the Miocene to the Pleistocene encompassing *P. major*, *S. darwini*, *S. meteai*, *G. bilaspurensis*, and *G. blacki*.
This is the best documented fossil ape lineage, but unless the Yeti is discovered it does not seem to have any living counterpart.

NEW SPECIMENS OF *RAMAPITHECUS*

The sample of *Ramapithecus* specimens have been greatly increased in recent
years. Although the actual quantity of material does not approach that of the
early Miocene pongids it is impressive by contrast to what there used to be.
Most of the material is late Miocene in age, but there are some middle Miocene
specimens from Turkey. These are very similar to *R. wickeri* from Kenya, which
is known from several new specimens from Fort Ternan (Andrews, 1971; Andrews
and Walker, 1976). In addition, a reconstruction of the dental arcades, based
on the probable association of a mandibular fragment with part of the symphysis
preserved and the type maxilla, provides new information on the morphology of
R. wickeri (Walker and Andrews, 1973). The main points distinguishing *R.
wickeri* from the dryopithecines are its robust mandible and maxilla, heavily
buttressed symphasis, anteriorly abbreviated mandible and premaxilla, raised

floor of the maxillary sinus, the lateral flare of the zygomatic arch, and
teeth with flattened occlusal surfaces, thick enamel, and enlarged M_1 relative
to M_3.   These points are all related to an increase in chewing stresses in the
region of the molars, and they have been combined in a single functional interpretation relating to increased lateral grinding in the chewing processes of *R.
wickeri* (Andrews, 1971).   The Turkish material consists of a mandible with P_3-
M_3 on the left side and P_4-M_3 on the right from Candir (Tekkaya, 1974) and
about 35 isolated teeth from Pasalar. Both sites have Vindobonian faunas and it
is possible that they are as old as or even slightly older than Fort Ternan in
Kenya. The mandible has been named *Sivapithecus alpani*(Tekkaya, 1974), but its
similarity to *R. wickeri* has been noted subsequently (Andrews and Tekkaya, 1976).
    The symphysis and body of the Candır mandible resemble those of *R. wickeri*
in being robust and shallow.   The plane of the symphysis is more erect and the
inferior transverse torus less developed than in *R. wickeri*, but the overall
dimensions are similar, and the body of the mandible is even stouter than in
R. wickeri   The anterior end of the mandible is greatly abbreviated and the
chin region is angled sharply at the anterior root of the P_3 so that the anterior end hardly projects beyond the level of the P_3 roots. This combination
of features can only be associated with a short orthognathous face such as
was predicted in the reconstruction of *R. wickeri* (Walker and Andrews, 1973).
The dental arcade of the Candır mandible is also similar to that reconstructed
for *R. wickeri*, with straight molar-premolar tooth rows that diverge only
slightly posteriorly, and with angles of divergence that are almost identical
(see Fig. 3). One difference from *R. wickeri* is the slightly greater overall
distance between the two tooth rows in the Candır mandible, giving a slightly
broader arcade. The dentition of the Candır mandible and of the series of isolated teeth are very similar indeed to that of *R. wickeri*, but some of the
teeth show definite traces of cingulum which led Tobien to postulate their
relationship with *P. africanus* (Becker-Platen et al., 1975). Their greatest
similarity, however, is with *Ramapithecus* from Kenya, and the position
adopted here is that they belong together in one species, *R. wickeri*.

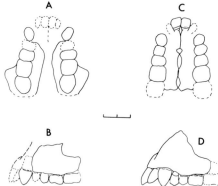

*Figure 3.   Outline drawings of the mandibles of R. wickeri from Fort Ternan,
           Kenya (A); R. wickeri from Candır, Turkey (B); and R. punjabicus
           from Pakistan (C). The Fort Ternan specimen is drawn from the re-
           construction made by Walker and Andrews (1973), but the other two
           specimens are complete as shown except for the P₄ crowns on the
           Pakistan mandible. The Fort Ternan mandible has a slightly narrow-
           er dental arcade than the Candır mandible, but the angle of diver-
           gence in both is the same. The Pakistan mandible has more diver-
           ging tooth rows, which are closer together anteriorly than in the
           other two mandibles and are further apart posteriorly.*

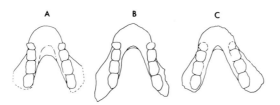

Figure 4.   Outline drawings of the palates of R. wickeri from Kenya (A,B)
and R. punjabicus from Hungary (C,D).   The R. wickeri palate is
based on the reconstruction of Walker and Andrews (1973);   the
R. punjabicus palate is based on a left maxillary fragment
(Rud 12) described by Kretzoi (1975), with a reversed tracing
to obtain the right side.   The two palates are extremely
similar both in shape and in size.

New material of *Ramapithecus punjabicus* has come mainly from Pakistan
and Hungary.   The Pakistan specimens have been recovered over the last
two years by Dr. David Pilbeam of Yale University, but as they have not
yet been described all that can be said here is that the best specimens
consist of two mandibles, one with both sides of the body intact and the
other more incomplete.   There are also a number of isolated teeth and
some postcranial elements, some of which almost certainly belong to *Ramapithecus*.   Measurements of some of the teeth are included on the bivariate
plots (Figs 1-2) and I thank David Pilbeam for permission to use them.
The Hungarian specimens have been assigned to a new genus and species,
Rudapithecus hungaricus (Kretzoi, 1975).   They consist of two maxillae,
three mandibles, and four isolated teeth (Kretzoi, 1975), but by far the
most important specimen is a left maxilla with I^1 and $C-M^1$ (Rud 12). This
specimen has much of the nasal process and nearly all of the palatal processes preserved on one side, and since the break in the palate is along
the median palatine suture it is possible to get a good idea of the shape
of the palate by putting a reversed print alongside a normal print so that
the two images of the suture come together.   The conformation of the palate
thus obtained (Fig. 4) is seen to be very similar indeed to the reconstruction
made for the Kenyan specimens.   The central incisor is small, approximately
the size of the I^1 of *Proconsul africanus*, but the crown is relatively reduced as well when compared with M^1 (80% crown size compared with 90-96%
for dryopithecines).   The canine has a crown similar in shape to that of
R. wickeri, but it is smaller and less projecting.   The premolars are nearly
equal in size, and neither they nor the M^1 have any trace of lingual cingulum.
The alveolar process of the maxilla is slightly more robust than in *R. punjabicus* and less robust than in *R. wickeri*, and the body of the mandible is
similarly intermediate.
The other maxillary fragment (Rud 15) is similar to the first, but it has
the crown of I^2 preserved.   Like the I^1 it is a small tooth, similar in size
to *P. africanus* specimens, and it lacks the sectorial specializations seen
on many of the dryopithecine specimens, a specialization associated with
long projecting canines.   The canine is unfortunately broken on this specimen,
but it would appear to have been larger than on Rud 12, as are the molars and
premolars as well.

   The mandible specimens are poorly preserved. No incisors are present.
The canine is present on two specimens (Rud 2 and 17) and appears remarkably
slender and bilaterally compressed, more so than in *R. wickeri*. The P_3 is a
large single cusped tooth and the P_4 is relatively smaller. The molars are
approximately equal sized but the M_3 is a broadly triangular tooth. The body
of the mandible is considerably less robust than in *R. wickeri*.
   In many respects the Hungarian specimens are intermediate between *R. wickeri*
and *R. punjabicus*. They appear more similar to the latter, but this raises the
question of the degree of difference between *R. wickeri* and *R. punjabicus*. The
differences noted earlier between *R. wickeri* and the dryopithecines are valid
also for *R. punjabicus* , and there is no doubt that they are closely related
and that they share the chewing adaptive complex described for *R. wickeri*
(Andrews, 1971), but the differences between them are of a comparatively minor
nature. The canine in *R. wickeri* is relatively longer than in *R. punjabicus*,
and the P_3 is large and preserves the primitive sectorial form of earlier apes.
In *R. punjabicus* the crown of P_3 may have been relatively smaller and closer
in size to P_4, although it is still basically one-cusped. The molars of *R.
wickeri* are slightly more elongated, and, particularly on the lowers, traces
of cingulum are often seen. The maxilla is considerably more robust in *R.
wickeri* than in *R. punjabicus* as shown by the distance between the alveolar
margin and the floor of the maxillary sinus (Andrews, 1971), and similarly
the mandible is also more robust. The body of the Candır mandible at the
level of M_3 has a greater thickness/depth ratio than any of the known specimens of *R. punjabicus* (Andrews and Tekkaya, 1976), and since the Kenyan specimens appear even more robust in the area of the symphysis than the Candır
mandible it seems likely that this is a general species character. Finally,
the dental arcade appears to be narrower and slightly less divergent in *R.
wickeri* than in *R. punjabicus*. The Fort Ternan and Candır mandibles have similar degrees of divergence, but the former is slightly narrower overall than
the latter, and both of them are narrower and less divergent than in the new
mandible from Pakistan. These differences are shown in Figs 3 and 4.

RELATIONSHIPS OF THE FOSSIL APES

   The relationships of the fossil apes are shown in Fig. 5. This is a time-
related cladogram: the time dimension is indicated by the length and position
of the solid horizontal lines beneath the species names; and the relationships
of the species, based on shared derived characters, are indicated by the
dashed lines. The recognition of shared derived characters, or synapomorphies,
depends on the recognition of the primitive condition both in anthropoids as
a whole and in the various subdivision within the suborder (Delson and Andrews,
1976). This will now be commented on briefly.
   The primitive condition in the anthropoid mandible was probably similar to
the moderate tooth row divergence of the dryopithecines, coupled with relatively deep mandibular body and symphysis, and an inferior transverse torus. This
is the condition in the monkey and gibbon lineages and in the Oligocene anthropoids from the Fayum. The early Miocene forms, *Proconsul* and *Limnopithecus*,
have lost the primitive inferior transverse torus, and, almost uniquely in
primates, have developed a clearly derived condition with a *superior* torus.
The question arises, therefore, whether the later dryopithecines, which
apparently have retained the primitive condition of the symphysis, could be
descended from the *Proconsul* group. The same applies to *Ramapithecus*, which
also has an inferior transverse torus. Two possibilities must be considered

for the relations of the early and middle Miocene dryopithecines according to
the mandibular evidence. Firstly, if the later dryopithecines and *Ramapithecus* evolved from the *Proconsul* group, their inferior transverse torus must be
secondarily derived, even though it is morphologically similar to the primitive
condition. This might be considered unlikely, but so is the alternative possibility, that the later dryopithecines and *Ramapithecus* evolved from an early
Miocene group which is entirely unknown and whose sole known attribute in that
it had an inferior transverse torus on the mandible. A possible candidate for
this group, if it exists, could be *Sivapithecus africanus*, only known at present from the maxilla.

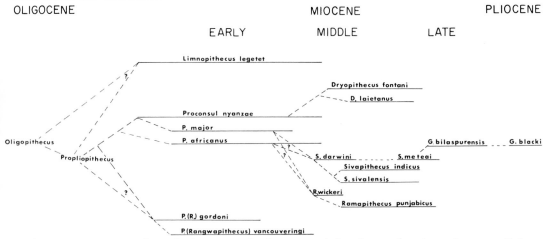

*Figure 5. Tentative phylogeny of the Dryopithecinae showing the possible
origin of Ramapithecus.*

The primitive condition of the anthropoid maxilla probably includes the
projecting snout, the narrow and shallow palate, the small maxillary sinuses,
the gracile zygomatic regions sloping posteriorly, gracile alveolar processes,
high narrow noses and the relatively large wide-set orbits. In most of these
characters the early dryopithecines retain the primitive condition similar to
other catarrhine groups such as the monkeys, hylobatids and the Oligocene apes.
They have a derived condition only in the reduction of the projecting snout.
Ramapithecus on the other hand has a derived condition in many of these features: the non-projecting face, reduced extent of the maxillary sinus, robust
zygomatic region with hardly any posterior slope, and extremely robust alveolar
processes. These distinguish it from the early dryopithecines, which are
basically primitive in maxillary structure, but the later dryopithecines, particularly *Sivapithecus* species, are intermediate in that they share some of
these derived characters, for instance the robust zygomatic processes and
small maxillary sinus, with *Ramapithecus*. In the maxilla, therefore, there is
an increasing trend of specialization of the maxilla leading directly from the
early Miocene *Proconsul-Limnopithecus-Rangwapithecus* group, though the middle
Miocene *Sivapithecus* group, to the middle to late Miocene *Ramapithecus* species.
Finally, the primitive condition in the dentition of the Anthropoidea
(Delson and Andrews, 1976) includes the following characters:
    relatively small incisors;
    relatively small and slender canines;
    sectorial or honing type of P_3, strongly bilaterally compressed;

upper premolars and molars relatively broad;
small M_1 and larger M_3, especially the elongated M_3; presence of molar
cingula.
Of these characters, dryopithecines have retained all except the sectorial
P_3. They still have one-cusped P_3's, but these no longer have the honing
function present (for instance) in monkeys and hylobatids. This derived character state, that is loss of honing function, is also shared with *Ramapithecus*.
Derived characters present in *Ramapithecus*, and shared by *Sivapithecus* and
Gigantopithecus, but not by other dryopithecines, include the development of
thicker enamel on the molars, the progressive increase in size of P_4 and M_1
relative to P_3 and M_3, the presence of a greater wear gradient from M_1 to M_3,
and the loss of the molar cingula. Finally, derived characters present in
Ramapithecus, but not in dryopithecines, include the oblique angle of the P ,
the slight reduction in size of M_3, and the slightly greater wear gradient. In
the dentition also, therefore, a trend is apparent from the primitive condition in the *Proconsul–Limnopithecus* – *Rangwapithecus* group to the more specialized condition in *Sivapithecus* and *Gigantopithecus*, with *Ramapithecus*
fundamentally similar to the latter group.

DISCUSSION AND CONCLUSIONS

The dryopithecines can be divided into two groups. The early Miocene
genera *Proconsul* and *Limnopithecus* are very similar and are considered to have
evolved from a common ancestor in the Oligocene. *Rangwapithecus*, formerly
assigned as a subgenus of *Dryopithecus*, is considered to be subgenus of *Proconsul* now that the latter has been reinstated as a valid genus. *Limnopithecus* and *Proconsul* share derived characters in the dentition, particularly
the loss of the primitive honing mechanism, in the mandibular superior transverse torus, and in the postcrania, and they distinguish these genera both
from the Oligocene taxa of *Propliopithecus* and *Aegyptopithecus* and from the
contemporary species formerly assigned to *Limnopithecus*, "L". *macinnesi*.
Dryopithecus fontani also appears fundamentally similar to the early Miocene
dryopithecines, although the lack of material makes its association of doubtful significance.
In contrast with this, the second group of dryopithecines is quite distinct.
The *Sivapithecus* species share with *Ramapithecus* and *Gigantopithecus* a number
of characters that are almost certainly derived, in contrast to the primitive
condition present in the other dryopithecine species. These characters include the development of thick enamel on the molars and premolars, the proportional increase in size of the fourth premolar and first molar relative to
P_3 and M_3, and the increase in robusticity of the mandibular and maxillary
bodies. These shared derived characters indicate a close relationship between *Gigantopithecus*, *Sivapithecus* and *Ramapithecus*, and they distinguish
them from the rest of the Dryopithecinae. Unfortunately, there is insufficient evidence to link this group with the hominids on the basis of this
character complex, and it is quite possible that it forms an adaptive lineage
parallel to a contemporary hominid lineage (contemporary because the *Sivapithecus* – *Ramapithecus* – *Gigantopithecus* adaptive radiation covers a time
span from the middle Miocene to the Pleistocene); but it is becoming increasingly likely, if only on negative evidence, that the hominids originated from
some part of this lineage. The least specialized members of the lineage are
the species of *Ramapithecus*, and these will be briefly considered to see what
light they could throw on hominid evolution.

The derivation of *Ramapithecus* from the early Miocene dryopithecines is still unknown. *R. wickeri* could conceivably be derived either from *Rangwapithecus*, with its relatively large and elongated molars, or from *Proconsul*; but the specimen originally described as *Sivapithecus africanus* (Le Gros Clark and Leakey, 1950), and "*Kenyapithecus africanus*" (Leakey, 1967) can still be regarded as closest to the probable ancestral condition of *R. wickeri*. The place of origin of *Ramapithecus* has long thought to have been Africa (Leakey, 1962), but there is now considerable uncertainty about this. The middle Miocene *Ramapithecus* from Turkey is almost certainly too early in time to be considered a derivative of the *Ramapithecus* from Fort Ternan, and in some respects also it is more primitive than the Kenyan form, particularly in the retention of traces of cingulum on the molars. In this and other features it resembles the earliest form of *Sivapithecus*, *S. darwini*, which is present in the same deposits in Turkey. It seems possible, therefore, that *Sivapithecus* emerged from an African stock some time in the early to earliest middle Miocene, migrating out of Africa towards Asia in the process and giving rise to later Asian and European species of the genus and to *Ramapithecus*. At a later stage of the middle Miocene, *Ramapithecus wickeri* re-entered Africa as a completely isolated species significantly different by then from the other African hominoid primates. Nothing is known of its subsequent development, but its isolated position in the African primate fauna could be of the greatest significance in explaining how it could have begun the remarkable specializations that led eventually to modern man.

REFERENCES

Abel, O. (1902). *S. Ber. Akad. Wiss. Wien, math.-nat. Kl.*, 111, 1171-1207.
Andrews, P. (1970). *Nature, Lond.*, 228, 537-540.
Andrews, P. (1971). *Nature, Lond.*, 231, 192-194.
Andrews, P. (1973). Miocene Primates (Pongidae, Hylobatidae) of East Africa, Ph.D. thesis, University of Cambridge.
Andrews, P. (1974). *Nature, Lond.*, 249, 188-190, and 680.
Andrews, P. and Tekkaya, I. (1976). *Proc. IX Cong. I.P.P.S.1-14*.
Andrews, P. and Walker, A.C. (1976). *In* "Human Origins", (G.L. Isaac and E.R. McCown, eds), pp. 279-304, Staples Press, London.
Becker-Platen, J.D., Sickenburg, O. and Tobien, H. (1975). *Geol. Jb.*, 15, 19-46.
Bishop, W.W. (1964). *Nature, Lond.*, 203, 1327-1331.
de Bonis, L., Bouvrain, G., Geraads, D. and Melentis, J. (1964). *C.r. Acad. Sc., Paris*, 278(D), 3063-3006.
de Bonis, L., Bouvrain, G. and Melentis, J. (1975). *C.r. Acad. Sc., Paris*, 281, 379-382.
Delson, E. and Andrews, P. (1976). *In* "Phylogeny of the Primates: a Multi-disciplinary Approach", (W.P. Luckett and F.S. Szalay, eds), pp. 405-446, Plenum, New York and London.
Fleagle, J.G. (1975). *Folia primatol.*, 24, 1-15.
Gentry, A.W. (1970). *In* "Fossil Vertebrates of Africa", (L.S.B. Leakey and R.J.G. Savage, eds), Vol. II, pp. 243-324, Academic Press, New York and London.
Hopwood, A.T. (1933). *Zool. J. Linn. Soc. London*, 38, 437-464.
Koenigswald, G.H.R. von (1935). *Kon. ned. Akad. Wet. Amsterdam*, 38, 872-879.
Koenigswald, G.H.R. von (1972). *Kon. ned. Akad. Wet.*, 75, 385-394.
Kretzoi, M. (1975). *Nature, Lond.*, 257, 578-581.

Lartet, E. (1856). *C.r. Acad. Sc., Paris,* 43, 219-223.

Le Gros Clark, W.E. and Leakey, L.S.B. (1950). *Q. Jl. geol. Soc. Lond.,* 105, 260-262.

Le Gros Clark, W.E. and Leakey, L.S.B. (1951). "Fossil Mammals of Africa", Vol. I (The Miocene Hominoidea of East Africa), 117 pp., British Museum (Natural History), London.

Leakey, L.S.B. (1962). *Ann. Mag. nat. Hist.,* 4, 689-696.

Lewis, G.E. (1937). *Amer. Jour. Sci.,* 34, 139-147.

Ozansoy, F. (1957). *Bull. Miner. Res. Explor. Inst. Ankara,* 49, 29-48.

Pandey, J. and Sastri, V.V. (1968). *J. Geol. Soc. Ind.,* 9, 206-211.

Pilbeam, D.R. (1969). *Bull. Peabody Mus. nat. Hist. New Haven,* 31, 1-185.

Pilgrim, G.E. (1915). *Rec. Geol. Surv. Ind.,* 45, 1-74.

Prasad, K.N. (1969). *Am. J. Phys. Anthrop.,* 31, 11-16.

Simons, E.L. and Pilbeam, D.R. (1965). *Folia primatol.,* 3, 81-152.

Simons, E.L. and Chopra, S.R.K. (1969). *Postilla (Yale Peabody),* 138, 1-18.

Tattersall, I. and Simons, E.L. (1969). *Folia primatol.,* 10, 146-165.

Takkaya, I. (1974). *Bull. min. Res. Exp., Ankara,* 83, 148-165.

Walker, A.C. and Andrews, P. (1973). *Nature, Lond.,* 224, 313-314.

AN APPRAISAL OF MOLECULAR SEQUENCE DATA AS A PHYLOGENETIC TOOL, BASED ON THE EVIDENCE OF MYOGLOBIN

K.A. JOYSEY

University Museum of Zoology, Downing Street, Cambridge CB2 3EJ, UK.

INTRODUCTION

This communication is based on joint work with my colleague A.E. Friday, and with our biochemical colleagues H. Lehmann and A.E. Romero-Herrera, formerly of the University Department of Clinical Biochemistry, Addenbrooke's Hospital, Cambridge. The lecture which I gave at the Sixth Congress of the International Primatological Society in Cambridge was developed from two other lectures given earlier in 1976 under the title 'Phylogenetic problems and parallel evolution in myoglobin'. One of these has been published in outline (Joysey et al., 1977) but personal circumstances prevailed against meeting the deadline for the publication resulting from the NATO Advanced Study Institute on Major Patterns in Vertebrate Evolution. Accordingly, the present paper has been tailored to include the essential points of both lectures. A full account of the joint work on which these lectures were based is now in press (Romero-Herrera, Lehmann, Joysey and Friday, 1978).

Despite the title of this paper I am not a biochemist; I am a whole animal zoologist interested in palaeontology and evolution, but during the last few years I have become involved with biochemical colleagues in an attempt to integrate the results of evolutionary studies at the anatomical and the molecular levels. I am going to discuss here the evidence of only a single molecule, but it has much to tell us about molecular evolution.

Professor H. Lehmann is a biochemist who has spent many years investigating the so-called 'abnormal' human haemoglobins which often produce profound clinical effects as a result of a single amino acid substitution in the molecule. Several years ago, jointly with Alex Romero-Herrera, he began to investigate the myoglobin of man, some other primates and a few other mammals. They determined the amino acid sequence of the myoglobins of several species and then sought from the zoologists a 'reliable' phylogenetic branching pattern showing the relationships of these species, together with 'accurate' dates of divergence. At that time they had no idea of the differences of opinion which divide our ranks and they were unaware of the problems of geological correlation and stratigraphical uncertainty which complicate the dating of a fossil — even after one has decided which fossil may be indicative that a particular divergence has already occurred. But we were interested to become involved because at that time some biochemists seemed to be claiming that protein sequence data held promise to solve most of our phylogenetic problems, and that 'molecular clocks' could be used to determine dates of divergence.

THE MYOGLOBIN MOLECULE

By the way of background it is necessary to mention that myoglobin is a
protein which is found as an intracellular component of red muscle. It combines reversibly with oxygen and it is believed to act as an oxygen carrier
and reservoir within the cells. It consists of a single chain of about 153
amino acids, combined with a single haem group. We are here concerned with
changes (or substitutions) in the primary structure of the molecule (i.e. the
sequence of amino acids), which represent fixed mutations. Some sections of
the peptide chain have a helical form (secondary structure), and the whole
chain is looped around and twisted back upon itself in a very definite pattern.
This conformation of the molecule (or tertiary structure) is maintained by a
combination of Van der Waal's forces between non-polar side chains, by 7 salt-
bridges and by 22 specific hydrogen bonds, some intra-chain and others inter-
chain in position (Fig. 1).

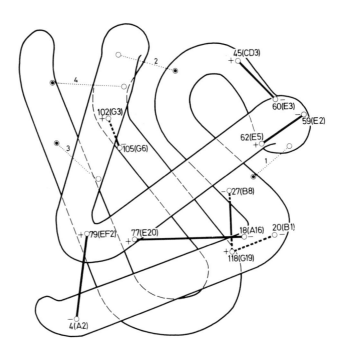

*Figure 1. An outline drawing of a myoglobin molecule, based on that of
sperm whale, showing the positions of four inter-segmental
hydrogen bonds (numbered 1 to 4), five inter-segmental and
two intra-segmental salt bridges.*

It is not difficult to envisage that some of the bonds and linkages which
determine the secondary and tertiary structure are of considerable functional
importance. Indeed, the residues present at about 40% of the positions have
been found to be invariant in all the myoglobins studied so far and it seems
likely that many of these are of importance either in the functional morphology of the myoglobin molecule itself or in its interaction with other molecules. At some positions where substitutions have been found the changes are
of a conservative nature; similarity in the properties of the residues

involved suggests that the changes are admitted under some constraint either at the molecular or the physiological level. Other positions seem to be able to accept amino acid substitutions quite readily and while one may be tempted to suggest that such positions are of less functional importance, it must be remembered that changes at any position are likely to have consequences upon adjacent regions of the molecule. In this context, adjacent may be taken to include parts of the molecule which are adjacent in the secondary and tertiary structure, although not necessarily so in the primary structure. Hence, the possible physiological and immunological consequences of any substitution may be relevant in terms of natural selection.

The haem group lies in an internal cavity formed by hydrophobic residues (the haem pocket), whereas the outside of the molecule in contact with the aqueous environment is mainly composed of hydrophobic residues. Thus, using sperm whale myoglobin as a model (Watson, 1969), 33 residues are internal and 120 can be described as external, and some of the internal residues are of particular importance as haem contacts. It is found that internal residues are less variable than external residues, that those external residues participating in salt bridges are less variable than the rest and that such positions usually admit only conservative changes. It is possible to rationalize the likely chemical consequences of particular substitutions among the other external residues, perhaps affecting the surface charge on the molecule, but direct experimental work on the physiological properties of the myoglobins of different species is mostly lacking.

MYOGLOBIN AS A PHYLOGENETIC TOOL

In our earliest collaborative study (Romero-Herrera et al., 1973), which was based on eighteen mammalian myoglobins, we found that the evidence of this single molecule provided some corroboration of generally accepted relationships based on the evidence of comparative anatomy. For example, palae-ontological evidence does not provide information on the sequence of branching of New World monkeys, Old World monkeys and Apes, but it is generally accepted that the two latter groups are more closely related to one another than is either to the New World monkeys. The evidence of myoglobin solved this particular trichotomy in the same way, there being several amino acid substitutions shared between the Old World monkeys and Apes, but absent in New World monkeys. On the other hand, we found that there were differences in the rate of fixation of mutations along two lines of descent arising from the same point, and hence fluctuations in the rate of fixation along a given lineage. Taken over quite long periods of time (80 Ma) there were still considerable differences in the overall rate of fixation of mutations in different lineages. Having some misgivings regarding the accuracy of this particular 'molecular clock' we suggested that 'average' rate was a more appropriate term than 'constant' rate in this context.

Encouraged by the fact that myoglobin seemed to be capable of solving problems in a manner concordant with conclusions based on comparative anatomy we then set out to obtain the opinion of myoglobin on some classical problems which have remained a matter for dispute. But when we sought to use myoglobin to settle the phylogenetic relationships of the tree shrew we found that, on criteria of 'cost' which will be discussed below, it fitted equally well either as a primate or as a non-primate (Lehmann et al., 1974).

Not discouraged, we sought to increase the total number of available myo-globin sequences, particularly among non-primates in the hope that this would

lay better foundations for solving such problems. In our 1973 paper, the 18
mammalian species included 10 primates, 3 cetaceans, 3 ungulates, 1 carnivore
and 1 marsupial. By 1975, the 30 vertebrate myoglobins available included 13
primates, 3 cetaceans, 4 ungulates, 4 carnivores, tree shrew, hedgehog, 2 marsupials and 2 birds.

Use of the difference matrix

It is possible to prepare a difference matrix which summarizes the total
number of amino acid differences between the myoglobins of pairs of species.
The difference matrix provides no information about the possible pathways of
evolution which have led to the present situation. It is tempting to investigate how much useful information can be extracted from such a matrix because
the data is comparable with that which is obtained through serological and
immunological studies, and such data is often easier to obtain than amino
acid sequence data. Using the myoglobin data Friday (1977) has explored
various procedures for their admissibility in constructing phylogenetic trees.
With suitable methods, it would be possible to place greater confidence in
phylogenetic results claimed on the basis of immunological studies alone. It
is sufficient to say here that there are various advantages and disadvantages
of the several clustering procedures available and that each procedure provided a slightly different answer, which is to be expected as each is based
on different assumptions.

The original Wagner method has been elaborated and presented in algorithm
form and it is particularly attractive because it is free from assumptions
concerning homogeneity of the rate of evolution (Farris, 1972). In the course
of development of the network it soon became apparent that, on accepted external criteria, the opossum was discrepant being assigned among the New World
monkeys and subsequently having the kangaroo linked in with it. The opossum
was left out and the procedure tried again but still the placings of several
animals were unexpected; for example, the harbour seal clustered with the
Cetaceans and the horse with the sportive lemur. These results are salutory.
Because we know that the opossum is a marsupial we do not read any phylogenetic inference into this similarity with the New World monkeys, but we
should bear this in mind when seeking hints on the affinities of such animals
as the hedgehog and the tree shrew where we do not know their phylogenetic
relationships.

Of the other procedures, the unweighted pair group method embedded the two
marsupials among the placentals, and the single linkage clustering method
widely separated the opossum and the kangaroo. The most effective clustering
method (i.e. the one which gave the 'best' zoological picture!) was a procedure
due to McQuitty which was originally devised for use in the social sciences;
nevertheless it placed the horse among the primates.

Certain groupings, such as the cetaceans, the apes and the artiodactyls do
remain stable through all the various clustering methods mentioned above and
in this sense might be regarded as 'reliable'. On the other hand, the repeated
clustering of the horse and sportive lemur remains a mystery and suggests pause
for consideration.

Reconstruction of amino acid substitutions

Turning now to the problem of reconstructing the possible pathways of evolutionary change in the form of a cladogram, it is usual to seek the most

economical (parsimonious) series of substitutions which will account for the
distribution of amino acids found in the living species. Clearly, the introduction of capricious change is open ended and common sense dictates that it
should be avoided. In effect one is reconstructing hypothetical myoglobin
chains of ancestral forms in such a way as to reduce the total number of subsequent changes to a minimum. When one sets out to fit the myoglobin data to
a given phylogenetic pattern, it is not surprising that alternative interpretations of amino acid substitutions can sometimes be reconstructed with equal
economy. Where it is necessary to exercise a preference it is sometimes possible to decide to accept back mutations in preference to parallel mutations,
or vice versa. Preference has always been given to those solutions which require single step changes rather than double hits at the codon level. Bearing
in mind that some amino acids may be coded in alternative ways, all interpretations of the pathway of change are undertaken by minimising the amount of
change at the codon level. It is to be expected that the introduction of new
species into the cladogram may cause some changes in the interpretation of the
nature of the ancestral residues (or their codons).

An alternative approach is to allow the phylogenetic branching pattern to
be changed in such a way as to produce the most economical solution. Under
these circumstances I follow the principle of parsimony with considerable
misgivings because the most economical solution may be unacceptable in zoological terms. For example, if the most parsimonious solution obtained
through a branch swapping procedure were to imply a close relationship between
the horse and the sportive lemur, I would hesitate to accept this on present
zoological evidence. It is possible that I might be wrong and that future
reappraisal of the evidence of comparative anatomy might suggest a close link
between Primates and Perissodactyla, and perhaps we should accept unexpected
molecular similarity as a hint to re-examine our preconceptions. On the other
hand, if the molecular evidence were to suggest that the opossum should be
linked with the New World monkeys I would seek some other explanation of the
molecular evidence.

On this basis, if it were found that the myoglobins of diving mammals were
similar to one another regardless of their systematic position, it would be
natural to seek an explanation in functional terms, and to regard such similarities as the product of parallel adaptation. In fact, there are some
parallel substitutions shared by some of the Cetacea and the harbour seal, and
there is one unusual substitution found only in the harbour seal and the
penguin.

Is it too far fetched to suggest that the unexpected similarity between the
myoglobins of the opossum and the New World monkeys, already detected by the
analysis of the difference matrix, may have some functional significance? The
geographical association of these two groups might even suggest parallel selection associated with some endemic disease which crosses the boundaries of the
mammalian orders, as does anthrax, but this is sheer speculation.

Alternative phylogenetic patterns

Based on the larger body of data (30 species instead of 18), the preliminary
results of our reconstruction of the possible pathways of evolution in myoglobin, with particular emphasis on primates, were presented to the Burg
Wartenstein Symposium 1975 (Romero-Herrera et al., 1976a). In Fig. 1 of that
publication (reproduced here as Fig. 2), we set out to emphasise the phylogenetic problem which we hoped to solve by showing five lines of descent

arising from an unresolved pentachotomy. In Fig. 2 of that publication (not reproduced here) we gave, as an example of our methods, one of the two most parsimonious solutions which we had found in resolving the pentachotomy, and we discussed in detail that part of the cladogram dealing with the Primates. We concluded that paper by drawing attention to our finding that parallel evolution is commonplace, about 50% of the substitutions being repeated elsewhere on the cladogram.

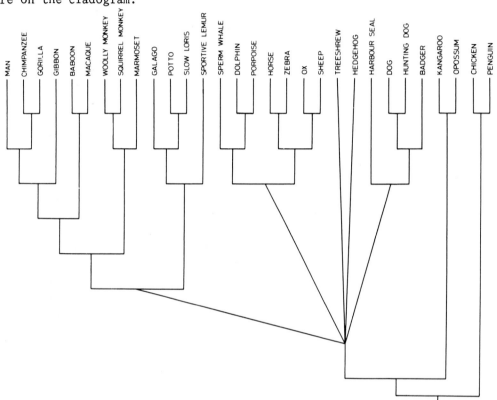

Figure 2. A phylogenetic pattern based on the evidence of comparative anatomy showing likely relationships between 30 living species. The pentachotomy represents the superimposition of four branching points, the order of which is controversial. An amino acid sequence is available for the myoglobin of each of these species.

Throughout our studies on myoglobin we have been reluctant to use branch swapping procedures which seek the most parsimonious solution regardless of its zoological acceptability. We have adopted an alternative approach of setting up a series of possible branching patterns and then seeking to discover which of these provided the most economical solution. In effect, we have tried to reach a compromise with the parsimony principle, but we have to some extent been disappointed by the results of our tactic.

Eight possible solutions to the pentachotomy of Fig. 2 are shown in Fig. 3. It may be seen that although these possible patterns of evolutionary relationship are very different from one another (e.g. hedgehog and tree shrew sometimes linked and sometimes separate, sometimes branching off before the carnivores and sometimes after, tree shrew sometimes included within the

Primates and sometimes excluded), when the myoglobin data is fitted to each
pattern the total number of fixed mutations required in each case is almost
the same. Of the eight possible phylogenies which we have explored, two were
equally parsimonious at a cost of 281 hits, one scored 282 hits and five were
equal at 283 hits; these differences cannot be regarded as convincing. Indeed,
on the evidence of myoglobin, there is little to choose between these different solutions. It is possible that the availability of more myoglobin sequences
will resolve this problem, but it is also possible that the evidence of myoglobin alone is incapable of resolving those same problems which have defeated
comparative anatomy, perhaps for similar reasons.

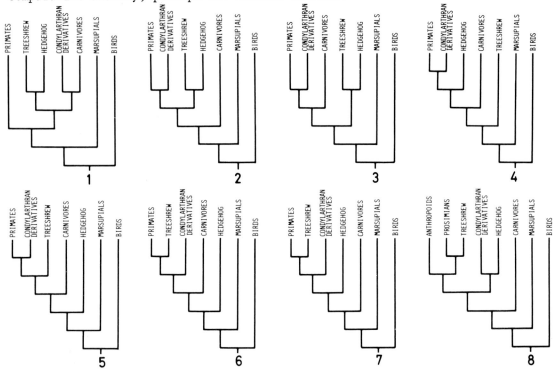

Figure 3. *Eight possible solutions of the pentachotomy shown in Fig. 2.
Condylarthran derivatives include the Cetacea, Artiodactyla and
Perissodactyla. The relationship between birds, marsupials and
placental mammals is fixed. Patterns 2 and 5 each cost 281 hits,
pattern 3 cost 282 hits and the other five patterns all cost
283 hits.*

Apart from the obvious difficulty of separating shared derived characters
from the parallel acquisition of characters (particularly when parallel evolution is commonplace) it is possible that the sequence of branching which we
were attempting to resolve might have occurred over quite a short period of
time relative to the known 'average' rate of fixation of substitutions in
myoglobin. For example, if the four branching events which are required to
resolve the pentachotomy all occurred within a period of 10 million years (as
is possible) then myoglobin, with an average rate of fixation of one mutation
in about 4 million years (in any given lineage), is unlikely to resolve the
problem. On the other hand, if one chose to use a molecule with a faster

rate of change, then after a period of some 80 million years the critical
changes are likely to have been over-written by subsequent change (including
back mutation). It seems possible that problems of this nature may be
solved in the future by combining the evidence of several molecules, each one
chosen with respect to its average rate of substitution, bearing in mind the
probable date of the unresolved events.

In a recent publication (Romero-Herrera et al., 1976b), we have shown that
the myoglobin of orangutan differs at one position (23 Ser) from the residue
shared by man, chimpanzee and gorilla (23 Gly), and it differs at another
position (110 Ser) from the residue shared by man, chimpanzee, gorilla and
gibbon (110 Cys). At each of these positions the residue present in orangutan
is shared with baboon and macaque. We found that the most parsimonious solution (Fig. 4) was obtained when the lineage leading to orangutan was placed so
that it branched off the lineage leading to man, chimpanzee and gorilla, before
that leading to gibbon. A more conventional pattern of relationship, with
the lineage leading to gibbon branching off before that leading to the orang-
utan, required either a back mutation (110 Ser) or a parallel mutation (110
Cys) within the hominoids, at the cost of one extra mutation. It was also
noted that the cost would be the same if the orangutan and the gibbon were
linked to one another on a common stem (i.e. accepted as being equidistant in
relationship to the other apes, although this pattern would still involve
either a back mutation or a parallel mutation).

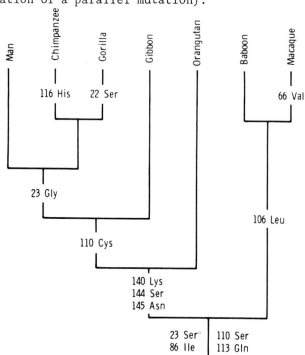

Figure 4.   A pattern which allows the most parsimonious solution of the
            amino acid substitutions leading to the known myoglobin
            sequences among catarrhines. (The asterisk indicates an
            ancestral condition brought forward from an earlier time.)
            It has recently been shown that the myoglobin sequences of
            gibbon and siamang are identical (Bruce et al., 1977).

It is evident that even within a well known group the acceptability of the most parsimonious solution is likely to be a matter of opinion and, of course, opinions may change as new lines of evidence become available. If concordant evidence from several unrelated molecules were all found to point in the same direction one might be persuaded to accept a new pattern of relationships among the great apes, but I hesitate to do so on the evidence of a single molecule, and prefer at present to relax the rule of parsimony.

RATES OF EVOLUTION

Estimates of the rate of molecular evolution are of interest because its supposed constancy has been one argument in favour of the hypothesis that a high proportion of fixed mutations are neutral, or nearly so, as far as selection is concerned. In trying to establish the best estimate of date for the various branching events on our cladograms we have, as a point of policy, rigorously demanded direct fossil evidence of the existence of a relevant new group and we have rejected all speculations about fossils not yet found, because such speculation about earlier origins is open ended and does not form a sound basis for phylogenetic discussion. In consequence, our dates are to be regarded as minimum dates, because many are likely to be later than the actual event. Even this statement needs to be qualified because the choice of a particular fossil as indicative of a particular event is often a matter of opinion. For example, in dating the origin of hominids we took *Ramapithecus* as the relevant fossil in our first paper (Romero-Herrera et al., 1973), and we used *Australopithecus* in a more recent paper (Joysey et al., 1977), not because of any particular commitment to either view but rather to explore the consequences of such differences of opinion on our investigation of rates of molecular evolution. A full justification of the dates we have adopted is given in Romero-Herrera et al.(1978).

The combination of these dates with the reconstructed fixed mutations derived from one of the two most parsimonious cladograms (Romero-Herrera et al., 1976a) indicates that there are considerable differences in the amount of change in different lineages (Fig. 5). For example, whereas one lineage (to gibbon) appears to have accepted no mutations during the past 20 Ma, another lineage (to ox) seems to have fixed 7 mutations during the last 18 Ma. Goodman, Barnabas, Matsuda and Moore (1971) have drawn attention to the apparently low rate of molecular evolution among higher primates; a similar observation applies to the myoglobin of the few Old World monkeys so far studied, but the myoglobins of the available New World monkeys and prosimians do not share this feature.

Sampling error will, of course, produce some fluctuations in rate but even over the relatively long period of about 80 Ma the fastest rate of fixation of mutations is about three times the slowest rate (based on 34 and 11 fixed mutations, respectively). Hence, we are inclined to discard the molecular clock as unreliable for dating divergences, at least within this span of time.

DISCUSSION

Given the constraints demanded by the functional morphology of the molecule itself and the constraints of the genetic code, it is to be expected that both will contribute to parallel change. One of the reasons that we are finding it difficult to assess the relative merits of phylogenetic patterns is that

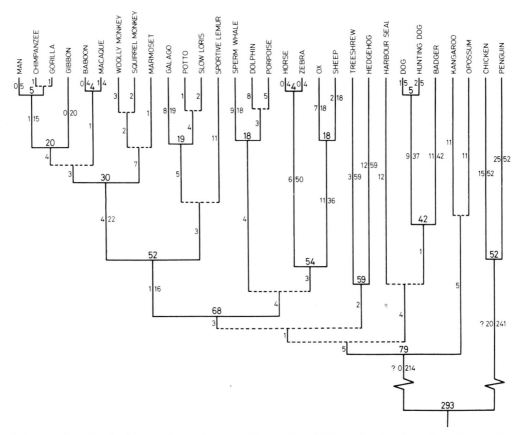

Figure 5. *A cladogram based on pattern 2 of Fig. 3 showing the date of*
divergence (where available) above each dichotomy, and to the
right of each branch the time span between successive dicho-
tomies. The number of mutations reconstructed along each
lineage between successive dichotomies is shown on the left of
each branch. Horizontal dotted lines represent dichotomies for
which no reliable date was available. The actual reconstruction
of the mutations is shown in Figure 2 of Romero-Herrera et al.,
(1976a).

parallel evolution is commonplace at the molecular level. It is evident that
there are limited possibilities for change, there being a limited number of
ways of remaining a functional myoglobin molecule. Functional morphology seems
to be of vital importance both at the anatomical and molecular levels. The
limitations of mechanics at the anatomical level and the limitations of, for
example, salt-bridges and hydrogen bonds at the molecular level place constraints on the possible direction that the evolution may take. Because the
laws of physics and chemistry have to be obeyed we find that there are a limited number of ways of making a living both at the anatomical level and at the
molecular level, and by inference at all intermediate levels of structure.
    Looking to the future, it seems that rigid application of the parsimony
principle is likely to produce unacceptable results if allowed to proceed to
a minimum solution regardless of all other lines of evidence. It could be
that, without necessarily proceeding to the minimum solution for any given

molecule, concordance of reasonably low solutions from several different molecules may deserve phylogenetic respect. It seems likely that procedures will be used to give less weight to substitutions which can be recognised, on other criteria, as being the product of parallel change, so giving relatively greater weight to those substitutions which are more likely to be monophyletic markers. Furthermore, for phylogenetic studies, the pooling of information from several molecules of comparable average rate may improve the resolution of the tool.

I am grateful to Adrian Friday for helpful discussion and criticism of this manuscript.

REFERENCES

Bruce, E.J., Castillo, O. and Lehmann, H. (1977). *FEBS Letters*, 78, 113-118.

Farris, J.S. (1972). *Am. Nat.*, 106, 645-668.

Friday, A.E. (1977). *In* "Myoglobin", (A.G. Schnek and C. Vandercasserie, eds), pp. 142-166, Editions de l'Université, Brussels.

Goodman, M., Barnabas, J., Matsuda, G. and Moore, G.W. (1971). *Nature, Lond.*, 233, 604-613.

Joysey, K.A., Friday, A.E., Romero-Herrera, A.E. and Lehmann, H. (1977). *In* "Myoglobin", (A.G. Schnek and C. Vandercasserie, eds), pp. 167-178, Editions de l'Université, Brussels.

Lehmann, H., Romero-Herrera, A.E., Joysey, K.A. and Friday, A.E. (1974). *Ann. N.Y. Acad. Sci.*, 241, 380-391.

Romero-Herrera, A.E., Lehmann, H., Joysey, K.A. and Friday, A.E. (1973). *Nature, Lond.*, 246, 389-395.

Romero-Herrera, A.E., Lehmann, H., Joysey, K.A. and Friday, A.E. (1976a). *In* "Molecular Anthropology", (M. Goodman and R.E. Tashian, eds), pp. 289-300, Plenum, New York.

Romero-Herrera, A.E., Lehmann, H., Castillo, O., Joysey, K.A. and Friday, A.E. (1976b). *Nature, Lond.*, 261, 162-164.

Romero-Herrera, A.E., Lehmann, H., Joysey, K.A. and Friday, A.E. (1978). *Phil. Trans. R. Soc. B.*, 283, 61-163.

Watson, H.C. (1969). *Prog. Stereochem.*, 4, 299-333.

RELATIVE GROWTH IN PRIMATES

B.A. WOOD

Department of Anatomy, The Middlesex Hospital Medical School, Cleveland Street, London W1P 6DB, UK.

INTRODUCTION

Although animals grow in an integrated way it is possible to study the progress of different parts of the animal by means of relative growth rates. Allowing for the effects of volume and area relationships, if relative growth rates are equal then animals will retain their shape and simply grow in size. If growth rates are not equal, then the shape of the animal will change as it grows in size.

Adult animal shape can be looked upon as the result of relative growth relationships during ontogeny. When Huxley (1932) synthesised his ideas on relative growth he concentrated on analysing relative growth rates, and the term 'allometry' was introduced to describe change of shape with increasing size (Huxley and Teissier, 1936). When it was recognized that the relationships of variables in 'grown' animals could be investigated, the study of allometry was subdivided into 'heterauxesis', which applied to longitudinal and cross-sectional growth studies, and 'allomorphosis', the study in adults of heterauxetic growth relationships (Huxley, Needham and Lerner, 1941). There is now a commonly agreed notation, $y = bx^{\propto}$, to describe the relationships between two variables, where y is the reference variable, x the variable being studied, b a constant and $\propto$, the allometry coefficient, is the slope of a plot of log y/log x.

Shape and size differences occur within species just as they do between them. In the Pongidae and Cercopithecidae, in particular, sexual dimorphism is a major source of intraspecific variability. Males in some species are not only larger than females but they also differ in shape; canine length and size, jaw size and limb proportions are three examples. The hypothesis that these sexual dimorphisms are due to allometric growth phenomena was tested in five groups of primates.

STUDY

A total of 99 adult male and 103 adult female skeletons of *Homo*, *Gorilla*, *Pan*, *Papio* and *Colobus* were examined and 90 measurements, cranial and postcranial, were made on each skeleton. Details of the samples and measurement protocols are given in Wood (1975, 1976). The size and shape differences between the sexes in each group were estimated using Penrose's size and shape distances and by computing the principal components of each data set. The percentage dimorphisms for each variable were combined to give a mean percentage dimorphism for each primate group. The eigenvalues of the first principal

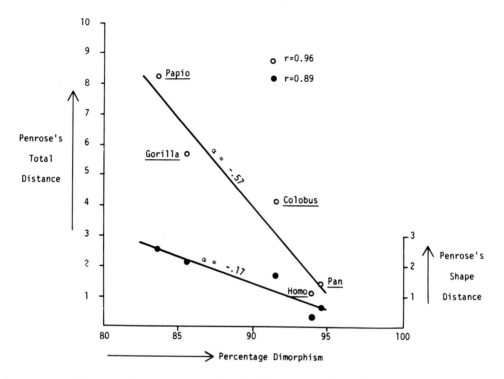

*Figure 1. Relationship between total distance, the shape component and
mean percentage dimorphism.*

components were correlated (r = 0.97) with the mean percentage dimorphism.
Similar correlation levels were found between mean percentage dimorphism and
both the total Penrose distance and its shape component (Fig. 1).

The allometry coefficients for each variable were calculated separately for
the males and females in each group, using femur length as the independent
variable in a least squares regression analysis. The significance of any difference between the male and female slopes was tested using the t-test for
means (Simpson, Roe and Lewontin, 1960). Although some allometry coefficients
were very different between the sexes, this was often due to poor correlation
between the variables and only in *Homo* did the number of significant differences
narrowly exceed the number that would be expected by chance alone.

In cases where there was no significant difference in the allometry coefficients slopes were calculated for the combined male/female data. Most of these
combined slopes deviated from unity so the covariance of these variables in the
adults indicated that growth had been allometric.

CONCLUSIONS

These results show that there are no consistent differences between the
growth patterns of males and females and it suggests that allometric growth may
account for shape differences between males and females in the primate groups
studied.

Huxley (1932) and Hersh (1934) had both suggested that comparison of allometry coefficients would be useful in taxonomy. They have been used in studies of living primates (Frick, 1959; Giles, 1956; Freedman, 1962; Bibus, 1967; Hemmer, 1964, 1967; Biegert and Maurer, 1972) and have also been employed in the analysis of fossil hominids (Kinzey, 1972; Leutenegger, 1973; Pilbeam and Gould, 1974). Results of earlier studies had suggested that allometry coefficients may be sexually dimorphic within a species group, but these latest results suggest that in primates intraspecific variations of the allometry coefficient are unlikely.

ACKNOWLEDGEMENTS

The support of the Royal Society, the Central Research Fund of the University of London and the Boise Fund are gratefully acknowledged.

REFERENCES

Bibus, E. (1967). *Folia primatol.*, 6, 92-130.
Biegert, J. and Maurer, R. (1972). *Folia primatol.*, 17, 142-156.
Freedman, L. (1962). *Growth*, 26, 117-128.
Frick, H. (1959). *Anthrop. Anz.*, 23, 64-71.
Giles, E. (1956). *Hum. Biol.*, 28, 43-58.
Hemmer, H. (1964). *Homo*, 15, 218-224.
Hemmer, H. (1967). *Fortschr. Evol.*, 3.
Hersh, A.H. (1934). *Amer. Nat.*, 68, 537-561.
Huxley, J.S. (1932). "Problems of relative growth", Methuen, London.
Huxley, J.S. and Teissier, G. (1936). *Biol. Zbl.*, 56, 381-383.
Huxley, J.S., Needham, J. and Lerner, I.M. (1941). *Nature*, 148, 225.
Kinzey, W.G. (1972). *Am. J. phys. Anthrop.*, 37, 442-443.
Leutenegger, W. (1973). *Folia primatol.*, 19, 9-17.
Pilbeam, D. and Gould, S.J. (1974). *Science*, 186, 892-901.
Simpson, G.G., Roe, A. and Lewontin, R.C. (1960). "Quantitative Zoology", Harcourt, Brace and World, New York.
Wood, B.A. (1975). "An analysis of sexual dimorphism in primates", Ph.D., University of London.
Wood, B.A. (1976). *J. Zool. Lond.*, 180, 15-34.

DENTOFACIAL RELATIONSHIPS IN SEXUALLY DIMORPHIC POPULATIONS

M.I. SIEGEL

Dept. of Anthropology, University of Pittsburgh, Pa 15260, USA.

The purpose of this paper is twofold: first, to describe some of the dento-facial correlates which exist in a sample of olive baboons and second, to point out the methodological pitfalls of lumping males and females when analyzing such sexually dimorphic populations.

In several earlier publications, the author investigated the relationship between tooth root length and facial proportions in carnivores (Riesenfeld and Siegel, 1970) and in nonhuman primates (Siegel, 1972). In both cases, it was found that the root dimensions of teeth in closest proximity to the palatal sutures were most highly correlated with measurements of the face. For these studies, males and females were treated as a total sample and not analyzed separately. It was the great degree of sexual dimorphism which provided the variability in facial protrusion, under consideration in the primate study. The lumping of the sexes also supplied an "adequate" sample size. Traditionally, when samples are small, and the sexing of individuals difficult because of their fragmentary nature, palaeontologists are forced to analyze such lumped samples. Sophisticated statistical treatments of the metric data are used to generate hypotheses concerning the affinities of these sexually lumped samples. Sexually dimorphic populations present unique problems when the sex of the individuals is not considered, and these problems can lead to spurious conclusions about the samples under consideration.

In the present study a sample of 40 olive baboons (*Papio cynocephalus panubis*), 19 males and 21 females, were used. 38 measurements were taken on each specimen: palatal length, palatal width, mandibular length, mandibular depth at three points, mesio-distal diameters of mandibular and maxillary teeth on one side and the root lengths for the same teeth. The overall sample and specific details of the measuring techniques are described fully by Siegel (1972).

The product moment correlation coefficient 'r' was computed between all dental dimensions and skeletal dimensions as well as between crown and root measurements.

For examples in this paper, twelve values of 'r', six for the maxillary and six for the mandibular dentition are reported. The entire matrix of over 1,000 correlation coefficients is more fully described in a longer paper now in preparation. All correlation coefficients are computed for the entire sample, for males alone and for females alone (Table I).

In the primate root length study, I reported (1972) an 'r' value of .728 for the correlation between maxillary canine root length and palatal length. This was used to strengthen my hypothesis that growth activity at the premaxillary-maxillary suture was responsible for the relationship between root development and facial growth, and that bony reduction could be related to some phylogenetic

TABLE I

Correlation Coefficients for Selected Dimensions

Mandibular Dentition	Correlation Coefficient and Level of Significance		
	Combined	Females	Males
P3 Length: Mandibular Length	.57 S*	.09 NS	.05 NS
C Length: Mandibular Length	.50 S	.20 NS	-.20 NS
M1 Length: Mandibular Depth	.58 S	-.05 NS	-.39 NS
M2 Length: Mandibular Length	.58 S	.45 S	.20 NS
M3 Root length: Mandibular Length	.52 S	.37 NS	.46 S
M3 Root length: Mandibular Depth	.62 S	.68 S	.19 NS
Maxillary Dentition			
C Root length: Palatal Length	.73 S	.16 NS	.35 NS
C Length: Palatal Length	.90 S	.15 NS	.17 NS
C Length: Palatal Width	.76 S	-.15 NS	.55 S
M1 Length: Palatal Length	.66 S	.24 NS	-.22 NS
M2 Length: Palatal Length	.74 S	.47 S	-.10 NS
M3 Length: Palatal Length	.72 S	.50 S	.04 NS

* S significant $P < .05$, NS not significant $p > .05$.

trends of dental reduction. When we look at the sexes separately (Table I) we see that the correlation is not significant for either males or females alone. This is also the case for five additional dimensions reported here, including those for mandibular P3, C, M1 and mandibular length as well as maxillary M1 and palatal length. Returning to the mandibular dentition, for M2 length correlated with mandibular length the 'r' value is significant for the total sample and for females alone, but not for males alone. For M3 root length, significant correlations are found with mandibular length and depth for the total sample, but for males we see a significant correlation with length and not depth, and for females just the opposite. With the maxillary dentition, further inconsistent findings are apparent. Palatal length is significantly correlated with M2 length and M3 length only for the total sample and for females, but not for males. The correlation of canine length with palatal width is significant for the total sample and for males, but not for females.

Even these few selected examples demonstrate the kinds of inconsistencies which exist when samples of males and females are lumped. Some relationships which appear to be significant for the total population are at best significant for one sex only. In these cases, or when the coefficients are significant for neither sex, the significant population value is an artifact of the two clusters of points in the distribution, which is extremely platykurtic or bimodal in nature. This is the technical problem posed by sexually dimorphic populations. Many metric variables may not be normally distributed. When this is the case, predictive correlations must not be based on lumped samples. Such was the case for our own conclusions concerning canine root length relations, which do not hold up when the sample is analyzed for males or females alone. Where size diagnoses of fossil primate species are made on dental dimensions of lumped samples (e.g. Gingerich, 1974) similar problems may occur. Such diagnoses imply that other dimensions of the animal are predictable from dimensions of a single tooth which has "special" properties. However, if for the total sample that tooth has dimensions which are bimodally distributed (i.e., if there is significant sexual dimorphism), the predictive value is lost and with it the ability to identify species.

It is therefore important to first determine if the dimensions under study are normally distributed and, if they are not, to separate the sample before analysis is begun. To ignore this important step is to run the risk of having someone point out the fallacies of one's research efforts. This investigator was fortunate enough to discover himself "the errors of his ways", even if four years late.

ACKNOWLEDGEMENTS

The skulls used in this study are from the primate skeletal collection of Dr. N.C. Tappen, University of Wisconsin-Milwaukee. Preparation and storage of the skulls were supported by National Institutes of Health Grant HD - 02033. I wish to express my appreciation to Dr. Tappen who made the specimens available. Travel to this conference was made possible through grants from the Provost's Research and Development Fund and the John Bowman Foundation.

REFERENCES

Gingerich, P.D. (1974). *J. Paleon.*, 48, 895-903.
Riesenfeld, A. and Siegel, M.I. (1970). *Am. J. Phys. Anthrop.*, 33, 429-432.
Siegel, M.I. (1972). *Acta Anat.*, 83, 17-29.

FUNCTIONAL MORPHOLOGY OF HOMINOID METACARPALS

R.L. SUSMAN*

*Dept. of Anthropology, University of Chicago,
1126 E. 59th Street, Chicago, Illinois 60637, USA.*

INTRODUCTION

Numerous recent functional and morphometric studies have focused on the hominoid wrist in attempts to explain structural-functional relationships in extant primates and to infer the behaviour of fossil Hominoidea (Lewis, 1969, 1972; Schön and Ziemer, 1973; Jenkins and Fleagle, 1975; Corruccini et al., 1975; Morbeck, 1975). These studies have reached different conclusions about the functional affinities of fossil forms.

The present study was undertaken to provide complementary data on the problem of the evolution of hominoid locomotor patterns by relating structure and function of hominoid manual rays. The study provides additional perspectives for modelling the biomechanics of the carpus and manual rays, and ultimately for inferring the behaviour of fossil Hominoidea.

MATERIALS AND METHODS

A series of measurements was made on individual metacarpals and phalanges of manual rays II-V in 255 adult hominoids (excluding *Pan paniscus*). Dissection of wet specimens and observations of captive animals (Tuttle, 1967, 1969; pers. obs.), provided a basis for determination of functionally significant variables. Measurements were taken on bones and from standardized lateral radiographs (see Susman, 1976 for details). Variables include length, shaft diameter, dorso-palmar head diameter, head diameter measured at the dorsal ridge, radio-ulnar head breadth, biepicondylar diameter, medullary cavity breadth, height of the dorsal articular ridge, and longitudinal curvature (see Table I).

Discriminant analyses were conducted on groups identified by gender and species. The results of statistical analyses are interpreted on biomechanical grounds on the basis of observed differences in locomotor hand posture. The following brief discussion focuses on analysis of metacarpal III, a primary weight-bearing ray in all of the extant apes.

---

* present address: Dept. of Anatomical Sciences, Health Sciences Center, S.U.N.Y. Stony Brook, Stony Brook, New York 11794, USA.

RESULTS

   Analysis of the third metacarpal reveals the distinctiveness of this segment
in each of the apes and man. Discriminant functions 1 and 2 account for 89% of
the variation between groups. Standardized coefficients of variables on functions 1 and 2 reveal the strong influence of features of the metacarpal head
(Table I).

TABLE I

*Standardized coefficients on the first two discriminant functions
(analysis of metacarpal III).*

Variable	Function 1	Function 2
Length	0.0155	0.9645
Radio-ulnar shaft diameter	0.1871	-0.3042
Dorso-palmar shaft diameter	-0.0745	-0.1915
Dorso-palmar head diameter	2.3559	0.3304
Radio-ulnar head breadth	0.2716	-0.4279
Biepicondylar (head) diameter	0.0911	0.2099
Medullary Cavity Breadth	-0.0903	-0.3304
Dorsal ridge height	0.9869	-0.0329
Longitudinal curvature	-0.0969	0.0875
Head diameter at the dorsal ridge	-2.5300	-0.0271

   On function 1 dorsal ridge development, the dorso-palmar head diameter and
head breadth are the most heavily weighted variables. On the second discriminant function, length, midshaft diameter and medullary cavity development
are prominent in addition to head breadth and diameter. This variation can be
explained in biomechanical terms that take account of the continuous differences
in hominoid hand posturing. Study has also revealed musculo-ligamentous adaptations in the manual rays that accommodate distinctive patterns of stress in
the metacarpus and phalanges.
   Gorilla and chimpanzee third metacarpals have: 1) deep metacarpal heads,
2) widened dorsal articular surfaces (in metacarpals III and IV), 3) pronounced
body ridges on the dorsal extent of the caput, and 4) a broad biepicondylar
diameter. These features reflect the increased range of movement of the proximal phalanges in the African apes, the buttressing of the metacarpal head
(Preuschoft, 1973), and a probable mechanism to help prevent bowstringing of
the collateral ligaments and intrinsic finger muscles across the metacarpophalangeal joints when the proximal phalanges are hyperextended. The dorsal expansion of the articular surface (in rays III and IV) reduces stress when the
hyperextended proximal phalanges are loaded. The African apes possess metacarpals with thin cortices and heavily trabeculated metacarpal heads.
   Metacarpal III in the Asian apes and man lacks strong development of secondary features of the shaft and head. Orang utan metacarpals are attenuated
and have thickened cortices compared with man and the other apes. The hylobatid apes possess long, thin metacarpals lacking strong secondary features
of the shaft and head. In the hylobatids the dorso-palmar extent of the distal
articular surface is reduced.
   The metacarpals of man are short and resemble *Pongo* in capitular morphology,

but resemble the African apes in cortical development and lack of longitudinal curvature.

The above variability reflects the marked differences in normal locomotor hand postures among the extant apes and man. In the African apes the fingers accommodate the unique digitigrade pattern known as knuckle-walking (see Tuttle, 1967, 1969). The proximal phalanx constitutes the principal load arm for the long digital flexors at the metacarpophalangeal joint. The moment of the long digital flexors is enhanced intrinsically by the combined depth of the metacarpal head, and the development of the palmar plate. In the African apes the power arm is increased in proportion to the load arm. In *Pongo* the intrinsic capacity of the metacarpophalangeal joints for load bearing is reduced by a short power arm and a relatively long load arm (= proximal phalanx). Other considerations of pongid rays in various suspensory modes reveal the increased capacity for efficient hook grasping of both vertical and horizontal supports in *Pongo* and the hylobatid apes.

CONCLUSIONS

The subtle, continuous differences in hominoid manual ray morphology require detailed quantitative observations of associated behavioural modes in order to properly explain the relationship between morphology and behaviour. Sound biomechanical inference then permits the explication of morphological patterns that have no extant counterpart (Rudwick, 1964).

The implications of the present work for interpreting the record of fossil hominoid hands are based on the employment of the hand in all manner of maintenance behaviour. The question of whether or not the initial hominid adaptive shift involved use of the hand for part- or full-time suspension, quadrupedalism, or whether it was free of a locomotor role awaits consideration of relevant fossils and their interpretation in a comprehensive comparative and functional framework.

ACKNOWLEDGEMENTS

I thank Professor Russell Tuttle for his help throughout this study and the American Museum of Natural History, the Cleveland Museum of Natural History, The United States National Museum, and the Field Museum of Natural History for the use of materials in their care. This work was supported by an NDEA Title IV and predoctoral fellowship and generous grants from the Henry Hinds Fund, Committee on Evolutionary Biology, University of Chicago.

REFERENCES

Corruccini, R.S., Ciochon, R.L. and McHenry, H.M. (1975). *Folia Primatol.*, 24, 250-274.
Jenkins, F.A. and Fleagle, J.G. (1975). *In* "Primate Functional Morphology and Evolution", (R.H. Tuttle, ed.), pp. 213-228, Mouton Pub., The Hague.
Lewis, O.J. (1969). *Amer. J. Phys. Anthrop.*, 30, 251-268.
Lewis, O.J. (1972). *Amer. J. Phys. Anthrop.*, 36, 45-58.
Morbeck, M.E. (1975). *J. Hum. Evol.*, 4, 39-46.
Preuschoft, H. (1973). *In* "Human Evolution", (M.H. Day, ed.), pp. 13-46, Soc. Study Hum. Bio., Symp. XI.
Rudwick, M.S. (1964). *Brit. J. Philos. Sci.*, 15, 27-40.

Schön, M.A. and Ziemer, A. (1973). *Folia Primatol.*, 20, 1-11.
Susman, R.L. (1976). "Functional and Evolutionary Morphology of Hominoid
    Manual Rays II-V", Doctoral thesis, University of Chicago.
Tuttle, R.H. (1967). *Amer. J. Phys. Anthrop.*, 26, 171-206.
Tuttle, R.H. (1969). *J. Morph.*, 128, 309-364.
Tuttle, R.H. and Basmajian, J.V. (1974). *In* "Primate Locomotion", (F.A. Jenkins
    ed.), pp. 293-347, Academic Press, New York and London.

ALLOMETRY AND ENCEPHALIZATION

H.J. JERISON

Neuropsychiatric Institute, The Center for the Health Sciences,
University of California, 760 Westwood Plaza, Los Angeles,
California 90024, USA.

Allometry is a phenomenon of growth that has also been used to describe animal species as if one species literally *grew* out of another. Encephalization is a phenomenon of brain evolution referring structurally either to the increased size of the brain relative to the rest of the nervous system, or to evolutionary enlargement of the brain in some species relative to others. The mathematical formulations of allometry and of encephalization are usually quite similar, which occasionally results in confusion of the phenomena or the assumption (possibly correct, though not accepted here) that the mechanism for encephalization involved the action of genetic systems (e.g. regulator genes) controlling allometric relations between brain and body.

The fundamentals of allometry as theory are derived from a problem of relative growth. Brain/body relations have often been used as an example, and the logic of the argument is as follows: the change in size of the brain dE should be related to a growth factor for the brain, a, and the size of the brain E at time t. This deads to a differential equation:

$$dE = a\ E\ dt$$

In a similar time span, the body's growth dP would be related to body size P according to a different growth constant b to lead to:

$$dP = b\ P\ dt$$

Combining these equations we have:

$$\frac{dE}{dP} = \frac{a}{b}\ \frac{E}{P}\ \frac{dt}{dt}$$

The dt term cancels, which is one objective of the analysis, i.e., to show the change in size of organs relative to one another during growth without the confusion of viewing the change in a graph in which three dimensions are needed: one for time and two more for the two organs in question. The ratio of growth rate factors for brain and body is designated alpha:

$$\alpha = \frac{a}{b}$$

and the equation can be arranged as:

$$\frac{1}{E}\ dE = \frac{a}{b} \cdot \frac{1}{P}\ dP = \alpha \cdot \frac{1}{P}\ dP$$

Integrating both sides of this equation gives:

$$\log E = \alpha \log P + \log k \qquad (1)$$

Log $\underline{k}$ is merely the constant of integration and is therefore of little importance in the analysis of growth, since its value will depend on the intervals
over which the integration is performed. Alpha is the constant of interest;
its magnitude relative to 1.0 provides a direct indication of growth. When
it exceeds 1.0, it signifies a higher growth rate for organ E (the brain, in
this instance) than for organ P (the entire body, in this instance). Values
less than 1.0 signify the opposite, while $\alpha = 1.0$ signifies equal growth rates.
Equation 1 has the additional advantage of being linear for logarithmic trans-
formations such as those provided by graphs in which ordinates and abcissae are
logarithmic.

The procedure may be applied to any sets of organ sizes, either in pairs for
bivariate allometry or in larger sets for multivariate allometry. The appli-
cations beyond problems of growth are usually descriptive rather than analytical and the theory just described, which follows Huxley (1932), is not directly
relevant. Efforts to develop relevant versions of the theory have been suggested by Gould (1971, 1975), and if these are successful they may be useful
to extend the analysis of encephalization that I propose.

Encephalization as a morphological problem involves the evolution of enlarged brains in some species relative to others, and a first step in a rational
analysis of this problem is to develop a theory of brain size that describes
the expected size of a brain in a species. Such a theory can be developed from
two fundamental ideas in the neuro-sciences: first the brain is constructed
in an unusually ordered way with the arrangement of elements at one level mapped
onto other levels with conservation of the pattern of the map (Woolsey, 1958;
Welker, 1973; Palay, 1967). (The degree of conservation is impressive, but
is of course not quite as simple as I assume for the purpose of theory-con-
struction.) The second point, less central for the theory but useful for
quantitative predictions, is that neural tissue in the brain is frequently
organized in such a way that "functional units" are of the order of 0.1 cm in
length, for example as in columnar units in sensory and motor neo-cortex. The
first statement in the theory is that the size of the brain, E, is determined
by the area of the mapped projection systems summed over the entire brain in
which mapping occurs. This statement is given dimensionally with the size
(volume) of the brain as having dimension L^3 and the summed projection areas
having dimension L^2. The source of the measures is given by subscript. Thus:

$$L_E{}^3 = f\ (L_p{}^2)$$

The function f must be balanced dimensionally, requiring multiplication of the
right-hand term by something with dimension L. That something can be the
depth of a functional unit L_d, giving us an exact statement:

$$L_E{}^3 = L_d \cdot L_p{}^2$$

In practice it may be impossible to measure all of the projection surfaces of
a brain, since all of these are probably not known, and in some instances there
is considerable overlap among surfaces. The $L_p{}^2$ term must, therefore, be estimated in some way, either from cortical mapped surface, total mapped surface,
or something related to the mapped surface. Among possible sources of the
measure one could take cross-sectional areas of cranial nerves and medulla,
or the actual areas of the sensory and motor surfaces of the body. These would

not be entirely adequate, because much of the projection system is amplified in the brain and the amplification occurs to different extents in different species. Whichever surface is taken, it will be necessary to multiply it by some amplification factor to make that surface exactly equal to the projection surface. The equation for brain size, dimensionally, thus becomes:

$$L_E^3 = m \cdot L_d \cdot L_p^2 \qquad (2)$$

The term m is a dimensionless constant representing the degree of amplification of the projection system relative to its provisionally estimated area L_p^2. An easy source for estimating L_p^2 is from the body size, especially since the specific gravity of the body is about 1.0 and one can equate weight and volume readily. If body size is used, we begin with dimensionality L^3, however, and we must therefore raise that to the $2/3$ power to insert it into an equation of the type of Eq. (2). An L_p^2 term is available from $(L^3)^{2/3}$, and the L_d term can be given its value of 0.1 in the centimetre-gram-second system of measurement. Going from the dimensional notation of Eq. (2) to the direct notation in which variables are named, as in Eq. (1), we can now rewrite Eq. (2) as:

$$E = 0.1\, m \cdot P^{2/3} \qquad (3)$$

or in logarithmic form:

$$\log E = \frac{2}{3} \log P + \log (0.1\, m)$$

The logarithmic form is obviously a version of Eq. (1), with $\alpha = 2/3$ and the "constant of integration" given by 0.1m cm. The theory presented here reached that result by an approach derived from the neurobiology of the brain, not from hypotheses about growth. The terms, therefore, have very different interpretations. Alpha is no longer the central source of information. It is, in fact, a constant derived from dimensional considerations. The "constant of integration" is no longer a constant. It is, instead, a term that includes an amplification factor m in addition to a dimensional constant reflecting a physiological process: the columnar organization of brain tissue. Most interesting, perhaps, is the fact that the amplification factor m is identical with the encephalization quotient EQ proposed several years ago (Jerison 1970, 1973). This is a rich source of further development of the theory of encephalization, which will be presented elsewhere (Jerison, 1977).

Before concluding it must be pointed out that encephalization is certainly not limited to brain tissue for which a "projection surface" of the type represented here as L_p^2 is appropriate. Much association cortex is of that type in most species (Masterton and Berkley, 1974), but at least some - such as the human speech and language areas - may not be analysable in these terms. Instead, one must assume that in the construction of the brain there is "added tissue" of the type that would function in brain/ body graphs in a way similar to the "extra neurons" parameter that I have used in the past (Jerison, 1973). In a complete theory, Eq. (3) would thus change to something like:

$$E = 0.1m\, P^{2/3} + A$$

This would result in families of curves with "A" as a parameter, which would be fitted by lines with slopes less than $2/3$ when regression analysis is applied to logarithmically transformed measures of brain and body size. The values of "A" would correspond to the computed values of k in such regression analyses in which Eq. (1) is used as the regression equation that is fitted. One then computes apparent values of alpha of the order of 0.2 or 0.3 (Hemmer, 1971) to find empirical "encephalization indices". These are only slightly related to alpha (including alpha in the Bauchot-Stephan progression index) but are in fact closely related to "A".

In conclusion, I would point out that equations that look the same do not necessarily mean the same thing. Allometry is the source of some of the empirical details - the curve fitting - used in the analysis of encephalization. Just as one must discover laws of growth to explain allometry as found for other systems, so must one discover laws of brain evolution to explain the similarity of the brain/body functions graphed on log-log paper to functions derived from allometric theory. I have shown how equations that were developed from dimensional consideration of the brain look like allometric functions but require very different interpretations.

REFERENCES

Gould, S.J. (1971). *Amer. Nat.*, 105, 113-136.
Gould, S.J. (1975). *Contrib. Primatol.*, 5, 244-292.
Hemmer, H. (1971). *Proc. III Cong. Primatol. Zurich*, 1, 99-107.
Huxley, J.S. (1932). "Problems of Relative Growth", Allen and Unwin, London.
Jerison, H.J. (1970). *Science*, 170, 1224-1225.
Jerison, H.J. (1973). "Evolution of the Brain and Intelligence", Academic Press, New York and London.
Jerison, H.J. (1977). *Ann. N.Y. Acad. Sci.*, 299, 146-160.
Masterton, R.B. and Berkley, M.A. (1974). *Ann. Rev. Psychol.*, 25, 277-312.
Palay, S.L. (1967). *In* "The Neurosciences: A Study Program", (G.C. Quarton, T. Melnechuk and F.O. Schmitt, eds), pp. 24-31, Rockefeller University Press, New York.
Welker, W.I. (1973). *Brain, Behav. Evol.*, 7, 253-336.
Woolsey, C.N. (1958). *In* "Biological and Biochemical Bases of Behaviour", (H.F. Harlow and C.N. Woolsey, eds), pp. 63-81, University of Wisconsin Press, Madison.

BRAIN SIZE AND INTELLIGENCE IN PRIMATES

R.E. PASSINGHAM

Dept. of Psychology, University of Oxford, South Parks Road, Oxford, UK.

Changes in the size of the brain during primate evolution have been thought
to imply changes in degree of intelligence. This can be justified by comparing the abilities of different living primates which vary in brain size.
Intelligence may be supposed to depend on the amount of brain tissue in excess of that needed to handle sensory information and control the movements
of the body. The amount of such tissue can be estimated by first establishing
what size of brain is needed to control a body of any particular weight.
Stephan (1972) does this by fitting a regression line to log/log plots of brain
and body size in insectivores. For any primate the difference in brain size
between that predicted by this line and the actual value provides an index of
excess tissue. However, it has long been known that the slope of the line relating brain to body weight is less steep for animals within a species and for
animals of closely related species. This appears to suggest that larger animals within a species must have less excess tissue than smaller animals, and
similarly for larger animals in a group of closely related species. Thus the
gorilla and orang-utan have low indices compared with the chimpanzee, and the
little talapoin has a much higher index than other cercopithecine monkeys
(Stephan, 1972).
A rank ordering of primates which avoids such paradoxical results can be
obtained by relating the size of the brain or the neocortex to the size of the
medulla (Passingham, 1975). On this ratio, the gorilla scores similarly to the
chimpanzee, and the talapoin to the other Old World monkeys. The rationale
for using the size of the medulla is that apart from the cranial nerves all
incoming and outgoing fibres pass through this region. The reason why comparisons with the medulla produce reasonable results can be seen by fitting a
least squares regression line through the data (N=38) provided by Stephan
et al. (1970) of medulla weight on body weight. The equation for the line is:
log medulla (mm^3) weight = 1.09 + 0.58 x log body weight (gms). If the deviation of each point from this line are calculated it is found that in 4/5
genera larger species score less favourably than smaller ones. The gorilla
has a much smaller medulla than expected for a primate of that size and the
talapoin a much larger one.
But the medulla gives only a very indirect measure of the inputs and outputs of the brain as it includes many nuclei. It would be better to compare
the size of the brain directly with the size of the spinal cord. This was
done by Krompecher and Lipack (1966). Using their data, a least squares regression line can be fitted to the spinal cord weight on body weight of eleven
mammals, five rodents, one rabbit, two carnivores and three primates including
man. The equation for the line is: log spinal cord weight (gm) = 4.10 + 0.73 x log

body weight (Kg). The value for the slope does not deviate far from the value of 0.66, which would be expected theoretically for the mapping of a surface area onto a volume. There are two further results of interest. First, the point for man lies very close to the regression line, showing that the size of the spinal cord in man is much as would be expected for a mammal of our body weight. Second the value for the orang-utan lies far below the line in the same way as did the value for the medulla of the gorilla in the previous plot. It seems likely that further data would show that the slope for log medulla on log body weight is less steep for animals in closely related species than for those in more distantly related ones.

Using the data of Krompecher and Lipack (1966), a least squares regression line can also be fitted for brain on spinal cord weight. Because the values for the three primates deviate far from those for other mammals a line was fitted only for ten mammals, five rodents, a rabbit, two carnivores and two ungulates. The equation is: log brain weight (gms) = 0.59 + 0.94 x log spinal cord weight (gms). The slope does not deviate far from a value of 1, indicating parallel changes in two volumes. The three primates lie on different intercepts, indicating much larger brain size for a particular weight of the spinal cord.

It appears that within a species or within closely related species the inputs and outputs of the brain do not increase as much with increase of body size as would be expected from data on more distantly related species. Why this should be so has not been explained. However, given that it is so it is clear why intelligence does not decrease with increasing body size within a species or within closely related species. Intelligence depends on the relation between the brain and the inputs and outputs, and both of these change less within related animals than might be expected from data on distantly related species. The relationship between them is not therefore affected in the way that might have been supposed if body weight were taken as an index of inputs and outputs to the brain.

This conclusion leads to an interesting corollary. Radinsky (1967) has suggested that the cross-sectional area of the foramen magnum would provide a useful measure of body size in fossil species. He therefore proposes that cranial capacity in such species be related to the area of the foramen magnum. The foramen magnum encloses an area of the medulla. Rather than using the cross-sectional area of the foramen magnum as an indirect measure of body size it might be better instead to regard it as an indirect measure of the inputs and outputs of the brain that is not likely to produce paradoxical results like some comparisons of brain with body weight (Passingham, 1975).

REFERENCES

Krompecher, S. and Lipack, J. (1966). *J. Comp. Neurol.*, 127, 113-120.
Passingham, R.E. (1975). *Brain, Behav. Evol.*, 11, 1-15.
Radinsky, L.B. (1967). *Science*, 155, 836-837.
Stephan, H. (1972). *In* "The Functional and Evolutionary Biology of Primates",
    (R. Tuttle, ed.), pp. 155-174, Aldine-Atherton, Chicago.
Stephan, H., Bauchot, R. and Andy, O.J. (1970). *In* "The Primate Brain",
    vol. I, (C.R. Noback and W. Montagna, eds), pp. 289-297, Appleton-Century-
    Crofts, New York.

FOSSIL EVIDENCE FOR PRIMATE VOCALIZATIONS?

J. WIND

Departments of Otorhinolaryngology and Human Genetics, Free University, Amsterdam, Netherlands.

The only hard evidence for ancestral primate vocalizations seems to be the somewhat scanty fossil remains. These, however, are insufficient for a reliable reconstruction of vocalizations, for interspecific differences in the latter are mainly determined by soft tissue anatomy. Accordingly most hypothesizing on this topic to date has been based on comparative observations of primate vocalizations, which, when supplemented by data from (for example) palaeoecology, may admittedly be adequate up to a certain level.

In recent years, however, a new method has been introduced by Lieberman and co-workers (see Lieberman, 1975). This is mainly based on the physical properties inferred from reconstructions of the oral and pharyngeal cavities, as deduced from fossil skull base anatomy. Having examined infant human, adult chimpanzee and Neanderthal skulls Lieberman and Crelin (1971) concluded that Neanderthal man could not have possessed human speech abilities because of his peculiar vocal tract.

The method has already been criticized on several grounds, viz. the skull base dimensions and flattening, and mandibular dimensions which, among others, were assumed to be decisive for speech abilities (LeMay, 1975), the alleged position of the hyoid (Falk, 1975), sphenoid and temporal bones (Burr, 1976), the alleged importance of chin morphology (Carlisle and Siegel, 1974; Wind, 1975), the underestimation of the large variation present in modern speaking humans (LeMay, 1975; Carlisle and Siegel, 1974; LeMay, 1976), the derivation of evidence from human newborns (Burr, 1976; Carlisle and Siegel, 1974; Wind, 1973; Wind, 1976), and the importance attributed to the information transferred by the vowels in modern human vocalizations (Fremlen, 1975). However, I think Lieberman's approach leaves a number of other questions unanswered:

1. The reconstruction of the casts, and especially the localization of the larynx, was largely based on the styloid process/skull base angle, as suggested by Wind (1970). Of the two Neanderthal skulls examined, however, only one shows - a small part of - the processes concerned (Fig. 1).

2. The direction of the styloid processes is in modern man determined by forces originating from the styloglossal, stylopharyngeal and stylohyoid muscles and those from the stylomandibular and the stylohyoid ligaments. These forces in Neanderthal are unknown, as is the case, for that matter, in modern man.

3. There is considerable variation in the anatomy of the human styloid process and its muscles and ligaments (von Lanz and Wachsmut, 1955). For example, a difference as large as 29.5° has been found in styloid process direction (Zivanovic, 1967), while often the processes curve half-way. In addition, in nonhuman primates, both modern (Zuckerman et al., 1962) and

fossil (Tobias, 1967), there is some intraspecific and considerable interspecific variation in styloid process direction. So, the direction of the
stylohyoid ligament in the Neanderthal skulls remains unknown.

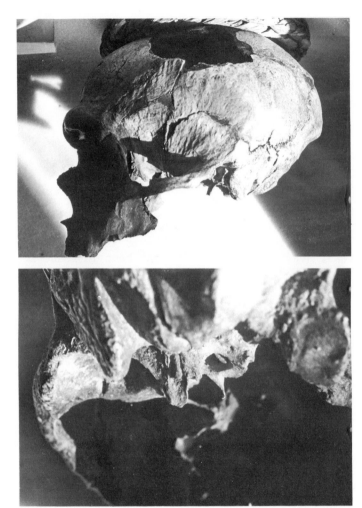

Figure 1. *A Neanderthal skull (La Chapelle-aux-Saints): a.(top) Left*
lateral view showing the vagina of the left styloid process.
b.(bottom) Detail of the left part of skull basis viewed
dorsally and caudally, showing the left styloid process, which
is broken off at 6mm from the basis, and its vagina protruding
5.5mm further (centre of picture). The remnant of the right
styloid process is even shorter, i.e., only 3mm; here the
vagina protrudes 10mm further. These meagre remnants are an
insufficient indicator for the direction of the original pro-
cesses. The other skull used by Lieberman and Crelin (1971)
for Neanderthal vocal tract reconstruction (La Ferassia) has
no styloid process or vagina at all. (Photographs made by the
author through courtesy of the Musée de l'Homme, Paris.)

4. Even if the direction of the stylohyoid ligament - and its length - were
known, it would simply indicate the position of the hyoid's lesser horn,
not that of the larynx. Yet, a large part of Lieberman's reasoning hinges
on laryngeal localization, which (in view of the additional absence of fossil
hyoid and laryngeal material) remains completely unclear.
5. Because of the absence of fossil evidence indicating the size, position
and mobility of the soft palate and epiglottis, conclusions about the inability
of earlier hominids to produce unnasalized sounds (Lieberman, 1975; Lieberman
and Crelin, 1971) remain speculative.
6. Even if the vocal tract reconstructions were correct, it can be questioned
what relevance this would have for the assessment of ancestral speech abilities.
The enormous functional redundancy of the human vocal tract (as apparent from
pathology) and the ability of nonhuman primates to produce various human-like
vocalizations (Hayes, 1951; Andrew, 1976) suggest that, if a modern human
were equipped with Lieberman's Neanderthal vocal tract, his speech would
be only slightly different from normal speech (e.g. its vowels would have
a slightly different timbre); and the encoding speed would not be significantly
lower than that of modern humans (Wind, 1976). This applies *a fortiori* when
such a creature's fellow-men would use the same vocalizations.

So, if Neanderthal man had mental properties similar to ours, as suggested
by their large brain size (Heim, 1970) and their cultural remains (Leroi-Gourhan,
1975; Solecki, 1975) the characteristics of their vocal communication must
have been similar to those of modern man.

Other arguments favouring the evolution of cerebral speech capacities
prior to that of modern vocal tract morphology are: (1) a new behaviour pattern usually precedes a new morphological trait; (2) human symbol decoding
capacity is clearly in advance of peripheral vocal encoding capacities, because decoding written language is much faster than decoding spoken language;
and (3) LeMay (1975) has suggested that Neanderthal man already had the cerebral assymetry associated with language capacities in modern man.

Another approach is offered by the *ear structures*, the characteristics of
which must have been shaped mainly by selective pressures resulting from
intraspecific communication and ecological conditions. Accordingly, in most
species the frequency range and the sensitivity of the auditory system is
geared to their vocalization frequency and amplitude in such a way that both
largely overlap. Therefore, if the hearing properties of extinct primates
can be deduced from their fossils, and if the then ecological conditions can
be reconstructed, some clues as to the character of their vocalizations might
emerge (Wind, 1976).

In the ecological niches occupied by our primate ancestors - forest, wood-
land and savanna - the presence of prey, predator or conspecifics must often
first have been detected by the clicks originating from breaking of dry
branches or leaves. Indeed, the maximum sensitivity of the human (Groves,
1965), the chimpanzee (El er, 1934) and the *Macaca* (Stebbins et al., 1966)
ear is in the 1 to 2 kHz range, where such clicks have most of their energy.
Interestingly, various speech consonants have their maximum energy in the
same range.

Hitherto, studies on the evolution of vertebrate hearing have mainly been
based on comparative observations (Manley, 1971, 1973). These show, in addition to the above audiological similarities within the primate order, that
chimpanzees are able to recognize differences in their fellows' calls which
the human ear considers as very small (van Lawick-Goodall, 1973; Beatty and
McDevitt, 1975), and that various animals including even chinchillas can

recognize small differences in human speech sounds (Kuhl and Miller, 1975).
   Hence, if the degree of mineralization and the rarity of fossil primates are
not too unfavourable, it is certainly worthwhile looking into the morphological
features of primate fossil middle and inner ears. One could assess, for example, by means of X-ray tomography, the length of the cochlea (providing an
idea of the width and localization of the frequency range), the drum/oval
window ratio (providing an idea about the sensitivity), and the microscopic
structure of the ossicles (providing an idea of the stiffness of the ossicular
chain and hence of the frequency range). Furthermore, fossil skull size may
provide clues concerning stereoacousia.
   In conclusion, the classical comparative approach still provides most of
the indications. It suggests (Wind, 1970; Wind, 1975b, 1976) that the primates, including their peripheral vocal and auditory organs, have been pre-
adapted to a large extent for speech long before *Homo sapiens sapiens* evolved,
and that it was cerebral reorganization that triggered speech-like communication. Perhaps these adaptations were followed by feed-back selective pressures which slightly modified the pharynx and larynx. Palaeoanthropology,
including Mousterian archaeology, suggests, however, that these pressures
have been much weaker than those resulting from a symbol-using speech-like
communication system, even if it contained sounds slightly different from
those of modern speech.

ACKNOWLEDGEMENTS

   I thank J-L. Heim for having provided the opportunity of examining the
Neanderthal skulls at Le Musée de l'Homme, and P. Andrews, G.W. Hewes, M.
LeMay, M.D. Leakey, G.G. Simpson, and R.W. Wescott for their support in pre-
paring the manuscript.

REFERENCES

Andrew, R.J. (1976). *Ann. N.Y. Acad. Sci.*, 280, 673-698.
Beatty, J. and McDevitt (1975). *Curr. Anthropol.*, 16, 668-669.
Burr, D.B. (1976). *J. Hum. Evol.*, 5, 285-290.
Carlisle, R.C. and Siegel, M.I. (1974). *Amer. Anthropol.*, 76, 319-322.
Elder, J.H. (1934). *J. Comp. Psychol.*, 17, 157-183.
Falk, D. (1975). *Amer. J. Phys. Anthrop.*, 42, 123-132.
Fremlen, J.H. (1975). *Science*, 187, 600.
Groves, J. (1965). *In* "Diseases of Ear, Nose and Throat", (W.G. Scott-Brown,
   J. Ballantyne and J. Groves, eds), pp. 283-329. Butterworths, London.
Hayes, C. (1951). "The Ape in our House", Harper, New York.
Heim, J.-L. (1970). *L'Anthropologie*, 74, 527-572.
Kuhl, P.K. and Miller, J.D. (1975). *Science*, 190, 69-72.
LeMay, M. (1975). *Amer. J. Phys. Anthrop.*, 42, 9-14.
LeMay, M. (1976). *Ann. N.Y. Acad. Sci.*, 280, 349-366.
Leroi-Gourhan, A. (1975). *Science*, 190, 562-564.
Lieberman, P. (1975). "On the Origins of Language: An Introduction to the
   Evolution of Human Speech", Macmillan, New York.
Lieberman, P. and Crelin, S. (1971). *Linguistic Inquiry*, 11, 203-222.
Manley, G.A. (1971). *Nature*, 230, 506-509.
Manley, G.A. (1973). *Evolution*, 26, 608-621.
Solecki, R. (1975). *Science*, 190, 880-881.
Stebbins, W.C., Green, S. and Miller, F.L. (1966). *Science*, 153, 1646-1647.

Tobias, P.V. (1967). "Olduvai Gorge", Vol. II: The Cranium and Maxillary
    Dentition of *Australopithecus (Zinjanthropus) boisei*, Cambridge University
    Press, Cambridge.
van Lawick-Goodall, J. (1973). "In the Shadow of Man", Collins, London.
Von Lanz, T. and Wachsmut, W. (1955). "Praktische Anatomie", Vol. XII,
    Springer-Verlag, Berlin.
Wind, J. (1970). "On the Phylogeny and the Ontogeny of the Human Larynx",
    Wolters—Noordhof Publishing, Grongingen.
Wind, J. (1973). *Curr. Anthropol.*, 14, 522.
Wind, J. (1975a). *Otorhinolaryngology*, 37, 58.
Wind, J. (1975b). *Acta Teilhardiana*, 12, 41-55.
Wind, J. (1976). *Ann. N.Y. Acad. Sci.*, 280, 612-530.
Zivanovic, S. (1967). *E. Afr. Med. J.*, 44, 298-302.
Zuckerman, S., Ashton, E.H. and Pearson, J.B. (1962). *Bibl. Primatol.*, 1,
    217-228.

EDITOR'S NOTE

The following comment was made by Dr. H.E. Heffner, Parsons Research Center,
University of Kansas:
"I found quite interesting your recommendation that we examine the morpho-
logy of fossil capacities. Several years ago we published a paper titled,
'The Evolution of human hearing' (Masterton, B., Heffner, H., and Ravizza,
R., *J. Acoust. Soc. Amer.*, 1969, 45, 966-985), in which we demonstrated that
the functional distance between the two ears is inversely correlated with
high-frequency hearing ability. This relation allows one to use the size of
a fossil mammal's head to obtain an estimate of its high-frequency hearing.
In addition, we are preparing a paper on hearing in the cohort Glires in
which we will show the existence of a positive correlation between high- and
low-frequency hearing. By taking advantage of these two correlations, it is
possible to estimate the hearing range of an extinct mammal given the size
of its head."
In response to a query about the ability of whales, which have large heads,
to hear high frequencies, Dr. Heffner added the curther comment.
"High-frequency hearing is correlated with the *functional* distance between
the two ears (i.e., the time it takes a sound to travel from one ear to the
other). Since the path of sound is through a whale's head instead of around
it and since the speed of sound is greater in the more dense medium of the head
than it is in air, the functional distance between the inner ears of a whale is
relatively small. Thus the Cetacea are not exceptional in their ability to
hear high frequency sounds."

LARYNGEAL 'PREADAPTATION' TO ARTICULATED LANGUAGE

B.R. FINK and E.L. FREDERICKSON

University of Washington School of Medicine, Seattle, Washington, 98195, USA and Woodruff Medical Center, Emory University, Atlanta, Georgia, 30322, USA.

Very little direct information is available on the behaviour of the larynx and laryngeal air sacs of the great apes. In this report we present some observations which may have bearing on the problem of the evolution of human language.

METHODS

The study was performed at the Yerkes Primate Center in 1975 and 1976, on three chimpanzees,, three gorillas, and three orangutans, lightly narcotized with intravenously administered ketamine. Respiratory rhythm was recorded continuously by means of a pneumograph. Motion pictures were made of the larynx visualized by direct laryngoscopy or with a fiberoptic laryngoscope.

Lateral radiographs of the laryngeal region were obtained in different phases of respiration or effort. The timing of the exposures was signalled on the respiratory record (paper speed 10 mm/sec.) by a marker pen actuated by the radiographer.

RESULTS

The radiographic profiles of the cervical air sacs indicated that in these young animals the sacs became partially deflated during inspiration and became reinflated during expiration or expiratory efforts (Fig. 1).

On the radiographs, the position of the larynx was more caudal in the inspiratory films, and the laryngeal vestibule enlarged and more translucent than in expiratory films, indicating that the vestibular walls underwent inspiratory attenuation by stretching.

The direct visual and motion picture observations showed the presence of cyclic inspiratory abduction and expiratory adduction of the arytenoids and attached laryngeal folds. It was noted that the vocal processes of the arytenoid cartilages remained parallel throughout the respiratory cycle. This finding contrasts with the respiratory behaviour in the human larynx, where the vocal processe are abducted in parallel in the first part of inspiration but may diverge during the later part of inspiration. In the apes, abduction in divergence was never observed.

DISCUSSION

The profile area of the sacs, and presumably their volume, was always larger in the expiratory and effort films than in those taken during inspiration.

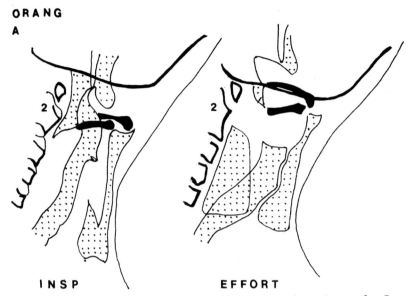

ORANG
A

INSP   EFFORT

Figure 1. *Tracings of lateral cervical radiographs of a male 3-year-old*
orangutan. Heavy lines indicate the vertebrae and lower jaw.
Solid black denotes ossified parts of the hyoid bone; stippled
areas denote translucent spaces containing air. EFFORT: the
area devoid of air below the hyoid bone marks the site of the
closed larynx. Below this the trachea is distended by the in-
creased air pressure, and sac profiles containing air are pre-
sent on either side of the trachea. Inspiration (INSP): the
hyoid bone and larynx are more caudal and the sac profiles are
smaller. In young chimpanzees and gorillas the profile change
was even greater.

Since an undefined interval of several minutes elapsed between successive
exposures, the films represent a random sampling of the condition of the sacs
during the respective phases of behaviour. So it seems likely that in young
pongids, at least, inflation of the sac with expiration or grunting effort,
and deflation with inspiration recurs with every cycle.

What the physiological significance of the tidal saccular air flow may
be is unclear. On morphological grounds the sacs have previously been thought
to function as resonance chambers (Starck and Schneider, 1960, 508). Negus
(1949, 50) believed that they were "for mixing and rebreathing of air".

The cyclic change in sac volume we appear to have demonstrated suggests
that flow into the sacs can also occur during phonation. This raises the
possibility that the saccular flow past the vestibular folds may enable
these folds to produce acoustic signals somewhat like a sail flapping in the
breeze. In this connection it is pertinent that in man the laryngeal mechanism that increases vocal ligament tension also increases the tension of the
vestibular ligaments, though without adducting the vestibular folds.

Our motion picture observations suggest that a phase of arytenoid abduction
in divergence occurs much more rarely in the respiration of great apes than it
does in man. In man, recent study of laryngeal biomechanics (Fink, 1975) indicates that the inspiratory abduction of the arytenoids is mechanically coupled

to the cranio-caudal-excursion of the larynx as a whole; abduction of the vocal processes in parallel is coupled to the early part of laryngeal descent while abduction in divergence is coupled to the later part of the descent.

Since the phase of abduction in divergence demands a greater vertical excursion of the larynx than does abduction in parallel, one may surmise that arytenoid abduction in divergence is infrequent or absent in hominoid apes, because the amplitude of the vertical laryngeal excursion of their larynx is insufficient.

Functionally, terminal abduction in divergence contributes a decrease in laryngeal resistance and a corresponding facilitation of respiratory air flow that is useful in situations calling on every ounce of reserve effort or endurance. It seems to follow that the marked cervical mobility of the larynx in man, which is responsible for respiratory arytenoid abduction in divergence, enables man to occupy an ecological niche employing protracted high respiratory air flows for minutes or hours, i.e. an ecological niche of high physical endurance, such as may be required in hunting big game and dragging the kill back to the home site to be shared by the female and young. Furthermore, it is noteworthy that in man, mobility of the larynx relative to the hyoid bone is exploited in producing effort closure of the larynx, an act which helps to convert the trunk into a rigid pillar that supports the efforts of the arms and legs and so improves the efficiency of these efforts. Finally, it is also likely that enhanced vertical mobility of the larynx would facilitate the development of articulated language, e.g. for planning and executing a hunt.

Thus the absence of accessory air sacs and the increased vertical mobility of the larynx in man would correlate with at least three adaptive characteristics:
1. increased respiratory inflow of air;
2. increased efficiency of exceptional physical efforts in utilizing the increased supply of oxygen;
3. facilitated use of language in planning the acquisition of food.
In what order these adaptations were acquired is unknown; the first or the second, or both, may have been preadaptive to language, but it is also possible that all three evolved concurrently and are integral parts of a single behavioural complex.

ACKNOWLEDGEMENTS

We gratefully acknowledge the help of Dr. Geoffrey H. Bourne, Director of the Yerkes Primate Center of Emory University, Atlanta, Georgia, and his staff. The work was supported by Grant HDO8511, National Institute of Child Health and Human Development, United States Public Health Service.

REFERENCES

Starck, D. and Schneider, R. (1960). *In* "Primatologia", Vol. III, part 2, (H. Hofer, A.H. Schultz and D. Starck, eds), Karger, Basel.
Negus, V.E. (1949). "The Comparative Anatomy and Physiology of the Larynx", Hafner, New York.
Fink, B.R. (1975). "The Human Larynx: A Functional Study", Raven Press, New York.

CAN EVOLUTIONARY KNOWLEDGE LEAD TO STABLE CLASSIFICATION? APPLICATION OF A NEW HYPOTHESIS TO THE STUDY OF SULAWESIAN MONKEYS

H. KHAJURIA

*Eastern Regional Station, Zoological Survey of India,
Risa Colony, Shillong 793003, Meghalaya, India.*

INTRODUCTION

Sulawesian (Celebese) monkeys form a very small group of related taxa well studied by a number of competent workers. The latest excellent monograph by Fooden (1969) recognizes only seven taxa on the basis of well-documented evidence. Despite exhaustive evolutionary knowledge and the very small size of the group, its classification is in great confusion because various workers are assigning very different ranks to these monkeys, e.g. subspecies of single species (Thorington and Groves, 1970), species of single genus (Fooden, 1969), or a number of different genera (see Fooden, 1969 and Hill, 1974). The example clearly shows that in the absence of good classificatory principles no agreement can be reached regarding ranks even in well studied very small taxa. In other words, it has not been possible to find even in a small taxon any solution to the serious problem of frequent changes in classification, which are resulting in tremendous wastage of human effort. The author (Khajuria, 1963) advanced a hypothesis that ranks can best be based on the characters of the most divergent taxa included in a group because, if microevolution is the rule, presence or absence of connecting links (which merely indicate extent of discovery) should not be allowed to alter the ranks. It was further pointed out (Khajuria, 1967a and b) that the most divergent taxa can easily be spotted by means of the values of their most divergent characters.

This principle is here applied to the study of Sulawesian monkeys in an attempt to stabilize their classification.

MATERIALS AND METHODS

The study is based on twenty-one specimens of Sulawesian monkeys present in the collection of Zoological Survey in India and on published literature. In addition, some skulls borrowed from British Museum (Natural History) and Bogor Museum, Java were studied.

The first step is to establish evolutionary lines in a group on the basis of study of holomorphs of included taxa (OTUS). If these lines are disputable, all taxa included in a group are to be arranged in all possible pairs, following the procedure adopted in numerical taxonomy, and the pair which shows maximum divergence is to be selected as the first dichotomy of the group. Other OTUS should be attributed to one of the branches of this dichotomy on the basis of proximity shown to a particular branch. Each branch is further divided by a similar procedure till the group is completely converted into a dendrogram. It may be noted, however, that not every dichotomy need be ranked and that a number of dichotomies may have to be included in one taxo-

nomic category. If the amount of divergence is in dispute, it should be
measured by the value (ratio of means) of the most divergent character in two
OTUS of the pair, as it appears that the total distance between any two OTUS
is related to the value of the most divergent character. This possibility is
suggested by the fact that there are infinite differences (characters) between any two OTUS and these can be arranged in a triangular matrix ranging
from minimum (approximating to zero) to the maximum (the most divergent
character).

CLASSIFICATION OF SULAWESIAN MONKEYS

   Three evolutionary lines can *with consensus* be distinguished, both morpho-
logically and geographically. (N.B. The scientific names mentioned here are
the ones given by the original describers of the taxon.)
   The first dichotomy is represented by: (1) *Macacus maurus* Schinz and (2)
the rest of the taxa with the most divergent taxon being *Cynocephalus niger*
Desmarest. The second line can be divided into: (a) *Papio* (*Inuus*) *hecki*
Matschie, *Papio nigrescens* Temminck and *C. niger* and (b) *Macacus tonkeanus*
Meyer, *Papio ochreatus* Ogilby, and *Papio* (*Inuus*) *brunnescens* Matschie (see
Fig. 1).

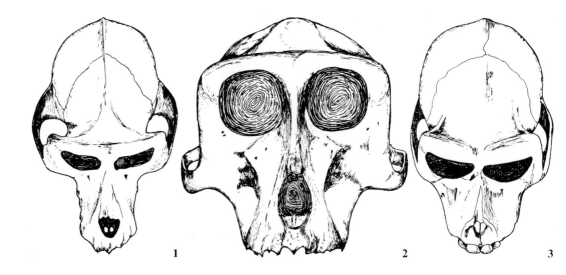

   *Figure 1. Skulls of 3 Sulawesian monkeys representing three principal
              evolutionary lines, both morphologically and geographically,
              and here ranked as separate genera:*
              1. Macaca maura, dorsal view (x 7/10)
              2. Cynomacaca ochreata brunnescens, front view (natural size).
              3. Cynopithecus niger, dorsal view (x 9/10).

   The most divergent taxa in latter dichotomy are: (a) *C. niger* and (b) *P.
ochreatus* or *P. brunnescens*.
   M. tonkeanus apparently represents a connecting line both geographically

and morphologically. In the first dichotomy, the most divergent taxa are distinguished by a number of non-overlapping characters such as maxillary ridges, ischial callosities, gluteal fields, coronal hairy crest, tail proportions, coloration, etc. and, thus, deserve generic distinction. In the second dichotomy the most divergent taxa are also distinguished by a number of similar non-overlapping characters. This dichotomy, therefore, also deserves generic distinction. For the first dichotomy the generic names available are *Macaca* Lacepede and *Cynopithecus* I. Geoffiroy; for the line represented by *P. ochreatus/P. brunnescens* the generic name available is *Cynomacaca* Khajuria. The taxa included in *Cynopithecus* line can be distinguished by a few non-overlapping characters and appear to be allopatric species while those included in *Cynomacaca* line are only distinguished by coloration, and, since they are also allopatric, they may better be considered as subspecies of *Cynomacaca ochreata*. The classification which emerges is as follows:

I.   Genus *Macaca*
      Macaca maura
II.   Genus *Cynopithecus*
      C. heckei
      C. nigresences
      C. nigra
III. Genus *Cynomacaca*
      C. ochreata tonkeana
      C. o. octreata
      C. o. brunnescens

REFERENCES

Fooden, J. (1969). *Bibl. Primatol.*, 10, 1-148.
Hill, W.C.O. (1974). "Primates: Comparative Anatomy and Taxonomy", Vol. VII, John Wiley and Sons, New York.
Khajuria, H. (1954a). *Rec. Indian Mus.*, 50, 301-305.
Khajuria, H. (1954b). *Rec. Indian Mus.*, 52, 101-127.
Khajuria, H. (1963). *Sci. Cult.*, 29, 256-257.
Khajuria, H. (1967a). *Proc. 54 Indian Sci. Congr.* (abstract).
Khajuria, H. 1967b). *Bull. Nat. Inst. Sci. India*, 34, 269-274.
Thorington, R.W. and Groves, C.P. (1970). *In* "Old World Monkeys", (J.R. Napier and P.H. Napier, eds), Academic Press, New York and London.

SECTION II

BEHAVIOURAL FACTORS IN PROSIMIAN EVOLUTION
Chairman and Section Editor: C.A. Doyle (Johannesburg)

INFLUENCES ON THE STRUCTURE OF VOCALIZATIONS OF THREE MALAGASY PROSIMIANS

LEE W. McGEORGE

Department of Zoology, Duke University, Durham, North Carolina, USA.

INTRODUCTION

The field of bioacoustics deals with the interaction between natural sounds and the environment. Two aspects of bioacoustics can be formulated into questions relevant to the study of acoustic communication in animals. First, how is the energy of a sound changed by the physical environment, and, second, how might other sounds present in the environment interfere with the sound under consideration.

Most of the research in bioacoustics has been applied to bird vocalizations (Morton, 1970; Ficken, et al., 1974; O'Neill, et al., 1976). Primates are also highly vocal animals and the principles of bioacoustics should apply to them as well. Madagascar is a particularly interesting place to study bioacoustically, not only because of the locally dense populations of noisy nocturnal and diurnal prosimian primates, but also because of the presence of highly vocal bird species. The great degree of endemism of the mammalian and avian faunas of Madagascar is likely to lead to a rewarding study of the co-evolution of behavioural features.

A decrease in the energy of a sound, called sound attenuation, always occurs as a sound is propagated away from its source. What happens to a sound in a particular habitat as it encounters wind and temperature gradients and obstacles, such as ground, tree trunks and leaves, characteristic of that habitat? If changes in sound energy occur nonrandomly and vary with respect to some feature of the structure of the sound, such as its frequency spectrum, one can put forth the hypothesis that the vocalizations of the species present in a particular environment will be characterized by structural features that would allow their best transmission through that environment (Morton, 1970).

Sounds present in the environment of an animal other than the sound signals of conspecifics could interfere with communication between conspecifics. Interference or masking of one sound by another actually occurs during the auditory processing of sounds received by an individual. Thus interference among the vocalizations of animals in a community cannot be fully predicted apart from knowledge of the neurophysiology of hearing in each of the animal species present. However, it can be said that some interference exists when a sound of one frequency overlaps in time with another sound having the same or nearly the same frequency (Morton, 1970). The probability that one particular sound will coincide with another changes daily and seasonally. For example, the species that call during the day are different from those calling at night, and all the calls overlap somewhat at dawn and dusk. Thus one can make the hypothesis that there are features of structure and timing of vocalizations of animals sharing a habitat such that the probability of one call jamming another is reduced.

In this paper I report the observations made on all the sound-producing animals in a gallery forest habitat of southern Madagascar during winter and spring of 1974. The prosimians *Lemur catta*, *Propithecus v. verrauxi* and *Lepilemur mustelinus leucopus* were present in high densities in the gallery forest, and both the breeding and the birth seasons of these animals occurred during the period of my study. *Microcebus murinus* and *Cheirogaleus medius*, the other prosimians occurring in this region, were seen rarely in the gallery and were never heard to vocalize during a formal observation. The other sound-producing animals of the forest included about 35 species of bird, the Malagasy fruit bat, several species of unidentified microchiropteran bats and several unidentified insects that I distinguish only on the basis of sound type. The structural feature of animal sounds I am concerned with in this paper is that of frequency, or what we hear as the pitch of a sound.

METHODS

In order to determine the effect of the physical environment on the transmission of sound I broadcast a tape of 10 pure tones and recorded the broadcast on a Sony 800 B recorder at distances of 10, 30 and 50 m from the source. The sound source and receiver were both 1.5 m off the ground. I did this at three sites in the gallery forest during the summer season. In the laboratory, using filters, a sound pressure level meter and a strip recorder, I determined the attenuation, measured in loss of decibels, sustained by each tone at the intermediate and far distances, taking the near distance as a reference point.

In order to characterize and quantify the sounds occurring in the gallery forest habitat I chose a permanent site in the forest and listened there for a total of 8,400 minutes over the winter and spring seasons of 1974. There were ten 24-hour observations in which I listened every other half hour during the day and night and for 2.5 continuous hours at dawn and dusk. A sound received a score of one if it was heard at least once during a five minute period. Thus for each half hour session a sound could receive a maximum score of six and during each 24-hour observation, a maximum of 168. On the basis of this scoring method I arbitrarily determined the major sounds of the environment. A type of sound had to be heard at least one tenth of the time over the duration of the study, i.e., receive a score of 168 or more, to be considered a major sound. If all the sound types emitted by one species of animal totalled a score of 168, each of those sounds is considered a major sound.

The spectrographic analyses of sounds were made from tape recordings taken in the field using a Sony cassette recorder or a Uher tape recorder. A Kay Sonagraph was used for the analyses.

RESULTS

The major sounds, excluding those given by most passerine birds, were given by a prosimian, *Lepilemur mustelinus*, the fruit bat, *Pteropus rufus*, and seven species of bird. Other major sound types that I am unable to identify according to species are two types of insect sounds and the twittering sounds of microchiropteran bats. Other vocal animals, whose several types of call only collectively met the major sound criterion, were the prosimian, *Lemur catta*, and one species of bird. Most of the passerines have been excluded from the present analysis because they have not yet been identified nor have their calls

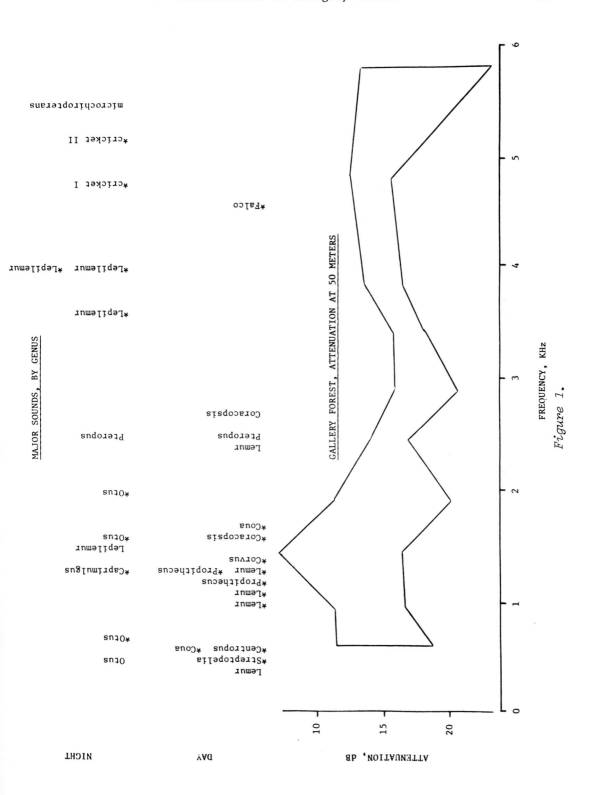

Figure 1.

been characterized spectrographically. However, I believe that the inclusion
of the passerines will not alter what I have to say about the prosimians, and
I will clarify this point later.

The top half of Fig. 1 shows the major sounds of the gallery forest, designated simply by the generic name of the animal producing them. They are segre-
gated according to whether they are heard primarily during the day or night.
The position of the calls along the horizontal axis corresponds to the frequency at which most of their sound energy is concentrated. The asterisk
indicates that the call is used in situations which would require communication over distances. I have included two types of call of *Propithecus verrauxi*,
although this prosimian does not meet my criterion for producing major sounds.

In the bottom half of Fig. 1 tone frequency is plotted against loss in
decibels for the 50 m receiver-source distance. Although the range of attenuation values for each tone over the three sites is wide, there are two sets
of frequencies that attenuate less than what is predicted from the computation
of the theoretical loss in sound energy at that distance. I will call these
sets "frequency windows".

DISCUSSION

Nine out of the 11 vocalizations of the three prosimians occur in or near
the frequency windows of sounds best propagated through the environment.
Eight out of these nine are calls that probably function in communication over
a distance, i.e., the calls usually elicit answers in kind from distant and
often visually isolated conspecifics. For example, the call of *Lemur catta*
whose major energy occurs at 1.1 KHz is the howl of the male that is usually
answered in kind by males of distant troops. The call of *Propithecus verrauxi*
that occurs at 1.2 KHz is the roar of the group that is set off by the appearance of a large bird of prey and sets off roars of neighbouring *Propithecus
verrauxi* groups. The two prosimian calls that do not occur within the frequency windows are two calls of *Lemur catta*, which seem not to function in
communication over distances. For example, the grunts of *Lemur catta*
occurring at .4 KHz are emitted by animals in a troop on the move, in which
the probability of animals seeing each other is high.

Calls of other animals seem to follow the pattern of the prosimian calls.
For example, at 1.7 KHz is the whistle of the parrot, *Coracopsis* sp., which
is usually responded to in kind by distant parrots. At 2.7 KHz is the squawk
of *Coracopsis* sp. that usually occurs during a close encounter between two
parrots and results in one parrot chasing the other.

When ordered according to emphasized frequency and categorized according
to whether given primarily during the day or primarily at night, the major
sounds show more separation within each category than between each category.
That is, if the two categories were superimposed on each other, the probability
is two to one that for a particular call the calls closest to it in frequency
are in the opposite or both categories rather than in its own category. For
example, the call of the Madagascar Scop's Owl, *Otus rutilus*, at .7 KHz is
closer in frequency to the calls of two diurnal birds than it is to another
nocturnal sound. The call of *Lepilemur mustelinus* at 1.3 KHz is equally close
to two diurnal sounds and one nocturnal sound.

There is some qualification to be put on the information shown in Fig. 1,
concerning the absence of calls of the passerine birds. I have spectrographically analysed only 10 out of about 30 types of frequently heard calls. These
fall outside the frequency range of all but one of the diurnal prosimian calls.

On the basis of these observations I make two suggestions that support the hypothesis that certain aspects of animal vocal behaviour reduce the probability of vocal interference. The first is that since day calls and night calls are separated temporally, overlap in frequency that is seen between them is of small consequence to the probability of one call jamming another. At least in the feature of frequency, the structures of day calls and night calls have evolved more or less independently of one another. I hesitate to say that they have evolved completely independently of one another, because day and night sounds do overlap temporally for short periods, i.e., at dawn and dusk. However, overlap of frequencies there is not perfect and still some separation is obtained. There is little overlap of any night sounds with any of the diurnal prosimian calls, with the exception of *Caprimulgus madagascariensis*, the nightjar. However, the vocalization of the nightjar is quite different in tempo from the vocalizations of the prosimians and probably would not be confused with them. Overlap in frequency between passerine bird calls and *Lepilemur mustelinus* calls probably occurs, but would offer little interference at dawn and dusk to the lepilemur, since only 12% of the total number of its calls in a 24 hour observation are given at those times.

My second suggestion, derived from the information given in Fig. 1, is that although the calls within the day or night categories occur together temporally, they are sufficiently separated according to frequency such that the probability of one call jamming another is minimized. This suggestion with respect to the day sounds is weakened by the absence of data on the frequencies of the passerine calls. However, as I mentioned earlier, this little affects the conclusion drawn about the calls of the diurnal prosimians, since the passerine calls analysed so far fall outside the frequency range of the prosimian calls, and thus overlap is avoided both during the day and at dawn and dusk.

A closer look at Fig. 1 will show a few instances of overlap in frequency within the day category. However, when there is overlap in frequency there are differences in other structural features of the calls that probably make them distinctive to the callers. For example, the calls of *Centropus toulou* and *Coua cristata* occur at the same frequency and in fact consist of repetitive short notes, but the rhythm of each call is different. For every one element of the *Centropus toulou* call, there are five in the *Coua cristata* call. A call of *Lemur catta* overlaps one of *Propithecus verrauxi*, but both calls serve to alert other members of the group and perhaps other groups to the presence of a ground predator. Any possible confusion resulting from frequency overlap would be advantageous rather than detrimental. However, in fact, the calls are quite different from one another, especially temporally, as are the calls of *Centropus toulou* and *Coua cristata*.

Within the category of night sounds separation according to emphasized frequency is very clear. The one instance of overlap in frequency occurs in two calls of *Lepilemur mustelinus*, whose energies are concentrated at 4 KHz. Obviously, there is no problem with interspecific interference in this case. One might expect that the similarity would render the functions of the two calls ambiguous. However, these two calls seem to have a similar function. One call from one animal is usually followed by the other call from a distant animal, and thus both serve to announce presence. Still they are not likely to be confused, since the one is a noisy call and the other is harmonically structured. The overlap here, near the midpoint of the upper frequency window, emphasizes the significance of that window for the calls in question. They are the most frequent calls of the lepilemur, a mostly solitary, territorial

animal   that puts a premium on knowing how many and how close are the conspecifics in the neighbourhood.

SUMMARY

I have reported on the acoustic characteristics of a forest habitat and on
the structure and temporal occurrence of the most common sounds emitted by
the animals of that forest. The analysis of the roles of each of the prosimian species lends support to the hypothesis that structure and timing of
the calls of animals inhabiting a particular environment show features that
maximize the chances of a sound signal reaching the appropriate receiver.

ACKNOWLEDGEMENTS

I especially thank Jay Russell for his encouragement and assistance in
Madagascar. The field work in Madagascar was supported by an NDEA Title IV
Fellowship, the Society of the Sigma Xi and the late Ms. Mamiedelle McGeorge.
Equipment for the data analysis was provided by the Departments of Otolaryngo-
logy, Psychology and Zoology at Duke University. Travel to the IPS Congress
in Cambridge was financed by the Graduate School and Zoology Department of
Duke University.

REFERENCES

Ficken, R.W., Ficken, M.S. and Hailman, J.P. (1974). *Science,* 183, 762-763.
O'Neill, W.E., Clark, C.W. and Vencl, F.V. (1976). The effects of foliage on
    the transmission and evolution of animal sounds in a northern deciduous
    woodland. Presented at the annual meeting of the Animal Behaviour Society,
    Boulder.
Morton, E.S. (1970). Ecological sources of selection on avian sound. Ph.D.
    thesis, Yale University, New Haven.

OLFACTORY COMMUNICATION, *GALAGO CRASSICAUDATUS*,
AND THE SOCIAL LIFE OF PROSIMIANS

ANNE B. CLARK

*Primate Behaviour Research Group, University of the Witwatersrand,
Johannesburg, South Africa.*

Olfactory communication is often cited as a 'primitive' feature of prosimian social behaviour. It is seldom studied as an integral part of an adaptive complex within which different communicative modes may complement each other in a non-additive fashion.

Here I focus on olfactory signalling in the largest galago species, *Galago crassicaudatus*, and attempt to relate it dynamically to known social behaviour. Since scent-marking has often been connected with territoriality and control of space (Mykytowyxz, 1970; Theissen, 1973; Charles-Dominique, 1974; see Eisenberg and Kleiman, 1972, for discussion), particular attention is given the space-usage system. Finally, olfactory signalling in prosimians and other ecologically comparable mammals is considered in more general terms.

Galago crassicaudatus has perhaps the most obvious elaboration of scent sources of all African galagines. The sources and behaviour patterns used in scent-marking with each have been well described (Bearder, 1974; Clark, 1975) and will only be summarized here.

Both sexes possess a glandular patch (size variable by sub-species) over the rostral sternum, which is rubbed against objects, particularly vertical supports. This patch is furred in juveniles less than 4-5 months old. Production of the yellow oily 'scent', not present at birth, can be detected on cotton wipings by twelve weeks by means of its yellow-white fluorescence under ultra-violet light (pers. obs.). Males generally chest-rub more frequently than females.

Enlarged apocrine and sebaceous glands are concentrated on the scrotal and circum-labial skin (Montagna and Yun, 1962). The secretion, also UV fluorescent, is deposited on branches by dragging the ano-genital area along them. 'Anogenital rubbing' often occurs in conjunction with chest rubbing.

The lips and muzzle contain large apocrine glands (Montagna and Yun, 1962) and are rubbed vigorously on objects, especially ends of supports. No UV-fluorescent secretion has been detected nor have responses to these marks been demonstrated. Face-rubbing may occur with chest and ano-genital rubbing and is provisionally classed as scent-marking (Clark, 1975).

All ages, beyond at least 112 days (pers. obs.), urine-wash in the manner described by several authors for this and other primate species (Eibl-Eibes-feldt, 1953; Andrew, 1964; Andersson, 1969; Andrew and Klopman, 1974). Urine is also deposited by means of 'rhythmic micturition' (Pinto, 1972). Urine is UV-fluorescent.

Both sexes use an aglandular patch on the hind sole, which is covered in 'short, corneal spines' (Montagna and Yun, 1962), in hind-foot rubbing (or leg rubbing, Bearder, 1974). Urine adheres to these epithelial spines and could

be deposited on a substrate. Rubbing can be quite noisy and, as it occurs
often in conjunction with other marking patterns, it might be an auditory
signal augmenting scent-marks.

Allomarking, as described for *G. senegalensis* (Andersson, 1969) and *Perodicticus potto* (Manley, 1974), has never been observed.

All potential scent sources, including the feet, are objects of olfactory
investigation on initial approach of one animal to another. Often a young or
subordinate individual approaches a fellow from below to sniff the hands first,
and then, if not repulsed, the lips and face. Males contacting females often
sniff the chest as well as the anogenital area (Clark, 1975).

Through a series of experiments involving presentation of perches scent-
marked by one or more galagos to a test galago in its home cage, a large amount
of information was shown to be present in scent marks (Clark, 1975). Sex,
individual identity and, in the case of males of *argentatus* and *crassicaudatus*
subspecies, subspecific identity could all be separately discriminated by
galagos on the basis of scent marks.

Age, or perhaps experience, appeared to improve the consistency of preferences for sex (overall preference being for female scent) and subspecies
(overall preference for 'own' subspecies' scent), as measured by time spent
sniffing and licking the perches. In the latter case particularly, older
males consistently directed most responses, including biting and chest rubbing, which have an aggressive context, to scent marks of their own subspecies.
Age of scent marker is also apparently reflected in scent. In three separate
experiments, galagos responded with consistently less sniffing when a young
(less than one year) male's scent was presented than when an older (two or
more years) male's scent was offered.

Discrimination of female reproductive condition, based on retrospective
analysis of sniffing and marking responses, was not demonstrable, but scent
marks of oestrous females were *licked* significantly more often than those of
diestrous or proestrous females. Licking has been connected with vomero-
nasal stimulation in mammals and, as is likely for many other groups (Estes,
1972), the well-developed prosimian accessory olfactory system (Stephan, 1965)
may be specialized to receive and process reproductive information.

The naturally marked perches carried some unspecified composite of galago
scent. Differentiation of information content and function of scents is still
under study. Urine and chest gland scent, both of which could be collected
free of contamination, were tested separately for individual identity information. No evidence of this information was found for urine which indeed
generally excites little interest. Chest gland scent was sufficient for
galagos to discriminate between some conspecifics (two females) but not others
(two males). In the latter case, scent from one galago was (unavoidably) collected as it exuded from a freshly cleaned gland, rather than from under a
pad worn by other scent donors. This result suggests that bacterial decay of
secretions may contribute the individual characteristics.

The function of scent marks and the circumstances under which galagos mark
is known primarily from Bearder's (1974) extensive ecological-behavioural
field study in South Africa, an observational study of a captive family
group (Clark, 1975), and the initial portion of a field study
on a large, fairly dense population (in progress). In addition, there have
been several social behaviour studies of groups introduced to each other as
adults (Jolly, 1966; Roberts, 1971; Tandy, 1974). Except for the latter
(which suffer from unnatural groupings of unfamiliar adults), there are no
real inconsistencies between studies in observed social behaviour and com-

munication. However, Bearder's (1974) field study concentrated for this data
on a single extended family group isolated from other groups. The temporo-
spatial distribution of reproducing adults, an aspect I will suggest is critical to understanding communication systems, is as yet unclear.

Bearder's (1974) observations and my own indicate a system of overlapping
male and female home ranges, and close association between particular males
and females and the females' young of up to a year, or more. The range overlaps seem more extreme than those reported for other lorisids (Charles-Dominique,
1971, 1972, 1974) and lemurids (Charles-Dominique and Hladik, 1971; Martin,
1972; Tattersall and Sussman, 1975); within my study area, large parts of
the home ranges of at least four males and two females coincide, though the
frequently-associated pairs and/or family groups are seldom seen together.
In this study, no instances of aggression between males have been observed.
In fact, upon meeting in the sleeping trees of a well-known female (♀5), two
adult males began grooming each other and continued for about twenty minutes.
Many galagos use the same feeding trees, especially the largest *Acacia karoo*
gum trees, sequentially, at different times of the night. Initial progression of known animals is almost always in the same direction along their riverine strip of trees. Early in the study, an observer caused small groups to
stop and watch for as much as thirty minutes; often a second or even third
group arrived until many more galagos were present than were ever normally
encountered.

In contrast with this apparent temporal spacing between groups, an individual may be alone while feeding. Then, when it settles to groom, its most
frequent associates (siblings, mother, 'mate') often arrive within minutes to
join in mutual grooming.

On the few occasions when agonistic vocalizations have been heard, the
'group', when located, consisted of ♀ 5, her three eight-month old young
plus the male twins of a second female (rarely seen herself), all within
♀ 5's home range. No mutual grooming between the twins and ♀ 5 or her triplets
has been seen.

From these and Bearder's (1974) observations, males seem to have large home
ranges which overlap both female and other male ranges. Females' smaller
ranges, of a size that sleeping trees are always within about thirty minutes'
travel time, are more exclusive of other females and their young. A female
associates mainly with one particular male, and her juveniles may move and
sleep with him as they become independent (Bearder, 1974). All galagos may
forage alone, at least out of sight of one another. However, groups organize
their resource use temporally as well as spatially so group members are likely
to be in the same area, and other groups more distant, at any time (pers.
obs.). Arboreal routes appear to be much the same for all groups.

What role does scent-marking play in coordinating this system? Again, observations on scent-marking in all studies are reasonably consistent, and all
are plagued by the lack of observed responses to scent marks. The experimental work (Clark, 1975) does provide insight on responses. Although the
conditions are far from natural, the test situation is most analogous to the
discovery of one or more branches recently and heavily marked by an unfamiliar
galago within a favoured portion of the home range.

Urine-marking remains enigmatic: as frequent as other patterns in the wild,
and more frequent under semi-natural conditions for all age-sex classes (Clark,
in prep.), it occurs in a bewildering variety of circumstances in this and
other species (for discussion, see Andrew and Klopman, 1974). It is most frequent at the beginning of activity and is also a usual response to sudden
noise, movements, etc. Trail-marking with urine, demonstrated for *Nycticebus*

(Seitz, 1969), has been discounted since animals do not regularly sniff along their routes and they often clearly orient visually (Bearder, 1974). A brief study of trail choice indicated that animals might prefer self-marked trails over unmarked ones when totally unfamiliar with the route (Clark, 1975).

While, as Doyle (1975) showed, the occurrence of urine-washing is non-random in a cage, no consistent spatial distribution has been noted in the wild (Clark, 1975). In the social context developed above, urine adhering to the feet, especially to the rough hind soles, could remain on branches and indicate to other animals that another had passed there recently. Even if lacking individual characteristics, the presence of fresh urine could alert a galago to other scent marks. A typical decay rate could contribute to its message value, especially as chest gland scent is very long-lasting and is unlikely to reflect accurately the time of marking. Urine-washing often occurs in grooming pauses, and is also likely to be present if an animal were slowed or diverted by an alarm. Thus it may be part of a non-specific temporal spacing system. That it has less social import than other scent is supported by the low interest it occasions during experiments and its very infrequent occurrence as a response to scent.

Chest-rubbing and anogenital-rubbing also occur when galagos are alarmed (in connection with growls, barks and hind-foot rubbing) or before and after agonistic encounters (Bearder, 1974; Clark, 1975). Chest-rubbing occurs repeatedly at certain points within the home range or cage (Bearder, 1974; Clark, 1975). Female 5 of the current field study chest-rubbed, anogenital-rubbed and sniffed both in the sleeping trees at the S extreme and, in two areas, at the N extreme of her range. Chest-rubbing was relatively less and anogenital-rubbing relatively more frequent in semi-natural conditions, difficult differences to interpret.

With overlapping ranges the rule, it is likely that these marking points are important information exchange posts between neighbours. Since chest- and anogenital-rubbing occur together frequently, the post would carry the individual identity of the marker together with the reproductive information as would likely be present in anogenital secretions.

Although chest-rubbing has been termed 'territorial' (Bearder, 1974), there is no evidence that boundaries are defended or respected. Certainly male ranges are not exclusive. Thus territorial defense is an unlikely function for such marking spots. But, for wide-ranging members of non-cohesive groups, the signalling of reproductive condition could be important. In semi-natural conditions, neither sex increased its marking just prior to female oestrus and females actually almost ceased marking during receptive periods which last an average of only 5.8 days(Eaton et al., 1973). However, this appeared to be compensated by increased locomotion by the female, frequent passages near the male as well as by his olfactory investigation of them during prolonged 'follows'. While a signal of approaching oestrus or even seasonal resumption of cyclic activity of females[1] may be needed to attract males or to synchronize their reproductive condition with that of females, copulation seems to be preceded by a period of close consorting when olfactory cues on a female's body would be more accurate signals.

---

[1] In the N hemisphere, no females cycled regularly from January to June and cycles became long and irregular by April. *G.c. argentatus*, the subspecies used in the study, do not exhibit periodic vaginal closure, even when acyclic, as most animals were by June (Clark, 1975).

In the experimental situation, galagos became quite excited and responded with all marking patterns except urine-washing. Marks were placed on presented scent-marks and cage walls; chest-rubbing especially was done over chest gland scent. Such marking over is a necessary basis for 'sign-posting'. Interestingly, females more than males marked over female scent, yet they sniffed it less than males. They used anogenital-rubbing more frequently and with a greater preference for rubbing on female-scented perches. This suggests, as do field observations, that females may tend to exclude females more than males exclude males.

G. crassicaudatus have a large number of vocalizations associated with contact, spacing and aggression. These, with one exception (a low, grating sound heard during male approaches to females and given several times by males when sniffing female scent) were not given in response to scent-marks. Nor do males in the wild simultaneously give their 'long call' and scent-mark. However, long calls are individually distinctive to my ears. Redundancy of some information between communicative modes would allow the integration of location and instantaneous mood reflected in calls with reproductive, etc., information left in chest-rubbing, at least.

Although brain size data from recent primate species show a consistent decrease in relative size of olfactory bulbs moving from prosimians through apes (Stephan and Andy, 1969), Jerison (1973) has stated that early mammals probably had olfactory bulbs no bigger than, if as large as, many recent species. Thus, well-developed olfactory structures and chemical signalling are less likely to be a primitive retention of early mammalian characteristics than a prosimian specialization. For mammals which evolve structures for social olfaction, chemical signals must represent part of a solution to communication problems posed by other ecologically necessary adaptations, e.g. nocturnal activity or spatial separation of individuals.

Charles-Dominique and Hladik (1971) have suggested that a population nucleus of overlapping, but separate, female home ranges, with central reproductive male home ranges, each overlapping those of one or more females, is the common and ancestral form of prosimian social organization. In evolving such a system, as boundaries are less important, individual relationships become more important. In particular, when spatial location of neighbours, with respect to a given individual, becomes less predictable, it would be more critical to signal instantaneous location, as well as its most frequently used areas (i.e. most likely location), its reproductive condition, etc., all in an individually-identifiable fashion.

While there seems to be some homogeneity within prosimian families as to what sources of scent, i.e. glands, are developed (see Table I), there is also a great variation in the number of such sources. The variation in number of vocalizations, also great, does not indicate that calls substitute for olfactory signals. Neither is there a trend to decrease the use of olfactory cues with increased social contact, as one might predict from generalizations about prosimian versus anthropoid sociality and communication. Consider, for comparative purposes, the relationship between signals and social structure in three closely-related species of phalangerid marsupials, all nocturnal, arboreal and ecologically similar to many prosimians.

Shoinobates volans, the greater glider, appears to be least gregarious and both sexes show a uniform, unclustered distribution through the habitat (Tyndale-Biscoe and Smith, 1969). Tyndale-Biscoe (1973) states that "Gliders are generally silent so if this spacing-out has a behavioural basis, it is likely to be olfactory, since paracloacal glands are well-developed, especially in males" (p. 168).

TABLE I

Family	Ref.	Urine	Anogenital Glands	Chest Glands	Gular Glands	Brachial Glands	Antebrachial Glands	Throat Glands	Head Gland	Facial Glands	Other
Lemuridae											
L. catta	31	+	+				+				Palmar marking
L. fulvus	18	+	+						+		
L. mongoz	36	+	+						+		
Indriidae											
I. indri	28	+	+		+						
P. verreauxi	29	+	+					+			
A. laniger					+						
Galagidae											
G. crassicaudatus	4,11	+	+	+						+	Hind-foot rubbing
G. senegalensis	1,5	+	+	+							
G. demidovii	6,7	+	+							+	
G. alleni	6	+	+								
Lorisidae											
P. potto	8,22	+	+								
A. calabarensis	22	+	+								
L. tardigradus	3,22	+				+*					
N. coucang	33,22	+				+*					

* Not same position as *L. catta's* brachial glands.

Scent sources of selected prosimian species by family (numbers refer to the following references:

1. Andersson, 1969.
3. Andrew and Klopman, 1974.
4. Bearder, 1974.
5. Bearder and Doyle, 1974.
6. Charles-Dominique, 1971.
7. Charles-Dominique, 1972.
8. Charles-Dominique, 1974.
11. Clark, 1975.
18. Harrington, 1974.
22. Manley, 1974.
28. Pollock, 1975.
29. Richard, 1974.
31. Schilling, 1974.
32. Seitz, 1969.
36. Tattersall and Sussman, 1975.)

Trichosurus vulpecula, the brush-tailed possum, maintains non-exclusive home ranges of 1-3 hectares with the larger, male ranges overlapping those of females. Sleeping is solitary; there is no mutual grooming or contact activity. Response to encounters is avoidance, although one study (Dunnet, 1964) reports male territorial defence against males. Six calls were identified and scent sources, best developed in males, include 'modified sebaceous glands on chin and chest as well as two sets of paracloacal glands' (Tyndale-Biscoe, 1973, citing unpublished data of J. Winter).

Pseudocheireus peregrinus, the ringtailed possum, is gregarious and groups (1 ♂, 1-2 ♀♀, plus young) cooperate in nest building. Home ranges of groups overlap; only nests are defended. Well-defined runways are used. Their scent, from anal glands, is distasteful to predators. Males, but not females, can identify sex from this scent. Vocalizations consist of a 'high-pitched chirruping twitter' given while feeding undisturbed (Thompson and Owen, 1964).

The largest potential number of olfactory and vocal signals is found in neither the most nor the least gregarious species, but in *Trichosurus* which shares space in a way reminiscent of many prosimians. Individual recognition between female neighbours could be important. Certainly, spatial location of possums does not ensure their separation from others and signalling past presence and reproductive condition would be essential to daily and seasonal coordination of individual movements.

Among prosimians, territorial species such as *Lepilemur* ensure separation of neighbours by spatial location only. Repetitive vocal territory advertisement suffices. Pottos and other lorisids contrast with *G. crassicaudatus* in their lack of vocalizations, while scent source elaboration is quite comparable. Again, because of their slow locomotion, scent marks may not need the added instantaneous location information of calls to be efficient at separating animals in space and time.

Lemur catta represent another extreme in that diurnality and cohesive groups select for even greater temporal precision in information transfer, such precision being ensured by visual signals. Here, olfactory cues seem to provide the context for, or intensify the effect of, visual cues and, in the words of Mertl (1976) "there is a fine intermeshing of information from the two sensory modalities". Thus, for territorial species, a neighbour's location is predictable by an individual as long as both are within their territories. Signals have only to be sufficient to advertise that. Increased overlap of ranges decreases predictability of neighbours' location and signals which inform both through space and time are required. For diurnal species in close contact, some temporal information will be most efficiently embodied in visual cues. For animals less able to see even their close associates, and only sometimes near them, an integrated combination of olfactory cues and vocal signalling seem to be the solution to the problem of finding or avoiding conspecifics on an individual basis.

In conclusion, I suggest that prosimians were early characterized by specialized use of olfaction. Further, an ancestral tendency to form social bonds based on individual relationships and tolerance (both ecological and social) of home range overlap would favour elaboration of chemical cues more than a shift into sociality via territorial individuals → territorial groups. This tendency is realized in the basic prosimian social system described by Charles-Dominique and Hladik (1971). *G. crassicaudatus*, in using a system whereby not only do individuals associate through home range overlap, but loose groups of associates share some parts of their ranges with other groups through temporal spacing, have both a large number of vocalizations for instantaneous location and a number of distinct chemical signals carrying sex, individual, subspecific information as well as advertising the use of the area. Some redundancy between modes, as is present between chest gland scent and long calls, allows for integration of temporal, spatial, and social information provided by each.

ACKNOWLEDGEMENTS

This work was supported by Ford Foundation Grant #67-375 to the Committee on Evolutionary Biology, University of Chicago; a Hinds Fund Grant, University of Chicago and, currently, by a University of the Witwatersrand Postdoctoral Fellowship with the Primate Behaviour Research Group. Special thanks are due to Dr. George B. Rabb, Brookfield Zoo, and Drs. J.A. Bergeron and P.H. Klopfer, not only for making facilities available, but for their advice and constant support.

REFERENCES

Andersson, A.B. (1969). Communication in the Lesser Bushbaby *Galago sene-galensis moholi*, unpublished M.Sc. Dissertation, University of the Witwatersrand.
Andrew, R.J. (1964). *In* "Evolutionary and Genetic Biology of the Primates", vol. II, (J. Buettner-Janisch, ed.), pp. 227-309, Academic Press, New York.
Andrew, R.J. and Klopman, R.B. (1974). *In* "Prosimian Biology", (R.D. Martin, G.A. Doyle and A.C. Walker, eds), pp. 303-312, Duckworth, London.
Bearder, S.K. (1974). Aspects of the Ecology and Behaviour of the Thick-tailed Bushbaby *Galago crassicaudatus*, unpublished Ph.D. Thesis, University of Witwatersrand.
Bearder, S.K. and Doyle, G.A. (1974). *In* "Prosimian Biology", (R.D. Martin, G.A. Doyle and A.C. Walker, eds), pp. 109-130, Duckworth, London.
Charles-Dominique, P. (1971). *Z. Tierpsychol.*, Beiheft 9, 7-41.
Charles-Dominique, P. (1974). *Mammalia*, 38, 355-379.
Charles-Dominique, P. (1974). *In* "Primate Xenophobia, Aggression and Terri-toriality", (R.L. Holloway, ed.), pp. 31-48, Academic Press, New York.
Charles-Dominique, P. and Hladik, C.M. (1971). *Terre et Vie*, 25, 3-66.
Clark, A.B. (1975). Olfactory communication by scent marking in a prosimian primate, *Galago crassicaudatus*, unpublished Ph.D. Dissertation, University of Chicago.
Doyle, G.A. (1975). *In* "Contemporary Primatology", (S. Kondo, M. Kawai and A. Ehara, eds), pp. 232-237, S. Karger, Basel.
Dunnet, G.M. (1964). *Proc. zool. Soc. Lond.*, 142, 665-695.
Eaton, G.G., Slob, A. and Resko, A. (1973). *Anim. Behav.*, 21, 309-315.
Eibl-Eibesfeldt, I. (1953). *Saugetierkundl. Mitt.*, 1, 171-173.
Eisenberg, J.F. and Kleiman, D. (1972). *A. Rev. Ecol. Syst.*, 3, 1-32.
Estes, R.D. (1972). *Mammalia*, 36, 315-341.
Harrington, J. (1974). *In* "Prosimian Biology", (R.D. Martin, G.A. Doyle and A.C. Walker eds), pp. 331-346, Duckworth, London.
Hladik, C.M. and Charles-Dominique, P. (1974). *In* "Prosimian Biology", (R.D. Martin, G.A. Doyle and A.C. Walker eds), ''. 23-37, Duckworth, London.
Jerison, H.J. (1973). "Evolution of the Brain and Intelligence", Academic Press, New York.
Jolly, A. (1966). Observations on *Galago crassicaudatus*, unpublished ms.

Manley, G.H. (1974). *In* "Prosimian Biology", (R.D. Martin, G.A. Doyle and A.C. Walker eds), pp. 313-329, Duckworth, London.

Martin, R.D. (1972). *Z. Tierpsychol.*, Beiheft 9, 43-89.

Mertl, A. (1976). *Folia primatol.*, 26, 151-161.

Montagna, W. and Yun, J.S. (1962). *Am. J. phys. Anthrop.*, 20, 149-165.

Mykytowyxz, R. (1970). *In* "Advances in Chemoreception", vol. I, (J.W. Johnston, D.G. Moulton and A. Turk, eds), pp. 337-360, Appleton-Century-Crofts, New York.

Pinto, D. (1972). Patterns of activity in three nocturnal prosimians: *Galago senegalensis moholi, Galago crassicaudatus umbrosus* and *Microcebus murinus murinus*, unpublished M.Sc. Dissertation, University of the Witwatersrand.

Pollock, J.J. (1975). *In* "Lemur Biology", (I. Tattersall and R.W. Sussman eds), pp. 287-312, Plenum Press, New York.

Richard, A. (1974). *In* "Prosimian Biology", (R.D. Martin, G.A. Doyle and A.C. Walker eds), pp. 49-74, Duckworth, London.

Roberts, P. (1971). *Folia primatol.*, 14, 171-181.

Schilling, A. (1974). *In* "Prosimian Biology", (R.D. Martin, G.A. Doyle and A.C. Walker eds), pp. 347-362, Duckworth, London.

Seitz, E. (1969). *Z. Tierpsychol.*, 26, 73-103.

Stephan, von H.   *Acta Anat.*, 62, 215-253.

Stephan, von H. and Andy, O.J. (1969). *Ann. N.Y. Acad. Sci.*, 167, 370-387.

Tandy, J. (1974). *In* "Prosimian Biology", (R.D. Martin, G.A. Doyle and A.C. Walker eds), pp. 245-259, Duckworth, London.

Tattersall, I. and Sussman, R.W. (1975). *Anthrop. Pap. Am. Mus. nat. Hist.*, 52, 193-216.

Theissen, D. (1973). *Am. Sci.*, 61, 346-350.

Thompson, J.A. and Owen, W.H. (1964). *Ecol. Monogr.*, 34, 27-52.

Tyndale-Biscoe, C.H. and R.F.C. Smith (1969). *J. Anim. Ecol.*, 38, 637-650.

Tyndale-Biscoe, C.H. (1973). "Life of Marsupials", Edward Arnold, London.

NECTAR-FEEDING BY PROSIMIANS AND ITS EVOLUTIONARY AND ECOLOGICAL IMPLICATIONS

R.W. SUSSMAN

Department of Anthropology, Washington University, St. Louis, Missouri 63130, USA.

INTRODUCTION

During July and August of 1973 in Ampijoroa, northwestern Madagascar, I and two other participants in this paper session (Mc George and Tattersall) observed *Lemur mongoz* (= *L.m. mongoz*) behaving in an unexpected manner. In this forest, the animals were nocturnal (although previously reported to be diurnal like other species of the genus *Lemur*), but even more unexpected was their feeding behaviour. Approximately 80% of the observed feeding behaviour of *L. mongoz* during this study was on the nectar-producing parts of four plants and 80% of this was on the nectar of the kapok tree (*Ceiba pentandra*) (see Tattersall and Sussman, 1975). In June, 1974, I returned to Ampijoroa and found *L. mongoz* still behaving in this unusual (at least for a primate) manner. In this paper, I will discuss the implications of this behaviour as it relates to inter-taxa interactions and to the ecological concepts of foraging guilds and co-evolution.

CEIBA PENTANDRA: A BAT-ADAPTED TREE

In Asia, Africa, and North and South America, the kapok tree is pollinated by plant-visiting bats and is adapted to attract these pollinating agents. The plant attracts bats with distinctive, easily accessible flowers and nutritious nectar. The bats then aid the plant in reproduction.

The kapok tree has coevolved with bats in a number of ways (Baker and Harris, 1959; Baker, 1965). The plants lose their leaves in the dry season before flowering occurs and leaflessness is prolonged until after the distribution of seeds. This not only allows free dispersal of seeds but also allows access to the flowers by flying animals larger than insects. The majority of flowers are produced, in dense clusters, on the terminal branches of the trees in the canopy. Flowering time is regular, with flowers opening about half an hour after dusk and remaining open during the night. Flowers produce a strong scent. The following day the corolla and stamens fall. Each flower produces five stamens which stand well above the corolla, and the pollen grains are sticky due to the secretion of oil drops in the anthers. Nectar is secreted at the base of the flower before it opens and disturbance of a cluster of flowers by wind or by visitors produces a rain of nectar when the flowers are open. Analyses of the nutritional content of the nectar of *Ceiba pentandra* indicate that it contains some sugars and a relatively large quantity of amino acids, with seven different amino acids represented (Baker and Baker, pers. comm.; see Sussman and Tattersall, 1977). Kapok trees are self-compatible as well as cross-pollinating.

There is only one plant-visiting bat in northwestern Madagascar, *Pteropus rufus*. This bat, however, is often destructive to *Ceiba pentandra* flowers and has not been considered a legitimate pollinator of this species. In this area of Madagascar, *Lemur mongoz* may fill a niche usually occupied by plant-visiting bats. It may serve as the major pollinator of the kapok tree and possibly other bat-adapted plants. As Tattersall (this volume) reports, the behaviour of *L. mongoz* is variable. In some areas *L. mongoz* is diurnal and, I suspect, has a very different diet.

NON-FLYING MAMMAL POLLINATORS

Rodents and marsupials

Whereas bats and birds are common vertebrate pollinators in the tropics, there are very few reports of pollination by non-flying mammals. A number of Australian marsupials and one Australian rat, *Rattus fiscipes*, are known to visit flowers for nectar and/or pollen (Rourke and Wiens, 1977) and the honey possum (*Tarsipes spencerae*) is a very specialized nectar feeder (Glauert, 1958; Vose, 1973). In Ceylon, the tree squirrel, *Funambulus palmar* is known to feed on the nectar of *Grevillea* and possibly *Ceiba* (Phillips, 1935 Walker, 1968). The interactions between these mammals and the plants they visit, or the indigenous species of birds and bats, have not been studied.

In Hawaii, introduced species of rats are known pollinators of *Freycinetia arborea*. In Asia, *Freycinetia* is pollinated by bats (van der Pijl, 1956) and shows many characteristics of a bat-pollinated tree (Faegri and van der Pijl, 1971). However, in Hawaii there are no plant-visiting bats and before the rat were introduced into Hawaii, *F. arborea* developed some ornithophilous traits and was probably pollinated by birds. It is likely that the retention of bat associated characters pre-adapted *F. arborea* to pollination by rats, which are highly aggressive and out-competed birds as pollinators (Rourke and Wiens, 1977).

Primates

The role of primates as potential pollinators has not been extensively studied, although a number of observations of flower-visiting by primates have been mentioned in the literature. Many primates eat blossoms or parts of blossoms but, in most cases, this is a very small part of their diet and the effect on the flower is destructive. Some of the nocturnal prosimians, however have been seen to feed on nectar without destroying the flowers. In these primate species, nectar may provide important sources of nutrients and the plant may benefit from the interaction.

In addition to *Lemur mongoz*, *Microcebus murinus* were observed at Ampijoroa licking nectar from the kapok tree without destroying the flowers. In southern Madagascar, Martin (1972) observed the mouse lemur feeding on flower of *Vaccinium emirnense* and of *Uapaca* trees. In western Madagascar, *Phanar furcifer* have been observed spending long periods of time licking clusters of flowers on the finest terminal branches of *Crateva greveana* trees. They also spend considerable time near the buds on the terminal branches of baobabs (*Adansonia*) (Petter et al., 1971). Outside of Madagascar, both of these species of plant are normally bat-pollinated. At Morondava, in western Madagascar, nectar from flowers of an as yet unidentified species of Caesalpiniac is the main food source of *Cheirogaleus medius* during the beginning of the ra season (Hladik, pers. comm.).

In the arid regions of southern Madagascar, during the driest part of the year, *Lepilemur* feeds mainly on the flowers of two endemic species of *Alluaudia*: *A. ascendens* and *A. procera* (Charles-Dominique and Hladik, 1971). Charles-Dominique and Hladik remark that it is the successive flowering of the two species that permits *Lepilemur* to survive the severest portion of the dry season. However, *Lepilemur* is a predator upon the flowers and these plants are normally insect pollinated (Hladik, pers. comm.).

The only description of   Lorisidae acting as a possible pollinating agent is that of Coe and Isaac (1965). They observed *Galago crassicaudatus* in East Africa visiting baobab trees and feeding on newly-opened flowers. The animals moved from flower to flower burying their faces within the flowers and around the sepals. The bushbabies lick the flowers, causing only superficial damage to them.

From these few accounts, it seems likely that nectar is an important potential source of nutrition for nocturnal primates. As more detailed observations of feeding behaviour are made on these animals, more examples of nectar feeding may come to light. However, in most of the cases of nectar feeding described, the primates are visiting plants that are normally bat-pollinated.

In areas where plant-visiting bats are rare or absent, the primates may be partially filling a void niche and the plants are able to take advantage of a potential pollinator. The adaptations of the plant to attract bats are sufficient to attract nocturnal primates, although I think that primates are not as adept at exploiting these resources as bats, nor are they ideal pollinating agents. It seems unlikely, but not impossible, that coevolution between plants and primates-as-pollinators has occurred. Much as *Daubentonia* fills in for woodpeckers in Madagascar (Cartmill, 1974), many of the nocturnal prosimians may be occupying (at least to some extent) the plant-visiting bat niche where bats are rare or absent.

EVOLUTIONARY AND ECOLOGICAL IMPLICATIONS

The role of primates in pollination ecology has brought to mind a number of related questions and problems which, I believe, have been relatively ignored in the primate literature. The first of these is the interaction between species of the major vertebrate taxa using the canopy of the tropical forest (e.g. birds, bats, squirrels, and primates). These interactions probably are among the major factors shaping present ecological adaptations and were extremely important in the evolution of the particular adaptive trends of these taxa. Related to this is the problem of defining major varieties of feeding and foraging adaptations of the primates and how these adaptive groups are apportioned in different tropical plant communities. Finally, to what extent is coevolution occurring between primates and the plants that they utilize? For the remainder of this paper, I would like to discuss these issues. The discussion will, because of time and length restrictions, be brief and, mostly, speculative. But my hope is, in the spirit of scientific inquiry, to throw out some ideas for the purpose of further thought and discussion and, perhaps more ambitiously, to stimulate further research in these areas.

Those vertebrate taxa that have the most species represented in the tropical forest canopy and that are most likely to be utilizing resources similar to those used by primates are bats, birds, and tree squirrels. All bats are nocturnal and most birds are diurnal. Although most tropical forest-dwelling mammals are nocturnal, most primates and all tree squirrels, except flying squirrels, are diurnal (see Charles-Dominique, 1975). Bats and birds avoid competition for many of their resources (especially insects and other prey, and nectar) simply by these differences in activity pattern. It seems likely that bats and nocturnal prosimians have been major or potential competitors si the Eocene. The success of bats in the tropics and the relative lack of succe of prosimians could very well be related. Existing nocturnal prosimians may u resources not normally exploited by bats, including gums and insect larva, as well as crawling insects, and catch much of their prey in the dense vegetation of the bush and canopy where it is difficult for bats to navigate. I believe more detailed investigations into bat and nocturnal prosimian interactions are warranted (and I wonder what *Aotus* is doing).

Besides basic differences in cycles of activity, there are patterns of resource partitioning and habitat selection within and between major taxonomic groups. Animals with parallel or convergent roles in similarly structured plant communities often share foraging and dietary adaptations, as well as certain general morphological adaptations. Groups of animals filling similar roles have been called "guilds"[1]. In each foraging guild are animals with adaptations that allow them to search, ingest, and digest certain types of available food items and that often share certain locomotor, sensory, dental, enzymic and gastrointestinal adaptations. Once foraging guilds are defined in primates, it may be possible then to attempt to relate various aspects of morphology, locomotion and individual and social behaviour to these dietary and foraging patterns. We expect to find some of these factors more similar and others more variable between members of similar guilds. A study of the relationships between similar guilds across taxonomic boundaries may also rais some interesting ecological and evolutionary questions.

Most research in this area has been done on birds (Root, 1967; Orians, 1969; Snow, 1971; Morton, 1973; Pearson, 1975; McKey, 1975) with a few papers on bats (Tamsitt, 1967; McNab, 1971; Fleming et al., 1972; Heithaus et al., 1975). In Table I, I list nine foraging guilds of tropical forest birds (from McKey, 1975, and Pearson, 1975). The majority of bird species inhabiting the tropical forest canopy are either exclusively insectivorous or partially frugivorous (usually feeding the young on insects and utilizing fruits heavily as adults). A very small percentage feed entirely on fruit (McKey, 1975; Pearson, 1975). The morphology, size, and behaviour of birds utilizing these resources is strongly determined by the particular type of foliage structure in which they typically forage. Furthermore, the prey species exploited by specific guilds have coevolved mechanisms to minimize predation or, in the case of plant species, to maximize efficiency of seed dispersal (Smythe, 1970; Snow, 1971; Morton, 1973; McKey, 1975). For example, fruits used by partially frugivorous birds usually attract a number of opportunistic dispersal agents, contain a great mass of small seeds, are sources of water and carbohydrates, and have short, displaced fruiting season. Fruits attracting specialized frugivorous birds are typically large, have one large seed, supply the birds with most of their lipids and proteins, as well

---

[1] "A *guild* is defined as a group of species that exploit the same class of environmental resources in a similar way. This term groups together species, without regard to taxonomic position, that overlap significantly in their niche requirements. The guild has a position comparable in the classifi- cation of exploitation patterns to the genus in phylogenetic schemes .... One advantage of the guild concept is that it focuses attention on all sympatric species involved in a competitive interation, regardless of their taxonomic relationship." (Root, 1967, p. 335.)

TABLE I

Foraging Guilds of Tropical Birds

Mainly Insect Foraging
    Glean insects
    Sally (both bird and prey on wing)
    Snatch (bird on wing, prey not)
    Peck and probe
    Glean and sally
    Glean and snatch
    Ant follower
Mainly Fruit Foraging
    "Opportunistic" frugivore
    "Specialized" frugivore

as carbohydrates, and have fruiting seasons spread over a long period (McKey, 1975). Species of trees utilizing the former seed dispersal strategy are most often found in the understory, in open vegetation areas or on the forest edge; those with the latter strategy are usually associated with the climax plant association in the canopy and emergent layers of forest (Snow, 1971; McKey, 1975). The birds utilizing these two types of fruits make up two frugivorous guilds differing in size, morphology, physiology, and behaviour.

Bats, squirrels, and primates also utilize these fruits and are seed dispersal agents, but specific information on fruit usage and seed dispersal in these animals is not extensive (however, see van der Pijl, 1957; Hladik and Hladik, 1967, 1969; Heithaus et al., 1975; and Smith, 1975). Fruit-eating guilds of birds, bats, squirrels and primates probably overlap in various ways. Pearson (1975), for example, found the proportion of fruit-eating birds, in two South American forests, to be inversely correlated with monkey population size, excluding the mainly leaf-eating species, *Alouatta seniculus*. He states that: "In general, it appears that a certain biomass of monkeys will offset a certain biomass of birds even though no effect can be detected at the level of species richness" (p. 464).

It is likely that leaf-eating is one of the major means by which diurnal primates have avoided extensive resource overlap and competition with birds and arboreal rodents. Besides a few specialized rodents, marsupials and edentates, primates are the only major group of vertebrates that feed on leaves in the forest canopy. Although some folivory may be beneficial to plants (Oppenheimer and Lang, 1969), in most cases it is detrimental (Ehrlich and Raven, 1964; Freeland and Janzen, 1974; Levin, 1976). Just as coevolution has occurred between plants, their pollinators and seed dispersal agents, mechanisms for protection from leaf predators and the strategies of folivory by mammals are interrelated (Janzen, 1970; Freeland and Janzen, 1974; Westoby, 1974; Atsatt and O'Dowd, 1976). Hence, among primates we may expect to find more than one pattern of folivory. For example, some primates are specialized leaf-eaters and have developed specialized gastrointestinal mechanisms for detoxifying plant secondary compounds (for example *Lepilemur*, Hladik et al., 1971, and many colobine monkeys). These species should use large amounts of several related toxic foods that are present in a year-round supply (Freeland and Janzen, 1974). Other, less specialized species of primate may adapt to a folivorous diet by using a particular foraging strategy such as: consuming a variety of plant foods at any one time, treating new foods with caution, ingesting small amounts on first encounter, sampling foods continuously, and feeding on leaves that contain small amounts of secondary compounds (see, for example, Glander, 1975, on *Alouatta seniculus*).

Extensive comparison and categorization of groups of species of primates that exploit the same class of resources in similar ways has not been done. However, recent research is pointing to some very interesting patterns of resource partitioning and utilization, especially in arboreal primates (see, for example, Clutton-Brock, 1975; Hladik, 1975; Struhsaker and Oates, 1975; Sussman, in press). I believe that further research into primate foraging guilds, inter-taxa interactions, and coevolution between primates and the plants they utilize will lead to a better understanding of primate morphology, behaviour and evolution.

ACKNOWLEDGEMENTS

This research was supported by National Science Foundation Research Grant No. BG - 41109 and Biomedical Research Support Grant RR - 07054 from the Biomedical Research Support Program, Division of Research Resources, National Institute of Health. Funds to attend the Conference were from a Washington University Summer Research Grant. I would like to thank the following people for their comments on various drafts of this paper: Linda Barnes, Marc Bekoff, J. Buettner-Janusch, Matt Cartmill, Jeremy Dahl, Bill D'Arcy, Steve Easley, Annette and Marcel Hladik, R.D. Martin, John McArdle, Stephen Molnar, Peter Raven, Bill Sawyer, Brian Suarez, Ian Tattersall and Steve Ward. I, of course, am responsible for the omissions and shortcomings.

Atsatt, P.R. and O'Dowd, D.J. (1976). *Science*, 193, 24-29.
Baker, H.G. (1965). *In* "Ecology and Economic Development in Tropical Africa", (D. Brokensha, ed.), pp. 185-216, Inst. Internat. Studs., Univ. California, Research Series 9, Berkeley.
Baker, H.G. and Harris, B.J. (1959). *Jl. W. Afr. Sci. Ass.*, 5, 1-9.
Cartmill, M. (1974). *In* "Prosimian Biology", (R.D. Martin, G.A. Doyle, A.C. Walker, eds), pp. 655-670, Duckworth, London.
Charles-Dominique, P. (1975). *In* "Phylogeny of the Primates", (W.P. Luckett and F.S. Szalay, eds), pp. 69-88, Plenum Press, New York.
Charles-Dominique, P. and Hladik, C.M. (1971). *Terre et Vie*, 25, 3-66.
Clutton-Brock, T.H. (1975). *Folia primatol.*, 23, 165-207.
Coe, H.J. and Isaac, F.M. (1965). *E. Afr. Wildlife J.*, 3, 123-124.
Ehrlich, P.R. and Raven, P.H. (1964). *Evolution*, 18, 586-608.
Faegri, K. and van der Pijl, L. (1971). "The Principles of Pollination Ecology, 2nd ed., Pergamon Press, Oxford.
Fleming, T.H., Hooper, E.T., and Wilson, D.E. (1972). *Ecology*, 53, 555-569.
Freeland, W.J. and Janzen, D.H. (1974). *Am. Nat.*, 108, 269-289.
Glander, K.E. (1975). *In* "Socioecology and Psychology of Primates", (R.H. Tuttle, ed.), pp. 37-57, Mouton, The Hague.
Glauert, L. (1958). *Austr. Mus. Mag.*, 12, 284-286.
Heithaus, E.R., Fleming, T.H., and Opler, P.A. (1975). *Ecology*, 56, 841-854.
Hladik, C.M. (1975). *In* "Socioecology and Psychology of Primates", (R.H. Tuttle, ed.), pp. 3-35, Mouton, The Hague.
Hladik, C.M., Charles-Dominique, P., Valdebouze, P., Delort-Laval, J., and Flanzy, J. (1971). *C. r. hebd. Séanc. Acad. Sci., Paris*, 272, 3191-3194.
Hladik, C.M. and Hladik, A. (1967). *Biol. Gabon*, 3, 43-58.
Hladik, A. and Hladik, C.M. (1969). *Terre et Vie*, 1, 25-117.
Janzen, D.H. (1970). *Am. Nat.*, 104, 501-528.
Levin, D.A. (1976). *Am. Nat.*, 110, 261-284.
Martin, R.D. (1972). *Z. Tierpsychol.*, Beiheft 9, 43-89.

McKey, D. (1975). *In* "Coevolution of Animals and Plants", (L.E. Gilbert and P.H. Raven, eds), pp. 159-191, University of Texas Press, Austin.

McNab, B.K. (1971). *Ecology*, 52, 352-358.

Morton, E.S. (1973). *Am. Nat.*, 107 8-22.

Oppenheimer, J.R. and Lang, G.E. (1969). *Science*, 165, 187-188.

Orians, G.H. (1969). *Ecology*, 50, 783-801.

Pearson, D.L. (1975). *The Condor*, 77, 453-466.

Petter, J-J., Schilling, A. and Pariente, G. (1971). *Terre et Vie*, 25, 287-327.

Phillips, W.W.A. (1935). "Manual of Mammals of Ceylon", Publ. Ceylon J. Sci., Dulau and Co., Ltd., London.

Root, R.B. (1967). *Ecol. Monogr.*, 37, 317-350.

Rourke, J. and Wiens, D. (1977). *Ann. Missouri Bot. Garden*, 64, 1-17.

Smith, C.C. (1975). *In* "Coevolution of Animals and Plants", (L.E. Gilbert and P.H. Raven, eds), pp. 53-77, University of Texas Press, Austin.

Smythe, N. (1970). *Am. Nat.*, 104, 25-35.

Snow, D.W. (1971). *Ibis*, 113- 194-202.

Struhsaker, T.T. and Oates, J.F. (1975). *In* "Socioecology and Psychology of Primates", (R.H. Tuttle, ed.), pp. 103-123, Mouton, The Hague.

Sussman, R.W. (in press). *In* "Primate Feeding Behaviour", (T.H. Clutton-Brock, ed.), Academic Press, London.

Sussman, R.W. and Tattersall, I. (1977). *Folia primatol.*, 26, 270-283.

Tamsitt, J.R. (1967). *Nature*, 13, 784-786.

Tattersall, I. and Sussman, R.W. (1975). *Anthrop. Pap. Am. Mus. nat. Hist.*, 52, 193-216.

van der Pijl, L. (1956). *Acta bot. neerl.*, 5, 135-144.

van der Pijl, L. (1957). *Acta bot. neerl.*, 6, 291-315.

Vose, H. (1973). *J. Mammal.*, 54, 245-247.

Walker, E.P. (1968). "Mammals of the World", 2nd ed., Johns Hopkins Press, Baltimore.

Westoby, M. (1974). *Am. Nat.*, 108, 290-304.

BEHAVIOURAL VARIATION IN *LEMUR MONGOZ* (= *L.M. MONGOZ*)

I. TATTERSALL

*American Museum of Natural History, Central Park West & 79th Street,
New York, N.Y. 10024, USA.*

Paradoxically, although students of animal behaviour have in general been much more keenly aware than morphologists of the balance between heredity and environment in the determination of the phenomena they study, it is the latter who have shown a greater sensitivity to the problems of variation. Given the far greater facility with which large samples of morphological characteristics can be measured and analysed, this is, of course, quite understandable: specimens in museum drawers are infinitely easier of access than are living animal populations. But it does mean that, until we have vastly more information on behavioural variability than is available at present for most species, we should be wary of employing methods analogous to those of the morphologists in the reconstruction of behavioural evolution. If we are not, we risk falling into the typological trap which engulfed the early morphological systematists. It may well be reasonable to suggest, as did Martin (1972), that it is permissible to discuss in these terms those behavioural characters that can be related directly to distinct morphological characteristics; but we cannot afford to overlook the fact that for the overwhelming majority of behavioural characters this is not the case. Quite simply, behaviour is not static in the sense in which, at least at any given point in time, morphology is.

Although, apart from the surveys of Petter (1962), almost half the extant species of Malagasy lemurs have been subject to more or less intensive behavioural investigation, very few have been studied in a variety of habitats or regions. Discussion of "species-specific behaviour", except perhaps in terms of very broad categories of locomotion, is thus somewhat premature where these animals are concerned. And the current impossibility of thus categorizing the behaviour of living species severely limits the precision with which we can interpret behavioural evolution, or even the interaction of morphology and behaviour, within the group. This is not to say that the effort is futile — clearly it is not — but rather that it should be undertaken with the limitations of the data constantly in mind.

Obviously, then, the study of individual prosimian species/subspecies in a variety of ecological settings is of paramount importance for our understanding of the nature of behavioural and morphological adaptation. The exemplary studies of Sussman (1974) and of Richard (1974) have begun to shed light on this vital area, and have already revealed that it is impossible to generalize about behavioural adaptation among the Malagasy primates. The primary purpose of this short note is to report the striking degree of behavioural plasticity shown by another lemur species: *Lemur mongoz.*

L. mongoz is represented by populations in northwest Madagascar and on two of the islands of the Comoro archipelago: Mohéli and Anjouan (Fig. 1). The

species was first studied at Ampijoroa in northwest Madagascar (Tattersall and
Sussman, 1975; Sussman and Tattersall, 1976),   particularly in regard to
its diet, activity pattern and group composition. I subsequently carried out
a survey of the Comorian populations between November, 1974 and January, 1975
(Tattersall, 1976a and b).   Sussman reports on dietary specialization in
L. mongoz elsewhere (this section); here I shall briefly discuss variations
in activity pattern and group composition within the species.

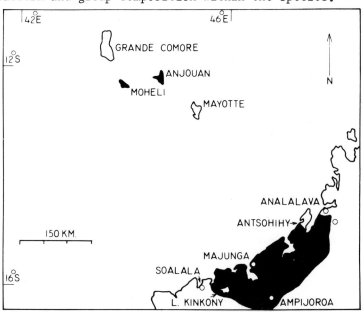

Figure 1. *Distribution of* L. mongoz *(solid black) in Madagascar and the
           Comoros. The southern and western limit of* L. mongoz *is not
           well known; Sussman and I have observed the species in the
           area of Lake Kinkony, but it has not been reported from
           Reserve Naturelle no. 8, some 20 km. due south of Soalala.
           The species does, however, occur both east and west of the
           Betsiboka River in the region of Ambato-Boeni, south of Ampi-
           joroa and on the same latitude as R.N.8. Distribution is not
           necessarily continuous within the areas shown.*

GROUP STRUCTURE

    Recensusing by R.W. Sussman in 1974 of three of the five groups of L.
mongoz censused in Madagascar in 1973 (Sussman and Tattersall, 1976:
Table 1) confirmed the original conclusion of Tattersall and Sussman (1975)
that pair-bonding is typical of the species, at least in the area
(Ampijoroa and Lake Kinkony) and season (June-August) of the study. No group
possessed more than four individuals, nor more than one fully adult animal
of each sex. Since L. mongoz breeds seasonally, and matures around two years
of age, up to two age-classes of immature individuals may be represented in
any given group. One group consisted in 1973 of an adult male and female, a
subadult female probably about 21 months old, and a juvenile male of around
nine months. The subadult female rested with the group during the day, but
frequently left it during the course of the night's activity. A year later,
this individual had departed from the group, while the young male, then aged

about 21 months, was still present. Two of the single pairs censused in 1973 possessed infants in 1974. On the basis of observations spread over two years, then, it seems reasonable to conclude that the pair-bonding observed is stable over time (although over exactly how long a time remains an open question).

TABLE I

Group compositions of Lemur mongoz

Male	Female	Inf/juv	No. groups	%
MADAGASCAR (1973; N=5)				
1	1		3	60
1	2		1	20
1	2	1	1	20
MADAGASCAR (1974; N=3)				
1	1		1	33
1	1	1	1	33
2	1	1	1	33
ANJOUAN (1974; N=26)				
1	1		6	31
1	2		2	8
2	1		2	8
1	2	1	5	19
2	1	1	4	15
1	1	1	4	15
2	2	1	1	4
	2		1	4
MOHELI (1974; N=22)				
1	1		6	27
1	2	1	2	9
1	1	1	3	14
2	2	1	4	18
2	2		4	18
3	2	1	2	9
4	2	1	1	5
	4		1	5

Table I also contains the results of censusing of *L. mongoz* on Anjouan (26 groups) and Mohéli (22 groups) in the period from mid-November 1974 to early

January 1975. Since no long-term observations were undertaken on any one
group, it was not always possible to discriminate between the adult-sized
although sexually immature 14-16 month-old subadults and the mature adults
(hence the categories are not differentiated in Table I); neither was it
possible to sex the infant/juveniles observed (around four weeks old in Mohéli
at the beginning of the study, and eleven weeks on Anjouan at the end), which
all still displayed the pelage colouration that persists unchanged in the
adult female.

On the reasonable assumption, confirmed in many cases by direct observation,
that the "third" adult-sized individual in any group was in fact immature, it
is clear that the groups of *L. mongoz* observed in Anjouan are best interpreted
as "family" units based on pair-bonding. Of the two cases which appear not
to fit this interpretation, one, a group consisting of an adult and subadult
female, is reliably reported to have lost its adult male through human interference shortly before censusing; the other, (two adult-sized males, two
adult-sized females and a juvenile) is most plausibly viewed as a result of
twinning in 1973. This would represent one twin birth in the thirty births
giving rise to the immature individuals censused (3.3%); Hill (1973) quotes
a figure of 8.2% twin births to *L. mongoz* in captivity. Interestingly, the
mean group size of 3.1 individuals is identical to that reported by Pollock
(1975) for *Indri indri*, which also forms pair bonds.

While observations in both Anjouan and Madagascar thus agree perfectly,
those from Mohéli present a rather different picture. Table I shows that
under half (45%) of the groups censused appear to have been normally-consti-
tuted "family" units. The compositions of six of the remainder might be explained by twin births in 1973, but only at the expense of invoking an abnor-
mally high twinning rate. The remaining six groups are clearly not in conformity with a pair-bonded group structure; one, for instance, contained six
adult-sized individuals (four males, two females), and one juvenile. The
possible significance of this variation in group structure will be discussed
below.

ACTIVITY RHYTHM

Malagasy *L. mongoz* was observed to be active exclusively at night. Onset
and cessation of activity corresponded to an extremely narrow range of ambient
light intensity. On Mohéli a similar pattern was observed, although the com-
mencement of activity was less tightly controlled by light-level; on two
occasions groups were observed to be active during the half-hour preceding
the official time of sunset, at light-levels of 115 and 6,300 Lux. All *L.
mongoz* observed on Mohéli between sunset and approximately 22.30 h were active;
again, this accurately reflects the pattern observed in Madagascar. The observation of nocturnality in Mohéli *L. mongoz* eliminates the possibility that
the nocturnality of *L. mongoz* at Ampijoroa was related to the night-flowering
of the kapok tree, its principal dietary resource at the time of the study.

L. mongoz groups in the coastal lowlands of Anjouan duplicated the activity
rhythm observed in the Mohéli population; in the central highlands of the
island, however, activity was observed throughout the hours of daylight. Of
the 15 groups located in the highlands during the daytime, only two were resting; not more than two others may have been moving as a result of the observer's approach. All the rest were normally active, and, indeed, intergroup
encounters were witnessed on two occasions: once around midday, and once
during the late afternoon.

DISCUSSION

Variation of this magnitude in two such basic aspects of the natural history of *L. mongoz* had not been expected. But at least in the case of activity rhythm there appears to be a clear-cut environmental correlation. Ampijoroa, Mohéli and the Anjouan lowlands possess a distinctly drier, brighter, warmer and more seasonal climate than do the highlands of Anjouan, which are cool, humid, semi-permanently cloud-covered, and support a montane evergreen rain forest. The fact that nocturnality is characteristic of a warm seasonal environment, whether in Madagascar, Mohéli or Anjouan, whereas diurnal activity is restricted to the only cool and consistently humid area surveyed, suggests very strongly that activity pattern is influenced by climatic factors. Possibly activity is metabolically uneconomic in the relatively severe conditions prevailing at night in the highlands of Anjouan.

But if there is a clear correspondence between climate and activity rhythm, whatever the actual basis of this correspondence may be, the environmental correlate of variation in group structure is considerably less evident. In Madagascar and in Anjouan, regardless of environment, pair-bonded groups were uniquely observed, whereas in Mohéli a mixture of social groupings, somewhat similar to the variety reported for *Propithecus verreauxi* by Richard (1974), occurred. It is possible that the latter reflected a seasonal shift in group composition among Mohéli *L. mongoz*. Group composition is likely to reflect, in addition to other factors, foraging strategy; optimal foraging strategy may change with seasonal shifts in the distribution and abundance of resources in the environment. Surveying was undertaken in Mohéli during the transitional period between the well-marked wet and dry seasons, which witness far-reaching phenological changes.

Observational evidence in support of this possibility is rather scanty, but on one occasion two groups were found resting together during the day; it was not realized, indeed, that separate groups were represented until they had moved off separately for the night's activity. Although the home ranges of *L. mongoz* groups are known in Madagascar to overlap extensively, encounters by neighbouring groups both there and in Anjouan involve much agitation and disturbance. The resting of totally separate groups in close proximity would be highly unexpected. It may therefore be most plausible to interpret the non-incident just described as representing a stage in the fission of a large group or in the coalescence of smaller ones. But the direction of the shift, if any, is unknown, and the hypothesis of seasonal change must, for the moment, remain no more than a possibility. Such factors as subtle habitat differences or population density might equally be involved in producing the mixture of group compositions seen in Mohéli.

In the context of this symposium it should be remarked, however, that it is precisely this inability so often to relate behavioural or morphological differences to environmental utilization or preference which should give us pause before we attempt generalizations about behavioural evolution, or before we apply gross adaptive models to the origin of taxa or to the actual fossil record. It is superfluous to say that behaviour is the critical factor in the evolutionary success or failure of organisms; that it is the crucial link between morphology and environment. But the conjunction of its enormous significance with its extreme lability should make us exercise the greatest care in propounding models of behavioural evolution.

ACKNOWLEDGEMENTS

   Fieldwork in Madagascar was financed by the Frederick G. Voss Anthropology
and Archaeology Fund of the Department of Anthropology, the American Museum of
Natural History, and in the Comoros by the National Geographic Society of
Washington, D.C. The studies could not have been carried out without the
cooperation and assistance of the Malagasy and Comorian authorities.

REFERENCES

Martin, R.D. (1972). *Phil. Trans. R. Soc. B.,* 264, 295-352.
Petter, J-J. (1962). *Mém. Mus. natn. Hist. nat. Paris, Série A.,* 27, 1-146.
Richard, A. (1974). *Folia primatol.,* 22, 178-207.
Sussman, R.W. (1974). *In* "Prosimian Biology", (R.D. Martin, G.A. Doyle and
   A.C. Walker, eds), pp. 75-108, Duckworth, London.
Sussman, R.W. and Tattersall, I. (in press). Cycles of activity, group com-
   position and diet of *Lemur mongoz mongoz* Linnaeus 1766 in Madagascar. *Folia
   primatol.,* 26, 270-283.
Tattersall, I. (1976a).   Note sur la distribution et sur la situation
   actuelle des lémuriens des comores. *Mammalia,* 40, 519-521.
Tattersall, I. (1976b). *Mammalia,* 40, 519-521.
Tattersall, I. (1976b). *Anthrop. Pap. Am. Mus. nat. Hist.,* 53, 367-380.
Tattersall, I. and Sussman, R.W. (1975). *Anthrop. Pap. Am. Mus. nat. Hist.,*
   52, 195-216.

FUNCTIONAL ANATOMY OF THE HIP AND THIGH OF THE LORISIDAE: CORRELATIONS WITH BEHAVIOUR AND ECOLOGY

J.E. McARDLE

Department of Anatomy, University of Chicago, USA.

INTRODUCTION

Investigations of primate locomotor behaviour and morphology (Ashton and Oxnard, 1963; Napier and Napier, 1967; Napier and Walker, 1967; Zuckerman, et al., 1973; Stern and Oxnard, 1973; Walker, 1974; and others) usually segregate the four genera of lorisines as a unique group with a distinctive locomotor behaviour. In these studies the lorisines are generally characterized as slow, deliberate, acrobatic, arboreal quadrupeds that neither leap nor run and which utilize quadrupedal suspensory postures as a significant part of their locomotor repertoires. Although the lorisines are capable of relatively rapid movement and only occasionally move in suspensory postures when feeding, grooming or moving between supports (Walker, 1969; pers. obs.), the basic description of lorisine locomotor behaviour is accurate. Previous studies of comparative morphology of prosimians (body proportions, myology or osteology) confirm the uniqueness of the lorisines as an overall group, but they have not examined the morphology of the individual species.

This apparent uniformity of the behavioural and morphological investigations has led to the practice of lumping the species of lorisine into a single group for studies of prosimian locomotion or comparative anatomy. This is in part a reflection of the small sample sizes that characterize many studies of prosimian morphology at the species level, but has also resulted from the tendency to utilize a 'typical' prosimian as if it were representative of all species within its particular taxon. Such lumping of species is useful for gross comparisons of major groups of prosimians already known to have different overall locomotor patterns. However, the finer, species-specific morphological and behavioural distinctions which may exist would be masked by lumping the species together. Such finer distinctions may more accurately reflect the incremental evolution of locomotor adaptations and behaviour in the lorisines specifically and the prosimians in general.

Recent investigations of the ecology and behaviour of wild populations of the lorisines indicate that there is considerable species-specific variability in locomotor behaviour, substrate preferences, dietary specializations, and various other ecological characteristics. *Perodicticus* inhabit the canopy in both primary and secondary forests, utilize substrates of various types and sizes including relatively large diameters, and are predominantly frugivorous (Jewell anq Oates, 1969; Charles-Dominique, 1971). *Arctocebus* are found in the understorey in both primary and secondary forests, prefer substrates of small diameter, and are mostly insectivorous (Charles-Dominique, 1971). *Loris* inhabit the understorey and terminal branches of the canopy in both primary

and secondary forests, move on small diameter supports but can climb larger
diameter substrates, and are almost entirely insectivorous (Petter and Hladik,
1970). *Nycticebus* have not been specifically studied in the wild, but some
observations indicate that they are most common in the canopy, utilize supports of larger diameter and are presumably frugivorous (D'Souza, 1974, pers.
comm.). Thus the lorisines consist of two basic ecological types. *Nycticebus*
and *Perodicticus* are found in the canopy associated with a wide variety of
substrates and a predominantly frugivorous diet. *Arctocebus* and *Loris* inhabit
the fine-branch niche in the understorey and the canopy associated with small
diameter substrates and an insectivorous diet.

MATERIALS

In order to determine the range of morphological diversity in the lorisine
hip and knee joints and its relationship to species-specific locomotor and
foraging behaviour, a detailed investigation of the comparative and functional
anatomy of the hip and thigh of each species was undertaken with additional
observations on other portions of the body. The data for this study were
derived from measurements of body proportions using both skeletons and cada-
vers (N = 21), from qualitative notes on and measurements of dissections (N =
24) and from comprehensive measurements of selected elements of the post-
cranial skeletons (N = 85). As a comparative sample, the same information
was collected for a much larger data set that represented all species of
prosimians, tarsiers and four species of tree shrews.

RESULTS OF INVESTIGATIONS

The initial visual impression of an animal is usually the relative proportions of the different body segments. In this regard lorisines are divisible
into two basic types:
1. Robust, large-bodied with relatively thick limbs (*Nycticebus* and *Perodicticus*);
2. Gracile, cylindrical, small-bodied with relatively long, thin limbs
(*Arctocebus* and *Loris*).
Homogeneity of lorisine morphology is indicated by examination of the brachial,
crural and intermembral indices. These three indices are essentially the
same for all four genera of lorisine, and the intermembral index separates
the lorisines from all other species of prosimian. However, examination of the
lengths of the entire limbs and individual limb segments in relation to body
size (i.e., trunk length) presents a distinct dichotomy. The limbs are relatively longer in the small-bodied genera.
Since all four genera of lorisine have the same mean number of presacral
vertebrae (Schultz, 1961), the difference in relative limb length probably
does not reflect a relative decrease in trunk length. The relative increase
in limb length is also indicated by examination of the relative proportions
of the femur. A major flexor of the thigh, the ilio-psoas muscle, inserts on
the lesser trochanter. Comparison of the position of the lesser trochanter,
relative to the maximum length of the femur, supports a division into two
groups with *Arctocebus* intermediate. This pattern may indicate either elongation of the femur or a relative proximal shift of the lesser trochanter and
a concomitant decrease in the moment arm of the ilio-psoas muscle in the small-
bodied genera. Comparison of the position of the lesser trochanter with a
dimension related to body size (the anterior-posterior diameter of the head of

the femur) indicates that there is no significant difference between the two groups. This implies that the small-bodied lorisines have relatively longer femora than the large-bodied species and that this elongation occurs in the diaphysis distal to the position of the lesser trochanter.

Measurements of the most distal point of insertion of muscle fibres for the principal extensors of the thigh, the adductor magnus and femorococcygeus muscles, and flexors of the knee, the semitendinosus and crural portion of the flexor cruris lateralis muscles, in relation to the maximum length of the femur and tibia respectively also show a division into two distinct groups, the same as that seen for the relative position of the lesser trochanter. The small-bodied species have relatively more proximally located insertions (see Figure 1). There are no significant intra-group differences in these data. Comparison of the small-bodied lorisines with the same measurements of muscle attachments from all other species of prosimian, tarsiers and some tree shrews indicates that the reduced muscle insertions onto the femur in the smaller lorises are more proximal than in any of the other species. The proximal reduction of the insertions onto the tibia and crural fascia is comparable to that seen in most of the leaping prosimians and tarsiers.

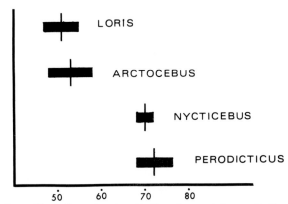

Figure 1. Relative distal extent of the insertion of the M. adductor magnus. Vertical line represents the position of the mean, and the horizontal bar equals one standard deviation unit.

Associated with the reduced distal extent of attachment for the fleshy, fascial and tendinous insertions of the major muscles inserting onto the femur and tibia is a decrease in muscle mass due to both a relative decrease in the area of insertion and increase in the length of muscle-free tendons. For example, the relative length of the tendon of insertion of the semimembranosus muscle expressed as a percentage of the total length of the muscle shows an increase in tendon length for the small-bodied lorisines due to a proximal shift in the muscle belly (see Figure 2). Correlated with this shift is the development of a modified 'pinnate' arrangement of the hamstring muscles in which the semimembranosus muscle receives diagonal fibres from both the flexor cruris lateralis and semitendinosus muscles in addition to its usual origin from the ischium.

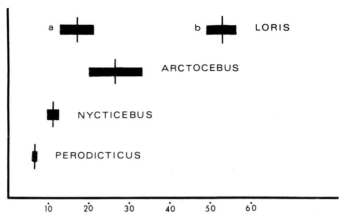

Figure 2. *Relative length of muscle-free tendon of insertion for the*
M. semimembranosus. Loris 'a' represents the completely
muscle-free tendon. Loris 'b' represents the length of tendon
minus the pinnate fibres from the other hamstring muscles.
These fibres are not present in the other genera.

SUMMARY AND CONCLUSIONS

The data for the lorisines indicate that they comprise two distinct morphotypes. This dichotomy is evidenced by examination of the body proportions,
muscle insertions, muscle tendons, muscle architecture and various dimensions
of the post-cranial skeletons of the respective species. The first type consists of large-bodied species with relatively short and stout limbs, extensive
and well-developed muscle insertions and relatively short tendons for the hamstring muscles. This type is represented in Africa by *Perodicticus* and in
Asia by *Nycticebus*. The second type consists of small-bodied species with
relatively thin and elongated limbs, reduced areas of muscle insertion, long
tendons for the hamstring muscles and the fibres of the hamstrings arranged
in a 'pinnate' manner. This type is represented in Africa by *Arctocebus* and
in Asia by *Loris*.

This same dichotomy is seen in the behavioural and ecological characteristics
of the four genera of lorisines. *Nycticebus* and *Perodicticus* inhabit the
canopy and frequently move on larger diameter substrates. *Arctocebus* and *Loris*
are more carnivorous and inhabit the fine-branch niche in which long, thin limbs
would allow a small animal to extend its reach in order to contact a greater
number of potential supports and to capture prey at greater distances from its
body while decreasing the amount of disturbance of the foliage and branches encountered in these movements.

A significant biomechanical problem associated with relative limb elongation
is an increase in the moment-of-inertia of the hindlimb around the hip and knee
joints and a concomitant increase in the muscle force needed to hold the limbs
in extended positions (a common posture for lorises). One response to this
problem is illustrated by the smaller lorises. Their muscles show a reduction
in mass by increasing the relative amount of muscle-free tendon and by decreasing the distal extent of the area of muscle insertions. This effectively
reduces the moment-of-inertia, but introduces the problems of a decrease in
the muscle force and the degree of excursion these muscles can produce. The
presence of a modified 'pinnate' arrangement for the muscle fibres of the hamstrings may be associated with the former problem and restriction to the fine-

branch habitat with its multiplicity of supports and abundance of prey may be associated with the latter. In all four genera of lorisines the arrangement of the quadriceps femoris muscles shows a pattern of insertion, tendon development and 'pinnation' similar to that for the hamstrings. This presumably reflects a response to a similar set of problems associated with the proximal segment of the hindlimb.

The problem of lumping species of prosimians into locomotor or higher taxonomic groups is not restricted to the lorises, but they are an excellent example of the extent to which species-specific differences in behaviour and ecology in respect to foraging, locomotor preference, habitat and biomechanics can significantly modify a basic locomotor morphology. Data such as those discussed in this paper have been collected for all species of prosimian, tarsiers and some tree shrews. This broader data set and such recent work as that of Manaster (1975) and Fleagle (1976) on the cercopithecines and colobines indicates that the taxonomic level at which one can distinguish morphological variability associated with specific behavioural or ecological characteristics is far finer than previously indicated.

ACKNOWLEDGEMENTS

This work was supported by the USPHS Anatomical Sciences Training Grant no. 5 - TO1 - GM00094; Hinds Fund, University of Chicago; Society of Sigma Xi Grant in Aid of Research; and National Science Foundation Grant GS 30508 to Dr Charles E Oxnard.

REFERENCES

Ashton, E.H. and Oxnard, C.E. (1963). *Trans. zool. Soc. Lond.*, 29, 553-650.
Charles-Dominique, P. (1971). *Biol. Gabon.*, 7, 121-228.
D'Souza, F. (1974). *In* "Prosimian Biology", (R.D. Martin, G.A. Doyle and A.C. Walker, eds), pp. 167-182, Duckworth, London.
Fleagle, J. (1976). *Amer. J. phys. Anthrop.*, 44, 178.
Jewell, P.A. and Oates, J.F. (1969). *Zool. Afr.*, 4, 231-248.
Manaster, B.J. (1975). "Locomotor Adaptations within the *Cercopithecus, Cercocebus,* and *Presbytis* Genera: a Multivariate Approach." Unpublished Ph.D. dissertation, University of Chicago, Chicago.
Napier, J.R. and Napier, P.H. (1967). "A Handbook of Living Primates" Academic Press, New York and London.
Napier, J.R. and Walker, A.C. (1967). *Folia primatol.*, 6, 204-219.
Petter, J.J. and Hladik, C.M. (1970). *Mammalia*, 34, 394-409.
Schultz, A.H. (1961). *Primatologia*, 4(5), 1-66.
Stern, J.T. Jr. and Oxnard, C.E. (1973). *Primatologia*, 4, 1-93.
Walker, A.C. (1969). *E. Afr. Wildlife J.*, 7, 1-5.
Walker, A.C. (1974). *In* "Primate Locomotion", (F.A. Jenkins Jr., ed.), pp. 349-381, Academic Press, New York.
Zuckerman, S., Ashton, E.H., Flinn, R.M., Oxnard, C.E., and Spence, T.F. (1973). *Symp. zool. Soc. London*, 33, 71-165.

SOLITARY AND GREGARIOUS PROSIMIANS: EVOLUTION OF SOCIAL STRUCTURES IN PRIMATES

P. CHARLES-DOMINIQUE

Laboratoire d'Ecologie Generale, Equipe de Recherche sur les Prosimiens, Museum National d'Histoire Naturelle, Brunoy, France, and Laboratoire de Primatologie, CNRS, Gabon.

Prosimians represent 30% of the existing Primates. Of great theoretical importance this primitive group contains 26 nocturnal species (13 genera), 3 nocturnal-crepuscular species (2 genera) and 9 diurnal species (4 genera). Their dietary specialisations and their social structures are even more varied than those of simian primates. Although the gregarious diurnal Lemurs attracted the initial attention of research workers, a growing body of research over the last ten years on the socio-ecology of nocturnal species now permits interesting comparisons to be made.

THE NOCTURNAL SOLITARY PROSIMIANS

Confusion exists in the literature concerning the term 'social', which is often opposed to the term 'solitary', even though the authors are referring in fact to 'gregarious' vs. 'solitary'. All of the solitary species that have been studied are social, but their social relationships are usually based on communication from a distance, using auditory and particularly olfactory signals (scent-marks) in which the message and its receipt are separated in time (deferred social communication). This general type of social relationship may vary in number and complexity but all share in common the fact that the general activity and, particularly, the movements of different individuals about their habitat are not synchronised. Other acoustic, visual, olfactory and tactile signals are also obviously used in short-distance communication (in male-female and mother-young relationships and in agonistic encounters, etc.).

General type Galago/Microcebus

Field studies of marked animals (following successive caputres and releases, direct observation and radio tracking) provide the basis for the interpretation of social structure in different species: *Galago demidovii, Galago alleni* (Charles-Dominique, 1972, 1977a, 1977b), *Galago senegalensis* (Bearder and Martin, in prep.), *Microcebus murinus* (Martin, 1972a), *Microcebus coquereli* (Pages, in prep.). The following scheme is given for these species:
a) Female-female and male-male territories overlap slightly (Fig. 1). A recent study of *Galago alleni* (Charles-Dominique, 1977b) shows a density of urine marking 8 to 10 times higher in these small overlapping zones compared to the other parts of the territory.

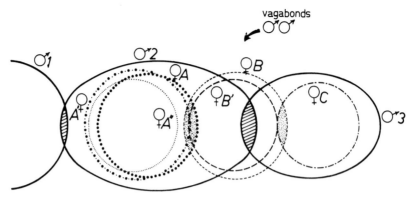

Figure 1. *Galago/Microcebus type of territorial organization: Although distributed in all directions, only territories disposed along a longitudinal axis are represented here. ♀ A, ♀ A' and ♀ A''; ♀ B and ♀ B' represent two matriarchies. ♂ 1 is bachelor; ♂ 2 is polygamous (♀♀ A, A', A'', B, B'); ♂ 3 is monogamous (♀ C). Male-male and female-female overlapping zones are generally independent (vocalizations and deposition of urine marks). Juveniles with their mothers are not represented separately.*

b) Male territories overlap those of females.
c) Male-female relations are permanent throughout the year, based on regular visits which involve direct contact (genital-flairing, allo-grooming), contact from a distance (vocalizations) and indirect contact (urine-marking).
d) Females retain privileged social ties with their female offspring which, when adult, may share the same territory in a matriarchal association.
e) Young males emigrate at puberty and often establish themselves at the periphery of the females' territories, being repulsed by the occupant male (*G. demidovii*, *M. murinus*), before assuming a central position when they are sufficiently strong.

Lorisine type (Perodicticus potto, Arctocebus calabarensis)

This second type of organization is derived from the first type, modified by certain eco-ethological contingencies which result in a type of social organization peculiar to the Lorisinae (Charles-Dominique, 1971, 1974, 1977a). These slow climbers use the strategy of dissimulation to avoid predators: slow progression, absence of sudden movements likely to attract the attention of a predator and complete immobilization at the slightest danger. The disposition of their territories (Fig. 2) is identical to that of *Galago* or *Microcebus* but there are no matriarchal associations, which contributes to a reduction in movement about the territory. Male-female relations are based exclusively on discrete visits by the male which 'controls' females' urine marks without meeting. The Lorisines have lost their loud calls, thus further reducing the possibility of detection by predators, though *Loris tardigradus* is reported to have a loud call (Petter and Hladik, pers. comm.).

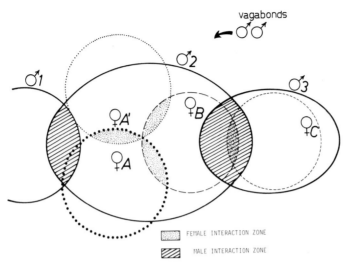

*Figure 2. Lorisine type (Perodicticus potto) of territorial organization:
Although distributed in all directions, only territories
disposed along a longitudinal axis are represented here. The
adult daughter ♀ A' and her mother ♀ A are separated (no matriarchies). ♂2 is polygamous (♀♀A, A', B); ♂3 is monogamous
(♀ C). Male-male and female-female overlapping zones are
generally independent (deposition of urine marking, no
vocalizations). Juveniles with their mothers are not represented separately.*

Lepilemur type (Lepilemur leucopus)

As with the Lorisines the Lepilemur type of social organization is similar
to that of the Galagines modified by certain eco-ethological considerations
(Charles-Dominique and Hladik, 1971; Hladik and Charles-Dominique, 1974).
Sportive Lemurs feed more exclusively on leaves and flowers, in very small
territories (50 to 60 m in diameter), resulting in very small borders. This,
together with the fact that the animals require little time for feeding (about
75 minutes per night) and that territories seldom overlap, allows them to
spend a great deal of time in direct mutual surveillance, exchanging visual
signals (movements of the head, 'heavy' leaping. . .) and engaging in vocal
duets. But for the absence of overlapping zones the disposition of territories
is of the Galaginae, Lorisinae or Cheirogaleinae type although, unlike these
groups, urine is not involved in social communication (Fig. 3).

Phaner furcifer type: pre-gregariousness

In this equally territorial species, the female is dominant to the male.
The male follows the female for a large part of the night, and eats after her
in gum trees which provides the bulk of their food (gums of *Terminalia* sp.
+ some insects). There is a contact vocalization related to movement about
the territory, reminiscent of the vocalizations of gregarious forest animals
(Lemurs, monkeys, some birds. . .). The further the male and female are
apart the louder are their contact vocalizations. Territorial relations, therefore, are based essentially on vocalizations, particularly in some small over-

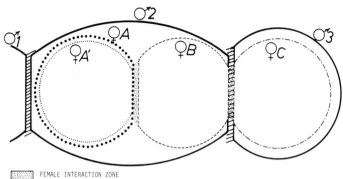

Figure 3.   *Lepilemur type (Lepilemur leucopus) of territorial organization:
Although distributed in all directions, only territories disposed along a longitudinal axis are represented here. ♀ A
and ♀ A' represent a matriarchy; ♂ 1 is bachelor; ♂ 2 is
polygamous (♀♀A, A', B); ♂ 3 is monogamous (♀ C): no overlapping zones; territorial confrontations take place along
the boundaries (vocalizations in duets, visual displays, no
urine markings). Juveniles with their mothers are not represented separately.*

lapping zones. These areas, similar to the overlapping zones of other nocturnal Lemurs (*Galago, Microcebus, Perodicticus*) involve as many as seven individuals which converge during the night and keep close together, engaging in
a 'concert'. After 10-20 minutes of calling they return to the centre of
their territories (Charles-Dominique and Petter, in prep.). In this species
vocal communication is predominant and, as in the case of *Lepilemur*, urine
is not involved in social communication. Male-female relations are complex,
characterized by almost permanent vocal contact and marking of the female
by the throat gland of the male. Nevertheless, territorial organization is
similar to that of other nocturnal Lemurs. This 'pre-gregarious' system of
social organization, based on numerous vocalizations, allows for identification and localization of each individual in the population.

    Although the social structure of the nocturnal Indriid, *Avahi laniger*, has
not yet been properly studied, preliminary observations suggest that *Avahi*
also has a kind of pre-gregarious social structure (Petter, 1962, Pollock,
pers. comm.).

    These three different types of social organization, characterizing the
solitary, nocturnal prosimians, apart from specific differences due to ecological peculiarities, share important characteristics:
1. The female territory is sufficiently large to allow feeding throughout
the year. This territory is signalled to conspecifics by urine-marks and/or
vocalizations; only adult females are repulsed. In some cases, the adult
daughter may more or less completely share the maternal territory (young
adult females are sedentary).
2. The male territory corresponds to the area occupied by his female(s)
and there is no relationship between territory size and dietary requirements.
Polygamy being frequent, males occupy much larger territories than females.
3. Young adult males emigrate at puberty. After their vagabond stage, they
establish themselves over 'vacant' female territories (or perhaps displace a

resident male).

4. After courtship a special social bond links the male to his female(s).
Male stability and territoriality depend, therefore, on the presence of females.

5. The male territory is signalled by urine marks and/or vocalizations. Only
adult males are repulsed though, in some cases, a small 'dominated' male may
be tolerated, as in *Galago demidovii*, and in some species males are more
tolerant than in others, as in *G. crassicaudatus* (Clark, this volume).

6. Territoriality is exhibited throughout the year, even when reproduction
is seasonal.

7. Female-female and male-male territories overlap slightly. These small
overlapping zones are of great importance in territorial relations.

8. Male and female territories overlap; male-female relations are permanent
throughout the year, based on regular visits by the male.

9. During the active period (night), occupants of overlapping territories
(male, females and infants) are solitary moving independently about their
territories. Social communication is based largely on direct auditory communication at a distance (lost in the African Lorisinae) and olfactory communication (urine) in which the 'message' and its 'receipt' are separated in
time (deferred social communication). This primitive mammalian way of communication is lost in *Lepilemur* and *Phaner*.

10. Direct communication and contact, which is rare during the night, is important during the daytime when animals gather together to sleep - female(s)
with infants and sometimes male(s).

11. There is a special call related to male-female relationships. At the beginning of the night the infant is deposited in the vegetation and recovered
later or at dawn (baby parking). In response to the mother's call, or spon-
taneously at dawn, the isolated infant emits a call.

12. The mother carries the infant in her mouth (except in the Lorisinae).

THE NOCTURNAL AND CREPUSCULAR GREGARIOUS LEMURS

Since the observations of Tattersall and Sussman (1975) and Petter et al.
(1976), *Lemur mongoz* and *Hapalemur griseus* may be compared to other Lemurs.
These two species are completely nocturnal in some regions, crepuscular or
partially diurnal in others. In all cases they are gregarious but the troop
is small compared to typical, diurnal Lemurs; there are very seldom two
adults of the same sex in one troop.

Apart from certain typical diurnal Lemur vocalizations and behaviours (contact vocalizations during progression, alarms in chorus, visual signals associated to alarm calls, etc.) *Hapalemur griseus* exhibit typically solitary
prosimian patterns such as 'oestrus call', baby parking and mouth transport
of the new born (Petter and Charles-Dominique, in press).

Nocturnal gregarious Lemurs can be considered as transitional forms between
nocturnal life and diurnal life. In the evolutionary past this transition has
probably passed back and forth through these two ways of life, and it is probably during a previous diurnal stage that gregariousness was established.

Madagascar, only colonized by a small number of zoological groups, presents
many vacant ecological niches. Given such conditions some species are able to
exploit the milieu during the night or day according to circumstances.

THE DIURNAL GREGARIOUS LEMURS

Most of the field data come from the observations of Petter (1962, Petter et al., 1976), Jolly (1966), Richard (1973, 1974), Sussman (1974), Pollock (1975a and b). *Propithecus verreauxi, Indri indri, Lemur catta* and *Lemur fulvus* are the best known species.
1. All diurnal Lemurs are gregarious; a troop may contain several adult males and females.
2. Female-female aggression is quite pronounced. The presence of several adult females in the same troop can be interpreted as a privileged tie persisting between mothers and their adult daughters (Paillette, in prep.). This situation recalls the matriarchies of nocturnal Lemurs.
3. Females dominate males, as in *P. verreauxi* (Richard and Sussman, 1974), or there is no male-female dominance relationship, compared to most simian species where the male is dominant. When a rank order is exhibited, it is usually a male-male and/or a female-female hierarchy, as in *Lemur catta* Jolly, 1966). This situation recalls those of solitary Lemurs where male territorial defence and female territorial defence are independent from one another.
4. During the mating period, male-female relationships may be established outside the troop, as in *Propithecus verreauxi* (Richard, 1974) or *Lemur catta* (Jolly, 1966). These occasional relationships are reminiscent of those of solitary prosimians where the male visits several matriarchies.

COMPARISONS BETWEEN DIURNAL AND NOCTURNAL PROSIMIANS

Nocturnal and diurnal prosimians form a relatively homogeneous phylum; anatomical comparisons are interesting for evolutionary considerations.
1. The eye of diurnal prosimians is of the nocturnal type: *Tapetum lucidum*, no *fovea*, little colour vision (Wolin and Massopust, 1970; Pariente, 1970, 1976).
2. Nocturnal and diurnal species present a similar external ear which folds and is completely orientable, particularly evident in nocturnal forms. In the Galaginae this mobility allows detection and localization of insects through an opaque screen (Charles-Dominique, 1971, 1977a). The audiogram is roughly similar in nocturnal and diurnal species, with a large spectrum in the ultrasonic range (Niaussat and Molin, these proceedings - Behaviour volume).
3. The hand is also similar in both nocturnal and diurnal prosimians with long, slender fingers, sensitive primarily on the palmar surface of the flattened terminal phalanges; dermatoglyphes arranged in parallel longitudinal ridges; 'reversed' articulation between the second and third phalanges permitting contact of only the terminal phalanges in opposition to the palm; poor neuro-muscular coordination of the fingers which cannot move independently (Bishop, 1964). In nocturnal species the capture of prey is accomplished by a sudden stereotyped grasp permitting capture of insects in flight. It is of interest to note that *Lemur fulvus, Lemur mongoz* and *Hapalemur griseus*, which never eat insects in the wild, can be trained to eat insects, to which they quickly become partial, in which case capture is accomplished by the same stereotyped movements observed in the nocturnal insectivorous prosimians. On the other hand nocturnal and diurnal Lemurs cannot manipulate or forage with their hands as do many apes and monkeys. The specialization of the Lemurs' hands seems incompatible with the capacity of complex movements of the finger.

These different arguments lend support to the consideration that diurnal Lemurs are derived from highly adapted nocturnal forms. One may describe the primitive Strepsirhine stock as follows: *Small arboreal nocturnal species, feeding upon insects, fruits and gums, and detecting by audition prey which are captured in flight.*[1] (Charles-Dominique and Martin, 1970; Martin, 1972b; Charles-Dominique, 1977a).

The nocturnal Strepsirhini hunting techniques have evolved correlatively with anatomical and physiological adaptation of hands and ears. Any diurnal species of Lemur can hunt animal prey by foraging in foliage and bark as do many species of ape and monkey. In fact, hunting techniques cannot be the same during the night and during the day when birds compete with mammals. The insectivorous monkeys find their prey largely by foraging and rummaging in the vegetation, with the aid of the hands in association with their excellent vision and greater intelligence. (For development of these arguments see Charles-Dominique, 1975). All diurnal Lemurs are frugivorous-folivorous and the Malagasy evolutionary radiation left unoccupied ecological niches equivalent to those elsewhere occupied by *Cercopithecus sp.* - *Miopithecus, Macaca, Cebus, Saimiri,* the Callithricidae, etc. In comparing diurnal Lemurs with monkeys it is of great importance to note that in both New World and Old World ecosystems there are insectivorous-frugivorous species, insectivorous-frugivorous-folivorous species and frugivorous-folivorous species living sympatrically (e.g. *Saguinus, Saimiri, Cebus, Ateles, Alouatta* in Panama; *Macaca, Presbytis entellus, Presbytis senex* in Sri Lanka; *Myopithecus, Cercopithecus sp., Cercocebus, Pan, Gorilla, Colobus* in Equatorial West Africa (Hladik, 1975; Gautier-Hion, these proceedings - Behaviour volume). Protein is rare in the pulpae of fruits and primates find protein in animal prey or in the green parts of plants, particularly buds and young leaves (Hladik and Hladik, 1969a). A highly specialized digestive gut, permitting degradation of cellulose- is needed in folivorous species for the extraction of protein from leaves; the extraction of animal protein in insectivorous forms is considered as primitive.

Apart from size, diurnal and nocturnal Lemurs are slightly different in terms of their anatomy, diurnal forms deriving from nocturnal forms. Comparison of social structures also suggests that diurnal forms derived from nocturnal forms: a 'cohesive tie' may be found in some nocturnal species (e.g. the pre-gregariousness of *Phaner*; the juvenile infant following his mother during a part of the night in some Lorisids - Charles-Dominique, 1972, 1976; Bearder, 1974). This cohesive tie seems to be highly developed in diurnal forms which find many advantages in gregariousness (visual location in diurnal conditions and a collective system of protection against predators). These diurnal species are generally larger than their nocturnal relatives; growth and maturation of infants is generally longer, resulting in an augmentation of the troop with juveniles of several generations.

Nevertheless, social structures of diurnal Lemurs cannot be only considered as the result of the cohesion of individuals whose territories overlap. A sexual dichromism is exhibited by several diurnal species as well as a number of visual signals (tail-waving, movements of the head, eyes screwing etc.) which do not exist in nocturnal species. In addition, numerous

---

[1] Martin (1972b) interprets the evolution of the procumbent 'tooth comb' of the Strepsirhini as an adaptation for the collection of gum. The grooming function of the tooth comb is probably a secondary adaptation.

cutaneous glands (perineal, brachial, antebrachial, throat) are involved in complex social interactions during which social partners mark each other and superimpose their markings on the same branches. Equivalent cutaneous glands don't exist in typical solitary species like Galaginae, Lorisinae, Chiero-galeidae or Lepilemuridae, and only *Phaner furcifer* and *Avahi laniger* males, which are pre-gregarious, have a cutaneous gland on their throat.

The bare patch on the sternum of the *Galago senegalensis* and *G. alleni* male is a non-differentiated area (Yasuda et al., 1961) rubbed on the support marked by urine (Doyle, 1974). However, *G. crassicaudatus* have a fully developed and functional sternal gland (Clark, this volume). The cutaneous scrotal gland is present- more or less developed, in all Lemur species.

GREGARIOUSNESS IN SIMIAN PRIMATES: COMPARISON WITH DIURNAL LEMURS

A large number of studies have been published on social structures of monkeys. In comparing monkeys and Lemurs, inter-individual relations and their modalities seem more important than the number of individuals involved in these interactions. The size and composition of troops are more often the result of ecological adaptations to food distribution; for example, *Presbytis entellus* and *Presbytis* (Hladik, 1975), *Lemur catta* and *Lemur fulvus* (Sussman, 1974), *Papio ursinus* (Hall, 1963). This aspect is not considered here.
1. Social communication among monkeys involves numerous visual signals in which colouration is important: faces and breeches of guenons, sexual skin of females, bottom of male Mandrills, etc. . .

In social communication, diurnal Lemurs use visual signals generally based on dark clear contrasts in the fur since their colour vision is poor (Pariente, 1976).
2. Social grooming in monkeys is based on fine manipulation of the fur (cor-related with foraging and search of prey).

In Lemurs, as in numerous mammal groups, allogrooming is mainly by licking (and use of the tooth combs), hands being used simply to grasp the partner's fur.
3. In monkeys, intra-group communication involves graded vocal signals in association with numerous visual signals, postures and facial mimics.

In Lemurs, the anatomy permits little mobility of the face which Andrew (1964) considers is correlated with retention of the snout. Communication is based largely on olfactory signals by means of the cutaneous gland and seems to play an important role in intra-group relations (group odour?). New World monkeys do urine-mark but this behaviour may be interpreted as the persistance of an archaic ancestral primate system of communication. Vocal communication of diurnal Lemurs is roughly equivalent to that of monkeys (Petter and Charles-Dominique, in press).
4. In monkeys, the males generally defend the group against predators and rival troops (cf. loud male call associated with these behaviours). Females engage little or not at all in territorial defence. In short, one observes a division of labour in the group. The females, more vulnerable, particularly when they carry infants, are less exposed to danger than the males. Injuries or death of some males are not of great importance. The function of the male is often correlated with sexual dimorphism, generally accompanied by great aggression. Intra-group relations are highly affected by this phenomenon and in most cases the males dominate the females (Eisenberg et al., 1972).

In contrast, sexual dimorphism is either absent or scarcely noticeable in the diurnal Lemurs and the whole troop participates in territorial advertisement and defence. An exception is *Lemur fulvus* in which males are more active than females during territorial interaction (Petter et al., 1976). The male-female hierarchy is not pronounced and when it exists females dominate males (Jolly, 1966; Richard and Sussman, 1974).

5. In monkeys, numerous complex behaviours exist; for instance, cooperation of males during attack, association of several males in competitions for dominance, leadership of one individual initiator in troop movements. Such behaviours have never been observed in Lemurs.

The small South American Callitrichidae have followed a different path of ecological and behavioural adaptation and, for this reason, only the Cebidae and Cercopithecidae are considered in these comparisons.

CONCLUSION AND DISCUSSION

Classically, diurnal Lemurs are considered as having reached, following ecological convergence, social structures that are roughly equivalent to those of simian Primates. However, this apparent similarity is based only on the observation of gregariousness; an ecological solution conferring many advantages in diurnal conditions (for example, defence against predators). Gregariousness is common to diurnal Lemurs and monkeys, but also to various zoological groups such as Equidae, Bovidae, Suidae, some birds, etc. With the exception of this aspect of their way of life, diurnal Lemurs and monkeys differ markedly in terms of their social relations (see Eisenberg et al., 1972).

Lemurs and monkeys represent two phyla separated probably since the Palaeocene/Eocene, each one primarily specialized, in terms of its physiology, to particular ecological conditions - nocturnal for Lemurs and diurnal for monkeys - with consequent subordination of social relations and systems of communication to these sensory specializations.

Diurnal gregarious Lemurs derive from a Strepsirhine stock highly adapted to nocturnal life conditions. Their sensory physiology, being of a nocturnal type, imposes limitations on the development of social organizations and systems of social communication compared to monkeys in similar conditions. To communicate within the group, diurnal Lemurs have developed other signals, compatible with their anatomy and sensory physiology - cutaneous glands and some visual and acoustic signals derived from nocturnal ancestors.

The study of social structures of several species of nocturnal prosimians shows a great homogeneity: individual territories; slight overlapping of male-male and female-female territories; overlapping of female territories and male territories; social communication based to a great extent on urine-marking (indirect communication) and on vocal communication (direct communication at a distance). These different characteristics are the same, in varying degrees, to those exhibited by other mammalian orders, particularly the more primitive families; for example, Viverridae (Charles-Dominique, in pres.), Tragulidae (Dubost, 1975), Manidae (Pages, 1975), Tenrecidae (Eisenberg and Gould, 1970). These social structures seem to be archaic features of primitive mammals, maintained quite unchanged in species living in ecological niches similar to those occupied by their ancestors.

Studies of the sociobiology of the larger diurnal species has progressed considerably in the last decade but a great deal remains to be done on the smaller nocturnal forms. Such studies may reveal elements for a more accurate interpretation of social evolution, from an ancestral stage largely hypothetical today.

REFERENCES

Andrew, R.J. (1964). *In* "Evolutionary and Genetic Biology of Primates", Vol.II (J. Buettner-Janusch, ed.), pp. 227-309, Academic Press, New York.

Bearder, S.K. (1974). Aspects of the behaviour and ecology of the thick-tailed bushbaby *Galago crassicaudatus*. Unpublished Ph.D. thesis, University of the Witwatersrand.

Bishop, A. (1964). *In* "Evolutionary and Genetic Biology of Primates", Vol.II, (J. Buettner-Janusch, ed.), pp. 133-225, Academic Press, New York.

Charles-Dominique, P. (1971). *Eco-éthologie des prosimiens du Gabon, Biol. Gabon, 1*, 121-228.

Charles-Dominique, P. (1972). *Z. Tierpsychol.*, Beiheft 9, 7-41.

Charles-Dominique, P. (1974). *Mammalia, 38*, 355-379.

Charles-Dominique, P. (1975). *In* "Phylogeny of the Primates", (W.P. Luckett and F.S. Szalay, eds), pp. 69-88, Plenum, New York.

Charles-Dominique, P. (1977a). "Behaviour and Ecology of Nocturnal Primates", Duckworth, London.

Charles-Dominique, P. (1977b).   *Z. Tierpsychol.*

Charles-Dominique, P., and Martin, R.D. (1970). *Nature, Lond., 227*, 5255; 257-260.

Charles-Dominique, P. and Hladik, C.M. (1971). *Terre et Vie, 1*, 3-66.

Doyle, G.A. (1974). *In* "Prosimian Biology", (R.D. Martin, G.A. Doyle and A.C. Walker, eds), pp. 213-231, Duckworth, London.

Dubost, G. (1975). *Z. Tierpsychol., 36*, 403-501.

Eisenberg, J.F. and Gould, E. (1970). *Smithsonian Contrib. to Zool., 27*, 1-137.

Eisenberg, J.F., Muckenhirn, N.A. and Rudran, R. (1972). *Science, 176*, 863-874.

Hall, K.R.L. (1963). *Symp. zool. Soc. Lond., 10*, 1-28.

Hladik, C.M. (1975). *In* "Socioecology and Psychology of Primates", (R.H. Tuttle, ed.), pp. 3-35, Mouton, The Hague.

Hladik, C.M. and Charles-Dominique, P. (1974). *In* "Prosimian Biology", (R.D. Martin, G.A. Doyle and A.C. Walker, eds), pp. 23-37, Duckworth, London.

Hladik, A. and Hladik, C.M. (1969). *Terre et Vie, 1*, 25-117.

Hladik, C.M. and Hladik, A. (1972). *Terre et Vie, 2*, 149-215.

Jolly, A. (1966). "Lemur behavior: a Madagascar field study", University of Chicago Press, Chicago.

Martin, R.D. (1972a). *Z. Tierpsychol.*, Beiheft 9, 43-89.

Martin, R.D. (1972b). *Phil. Trans. R. Soc., B., 264*, 295-352.

Pariente, G. (1970). *C. r. hebd. Seanc. Acad. Sci., Paris, 2701*, 1404-1407.

Pariente, G. (1976). Etude éco-physiologique de la vision chez les Prosimiens malgaches. Thèse Doctorat d'Etat, Montpellier (France), A.O. 12352 C.N.R.S

Pagès, E. (1975). *Mammalia, 39*, 4; 613-641.

Petter, J.J. (1962). *Mem. Mus. natn. Hist. nat., Paris, Série A, 27*, 146.

Petter, J.J., Albignac, R. and Rumpler, Y. (1976). "Faune de Madagascar: Lémuriens", Centre National de la Recherche Scientifique, Paris.

Petter, J.J. and Charles-Dominique, P. (in press). *In* "The Study of Prosimian Behavior", (G.A. Doyle and R.D. Martin, eds), Academic Press, New York..

Pollock, J.I. (1975a). The social behaviour and ecology of *Indri indri*. Unpublished Ph.D. thesis, London University.

Pollock, J.I. (1975b). *In* "Lemur Biology", (I. Tattersall and R.W. Sussman, eds), pp. 287-311, Plenum Press, New York.

Richard, A.F. (1973). The social organization and ecology of *Propithecus verreauxi*. Unpublished Ph.D. thesis, London University.

Richard, A.F. (1974). *In* "Prosimian Biology", (R.D. Martin, G.A. Doyle and A.C. Walker, eds), pp. 49-74, Duckworth, London.

Richard, A. and Sussman, R.W. (1974). *In* "Primate Aggression, Territoriality and Xenophobia", (R.L. Holloway, ed.), pp. 49-76, Academic Press, New York.

Sussman, R.W. (1974). *In* "Prosimian Biology", (R.D. Martin, G.A. Doyle and A.C. Walker, eds), pp. 75-108), Duckworth, London.

Tattersall, I. and Sussman, R.W. (1975). *Anthrop. Pap. Am. Mus. nat. Hist.*, 52, 193-216.

Wolin, L.R. and Massopust, L.C. (1970). *In* "Advances in Primatology", Vol. I, (C.R. Noback and W. Montagna, eds), pp. 1-27, Academic Press, New York.

Yasuda, K., Aoiki, J. and Montagna, W. (1961). *Am. J. phys. Anthrop.*, 19, 23-33.

DISCUSSION OF BEHAVIOURAL FACTORS IN PROSIMIAN EVOLUTION

G.A. DOYLE

*Primate Behaviour Research Group, University of the Witwatersrand,
Johannesburg, South Africa.*

Since the time of Darwin and the early ethologists it has been argued that behaviour must evolve in the same way as other biological characteristics like morphology and anatomy. It should, therefore, be possible, on the basis of behavioural similarities, to determine evolutionary sequences of behaviour and thus to construct phylogenetic trees. This has been done with numerous species listed by Hinde (1970, pp. 664-674), Klopfer (1973, p. 114), Moynihan (1973, p. 22) and others. Most of these studies, particularly the earlier ones, have been on insects, fish and birds and have been largely concerned with displays. Some more recent studies have been concerned with the evolution of facial expressions and vocalizations in simian primates (Van Hooff, 1962, 1967; Andrew, 1963a, b) and Hinde (1970) notes that the principles involved in these studies are the same as those for insects and lower vertebrates.

With the exception of the Order Primates mammals do not feature prominently in this list largely because, unlike many orders of insect, fish and bird, there are no mammalian orders in which a significant proportion of the extant members appears to represent a graded series approximating an evolutionary sequence. In the Order Primates selected extant members do appear to constitute such a sequence from early ancestral prosimian primates to man, as Huxley (1863) first pointed out. This should permit us to use the living primates as a series of models for determining the course of behavioural evolution. Caution must be exercised here since, as Doyle and Martin (1974, p. 3) point out: "The living primates are, of course, the living end-products of an evolutionary radiation in which all species have, without exception, undergone some modification from the ancestral primate condition. This must be borne in mind when any attempt is made to reconstruct an actual evolutionary sequence of primate behaviour."

If behaviour is to be regarded as an index of taxonomic affinity, no more and no less important than other criteria, three problems arise. Firstly, as many research workers, and more recently Klopfer (1973), have pointed out, similarities between organisms may be due to either homology or analogy, and very often the distinction between the two depends on our knowledge of phyletic relationships already established on the basis of other factors. Homologous similarities are very difficult to prove particularly when the structure or function in question is clearly adaptive.

Secondly, and perhaps more importantly, there is the assumption made quite explicit by Ewer (1968), for instance, that species-typical behaviours are endogenous or genetically encoded or, at least, it is assumed that they are highly circumscribed by the structures subserving them which are, in turn, genetically encoded. Such assumptions do not accord well with the wealth of evidence on behavioural plasticity on which both Sussman's and Tattersall's papers provide further evidence. The effectiveness of genetic constraints, and no one will argue against their existence, will depend on the degree of order-liness and homogeneity of the environment in which the organism is called upon to function. The more invariant, orderly and homogeneous the environment the more similar and invariant the behaviour. Conversely, the more variable the habitat the more variable will be the behaviour. The behaving organism acts like a continual feed-back system sensitive to changes in its environment and continuously affected and modified by its own output.

However, we have good grounds for believing that this does not apply to the same extent to all behaviours. Some behaviours are extremely plastic while others are resistant to modification by the environment.

Thirdly, behaviour is a changing process and not an invariant structure: it appears only under certain conditions, at intervals of time and only a few at a time, as Kummer (1970) points out. Morphological features, which figure prominently in taxonomic assessment, are not only present all the time but many survive death by fossilizing; behaviour does not.

The above constitutes good grounds as to why behaviour has, thus far, played only a minor role in taxonomic classification. Nevertheless, data are emerging which are helping substantially to clarify our thinking about the evolution of behaviour in general. Despite the fact that none of our contributors, as far as I know, had this problem specifically in mind when they did the research they reported on today, they have, nevertheless, presented their papers for discus-sion within this context and, in doing so, have contributed greatly to some of the more vexing problems of prosimian phylogeny.

Rather than summarize the papers given this morning some salient points, some of which were discussed and all of which merit further discussion, will be selected within the framework of the title of our session.

Clark showed that *Galago crassicaudatus argentatus* and *G.c. crassicaudatus* are able to distinguish one another on the basis of a composite of scents, including chest-gland scent and, possibly, urine, collected from perches in their cages. Doyle posed the question as to whether this implied sub-specific recognition of scent in this species or whether it constituted further evidence that *G.c. argentatus* and *G.c. crassicaudatus* should be separated taxonomically at the sub-specific level, as many research workers believe should be the case on other grounds (morphological and biochemical, for instance). Clark pointed to other differences between the two sub-species. Morphologically *G.c. argen-tatus* have a much larger chest-gland patch than *G.c. crassicaudatus* and, more importantly, she noted that their calls are spectrographically different.

A number of discussants drew attention to other research in which individual recognition on the basis of scent had been investigated in various species. Schilling, for instance, observed that male *Microcebus coquereli* are able to recognise other males on the basis of urine while Bailey observed that *Lemur catta* pay very little attention to urine. Manley pointed out too that male *Galago senegalensis* and a number of lorisine species recognise females in oestrus on the basis of urine. Male *G. crassicaudatus* likewise appear to re-cognise females in oestrus in that they lick the perches of oestrous females, on which a composite of scent has been deposited, more frequently than the perches of anoestrous females.

However, only Clark and Harrington (1974) have specifically tested for sub-specific recognition of scent. Harrington's results were different from Clark's. He showed that three *Lemur fulvus* sub-species were not able to dis-tinguish one another's scent collected in a manner similar to that in Clark's study.

Marler raised an important methodological problem following McGeorge's presentation. He pointed out that the time of day when sounds are recorded is very important since the transmission properties of the environment may vary considerably between dawn and dusk. For example, at dawn, after a cool night, the forest canopy warms rapidly resulting in a very steep temperature gradient at the canopy which has the effect of refracting sound. Because of this, transmission distance may be far greater at dawn than at midday. An analogous situation obtains at dusk as a result of the canopy cooling more rapidly than the forest floor. McGeorge recalled that there is a concentration of calls at dawn and dusk for both nocturnal and diurnal forms.

The nocturnal calls, Findley noted, appear to be concentrated in the higher frequency windows while the calls of the diurnal forms occupy the lower frequency windows. The calls of the nocturnal forms were also spread out over a greater frequency range, and this would be a more critical criterion than the actual frequencies themselves in minimising the possibility of masking, understandable in view of the fact that auditory communication is generally considered to be more important in nocturnal forms than in diurnal forms. Frequency range is also related to size, Charles-Dominique pointing out that nocturnal forms tend to be smaller than diurnal forms.

Subtle but strong constraints must have existed to ensure the evolution of characteristic patterns of vocalization each with a maximum capacity for effective communication within a community of unrelated sympatric forms with consequent pressures to avoid temporal, frequency and other forms of overlap.

At first sight Sussman's paper on the evolutionary and ecological implications of nectar feeding in *Lemur mongoz mongoz*, within guilds of unrelated but sympatric species of animal inhabiting the forest canopy, seems analogous to McGeorge's paper on patterns of vocal communication evolved within an acoustic environment of calls emanating from at least as large a community of unrelated, sympatric forms.

With the presentation of the paper and the discussion which followed the analogy turns out not to be a good one. For one thing Sussman emphasized at the beginning of his paper that the feeding behaviour of the *L. mongoz* he observed in northwestern Madagascar was peculiar, i.e., uncharacteristic of *L. mongoz* generally and, therefore, an adaptation to very localized conditions. Following Struhsaker's question on the possibility of *L. mongoz* having coevolved with the kapok tree, *Ceiba petandra*, the analogy seems to break down even further. Sussman was of the opinion that *Ceiba*, an exotic species introduced into Madagascar, was already adapted to large pollinating agents like fruit bats in Africa, Asia and South America, and that in Madagascar *L. mongoz* were simply taking advantage of a tree already adapted to large pollinating agents. In contrast to South America, for instance, where there are many bats, Madagascar has many prosimian forms using flowers as food sources but only one plant-visiting bat, *Pteropus*, which Sussman said was often, but not always, destructive of *Ceiba*. *L. mongoz*, in northwestern Madagascar at least, were probably simply filling a niche left vacant by bats just as *Daubentonia* filled a niche left vacant by woodpeckers though without any apparent morphological and anatomical changes analogous to those which have taken place in *Daubentonia* enabling it to fill this niche.

Sussman gave examples of other Madagascan prosimian forms visiting flowers to feed on nectar without destroying them and thus acting as potential pollinators. *Galago crassicaudatus* was given as an example from Africa. The composition of the nectar of many of these plants is highly nutritious containing sugars, amino acids and, in some cases, small amounts of lipids and proteins

(Baker and Baker, 1975) sufficient to sustain some species of prosimian, at
least during lean periods, and possibly even a large bat like *Pteropus* which,
according to Sussman, uses a lot more energy than *L. mongoz*. It was also
pointed out that even purely or partly insectivorous prosimian species, which
visit plants in search of insects attracted to them, could act as pollinators
provided that, in catching insects, their mouths came into contact with pollen

Both Tattersall's and Sussman's papers suggest that for most behaviours,
and certainly those that their two papers deal with, the range of possibilitie
is so wide that it is difficult to characterize a species exactly in terms of
its behavioural repertoire. Wilson (1975) emphasizes the point strongly with
respect to a number of patterns of behaviour, including social behaviour in
some species of primate, where he deals briefly with the concept of 'behavioural scaling'; differences in behaviour between groups of the same species
are not permanent nor less are they genetically determined differences but are
adaptations to the demands of peculiarities in the immediate environment and,
therefore, such groups must be regarded as temporarily at different points
on the same behavioural scale. Gartlan (1973), reviewing studies of social
structure in African cercopithecines, shows how dependent is social organization within a species on environmental factors like population density and
seasonal changes in food supply, amongst other things. In two very different
habitats in Uganda, the one described as disturbed and deteriorating with a lc
population density of *Cercopithecus aethiops,* and the other as rich and regene
ting with a high population density of the same species, there were profound
differences in social behaviour between groups of the same species. In the
former, inter-group grooming was observed, intra-group grooming was infrequent
territorial behaviour was absent and aggressive intra-group encounters were
rare. In the latter, inter-group grooming was never seen, intra-group grooming was more frequent, territories were vigorously defended against intruding
conspecifics and aggressive intra-group encounters were more common.

Tattersall, as Wilson (1975) does, draws attention to the need to study
the same species in a variety of different habitats. Tattersall reported on
two other aspects of behavioural plasticity in *Lemur mongoz* - patterns of acti
vity and group composition. Jerison and Petter raised the question about vari
bility across the nocturnality/diurnality dimension in other species of prosimian, Petter pointing out that *Lepilemur*, specialized for nocturnal activity
may awaken during the day, if cold, seek the sun and eat; and Jerison pointir
out that *L. mongoz,* in captivity, may be entrained by social and other stimuli
to reverse its activity pattern. Charles-Dominique (this section) makes similar observations with respect to *Hapalemur* and notes further that both *Hapalemur* and *L. mongoz,* the only two fully gregarious, nocturnal, prosimian
species, represent forms transitional between nocturnal and diurnal life and
that, depending on their habitat, they may be nocturnal, crepuscular or partl
diurnal.

Jerison also expressed an interest in physiological factors related to
activity cycles drawing attention, for instance, to the fact that nocturnal
species are generally deficient in colour vision. Tattersall confirmed this
view citing the behaviour of both *L. mongoz* and the closely related *L. fulvus.*
He noted that in Anjouan, where *L. mongoz* are found, central massifs may rise
to as high as 1500 m and that at about 700 m *L. mongoz* begin to show a shift
in activity. On Mayotte, where *L. fulvus* are found and which is much lower
than Anjouan with forested areas rising to about 650 m, *L. fulvus* are never
found above 400 m and rarely above 350 m. Physiological differences between
the two species must account for this fact.

Tattersall's paper shows that while there is an obvious relationship between climate and shifts in activity rhythm, environmental correlates of variation in group structure within the same species are much less obvious. Perhaps Wilson's concept of behavioural scaling can be invoked here. Jolly (1972),
for instance, showed how both group size and inter-group behaviour in *Lemur
catta* at Berenty, Madagascar, changed markedly over a six-year period, probably due to long-term population effects or seasonal food shortages. This
would accord well with Gartlan's (1973) observations on *C. aethiops*. In the
first case we have changes in social behaviour within the same animals over
time, and, in the second, differences between troops of the same species at
the same time.

Clearly then behaviour is not static in time as morphology is and as Tattersall points out. The term species-typical behaviour appears to apply only in
the broadest sense, at least with regard to many behaviours. We need not only
to study the same species in a variety of different habitats and under a
variety of conditions but also to study the same animals over long periods of
time.

McArdle's detailed examination of the functional anatomy of hip and thigh
confirms the close taxonomic affinity of the four lorisine species, *Perodicticus potto, Nycticebus coucang, Arctocebus calabarensis* and *Loris tardigradus,*
and confirms further that the latter two more gracile forms are more similar
to one another than the two more robust forms. More importantly McArdle
shows how these differences relate to their respective ecological niches, on
which Charles-Dominique (1977) has provided a great deal of evidence elsewhere.
Does this mean that we can infer locomotor behaviour from the sort of anatomical details provided by McArdle's paper? It might appear so from a consideration of the four lorisid species alone but what of *Galago crassicaudatus,* anatomically clearly a galagine but with respect both to its locomotor behaviour
and its feeding habits, which are bound together, as Bearder (1974) noted,
it seems to represent a compromise between lorisine and galagine. Structural
correlates subserving fine differences in locomotor behaviour are not apparent,
though, in the discussion, McArdle noted that the foot of both *G. crassicaudatus* and *G. elegantulus* is shorter than in other galagines. The relationship between structure and function is clearly not as close in some species
as it is in others and the point was made that the important factor is the
study of structure in action.

Charles-Dominique's interesting and comprehensive paper dealt with the
social evolution of the diurnal and gregarious lemuriformes from ancestral,
nocturnal, strepsirhine stock. Only in Madagascar are the diurnal, gregarious
lemurs found where, unlike the nocturnal prosimians of Africa and Asia, they
have evolved without competition from and in parallel with the simian primates.
As a result of this he notes their greater degree of flexibility, compared to
the lorisids, in activity rhythms, social behaviour and structure and feeding
behaviour and their consequent ability to exploit niches which in the New
and Old World are occupied by simian primates. *Lemur mongoz,* the subject of
both Tattersall's and Sussman's papers, together with *L. fulvus* and *Hapalemur*
are given as examples of this flexibility characteristic of the gregarious
Lemuriformes. Having developed in parallel with the simian primates they
show many examples of convergent evolution of social structure with simian
primates, though limited in this evolutionary development by sensory specializations still largely characteristic of the nocturnal and more solitary forms
and, therefore, less complex in social behaviour and communication.

Jolly expressed surprise that there should apparently be less in the way of olfactory marking in the nocturnal forms compared to the diurnal forms. By way of explanation Charles-Dominique, contrasting the diurnal prosimians with the simians, suggested that olfactory marking in the diurnal prosimians probably compensates for the absence of facial gestures and other visual signals so characteristic of the simian primates. The diurnal prosimians possess numerous cutaneous glands poorly developed in the nocturnal prosimians. One must not forget too that many of the elaborate forms evolved for depositing scent probably have a strong visual component as Jolly (1966) and others have noted.

Jolly asked further about the presence of visual signals in the nocturnal forms noting the use of visual signals in both *Phaner* and *Lepilemur*. Charles-Dominique noted that the nocturnal forms often move about where the night light is best, and drew attention to the distinctive marking on the head of *Phaner*, which probably serves as a visual recognition signal and he noted that the 'heavy' way in which they jump probably constitutes a mixture of visual and auditory communication.

Manley cautioned against thinking that because a species is nocturnal it must rely little on visual signals pointing out that, in the lorisids, for instance, the defence posture in *Potto* and *Galago* contains clear visual components effective particularly over short distances.

Clearly some behaviours are more resistant to modification by the immediate environment than others, patterns of communication, particularly vocal communication, being good examples and conforming, therefore, more than others, to what are characteristically called species-typical behaviours. Conversely other behaviours, like social behaviour, activity rhythms and feeding behaviour, are clearly very plastic. Some behaviours too can be related reasonably closely to morphological characteristics, like locomotor behaviour, while others cannot, and again those same behaviours shown to be most susceptible to modification by the immediate environment, particularly social behaviours, fall into this category.

To the extent that behaviour studies can contribute to the problems of systematics they must obviously concentrate on those behaviours most resistant to modification by the immediate environment like the characteristics of vocal and visual communicative patterns, as Kummer (1970) pointed out, as well as olfactory communication. Although Gartlan (1973) notes that the vocal repertoires of species may differ markedly from one area to another as a function of such environmental factors as population density and population pressure, their essential characteristics are, nevertheless, remarkably stable. This would apply too to olfactory communication. Studies should also concentrate on those patterns of behaviour that can be most closely related to the structures subserving them, like locomotor behaviour. This does not mean that we should ignore other behaviours. We need to know more about social behaviour and social structure and feeding behaviour, for instance, and how and to what extent they are susceptible to the influences of the immediate environment and to this end we need to study the same species both in a variety of habitats and in a particular habitat over an extended period of time. These behaviours should eventually be able to make significant contributions to our understanding of the process of evolution by providing concrete and immediate information to help explain many biological phenomena which are themselves the causes of evolutionary change.

But, however important they are, and given both our present state of knowledge and the rather unrefined techniques at our disposal, I can do no more than emphasize Tattersall's note of caution that great care must be exercised in using them as a basis for propounding models of behavioural evolution.

ACKNOWLEDGEMENTS

I would like to acknowledge the support of the Council Research Committee
of the University of the Witwatersrand and the Human Sciences Research Council
of the Republic of South Africa.

REFERENCES

Andrew, R.J. (1963a). *Behaviour*, 20, 1-109.
Andrew, R.J. (1963b). *Symp. zool. Soc. Lond.*, 10, 89-101.
Baker, H.G. and Baker, I. (1975). *In* "Coevolution of animals and plants",
    (L.E. Gilbert and P.H. Raven, eds), pp. 100-140, Univ. of Texas Press,
    Austin.
Bearder, S.K. (1974). Aspects of the ecology and behaviour of the thick-
    tailed bushbaby, *Galago crassicaudatus* , unpublished Ph.D. thesis, Uni-
    versity of the Witwatersrand.
Charles-Dominique, P. (1977). "Ecology and Behaviour of Nocturnal Primates",
    Duckworth, London.
Doyle, G.A. and Martin, R.D. (1974). *In* "Prosimian Biology", (R.D. Martin,
    G.A. Doyle and A.C. Walker, eds), pp. 3-14), Duckworth, London.
Ewer, R.F. (1968). "Ethology of Mammals", Logos Press, London.
Gartlan, J.S. (1973). *In* "Precultural Primate Behaviour", (I.W. Menzel, Jr.,
    ed), vol. I, pp. 88-101, S. Karger, Basel.
Harrington, J. (1974). *In* "Prosimian Biology", (R.D. Martin, G.A. Doyle and
    A.C. Walker, eds), pp. 331-346, Duckworth, London.
Hinde, R.A. (1970). "Animal Behaviour: A Synthesis of Ethology and Compara-
    tive Psychology", 3rd ed., McGraw-Hill, Tokoy.
Huxley, T.H. (1863). "Evidence as to Man's Place in Nature", Williams and
    Norgate, London.
Jolly, A. (1966). "Lemur Behavior: A Madagascar Field Study", Univ. of
    Chicago Press, Chicago.
Jolly, A. (1972). *Folia primatol.*, 17, 335-362.
Klopfer, P.H. (1973). *Ann. N.Y. Acad. Sci.*, 223, 113-119.
Kummer, H. (1970). *In* "Old World Monkeys", (J.R. Napier and P.H. Napier, eds),
    pp. 25-36, Academic Press, New York.
Moynihan, M. (1973). *Breviora*, 415, 29 pp.
Van Hooff, J.A.R.A.M. (1962). *Symp. zool. Soc. Lond.*, 8, 97-125.
Van Hooff, J.A.R.A.M. (1967). *In* "Primate Ethology", (D. Morris, ed.),
    pp. 7-68, Weidenfeld and Nicolson, London.
Wilson, E.O. (1975). "Sociobiology: the new synthesis", (pp. 19-21), Harvard
    U.P., Cambridge, Mass.

SECTION III

SOME OTHER ASPECTS OF PROSIMIAN BIOLOGY
Section Editors: D.J. Chivers (Cambridge)
and K.A. Joysey (Cambridge)

LORISIFORM HANDS AND THEIR PHYLOGENETIC IMPLICATIONS:
A PRELIMINARY REPORT

H.-U.F. ETTER

*Anthropologisches Institut der Universität Zürich,
Künstlergasse 15, 8001 Zürich, Switzerland.*

The hand structure of all primates is directly influenced by their habitat and by their behaviour. Consequently the hand, involved in both locomotor and manipulating activities, provides an excellent indication of an animal's way of life. It is reasonable, therefore, to assume that a detailed analysis of the hand skeleton of the prosimians will add something new to the recent discussions about Lemuriform/Lorisiform phylogeny (Cartmill, 1975; Charles-Dominique and Martin, 1970; Groves, 1974; Szalay and Katz, 1973; Tattersall and Schwartz, 1975).

MATERIALS AND METHODS

This study was carried out on the hand skeletons of 99 adult prosimian primates (Table I). The measurements were taken according to Martin's methods (1914). The length of the skeletal trunk is taken after the methods described by Biegert and Maurer (1972).

RESULTS

The hands of the Lorisinae are the shortest when compared with the hand lengths of other prosimians (Fig. 1). The small bodied species, *Loris tardigradus* and *Arctocebus calabarensis*, have the shortest hands. The Galaginae have slightly longer hands than the Lemuridae. The regression lines of the Galaginae and the Lemuridae are almost parallel. The differences in the hand lengths between the different genera in the Lorisinae are remarkable, whereas the correlation in the Lemuridae and the Galaginae is highly significant.

The carpus of the Lorisinae is the longest of all prosimians (Fig. 2). In spite of slightly elongated hands, the Galaginae have relatively longer carpi than the Lemuridae. The regression lines are more or less parallel and the correlations in all three groups are highly significant.

The thumb ray is longer in the Lorisinae than in the Lemuridae, the Galaginae laying in between (Fig. 3). The regression lines are more or less parallel and the correlations are significant in the Galaginae and in the Lemuridae.

In the Galaginae, the large bodied *G. crassicaudatus* is closest to the Lorisinae in handlength, carpus length and thumb-ray length.

The second ray is especially short in the African Lorisinae, whereas in the Asian Lorisinae it is only slightly shorter than in the other prosimians (Fig. 4). The Lemuriformes and Galaginae do not differ in the relative length of their second rays. The regression lines are more or less parallel and the correlation is highly significant in the Lemuridae, the Galaginae and the Indriidae.

TABLE I

Species	skeletal trunk**		longest ray		sec. ray		thumb ray		carpus	
	n	x̄	n	x̄	n	x̄	n	x̄	n	x̄
Arctocebus calabarensis	19	164.4	19	25.5	19	11.1	19	18.6	4	5.3
Perodicticus potto	25	208.5	25	47.4	25	19.4	25	27.2	13	9.9
Loris tardigradus	3	112.9	4	23.7	4	16.2	4	14.6	1	5.0
Nycticebus coucang	7	188.4	7	38.2	7	24.1	7	20.6	2	9.0
Galago demidovii	4	75.5	4	19.9	4	13.9	4	11.2	3	3.3
Galago senegalensis	6	115.7	6	33.8	6	25.6	6	18.7	4	6.0
Galago elegantulus	13	125.7	13	36.2	13	28.0	13	20.0	11	6.4
Galago crassicaudatus	1	168.2	1	43.1	1	32.8	1	23.4	1	8.5
Propithecus verreauxi	3	265.0	3	82.2	3	59.6	3	41.4	4	12.0
Indri indri	0	-	1	136.0*	1	103.2*	1	69.2*	0	-
Avahi laniger	1	164.7	1	61.7	1	45.7	1	22.6	0	-
Lepilemur leucopus	3	157.0	3	41.3	3	31.2	3	18.9	2	6.5
Hapalemur griseus	2	182.9	2	46.3	2	33.5	2	23.2	1	7.0
Lemur albifrons	1	257.2	1	59.8	1	50.4	1	30.3	2	9.5
Lemur catta	1	239.9	1	53.3	1	45.8	1	28.3	2	11.8
Microcebus murinus	2	77.9	2	16.9	2	13.3	2	8.7	1	2.0
Daubentonia madagascariensis	2	233.0	3	114.7	3	74.0	2	45.6	1	10.5
Tarsius syrichta	2	72.4	2	33.4	2	30.1	2	18.1	1	4.0
Tarsius bancanus	1	77.7	1	37.9	1	32.5	1	19.2	1	4.5

*   The length of the terminal phalanges in *Indri indri* were estimated.
** Data on skeletal trunk length are partly from Stettler (1970).

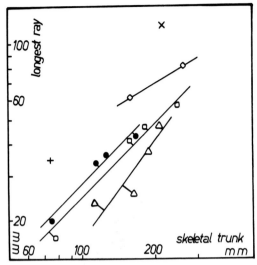

Figure 1. *The length of the longest hand ray is plotted against the ske-*
letal trunk length. Every plot (excl. Tarsius and Lemur) shows
the mean value of a prosimian species in a double logarithmic
system.

Correlations	Lemuridae	Galaginae	Lorisinae
r	0,98	0,98	0,89

Symbols (species within a symbol grouped after increasing
lengths of longest hand rays):

  : *L. tardigradus, A. calabarensis, N. coucang, P. potto*
  : *M. murinus, L. leucopus, H. griseus, L. catta and L.*
    albifrons
  : *G. demidovii, G. senegalensis, G. elegantulus, G. crassi-*
    caudatus
  : *A. laniger, P. verreauxi, (I. indri)*
  : *T. syrichta and T. bancanus*
  : *D. madagascariensis*

Concerning the morphology of the carpus, the Lorisinae show a scaphoid
and a lunate which are broad, whereas the triquetrum and the pisiforme are
reduced, so that the articulatio ulnacarpea is orientated in a more ulnar
and dorsal direction (Fig. 5). The base of the second ray, the multangulum
minus, is also reduced and the capitatum is relatively large. The articulatio carpo-metacarpea pollicis is distinctly inclined to radial.
    The distinct rotational movement of the multangulum maius has resulted in
a strongly bowed carpus, and marginal elevations are prominent (Fig. 6). The
carpal tunnel thus becomes high. In *Loris* and *Potto*, this carpal tunnel is
closed by an extra bony plate, incorporated in the ligamentum transversum
carpi, whereas in *Arctocebus* the hamulus of the hamatum is very long, robust
and strongly curved on the volar side, so that here also the carpal tunnel
is almost closed. Subba-Rau and Sundaresan described this bony plate in
Loris in 1931, naming it os ventrale. The articulatio carpo-metacarpea
pollicis is long, narrow, and saddle-shaped. The metacarpals have undergone
torsion with the result that the flexion planes of digits 3 to 5 are more or
less parallel.

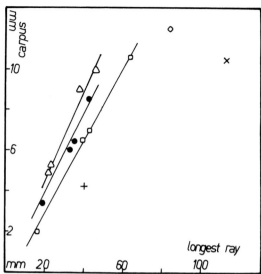

Figure 2. *The length of the carpus is plotted against the length of the*
longest hand ray. Every plot shows the mean value of a pro-
simian species (excl. Lemur and Tarsius). For symbols see
Fig. 1.

Correlations r	Lemuridae	Galaginae	Lorisinae
r	0,99	0,99	0,98

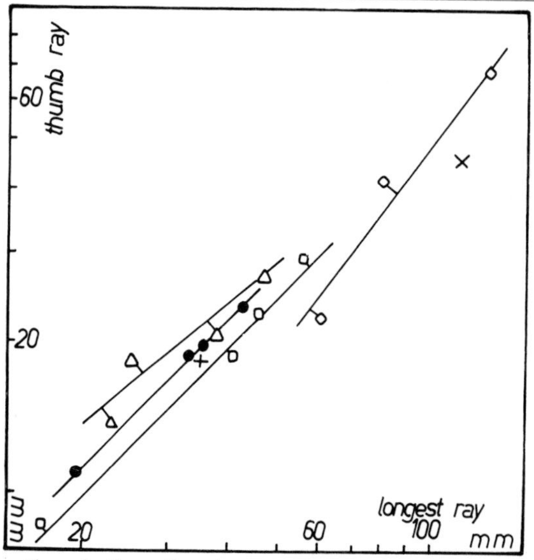

Figure 3. *The length of the thumb ray is plotted against the length of the*
longest hand ray. Every plot shows the mean value of a pro-
simian species (excl. Lemur and Tarsius) in a double logarith-
mic system. For symbols see Fig. 1.

Correlation	Lemuridae	Galaginae	Lorisinae	Indriidae
r	0,99	0,99	0,95	0,99

100
II ray
60
20
mm
20
60
100
longest ray
mm

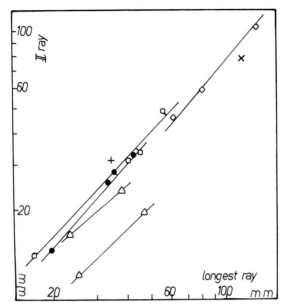

Figure 4. *The length of the second ray is plotted against the length of the longest hand ray. Every plot shows the mean value of a prosimian species (excl. Lemur and Tarsius) in a double logarithmic system. For symbols see Fig. 1.*

Correlation	*Lemuridae*	*Galaginae*	*Indriidae*
r	*0,98*	*0,99*	*0,99*

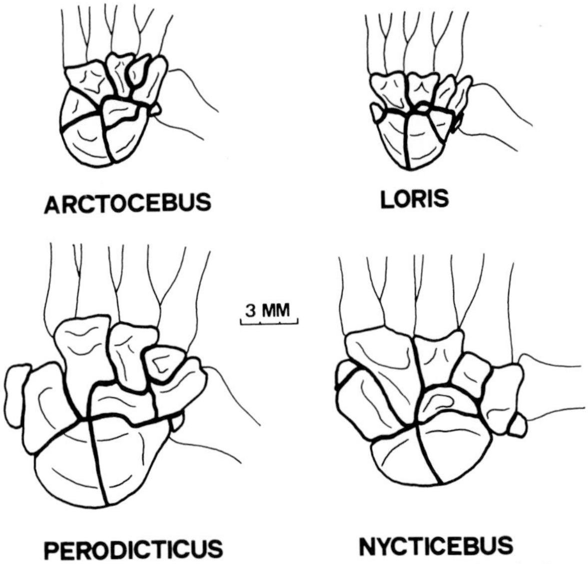

ARCTOCEBUS
LORIS
3 MM
PERODICTICUS
NYCTICEBUS

Figure 5. *Dorsal view of the left carpi of the 4 lorisine generas. Note the large multangulum maius, capitatum and hamatum, and the reduced multangulum minus.*

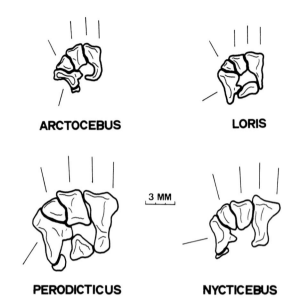

Figure 6. *Distal view of the left carpi of the 4 lorisine generas. The straight lines indicate the dorso-volar plane of the capitula of the metacarpalia I to V. Note the shape of the multangulum maius and of the hamatum and the bony closure of the carpal tunnel in Loris, Potto and Arctocebus.*

The pincer-like, strong hands with the closed bony carpal tunnel of the Lorisinae appear to be best adapted for their functions (Bishop, 1964). However, detailed analysis of correlations between the morphology of the hand skeleton and its proper functions remain to be made on an intergeneric basis.

The hand skeleton of the Galaginae show in many respects lorisine features (Figs 7, 8) like the tendency towards a large and robust, clearly abducted thumb, a broad capitatum and lunatum, an articulatio ulnacarpea which is distinctly ulnar and dorsal orientated, a long, narrow saddle-shaped articulatio carpo-metacarpea pollicis and the parallel capitula of the metacarpalia III to V as a result of their torsion.

These features in the hand skeleton common to the Lorisiformes, show quantitative differences within the different species but serve to distinguish them from all other primates.

Compared with the hand skeletons of the Lemuridae, the *Galago* hand skeletons resemble those of the most generalized sub-family of this group, the Cheirogaleinae. Except for lorisine features of the carpus, all other features are similar to those of *Microcebus*, especially the morphology of the metacarpalia and phalanges.

The terminal phalanges are identical (Fig. 9). They are broad, robust, with exceptionally robust, broad and big nail bases. The same kind of terminal phalanges are found in the other Lemuridae and Indriidae and also in *Tarsius*.

Concerning the hand skeleton, the Galaginae differ from the Lemuridae in those features, which are most pronounced in the Lorisinae. In the carpus and in the proportions of the hands, the Galaginae are closest to the Lorisinae, whereas in the morphology of the metacarpalia and phalanges, the Galaginae are closer to the Lemuridae and especially to the Cheirogaleinae. However, morphological and metrical differences between the 4 lorisine species are remarkable; their common features are more pronounced in the African species than in their Asian counterparts. In the smaller animals, *Arctocebus calabarensis* and *Loris tardigradus*, they are more pronounced than in the larger ones. Concerning hands, *Nycticebus coucang* is the least specialised lorisine primate. In the Galaginae it seemed that the larger-bodied species, *G. crassicaudatus*, is closer to the Lorisinae than the smaller-bodied.

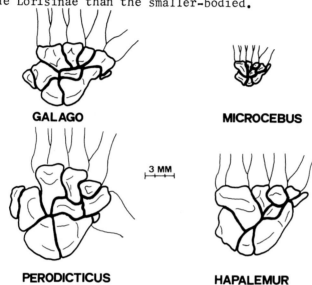

Figure 7. Dorsal view of the left carpi of 4 prosimian generas. The large capitatum, hamatum and multangulum maius in the galagine carpus are lorisine features.

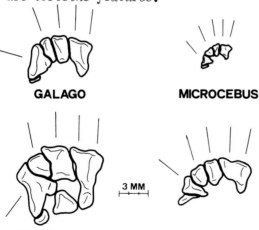

Figure 8. Distal view of the left carpi of 4 prosimian generas. The straight lines indicate the dorso-volar plane of the capitula of the metacarpalia I to V. The position and the shape of the multangulum maius and the bowing of the carpus in Galago are lorisine features.

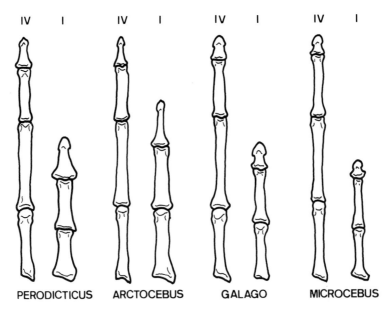

Figure 9. *The first and the fourth ray of four prosimian generas. The
fourth rays of all 4 generas are reduced to the same length and
the first rays are in proportion to the fourths. The terminal
phalanges in Galago are very similar to those in Microcebus.*

DISCUSSION

These findings are in agreement with the results Biegert (1959) arrived
at in his studies of the palms of prosimians, the galagos being in many respects intermediate between Lorisinae and Lemuridae. As mentioned by Bearder
and Doyle (1974) locomotion of *G. crassicaudatus* is midway between the active-
leaping and the slow-climbing locomotion of Lorisinae. Assuming that the structure and function of the hands of the Lorisinae are closely related, there are
sound reasons to suppose that the different galagos show different degrees of
lorisine specialisations in both locomotor behaviour and structure of the hands.

The broad and robust form of the terminal phalanges is common to Galaginae
and Cheirogaleinae and connected with their broadened, discus-shaped fleshy
pads (Biegert, 1961). It seems highly probable that this type of finger tip
provides a maximum of friction and adaptability to any surface for control of
gripping, landing, holding fast, etc. This feature is an advantage and especially useful in leaping locomotion. Napier and Walker (1967) even suggest
that the dominant ulnar border of the prosimian hand represents an adaptation
to the vertical resting position associated with clinging and leaping locomotion.

These results agree with those of Charles-Dominique and Martin (1970) which
showed that there are great ecological similarities between *Microcebus murinus*
and *Galago demidovii*. Charles-Dominique and Martin concluded that, in this
respect, Cheirogaleinae and Galaginae may show ancestral primate features,
including their mode of locomotion: a mixture of horizontal branch-running
and vertical clinging and leaping in trees, and hopping on the ground.

Napier and Walker (1967) suggest that"... Eocene representatives of this
locomotor group could have been capable of providing the basic stage from which

other primate locomotor groups have evolved" (p. 216). Walker (1970, 1974)
also deduced from fossil remains of prosimians of the Eocene and Oligocene
from Europe, Africa and America, that leaping locomotion was the dominant
locomotor habit at this time. He concludes from the early Miocene lorisid
postcranial material that "... although the dental and cranial remains show
quite eclectic resemblances to different galagines and lorisines all post-
cranial elements are clearly from animals that had a vertical clinging and
leaping locomotion like that of modern galagos" (p. 254, 1970). Hence he assumes that post-cranial specializations of the Lorisinae seemed to have evolved from a clinging and leaping type of locomotion and most probably later
than the Miocene.

The analyses of the structures and proportions of the hand skeletons of
the Lorisiformes are also in agreement with the assumption that the recent
galagos represent a group of animals that derived from a stock of clinging
and leaping prosimians, similar to the recent Cheirogaleinae. The specialized
hind-limb structures and the slightly elongated hands are adaptations to an
even more advanced type of leaping locomotion. The differing lorisine features in the hands and the differing type of locomotion within the Galaginae
are arguments for the assumption that some galagos may then have showed tendencies to adapt gradually to a more lorisine mode of locomotion. From this
locomotor and structurally diversifying group of galagos the different lorisine genera split off probably at different times. The remarkable differences
within the Lorisinae makes this probable. Thereafter the African and the
Asian lorisine species adapted to similar advanced slow-climbing locomotor
habits in similar ecological niches, both separated into two different habitats: the large bodied animals occupying rather high vegetational strata, preferring fruit and gums, and the small-bodied animals utilizing the undergrowth
and feeding more on animal prey (Charles-Dominique, 1974; D'Souza, 1974). The
generally more specialized hand of the small bodied Lorisinae and the rather
remarkable differences in the hand-structure within the Lorisinae may be in
correlation with their differences in ecology and habitat.

ACKNOWLEDGEMENTS

The author wishes to thank Dr. J. Biegert for permission to examine the
prosimian material of the A.H. Schultz-Collection and of the primate collection at the Institut of Anthropology, University of Zürich as well as for the
many helpful discussion we had. Thanks for additional prosimian material goes
to Dr. H. Burla (Zoological Institut, University of Zürich), Dr. H. Schäfer
(Museum of Natural History, University of·Basel) and Dr. H. Huber (Museum of
Natural History, University of Berne).

REFERENCES

Bearder, S.K. and Doyle, G.A. (1974). *In* "Prosimian Biology", (R.D. Martin,
    G.A. Doyle, A.C. Walker, eds), pp. 109-130, Duckworth, London.
Biegert, J. (1959). *Z. Morph. Anthrop.*, 49, 316-409.
Biegert, J. (1961). *In* "Primatologia", (H. Hofer, A.H. Schultz, D. Starck,
    eds), Vol. 2, pp. 1-326, Karger, Basel and New York.
Biegert, J. and Maurer, R. (1972). *Folia primatol.*, 17, 142-156.
Bishop, A. (1964). *In* "Evolutionary and genetic biology of primates", (J.
    Buettner-Janusch, ed.), Vol. 2, pp. 133-226, Academic Press, New York and
    London.

Cartmill, M. (1975). *In* "Phylogeny of the primates", (W.P. Luckett, F.S.
    Szalay, eds), pp. 313-354, Plenum Press, New York.
Charles-Dominique, P. (1974). *In* "Prosimian Biology", (R.D. Martin, G.A.
    Doyle, A.C. Walker, eds), pp. 131-150, Duckworth, London.
Charles-Dominique, P. and Martin, R.D. (1970). *Nature, Lond.*, 227, 257-260.
D'Souza, F. (1974). *In* "Prosimian Biology", (R.D. Martin, G.A. Doyle, A.C.
    Walker, eds), pp. 167-182, Duckworth, London.
Groves, C.P. (1974). *In* "Prosimian Biology", (R.D. Martin, G.A. Doyle, A.C.
    Walker, eds), pp. 449-473, Duckworth, London.
Martin, R. (1914). "Lehrbuch der Anthropologie", Fischer, Jena.
Napier, J.R. and Walker, A.C. (1967). *Folia primatol.*, 6, 204-219.
Stettler, M. (1970). Skelettproportionen bei prosimischen Primaten,
    (unpublished), Zürich.
Szalay, F.S. and Katz, C.C. (1973). *Folia primatol.*, 19, 88-103.
Subba-Rau, A. and Sundaresan, K. (1931). *Half yearly Journ. Mysore Univ.*,
    5, 100-190.
Tattersall, I. and Schwartz, J.H. (1975). *In* "Phylogeny of the primates",
    (W.P. Luckett and F.S. Szalay, eds), pp. 299-312, Plenum Press, London.
Walker, A.C. (1970). *Amer. J. phys. Anthrop.*, 33, 249-261.
Walker, A.C. (1974). *In* "Primate Locomotion", (F.A. Jenkins, ed.), pp. 349-
    381, Academic Press, New York and London.

HOME RANGE, BEHAVIOUR AND TACTILE COMMUNICATION IN A NOCTURNAL MALAGASY LEMUR *MICROCEBUS COQUERELI*

ELISABETH PAGES

Laboratoire d'Ecologie Générale - Museum National d'Histoire Naturelle, 4, avenue du Petit Chateau, 91800 Brunoy, France.

In June and July 1974, during the dry season, six *Microcebus coquereli* (2 ♂♂, 2 ♀♀, 2 juveniles) were followed by radio-tracking, in a preliminary study, and seven others were captured and marked. This study was carried out near Morondava in western Madagascar.

This region undergoes pronounced seasonal changes; rain falls only during summer, while 5-6 months are dry and relatively cold at night. The forest type is tropical dry deciduous. The dry season is characterized by an absence of flower, fruits, buds, and of leaves in the higher strata, and by a scarcity of insects, reptiles, and mammals (Hladik and Abraham, in press; Hladik, 1977; Hladik et al., in press; Pariente, 1974; Petter, 1977).

During the wet season, *M. coquereli* have an omnivorous diet including insects, fruits, flowers and probably small mammals, chameleons, eggs and frogs. In contrast, they are extremely specialized during the drastic dry season, feeding mainly on secretions of colonies of Homopteran larva.

Adult *M. coquereli* weigh about 300 grams, and, during the study period, they confined most of their activity to the remaining zones of foliage (1 to 6 metres above the ground). Scent, sight, hearing and touch, all play an important role in intra-specific communication (Schilling, 1977; Pariente, 1977; Petter and Charles-Dominique, 1977), although the importance of each depends on the season.

Mating takes place in October; gestation lasts three months and normally two young are born.

HOME RANGE AND SOCIAL ENCOUNTERS

The home range of adults of both sexes are partially exclusive, within which several concentric zones may be distinguished (Figs 1 and 2). The central region (zone I and II, 6, Fig. 2) constitutes a small percentage of the total area, but the animal spends a large percentage of its time there (see legend, Fig. 2). Most activities are concentrated in this region, such as sleeping and resting, feeding, scent marking, vocalizations, grooming and mutual manipulative play (Pages, in press).

The male surveys this region from the top of the higher bared trees, in the canopy, and seems to defend it (two observations).

The rest of the home range (zone III) is used chiefly during the second half of the night. Its limit depends on encounters with the neighbouring conspecifics.

The animal may travel long distances within this zone, the male usually moving in straight lines up to 400 m, while the female circles around the periphery. The maximum distance covered in a single night may reach 1.5 km.

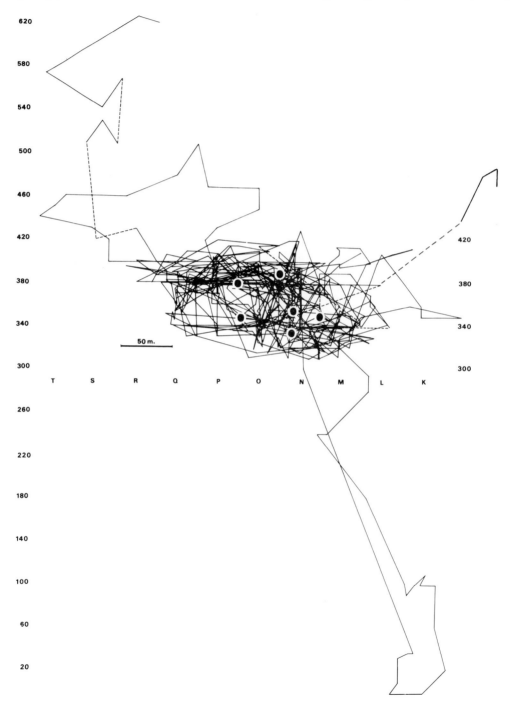

Figure 1. *Nests and movement of ♂6 in 100 hours of observations. The numbers indicate parallel paths 40 metres apart.*

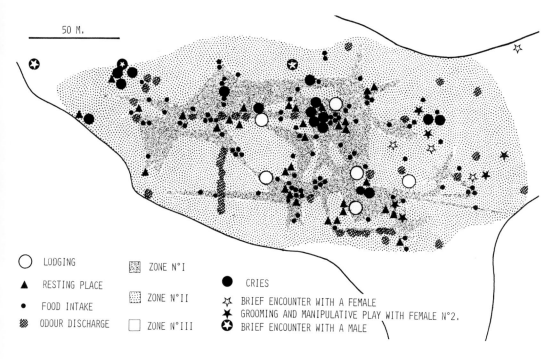

LODGING   ▨ ZONE N°I

▲ RESTING PLACE   ⬚ ZONE N°II   ● CRIES

• FOOD INTAKE   ▨ ZONE N°II   ☆ BRIEF ENCOUNTER WITH A FEMALE
                                     ✦ GROOMING AND MANIPULATIVE PLAY WITH FEMALE N°2.
▨ ODOUR DISCHARGE □ ZONE N°III   ✪ BRIEF ENCOUNTER WITH A MALE

Figure 2.   Home range, zones of regular use, and activity localization (100
hours of observation of ♂6):
zone I: 10% of total home range area: 60% of the animal's
total activity takes place in this zone, 80% of which occurs
before midnight.
zone II: 20% of home range area and 25% of total activity, 55%
of which occurs before midnight.
zone III: 70% of home range area and 15% of activity, 18%
of which occurs before midnight.

   The home range of the female is larger than that of the male. The central
region (zone I and II) measures about 0.03 km² for the female and 0.02 km² for
the male; the total home range during the study period was 0.1 km² and 0.08
km² for female and male respectively.
   The home range of the young are included within that of the mother, although
they forage separately. The mother and young sleep together and use loud calls
to relocate each other, and to return to the nest.
   During the period of observation, encounters between conspecifics were rare,
and two adults were never seen to sleep together. Male n° 6 only met a conspecific once every two nights, during which he encountered 2 ♂♂, 2 ♀♀ and 2
neighbouring juveniles. Prolonged contacts, which consisted of resting side
by side, and of mutual grooming and manipulative play while hanging suspended

by the feet, occurred only every 3 or 4 days, and only with ♀ n° 2.

There was a close association between two animals of the opposite sex, and the most frequently associated pairs (♀2 - ♂6 and ♀ 10 - ♂ 13) may indicate a loose pair-bonding.

RHYTHM OF NOCTURNAL BEHAVIOUR

M. coquereli are strictly nocturnal (17.15-18.25 to 0540-0610 hours).

Routes of travel, activities and social behaviour differ considerably from one night to another, but, in general, there is an asymmetrical division of activities in the course of the night (Fig. 3).

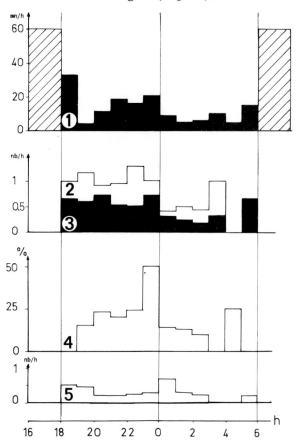

Figure 3. *Rhythm of nocturnal behaviour (100 hours of observation of ♂6)*
A) Activities which are more numerous during the first half
of the night:
1 Resting and self-grooming
2 Total number of food intakes seen
3 Homopteran secretion intake
4 Percentage of total observations in which animal was seen
higher than 15 m
5 Odour discharges

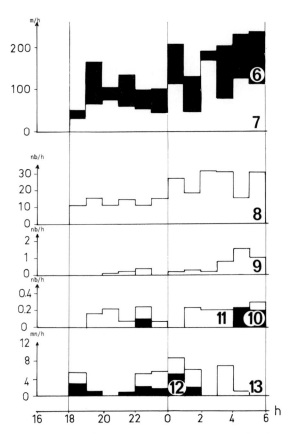

Figure 3. B) *Activities which are more numerous during the second part of the night:*

  6   *Distance covered every hour*
  7   *Distance "as the crow flies" between animal's position every hour*
  8   *Powerful jumps*
  9   *Cries*
 10 *Encounter with a male*
 11 *Encounter with a female*
 12 *Play with a female*
 13 *Social contact with a female*

    Certain maintenance activities, such as feeding, self-grooming, resting, odour discharges, inspection of the home range, are more numerous during the first half of the night, and take place only in the central region of the home range.

    Social activities, which permit or depend on contact with a conspecific, such as vocalizations, encounters, mutual grooming and play, are more frequent during the second half of the night, and take place in the intermediate or outer zones of the home range.

TACTILE COMMUNICATIONS AND REPRODUCTION IN CAPTIVITY

Both the general activity of *M. coquereli* and the frequency, duration and type of social contact are profoundly dependent on the season and oestrous cycle. Sexual receptivity is determined by the seasonal light cycle (Petter-Rousseaux, 1968, 1974;   Actograms have been recorded and analysed for two years, as well as recording of observations using video-tape and a tape recorder.

In captivity, at the end of winter and before the beginning of sexual activity, the pair-bonds strengthen, as evidenced by a progressive increase in the length of tactile contact (resting side by side, sniffing, passing over the partner while rubbing it, reciprocal grooming and play), which is abruptly reduced following copulation. Tactile interrelations, which last all year, but at a low level, permit at last a minimum of contact between ♂ and ♀ during the period of sexual quiescence and facilitate reproduction by the progressive increase of contact preceding copulation.

The prolonged and repeated tactile interrelations seem to function to maintain contact and reduce aggression before and during oestrus.

On the nights preceding oestrus, the female is less tolerant towards male's approach; the male, already strongly stimulated by the female's odour, follows and attempts to copulate with her. However, she only tolerates contact in the form of grooming and manipulative play.

During oestrus, there is a great deal of grooming, especially self-grooming, but no manipulative play. The female is sexually receptive for one night; other males are attracted by the odour and the particular loud call of the female, but the presence of the familiar ♂ excludes the neighbouring males. Experiments show that reproduction only occurs successfully if one male is alone with the female (observations based on 14 oestrous cycles of two ♀♀, showed that only 6 conceptions occurred). The presence of other male adults or even young of the previous litter, prevents fertile mating, even if they do not openly challenge the male. In the presence of the other males, a female may vigorously attack and pursue her usual companion, or simply refuse his approach (2 instances). Even if she accepts normal copulation (6 instances) she may not conceive, in which case another oestrous cycle will follow after 22-23 days (the normal cycle). In one case, a spontaneous abortion occurred, followed by a delayed oestrous cycle. A pair-bond based on repeated tactile relationships seems therefore to be necessary to permit mating. This social structure seems to be intermediate between those of the more solitary and the more strictly gregarious prosimians (Charles-Dominique, this volume).

Observations in the field (and in captivity) indicate that social behaviour occurs at specific times during the night, at a specific area within the home range and at varying levels throughout the year.

Tactile relationships between male and female, together with other forms of communication, establish and maintain a pair bond which permits successful reproduction at the period of sexual activity.

REFERENCES

Hladik, C.M. (1977). *In* "The Study of Prosimian Behaviour", (G.A. Doyle and R.D. Martin, eds), Academic Press, New York and London.

Hladik, A. (in press). Phenology and climate of the dry forest of the West coast of Madagascar. *In* "Contributions to Primatology", (P. Charles-Dominique, ed.), Karger, Basel.

Hladik, C.M., P. Charles-Dominique and J.J. Petter (in press). Feeding strategies of five nocturnal Lemurs of the dry forest of the West Coast of Madagascar. *In* "Contributions to Primatology", (P. Charles-Dominique, ed.), Karger, Basel.

Pariente, G.F. (1974). *In* "Prosimian Biology", (R.D. Martin, G.A. Doyle and A.C. Walker, eds), pp. 183-198, Duckworth, London.

Pariente, G.F. (1977). *In* "The study of Prosimian behaviour", (G.A. Doyle and R.D. Martin, eds), Academic Press, New York and London.

Pages, E. (in press). Ecology and social behaviour of *Microcebus coquereli*. *In* "Contributions to Primatology", (P. Charles-Dominique, ed.), Karger, Basel.

Pages E. and A. Petter-Rousseaux (in press). Annual variations of circadian rhythms of five sympatric nocturnal Lemurs observed in artificial conditions. *In* "Contributions to Primatology", (P. Charles-Dominique, ed.), Karger, Basel.

Petter-Rousseaux, A. (1968). Cycles génitaux saisonniers des Lémuriens malgaches. Entretiens de Chizés, n° 1 (R. Canivenc, ed.), pp. 11-12, Masson, France.

Petter-Rousseaux, A. (1974). *In* "Prosimian Biology", (R.D. Martin, G.A. Doyle and A.C. Walker, eds), pp. 365-373, Duckworth, London.

Petter, J.J. (1977). *In* "Recent Advances in Primatology: Behaviour", (D.J. Chivers and J. Herbert, eds), Academic Press, London.

Petter, J.J. and P. Charles-Dominique (1977). *In* "The Study of Prosimian behaviour", (G.A. Doyle and R.D. Martin, eds), Academic Press, New York and London.

Petter, J.J., A. Schilling and G.F. Pariente (1971). *Terre et Vie*, **3**, 287-327.

Schilling, A. (1977). *In* "The Study of Prosimian behaviour", (G.A. Doyle and R.D. Martin, eds), Academic Press, New York and London.

PHYSIOLOGICAL ASPECTS OF *PERODICTICUS POTTO*

M. GOFFART

Dept. of Physiology, University of Liège, Belgium.

Field studies in Gabon have informed us about the ecology and social relations of pottos (Charles-Dominique, 1971) but these prosimians raise interesting physiological problems.

STANDARD METABOLISM AND THERMOREGULATION

The body temperature of this furry species is highly variable, 2-3°C below that of most mammals. Its resting metabolic rate is 45% below the standard value for a 1 Kg primate (Hildwein and Goffart, 1975). The low metabolism could not be related to the status of the thyroid gland (Lemaire and Goffart, 1974) nor to the rate of output of catecholamines (Canguilhem et al., 1975), which are both normal. The tropical potto does not stand well a warm environment in laboratory conditions: although the skin conductance doubles and the heat-loss by vaporization increases (polypnea and salivation) the core temperature rises. The potto is cold tolerant, however: it can stand ambient temperatures 15°C below the lowest temperature encountered in the tropical forest. In the cold the potto does not shiver and its metabolic rate is raised by an increased secretion of noradrenaline and 17-OH corticosteroids (Canguilhem et al., 1976). The thermoregulation of the potto can hardly be regarded as "primitive" with respect to that of higher primates.

MOTOR ACTIVITY

The usual behaviour of the arboreal potto is sluggish, but it can speed up considerably. The isometric contraction-time, the half-relaxation time and the fusion frequency of five different hind limb muscles have been shown to be intermediate between those of "fast" and "slow" muscles in the cat.

The velocity constant derived from the force-velocity relationship is similar to that of cat soleus. Histo-enzymological studies showed that potto's striated muscles are made of a regular chequerboard distribution of "fast" and "slow" fibres in approximately equal proportions. The maximal force developed by the muscles per square centimetre cross-section is high. The muscles are thus adapted to the general behaviour of the species (Marechal et al., 1976). The slow movements cannot be attributed to a deficiency in the extrapyramidal motor system for the dopamine, the dopac and the tyrosinehydroxylase contents of the caudate nucleus are similar to those in *Macaca mulatta* and *Macaca fascicularis* (Gerardy et al., 1975).

CEREBRAL CORTEX

The cortical sensory projections of somatic, auditory and visual origin have been mapped and their precise relations to the sulci determined. No heterosensory potentials (visual or auditory) can be recorded from the soma-tomotor area nor from any other part outside their primary projection areas. In that respect the sensory projection system of the potto seems to be less developed than in the cat (Boisacq-Schepens et al., 1977).

VISION

Contrary to common teaching, the nocturnal potto has a duplex retina with about 1 cone for 300 rods. The electroretinogram of the dark adapted potto resembles that of the night monkey (*Aotus*), and its time course is slower than in guinea pig, cat and man. Red sensitive cells respond to wavelengths be-yond those which affect visual purple (> 650 mμ) and give rise to an ERG where the "a" wave vanishes, and the "b" wave is delayed when compared with the ERG obtained when the retinal elements are stimulated by white or blue light. Nevertheless the fusion frequency in deep red light does not exceed 9 per sec. After strong illumination the retina of the potto may remain refractory for 5 to 10 sec. Red sensitive elements recover faster than the others and com-plete recovery lasts as long as 40 min. Because the few cones in potto's retina activate a great number of bipolar cells it is suggested that they subserve day vision (Faidherbe and Goffart, 1975; Goffart et al., 1976).

DIGESTION

The omnivorous potto can digest insects by means of the gastric secretion of a very active chitinase, the specificity of which has been assessed by a new method (Cornelius et al., 1976). No cellulolytic enzymes were found in the digestive tract but the pancreatic and intestinal amylases dispose of the carbohydrates. The conical coecum is not homologous to the human appendix (Beerten-Joly et al., 1976).

REFERENCES

Beerten-Joly, B., Piavaux, A. and Goffart, M. (1974). *C.R. Séances Soc. Biol. Filiales,* 168, 140-143.
Boisacq-Schepens, N., Gerebtzoff, M.A. and Goffart, M. (1977). *Primates,* 18, 401-410.
Canguilhem, B., Hildwein, G., Juchmès, J. and Goffart, M. (1975). *C.R. Séances Soc. Biol. Filiales,* 169, 695-700.
Canguilhem, B., Hildwein, G. and Goffart, M. (1976). (unpublished).
Charles-Dominique, P. (1971). *Biologia gabonica,* 7, 121-228.
Cornelius, C., Dandrifosse, G. and Jeuniaux, Ch. (1976). *Int. J. Biochem.,* 7, 445-448.
Faidherbe, J. and Goffart, M. (1975). *C.R. Seances Soc. Biol. Filiales,* 169, 1641-164
Gerardy, J., Dresse, A. and Goffart, M. (1975). *C.R. Séances Soc. Biol. Filiales,* 169, 706-709.
Goffart, M., Missotten, L., Faidherbe, J. and Watillon, M. (1976). *Arch. internat. Physiol. Biochim.,* 84, 493-516.
Hildwein, G. and Goffart, M. (1975). *Comp. Biochem. Physiol.,* 50 A, 201-213.
Lemaire, M. and Goffart, M. (1974). *Arch. internat. Physiol. Biochim.,* 82, 149-
Maréchal, G., Goffart, M., Reznik, M. and Gerebtzoff, M.A. (1976). *Comp. Bioch Physiol.,* 54 A, 81-93.

THERMOREGULATION AND BEHAVIOUR IN TWO SYMPATRIC GALAGOS: AN EVOLUTIONARY FACTOR

F. VINCENT

Laboratoire de Psychophysiologie, Université de Paris X, France.

Contradictory conclusions were drawn from the works on the thermoregulation of Afro-malagasy prosimians. Suckling (1969), Suckling and Suckling (1971), Suckling et al. (1969) all found that the internal temperature of the Potto was independent of the ambient temperature, and that the animal's metabolic expenditure was comparable to that of mammals of the same size.

On the other hand, Bourlière et al. (1956), Bourlière and Petter-Rousseaux (1953, 1966), Hildwein (1972), and Goffart and Hildwein (1972, 1975) concluded that the central temperature was variable in five species (the Potto and four Malagasy prosimians), and that it was dependent on the external temperature. Further, Goffart and Hildwein (1972) calculated that the metabolism of the Potto at rest was lower by 45% than in mammals of the same size. Vincent (1969) has already mentioned the relative instability of the rectal temperature of Demidoff's *Galago* (35-40°C; Congo, 4°S).

All these observations, made on semi-captive animals, show that the problem of the more or less rigorous homeothermy of Prosimians is thus worth going into. The present paper concerns only free-ranging prosimians; it tries to throw light on the evolutionary importance of the physiologico-behavioural strategies of thermoregulation in differentiating ecological niches, in this case between two close-related sympatric primates: *Galago alleni* (Waterhouse, 1837) and *Galago (Euoticus) elegantulus* (Le Conte, 1857).

METHODS

In 1970-73 nocturnal prosimians were collected at night close to the village of Nkolngem I (4°N) in the forests of Southern Cameroon (cf. Heim de Balzac, 1969; Vincent, 1972 and Molez, 1975-76).

The ambient temperature was recorded 0.8 m above ground level at the time of the animals' capture. Their rectal temperature was taken less than 4 minutes after death, with a Wesco thermometer (dia. 2-4 mm) inserted by 3 cm. This thermometer was cleaned (95° alcohol), dried and wiped after each temperature recording.

RESULTS

22 rectal temperature recordings concern *G. alleni* and 31 for *G. elegantulus* (total 53).

Specific variability: the extreme values of the central temperature are much closer among the *G. elegantulus* than among the *G. alleni* (Table I): whilst the average values (37.2 and 37.8°C respectively), and the maximum values

TABLE I

*Rectal temperature variability in two sympatric African forest
prosimians in their natural habitat*

			Rectal Temperature		
			Mean	Maximum	Minimum
Galago alleni (N=22)	ADULTS (N=19)	♂ (N=6)	37.9°C	38.6°C	37.4°C
		♀ (N=13)	37.8°C	38.5°C	37.5°C
		♂ + ♀ (N=19)	37.8°C	38.6°C	37.4°C
	SUBADULTS 150-200 g (N=3)		38.1°C	38.6°C	37.6°C
Galago (Euoticus) elegantulus (N=31)	ADULTS	♂ (N=18)	37.5°C	38.6°C	35.2°C
		♀ (N=10)	37.0°C	38.0°C	34.0°C
		♂ + ♀ (N=28)	37.2°C	38.6°C	34.0°C
	SUBADULTS 180-220 g (N=3)		36.8°C	38.2°C	35.0°C

(38.6°C) are very similar for the two species, the minimum temperatures vary
considerably from one species to another, both among adults (34 and 37.4°C) and
subadults (35 and 37.6°C).

Sexual variation: no significant difference between males and females,
either among the *G. alleni* or the *G. elegantulus*.

Circannian variation: 45% of the *G. alleni* were captured during the dry
season. The variation in their rectal temperature (37.4 to 38.6°C) is similar
to that of the year as a whole. Captures of *G. elegantulus* were more regularly
distributed throughout the year, with no significant difference between two
seasons. Consequently, the rectal temperature does not appear to have circannian variations.

Circadian variation (period of activity only): no significant relationship
between the rectal temperature and the time of the animals' capture (1900 to
0100 hours for *G. alleni*; 1900 to 0300 hours for *G. elegantulus*).

Central temperature and height of the bushbaby in relation to ground-level
(Fig. 1): no significant difference with *G. alleni* ($r = 0.02$); *G. elegantulus,*
however, had a significant slight dependency ($r = 0.042$ at a two-way probability of 5%).

Variation in relation to ambient temperature: no correlation between the
two temperatures ($r = 0$) for 17 adult *G. alleni*; however, there was a significant correlation ($r = 0.4$) for 27 adult *G. elegantulus*. There is no reason
to assume that the rectal temperature increases as ambient temperature decreases; one can thus reason unilaterally, and $\alpha = 2\%$ (Fig. 2).

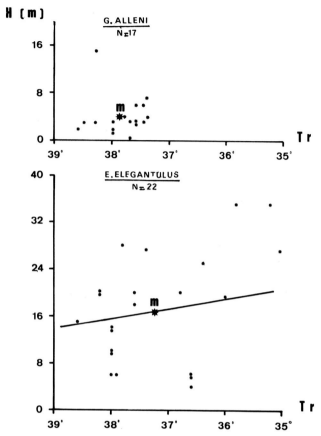

Figure 1. *Compared variation of adult rectal temperature (Tr, in °C), in relation to height (H, in metres above ground-level) of the location occupied by the animal at the time of its capture, between 2 African forest prosimians (sympatric), in their natural habitat. m = mean.*
Slope of the regression line: 0.04

DISCUSSION AND CONCLUSIONS

Thus we have two species of similar size and morphology, with almost the same area of geographical distribution (Fig. 3), captured on the same station and under the same conditions. The rectal temperature of *G. alleni* hardly varies in relation to height of the bush-baby above ground-level, or to ambient temperature at ground-level at the time of capture: this animal thus appears to be "perfectly homeotherm". The central temperature of *G. elegantulus* varies with the two factors mentioned above: the homeothermy is not so strict in this case (Fig. 4). Now, in forests, the ambient temperature is less stable near the canopy than at ground-level: the nocturnal temperature of the canopy strata B (Richards, 1964), very discontinuous on our station (Molez, 1975), decreases to a greater extent and faster than those of stratas C to E. Had we been able to measure the ambient temperature at the exact altitude of capture, we would probably have recorded wider thermal variations and we would no doubt have brought to light a better correlation between

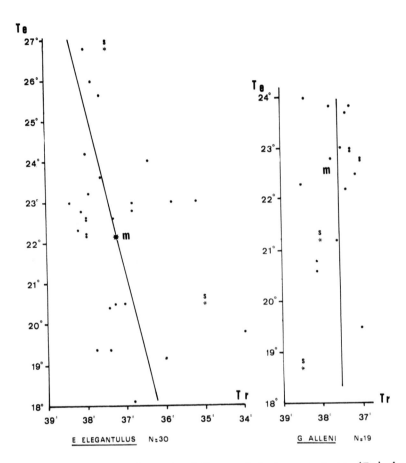

Figure 2. *Compared variation of adult rectal temperature (Tr) in relation*
to the external temperature (Te) recorded 0.8m above ground-
level at the time of the capture. m = mean.
Slope of the regression line 0.4. Subadult values have been
added on figure.

external and internal temperatures. Further, if the "altitude of the bush-
baby/central temperature" correlation proves to be inferior, this is because
(i) our evaluation of the altitude was only accurate to within 15%; (ii) and
also because the animal was in the process of moving.

It is interesting to correlate these differences in physiological strategi
with behavioural differences.
1. *vertical distribution* (Molez, 1975-76): *G. alleni* lives rather in the
lower stratas of the forest (relatively stable ambient temperature: 18-27°C),
and *G. elegantulus* more in the upper stratas (wider thermal variations:
13-40°C).

An experimental study (Vincent and Lemaho, unpublished) confirms that *G.*
alleni would retain the same thermal variability within the range of external
temperatures of the other species.

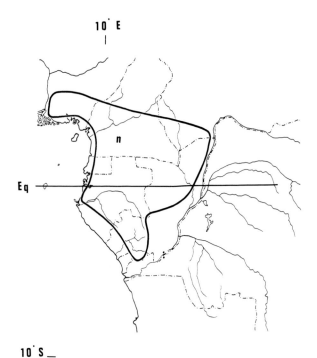

Figure 3a. West central Africa: area of geographical distribution of
G. alleni. n = station of capture.

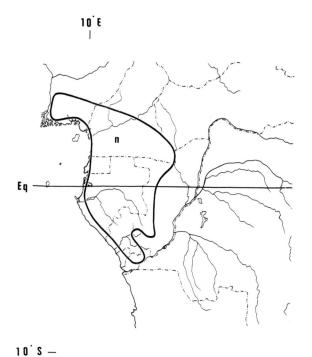

Figure 3b. West central Africa: area of geographical distribution of
G. elegantulus. n = station of capture.

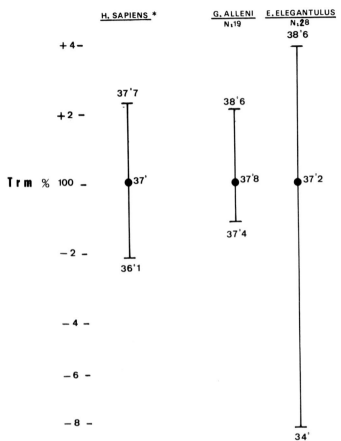

Figure 4. *Variability(%) of adult rectal temperature in 2 African forest prosimians (sympatric), in their natural habitat. Compare with human variability (from Fabre and Rougier, 1954). Trm = mean rectal temperature.*

2. *diet* (Charles-Dominique, 1971, in primary forest; Molez, 1975-76, in secondary forest): *G. elegantulus* is clearly insectivorous, *G. alleni* more frugivorous.

3. *feeding strategies* (Molez, 1975-76): *G. alleni* first eats fruits (a quickly assimilable energy source) in the first part of the night, before beginning to hunt insects.

4. *Choice of sleeping sites:* I observed *G. elegantulus* sleeping on branches, with or without a bedding of leaves, when *G. alleni* often places leaves in a hole in a tree and sleeps there.

   If the "perfect homeothermy" of *G. alleni* is possible because he lives in a buffered thermal environment, this regulation appears to be energetically very expensive. He must avoid heat loss during sleep, and restore energy immediately on awakening, more so than in *G. elegantulus.* So the thermal environments have probably exerted a non-negligible evolutionary pressure on the specific (physiological and behavioural) differentiation of the choice of ecological niche; this may be the case not only for these two sympatric prosimians, but also for other forest dwellers.

REFERENCES

Bourlière, F., Petter, J.J. and Petter-Rousseaux, A. (1956). *Mêm. Inst. Sc. Madagascar*, A, 10, 303-304.
Bourlière, F. and Petter-Rousseaux, A. (1953). *C. R. Soc. Biol.*, 147, 1594.
Bourlière, F. and Petter-Rousseaux, A. (1966). *Folia Primatol.*, 4, 249-256.
Charles-Dominique, P. (1971). *Biologia gabon.*, 7, 121-228.
Fabre, R. and Rougier, G. (1954). "Physiologie médicale", Maloine, Paris.
Goffart, M. and Hildwein, G. (1972). *C.R.S. Soc. Biol.*, 166, 10, 1382-1384.
Goffart, M. and Hildwein, G. (1975). *J. comp. Biochem. Physiol.*, 50, A, 201-213.
Heim de Balzac, H. (1969). *Ann. Fac. Sc. Cameroun*, 2, 49-58.
Hildwein, G. (1972). *Arch. Sc. physiol.*, 26, 279-385 and 387-400.
Le Conte, J. (1857). *Proc. Acad. nat. Sc. Philadelphia*, 9, 10-11.
Molez, N. (1975). "Adaptation alimentaire du Galago d'Allen (primate africain) aux milieux forestiers secondaires", Thèse 3° cycle, Univ. de Paris.
Molez, N. (1976). *Vie et Milieu*, in press.
Richards, P.W. (1964). "The tropical rain forest", Cambridge University Press.
Suckling, J.A. (1969). "The retia mirabilia of *Perodicticus potto*. A functional study", unpublished Ph.D. Thesis, University of East Africa.
Suckling, J.A. and Suckling, E.E. (1971). *In* "Research in Physiology", (F.F. Kao and K. Koizumi, eds), pp. 661-672, Aulo Gaggi, Bologne.
Suckling, J.A., Suckling, E.E. and Walker, A. (1969). *Nature*, 221, 379.
Vincent, F. (1969). "Contribution à l'étude des Prosimiens africains, le Galago de Demidoff: reproduction (biologie, anatomie, physiologie) et comportement", C.N.R.S., A.O. 3575.
Vincent, F. (1972). *Ann. Fac. Sc. Cameroun*, 9, 155-167.
Waterhouse, G.R. (1837). *Proc. zool. Soc. London*, 5, 87-88.

THE EFFECT OF THALIDOMIDE ON THE GREATER GALAGO (*GALAGO CRASSICAUDATUS*)

H. BUTLER

*Department of Anatomy, University of Saskatchewan,
Saskatoon, Saskatchewan, Canada.*

The typical human thalidomide syndrome of amelia and phocomelia can be regularly reproduced in seven species of Anthropoidea (Hendrickx, 1973). Thalidomide acts very specifically during the period of early limb bud formation between days 22 and 30 of pregnancy. Eighteen greater galagos (*Galago crassicaudatus*) were treated with presumably teratogenic doses of thalidomide between days 16 to 30 of pregnancy (Wilson, 1969). This period of pregnancy was selected on the assumption that the greater galago followed the same timetable of development as the rhesus monkey (*Macaca mulatta*). All animals gave birth to normal infants. Butler (1972) showed that the development of both the lesser galago (*Galago senegalensis*) and the greater galago lags about 10 days behind that of the Anthropoidea. Hence, it is to be expected that limb bud formation occurs between days 30 to 40 of pregnancy.

Four greater galagos were given 20 mg of thalidomide per kg of body weight on days 30 to 40 of pregnancy. The foetuses were removed by Caesarian section on about the 115th day of pregnancy. All were normal.

Adding these experiments to those of Wilson (1969) shows that 22 pregnant greater galagos have been treated with doses of thalidomide between days 16 to 40 of pregnancy. The dose used would have been highly teratogenic in the rhesus monkey and was administered during the period of early limb bud formation. Clearly, thalidomide is not teratogenic in this particular prosimian. In considering the different response between the seven species of Anthropoidea and the prosimian species, the following factors have to be considered:

1. *Time of administration.* The recent experiments took place during the period of limb bud formation which is the only time thalidomide produces limb deformities.

2. *Placental factors.* Anthropoidea have a haemochorial placenta whereas the Prosimii have an epitheliochorial placenta. The Prosimii have a very large yolk sac which forms a functional yolk sac placenta which is later replaced by an allantoic placenta. The yolk sac of Anthropoidea is vestigial and never forms a functional yolk sac placenta. The transmitting ability of allantoic and yolk sac placentas are very different (Brambell, 1970). It is noteworthy that the prosimian embryo is almost entirely dependent on the yolk sac placenta during the period of early limb bud formation.

3. *Metabolic factors.* These include maternal, foetal and placental metabolism and, unfortunately, little is known in respect of non-human primates. There are clear-cut differences in the manner in which Prosimii and Anthropoidea metabolize phenoxyacetic acid (Lister, 1974).

The present evidence indicates that the greater galago is not sensitive to thalidomide. All Prosimii have an epitheliochorial placenta and a large functional yolk sac placenta during early development. Consequently, it is most likely that all of them are insensitive to thalidomide. Hence, Prosimii, which are economical to keep and breed well in captivity, are not suitable models for the study of the teratogenic abilities of drugs used on humans.

REFERENCES

Brambell, F.W.R. (1970). "The Transmission of Passive Immunity from Mother to Young", North-Holland Publishing Company, Amsterdam.
Butler, H. (1972). *Folia primat.*, 18, 368-378.
Hendrickx, A.G. (1973). *J. Med. Prim.*, 2, 267-276.
Lister, R.E. (1974). *Biochem. Soc. Trans.*, 2, 695-699.
Wilson, J.G. (1969). "Methods for teratological studies in experimental animals and man", Igaku Shoin Ltd., Tokyo.

SECTION IV

PHYLOGENY OF *TARSIUS*
Chairman and Section Editor: M. Cartmill (North Carolina)

PHYLOGENY OF *TARSIUS*: INTRODUCTION

M. CARTMILL

Department of Anatomy, Duke University Medical Center,
Durham, North Carolina, USA.

Twice in this century, an unorthodox idea about the place of *Tarsius* in
primate evolution has led to a storm of argument and spurred primatologists
to more detailed investigations of tarsier biology. The first of these was
the "tarsioid" theory of human origins propounded by F. Wood Jones in a series
of papers and books beginning with the publication of "Arboreal Man" in 1916,
which initiated two decades of controversy between Jones and other major students of primate evolution. The second such idea is the lemuroid theory of
anthropoid origins proposed by P.D. Gingerich, which is now beginning to have
a similar stimulating effect on European and American primatology.

Jones' work prompted the Zoological Society of London to organise its 1919
symposium on tarsier affinities, published in the Society's *Proceedings* for
that year. That symposium, which could profitably be read as a preface to the
present collection of papers, marked a watershed in scientific thought about
tarsiers. Some of its participants treated *Tarsius* in an essentially pre-
Darwinian fashion; they analysed its special resemblances to several diverse
groups of primates as "annectant" features which placed it in some sort of an
intermediate or transitional position between these groups, as though it represented a crucial link in the Great Chain of Being. This mode of thought is
exemplified by the statement, which to us today is almost meaningless, with
which Woollard concluded his 1925 monograph on *Tarsius*: "... there can be no
doubt that *Tarsius* is a Lemur of the Lemurs and is annectant to the early Eocene
primitive placentals, and that standing at the base of the Primate stem it
reaches forth to the Simian forms and is annectant to the Anthropoidea."
(Woollard, 1925, *Proc. Zool. Soc. Lond.*).

Other participants in the 1919 symposium invoked cladistic principles of
the sort widely accepted today and perhaps too exclusively associated of late
with the name of Willi Hennig. In his trenchant attack on Wood Jones, Chalmers
Mitchell clearly expressed the basic principle which distinguishes this approach
from the older one: "Characters have to be judged as well as counted, if it be
intended to use them for estimating the relative degree of affinity between
animal types.... Primitive characters may be useful for the description or definition of a group — they have no value for assigning degrees of affinity."
(Mitchell, 1919).

In the present symposium, biochemical and other new sorts of data have been
brought into the argument, and new fossil information has complicated the picture and engendered a diversity of opinion which did not exist in 1919. But
as in 1919, the most important areas of disagreement concern the use of biological data in systematics. Do fossils have epistemological priority over
living animals in reconstructing phylogeny? Are two structures homologous

by virtue of their intrinsic structure, or their extrinsic relationships, or their developmental history? Does Chalmers Mitchell's principle beg the question it pretends to answer? The problem of tarsier affinities is particularly significant because it confronts us with such questions in a sharp, unambiguous way. We are grateful to Gerald Doyle and David Chivers for making it possible for us to come together and discuss these questions in Cambridge.

IF *TARSIUS* IS NOT A PROSIMIAN, IS IT A HAPLORHINE?

J.H. SCHWARTZ

Department of Anthropology, University of Pittsburgh, Pennsylvania, USA. and Section of Vertebrate Fossils, Carnegie Museum of Natural History.

INTRODUCTION

The phyletic and taxonomic position of *Tarsius* and its allies has, from time to time, been the centre of much debate and controversy. There are, at present, two major schools of thought: 1) *Tarsius* is allied with lemurs and lorises and all should be classified in the Suborder Prosimii (e.g. LeGros Clark, 1962; Simons, 1972; Simpson, 1945); and 2) *Tarsius* is more closely related to Anthropoidea and, with them, should form the Suborder Haplorhini (e.g. Hill, 1955; Luckett, 1974a, b; Martin, 1972; Szalay, 1975). In general, *Tarsius* is retained within Prosimii or allocated to Haplorhini primarily on the basis of grade or degree of morphologic (i.e. phenetic) similarity. Advocates of the haplorhine status of *Tarsius* commonly rely on neontological evidence and rarely appear cognizant of the ramifications of the fossil evidence. Conversely, reliance on the notion that tarsier-related fossils are inherently primitive forms the basis for retention of the Prosimii-Anthropoidea dichotomy. Generally shared by each school, however, is the belief that *Tarsius* represents a stage of organization intermediate between that of living prosimians and Anthropoidea (e.g. LeGros Clark, 1962; Luckett, 1974a, b) and that the omomyids (*sensu* Gazin, 1958) are generalized tarsioids from which higher primates evolved. Contrasted with either view is that proposed by Gingerich (1974, 1975a, b), who has argued that dental similarities warrant association of plesiadapiforms and tarsiiforms (=Plesitarsiformes), on the one hand, and non-tarsier prosimians and Anthropoidea (=Simiolemuriformes), on the other.

CRITERIA FOR PHYLOGENETIC ASSESSMENT

Since our concern should be with the origin of taxa, it is necessary to distinguish those characters which reflect recency of origin from those which are merely primitive retentions; the former are synapomorphic (Hennig, 1966) or shared derived character states. In other terms, synapomorphic characters are "shared derived homologues" which, at low levels of universality, can be contrasted with retained primitive homologues (i.e. homologous characters at high levels of universality) (Wiley, 1975). If one constructs a phylogeny on the basis of shared (derived) homologous character states, this can be tested because "... homologies are *potential falsifying hypotheses* of other independent homologies and the phylogenies with which these homologies are associated (Wiley, 1975: 240)."

The question I shall entertain is, "If *Tarsius* is not a prosimian, is it a

haplorhine," I suggest that the identification of the teeth of *Tarsius* and
other primates should be revised and that the resultant homologies indicate,
on the one hand, a plesiadapiform-tarsiiform clade and, on the other, a
Strepsirhini-Anthropoidea sister-group. Data from other studies will be discussed with regard to homologies and used to test this hypothesis.

TARSIUS: DENTAL MORPHOLOGY

   The dental formula of *Tarsius* is commonly accepted as 2.1.2.3./1.1.3.3
(e.g. Le Gros Clark, 1962; Simons, 1972). As generally described (e.g. Gregory, 1922; Hill, 1955; Le Gros Clark, 1962; Simons, 1961), the upper central
incisor is large, trenchant and caniniform and overwhelms the small, simple
premolariform lateral incisor; the single lower incisor is similar to the upper
lateral incisor. The C_1^1 are larger and more salient than the neighbouring incisors or the following P_2^2 but are, nonetheless, more premolariform than otherwise; the upper canine is demonstrably less caniniform and trenchant than the
upper central incisor. P_2^2 are small and thus distinguished from the posterior
two of the recognized premolar set. M_1^1-M_3^3 are of relatively simple morphology
and quite similar to those of omomyines, especially *Omomys* (cf. Gazin, 1958;
Simpson, 1940; McKenna, pers. comm.).

FEATURES OF DENTAL DEVELOPMENT AND ERUPTION

   In an earlier work, I (Schwartz, 1977) concluded that the ancestral primate
stock had possessed five premolars in its dental complement, and that this configuration had been retained in *Tarsius* as well as many tarsiiforms and plesiadapiforms. I further concluded that the anteriormost caniniform tooth in
these primates was a canine and not a transmuted incisor. This would be seen
in the upper and lower jaws of plesiadapiforms and tarsiiforms, except that
Tarsius had lost the lower canine.
   I based these conclusions on study of dental development and eruption in
term and older specimens of *Tarsius* which do not exhibit replacement at the loci
of the lower "incisor", "I^2" or "P_2^2". I thus interpreted the dental formula of
Tarsius as C dp_1^1 P_2^2 dp_3^3 P_4^4 P_5^5 M_1^1 M_2^2 M_3^3. In fetal tarsiers, however, precursors
to these deciduous teeth have been noted in some specimens (Greiner, 1929;
Leche, 1896). These precursors are resorbed and not functional postnatally.
This is similar to the condition Thomas (1894) described for the premolar region of a fetal specimen of the rodent *Thryonomys swinderianus*. This raises
some interesting interpretations of the evolution of mammalian dental replacement, which will be dealt with elsewhere (Schwartz and Krishtalka, in prep.).
For the time being, however, while the loci I suggested retained deciduous
teeth may in fact contain permanent teeth, five premolar loci are, nevertheless,
identifiable.
   One of the most fascinating aspects of tarsier development is that the lower
"incisor", "I^2" and "P_2^2" develop and erupt quite a bit earlier than the undis-
puted deciduous teeth and yet are retained and function not only with these
teeth but with their permanent successors (Dahlberg, 1948; Schwartz, 1974,
1975a). An initial interpretation of premolar development and eruption gives
the sequence "P_2^2" → "P_4^4" → "P_3^3" (ibid.). Such a sequence is also seen in the
parapithecids, a plethora of prosimians and some ceboids (cf. Schwartz, 1975b,
1977, and references therein). However, the anterior-most premolariform tooth
of these primates does not come into place, relative to "P_3^3" and "P_4^4", as early
as does "P_2^2" of *Tarsius*. In *Tarsius*, it is the premolariform "canine" which

contributes to a sequence of premolar appearance most like that of the other primates. This suggests that the tarsier "canine", which is more premolariform than caniniform, is homologous with the "$P\frac{2}{2}$" of these other primates.

In addition to the above, the majority of lemurs and lorises display a co-ordinated appearance of the upper canine and anteriormost upper and lower pre-molariform teeth (Schwartz, 1974). This integrity continues not only from de-velopment through eruption, but through the functional life of these three teeth. While the upper "canine" and "$P\frac{2}{2}$" of *Tarsius* do not appear in such inte-gration, the upper "central incisor" and upper and lower "canines" do (Schwartz, 1974, 1977). Therefore, not only can the "canine" of *Tarsius* again be homo-logized with "$P\frac{2}{2}$" of other primates, but the tarsier upper "central incisor" — which is demonstrably the most caniniform tooth — can be seen as homologous with the upper canine of other primates. With such congruity of morphology and dental development and eruption, I, therefore, suggest that the upper and lower jaws of *Tarsius* contain five premolar loci — $P\frac{1}{1}$ $P\frac{2}{2}$ $P\frac{3}{3}$ $P\frac{4}{4}$ $P\frac{5}{5}$ — as well as three molars; the anteriormost upper tooth is the canine. I further suggest that the caniniform anteriormost tooth of other tarsiiforms as well as plesiadapiforms should also be identified as a canine, behind which are maximally five premolar teeth (see Schwartz, 1977, for fuller discussion).

These suggested homologies — that a canine is the first tooth not only in the lower but also the upper jaw — were also proposed by Matthew (1915) for *Tetonius* (=*Pseudotetonius* (Bown, 1974)) and McKenna (1963) for apatemyids. How-ever, most scholars continue to identify teeth on the traditional basis of tooth position and occlusion. But, why cannot a large, caniniform tooth at the front of the jaw be homologous with a caniniform tooth in its "expected" posi-tion? There is no rationale in the literature as to why the anteriormost tooth in the jaw can only be an incisor. Nevertheless, there is almost a religious adherence to the belief that, short of being lost, canines, incisors and pre-molars will always be in the same positions in the jaws of all mammals. With this rigidity, we are left with invocations of "adaptation" and "selection" and elaborate explanations of why a tooth or teeth have come to look and function like another.

There is no reason why incisors cannot be lost and the canine developed as the first tooth in the jaw. The premaxillary-maxillary suture is not a struc-ture which has bearing on the segmentation and differentiation of tooth sets. Indeed, as demonstrated, for example, in the mouse and *Homo sapiens*, the meso-dermal determinants of both tooth formation and morphology are established well before the onset of ossification of facial and mandibular bone (Kollar and Baird, 1971; Miller, 1971; Tonge, 1971). Therefore, it is bone differentiation which occurs after tooth position and which tooth in which locus has been determined, and not *vice versa*.

With regard to the above-argued dental homologies, *Tarsius* is obviously con-trasted with strepsirhines and Anthropoidea. The latter two groups, with the exception of *Daubentonia*, possess incisors and canines in their expected positions while *Tarsius* is most like other tarsiiforms and plesiadapiforms in that (from available evidence) caniniform teeth are anteriormost in the jaw. I suggest that "loss" of incisors and concomitant development of the canine at the front of the jaw is a shared derived character of plesiadapiforms and tarsiiforms and that this clade is a sister-group of a strepsirhine-anthro-poid group. Derived characters which would have characterized the ancestral strepsirhine-anthropoid stock include tooth "loss" at the $P\frac{3}{3}$ locus.

OTHER CHARACTERS

The sister-groups suggested here (also in Schwartz, 1977) are the same as those Gingerich (1974, 1975a, b) proposed from study of dental morphology and are further supported by Butler's (1973) work on homologous wear facets. Studies of the ontogeny of the auditory bulla indicate that strepsirhines and Anthropoidea possess a totally petrosal bulla (Cartmill, 1975; Kampen, 1905; Major, 1899; Starck, 1975), whereas *Tarsius* has an entotympanic contribution (Starck, 1975; Martin, pers. comm.; Schwartz, ms.). Accepting a totally entotympanic bulla as primitive (cf. McKenna, 1966), the polarity of the developmental morphocline again suggests that non-microsyopid plesiadapiforms and tarsiiforms as well as Strepsirhini and Anthropoidea, respectively, are sister-groups. Thus, studies of dental morphology, development of the auditory bulla and homologies based on the correlation of dental development and eruption with dental morphology all lead to the same general interpretation of the relationships of higher categories of primates.

On the other hand, study of the placenta and fetal membrances in primates has led Luckett (1974a, b) to suggest that *Tarsius* is more closely related to Anthropoidea than to strepsirhines. The major similarities cited in support of a *Tarsius*-Anthropoidea sister-group are: the development of a mesodermal body stalk; a rudimentary allantoic vesicle; a small, free yolk sac; the presence of a primordial amniotic cavity; lack of a choriovitelline placenta; and the development of a discoidal, hemochorial placenta. These characters are considered synapomorphic for *Tarsius* and Anthropoidea and derivable from the primitive Eutherian condition which, for the most part, is retained in lemurs and lorises: i.e. amniogenesis entirely by folding; an initially large, free yolk sac; development of a choriovitelline placenta prior to the chorioallantoic placenta; a large, permanent allantoic vesicle; and development of a diffuse, epitheliochorial placenta. However, the question arises whether the similarities between *Tarsius* and Anthropoidea indicate retention from a common ancestor or are autapomorphic.

Contrasted with that of strepsirhines and Anthropoidea is the process of amnion formation in *Tarsius*. As in sauropsids and most mammals, amniogenesis in strepsirhines occurs by folding; in these primates, Rauber's layer is lost prior to implantation. Amnion formation in anthropoids proceeds by cavitation — creating a primordial amniotic cavity — Rauber's layer remaining intact. In *Tarsius*, on the other hand, Rauber's layer is lost prior to implantation and, while amniogenesis begins with cavitation — at this stage a primordial amniotic cavity is formed — definitive amniogenesis occurs by folding. Luckett (op.cit.) has suggested that the initial period of amniogenesis by cavitation in *Tarsius* is homologous with the uninterrupted process in Anthropoidea and that this character is one of the derived states between these primates. Taken as a complex, however, the entire process of amnion formation in *Tarsius* is unique among primates and, thus, should more plausibly be interpreted as an independently derived character state.

Luckett (1974a, b) also suggests that discoidal, hemochorial placentation is a shared derived character of *Tarsius* and Anthropoidea. The development of this type of placentation, however, is preceded by the implantation of the blastocyst to the uterine endometrium which is different in *Tarsius*: i.e. the blastocyst attaches by the paraembryonic trophoblast to the mesometrial pole whereas, in anthropoids, attachment is by the embryonic trophoblast to the orthomesometrial pole of the uterine endometrium. Thus, as has also been suggested for the presence of hemochorial placentation in macroscelidids, erinaceids, dasypodids, Dermoptera and many rodents (ibid.), it would appear that *Tarsius* and Anthropoidea have independently developed this morphology. Regardless of this, Luckett (1974a) maintains "...that the differences in initial attachment and amniogenesis between Tarsioidea and Anthropoidea are the result of the development of a simplex uterus in Anthropoidea (p. 221)." Luckett (1974a, b) further suggests that, if *Tarsius* had a simplex rather than a bicornuate uterus, the features of blastocyst implantation would be the same as in Anthropoidea. However, while *Tarsius* does not have a simplex uterus, it has, by different means of blastocyst implantation, developed hemochorial placentation. Additionally, many of the taxa which Luckett cites as convergently developing hemochorial placentation have bicornuate uteri (Vaughan, 1972) and also display the same characters of blastycyst implantation as do Anthropoidea (Luckett, op.cit.). Therefore, contrary to Luckett's insistence, it would appear that *Tarsius* is not intermediate between strepsirhines and Anthropoidea but, in its derived characters, is unique among extant primates. Furthermore, since hemochorial placentation only occurs by the precocious establishment of a chorioallantoic placenta which bypasses a transitory choriovitelline stage and this, in turn, is correlated with rudimentary development of an allantoic diverticulum and precocious differentiation of a mesodermal body stalk (as evidenced in *Tarsius*, Anthropoidea, dasypodid edentates and various rodents) (ibid.), these characters cannot be considered separately in determining phylogenetic relationships. Thus, the discernable derived characters of *Tarsius* are not homologous with those of Anthropoidea and do not support a sister-group relationship between these two taxa.

Immunodiffusion and electrophoretic studies on blood serum proteins, in recent years, have been offered as the ultimate means of deciphering the "real" phylogenetic relationships of organisms. This is based on the argument that the genome is the basic unit of inheritance and, thus, since the genotype determines the phenotype, unravelling the genetic code will reveal the pathways of evolutionary change (Goodman, 1975; Zuckerkandl, 1963). Indeed, the title of Lewontin's (1974) recent work, *The genetic basis of evolutionary change*, reflects this general belief. However, perhaps it would be more fruitful and realistic for these studies to approach the problem of phylogeny reconstruction from a different point of reference, i.e. with regard to "the evolutionary basis of genetic change" (Eldredge, pers. comm.). For, after all, proteins, DNA, genes, etc. are morphological complexes which should be subject to the same rigors of interpretation as "phenotypic" morphologies.

The morphological complexes would embody sequence data on proteins and especially DNA about which, however, very little is known (Baba et al., 1975; Kohne, 1975). Therefore, phylogenies derived from molecular data are at present primarily based on phenetic (antigenic) similarities of assumed homologous sites rather than on the results of isolating shared primitive from shared derived *known* molecular homologues.

Inherent in the problems of using antigenic similarity for phylogeny reconstruction is the dilemma of sorting out convergent molecular character states (cf. Cook and Hewett-Emmett, 1974). Furthermore, as Goodman (1963) has argued, it appears that the differentiation of certain molecular structures may be correlated with the types of placentation and, as a result, the accessibility of the maternal immunological system. Thus, the resemblances, for example, between *Tarsius* and *Homo sapiens* in hemoglobin and adenylate kinase banding patterns (Barnicott and Hewett-Emmett, 1974), if not primitive retentions, may well be convergences correlated with the independent (as argued above) development of hemochorial placentation. In fact, if a consequence of hemochorial placentation is selection against "...divergence from the ancestral genetic code (Goodman, 1963: 212)", perhaps many of the molecular similarities between *Tarsius* and Anthropoidea are merely primitive retentions (cf. Baba et al., 1975: 97-98). Nevertheless, molecular — and I would also include kary-ological — studies do not, at present, offer the kind of data which are neces-sary to falsify hypotheses concerning the phylogenetic relationships of taxa (also see discussion of Thorington, 1970).

CONCLUSION

   The two hypotheses I question in my title are basically rewordings of the same hypothesis: *Tarsius* is intermediate between strepsirhines and Anthropoi-dea. For years, systematists have been working under the assumption that this evolutionary schema is real and, thus, the task has been to push the pendulum — with *Tarsius* clinging to its end — closer to one or the other group of pri-mates. Similarity has been taken to represent relatedness and the "principle of parsimony" is the guiding light.

   Not only has the hypothetical intermediacy of *Tarsius* become a truism, but the method of determining dental homologies — upon which phylogenetic rela-tionships are a good deal based — has not been subjected to even the slightest questioning. I have attempted to do this with developmental data on primates and believe that this age-old hypothesis — that teeth will be in the same positions in the jaws of all mammals — has been falsified. The next step, obviously, is to try to falsify this new hypothesis: that teeth other than incisors can develop in the premaxilla and as the anteriormost tooth in the mandible. However, to reject this hypothesis through reasoning which is based on the traditional approach — e.g. because of the belief that the lower canine is not supposed to occlude behind the upper canine, or because the caniniform anteriormost tooth of other mammals is still identified as a trans-muted incisor — is circular and not a proof of falsification.

   Concern with "trends", "adaptive arrays", and the pursuit of ancestor-descendent relationships has led to much confusion in systematic studies — confusion which also results in the relegation of certain taxa to wastebasket status. The now oft-quoted phrase "primates of modern aspect" reflects a bias which obscures rather than clarifies attempts to understand phylogenetic relationships between taxa whether they be fossil or extant. Obviously, extant taxa and their morphologies are modern, in the sense that they are around now. However, this is a truncated view of an ongoing phenomenon — it is fortuitous which taxa will remain to be seen as "modern".

   An objective evaluation of the characters of all living and fossil forms should result in synapomorphies which indicate not only monophyly but the cohesiveness and reality of higher taxonomic groups. Thus, within mammals, one can discern the derivedness of molar morphology which delineates the Order Primates (cf. Gregory, 1922; Simpson, 1935, 1940) and would also have charac-terized the ancestral primate stock. Within this group of "dental primates", the ancestral stock of microsyopids, plesiadapiforms and tarsiiforms would have been characterized by the loss of incisors and the development of the canine at the front of the jaw while the ancestral strepsirhine-anthropoid stock would have been characterized by the reduction from five to four premolars, the loss of the medial branch of the carotid artery as well as the development of a post-orbital bar, a petrosal bulla and nails at least on the hallux and pollex. In the ancestral plesiadapiform-tarsiiform stock, the medial branch of the carotid artery would not have been present and the auditory bulla would have been com-

pound. And, finally, the ancestral tarsiiform stock would have been characterized by the presence of a postorbital bar and nails on at least the hallux and
pollex.

ACKNOWLEDGEMENTS

I thank Drs. M. Cartmill, N. Eldredge, L. Krishtalka and I. Tattersall for
discussion and criticism of the manuscript. Research and travel were funded
in part by The Wenner-Gren Foundation for Anthropological Research and a John
G. Bowman Faculty Grant, University of Pittsburgh.

REFERENCES

Baba, M.L., Goodman, M., Dene, H. and Moore, G.W. (1975). *J. Human Evo.*, 4,
    89-102.
Barnicott, N.A. and Hewett-Emmett, D. (1974). *In* "Prosimian Biology", (R.D.
    Martin, G.A. Doyle and A.C. Walker, eds), pp. 891-902, Duckworth Press,
    London.
Bown, T.M. (1974). *Contr. Geol.*, *Univ. Wyo.*, 13, 19-26.
Butler, P.M. (1973). *In* "Craniofacial Biology of the Primates", (M.R. Zingeser,
    ed.), Symp. IVth Int. Congr. Primat., vol. 3, pp. 1-27, Karger, Basel.
Cook, C.N. and Hewett-Emmett, D. (1974). *In* "Prosimian Biology", (R.D. Martin,
    G.A. Doyle and A.C. Walker, eds), pp. 937-958, Duckworth Press, London.
Dahlberg, A.A. (1948). *Amer. J. Phys. Anthrop.*, 6, 239-240.
Gazin, C.L. (1958). *Smithsonian Misc. Coll.*, 136, 1-112.
Gingerich, P.D. (1974). Ph.D. thesis, Yale University.
Gingerich, P.D. (1975a). *Univ. Mich. Contr. Mus. Paleon.*, 13, 135-148.
Gingerich, P.D. (1975b). *Nature*, 253, 111-113.
Goodman, M. (1963). *In* "Classification and Human Evolution", (S.L. Washburn,
    ed.), pp. 204-234, Aldine Publishing Co., Chicago.
Goodman, M. (1975). *In* "Primate Functional Morphology and Evolution", (R.H.
    Tuttle, ed.), pp. 193-199, Aldine Publishing Co., Chicago.
Gregory, W.K. (1922). "The Origin and Evolution of the Human Dentition",
    Williams and Wilkins Co., Baltimore.
Greiner, E. (1929). *Z. Anat. Entw.*, 89, 102-122.
Hennig, W. (1966). "Phylogenetic Systematics", University of Illinois Press,
    Urbana.
Hill, W.C.O. (1955). "Primates, Comparative Anatomy and Taxonomy, vol. II,
    Haplorhini: Tarsioidea", University Press, Edinburgh.
Kampen, P.N. van. (1905). *Morph. Jn.*, 34, 321-722.
Kohne, D. (1975). *In* "Phylogeny of the Primates", (W.P. Luckett and F.S.
    Szalay, eds), pp. 249-261, Plenum Press, New York.
Kollar, E.J. and Baird, G.R. (1971). *In* "Dental Evolution and Morphology",
    (A.A. Dahlberg, ed.), pp. 15-30, The University of Chicago Press.
Leche, W. (1896). *Festschr. für Gegenbaur*, Leipzig, 3, 127-166.
Le Gros Clark, W.E. (1962). "The Antecedents of Man", (2nd ed.), Edinburgh
    University Press.
Luckett, W.P. (1974a). *In* "Reproductive Biology of the Primates", (W.P.
    Luckett, ed.), Contr. Primat., vol. 3, pp. 142-234, Karger, Basel.
Luckett, W.P. (1974b). *In* "Prosimian Biology", (R.D. Martin, G.A. Doyle and
    A.C. Walker, eds), pp. 475-488, Duckworth Press, London.
Major, C.I.F. (1899). *Proc. Zool. Soc. Lond.*, 1899, 987-988.
Martin, R.D. (1972). *Phil. Trans. R. Soc. Lond.*, 264, 295-352.

Matthew, W.D. (1915). *Amer. Mus. Nat. Hist. Bull.*, 34, 429-483.
McKenna, M.C. (1963). *Amer. Mus. Nov.*, no. 2160, 1-39.
McKenna, M.C. (1966). *Folia primatol.*, 4, 1-25.
Miller, W.A. (1971). *In* "Dental Evolution and Morphology", (A.A. Dahlberg, ed.), pp. 31-44, The University of Chicago Press.
Schwartz, J.H. (1974). Ph.D. thesis, Columbia University.
Schwartz, J.H. (1975a). *Folia primatol.*, 23, 290-307.
Schwartz, J.H. (1975b). *In* "Lemur Biology", (I.M. Tattersall and R.W. Sussman, eds), pp. 41-63, Plenum Press, New York.
Schwartz, J.H. (1977). *Spec. Pub. Carnegie Mus. Nat. Hist.*, in press.
Simons, E.L. (1961). *Postilla*, no. 54, 1-29.
Simons, E.L. (1972). "Primate Evolution", MacMillan Co., New York.
Simpson, G.G. (1935). *Amer. Mus. Nov.*, no. 816, 1-30.
Simpson, G.G. (1940). *Bull. Amer. Mus. Nat. Hist.*, 77, 185-212.
Simpson, G.G. (1945). *Bull. Amer. Mus. Nat. Hist.*, 85, 1-350.
Starck, D. (1975). *In* "Phylogeny of the Primates", (W.P. Luckett and F.S. Szalay, eds), pp. 357-404, Plenum Press, New York.
Thomas, O. (1894). *Ann. Mag. Nat. Hist.*, 13, 310-322.
Thorington, R.W., Jr. (1970). *In* "Old World Monkeys", (J.R. Napier and P.H. Napier, eds), pp. 3-15, Academic Press, New York.
Tonge, C.H. (1971). *In* "Dental Evolution and Morphology", (A.A. Dahlberg, ed.) pp. 45-58, The University of Chicago Press.
Vaughan, T.A. (1972). "Mammalogy", W.B. Saunders Co., Philadelphia.
Wiley, E.O. (1975). *Syst. Zool.*, 24, 233-243.
Zuckerkandl, E. (1963). *In* "Classification and Human Evolution", (S.L. Washburn, ed.), pp. 243-272, Aldine Publishing Co., Chicago.

DISCUSSION

Joysey (Cambridge): You commented, quite rightly, that one can't use traditional evidence to falsify your new dental homologies. What sort of evidence
might falsify them? Have you any other criteria to offer? Unless there is a
second method, it is impossible to falsify the first.

Schwartz: I think we need to know more precisely what's going on in the actual
differentiation of facial and dental structures, and this is of course being
worked on by others. What I'm starting to do is to look at sequences of tooth
development and eruption in various insectivores with caniniform anterior teeth
to see if similar arguments apply in those cases.

Cartmill (Duke University): What is the evidence for your suggestion that
Tarsius retains an entotympanic element in the bulla?

Schwartz (projecting a slide): This is the skull of a juvenile *Tarsius* which
shows a bilateral opening on the medial aspect of the hypotympanic sinus,
anterior to the carotid foramen. The edge of the opening represents a separate entotympanic element that was either not yet ossified or else lost in
preparation. Some other specimens show vestiges of a suture delimiting this
area.

Cartmill: Bob Martin has also suggested that *Tarsius* retains an entotympanic.
Bob, is this the same element you identified?

Martin (Wellcome Institute): No. The entotympanic that I would describe is
more lateral, anterior to the constriction in the bulla marked by the carotid
foramen. We have a juvenile specimen in the British Museum of Natural History
that apparently has a separate ossified element here. With a scanning elec-
tron microscope, you can find a suture at the posterior margin in some adult
tarsiers. That does not necessarily indicate the presence of two separate
bones; it might be a petroso-petrosal suture, like the palato-palatine suture
found in the orbit of *Tupaia*, where the palatine grows back to fuse with itself. The scepticism we may feel here about the presence of an entotympanic
element (or perhaps even two different entotympanics) in the tarsier bulla is
a salutary lesson for the interpretation of fossil material, where it is even
more difficult to say whether an entotympanic was present.

Cartmill (projecting a slide): This is the skull of a newborn *Leontopithecus
rosalia*. It appears to have a separate entotympanic element, which (as you
can see from its spiculate front edge) is beginning to ossify the floor of the
bulla's anterior chamber (the "hypotympanic sinus"), in front of the transverse septum that contains the carotid canal. This element is in much the
same position as that described by Jeff Schwartz. If this is a separate element in both animals, it might represent a haplorhine symplesiomorphy rather
than a plesiomorphy unique to *Tarsius*. But we can't be sure it's a separate
element until we look at sectioned material.

Maier (Frankfurt): I've studied virtually all genera of New World monkeys by
serial section, and found not the slightest evidence for any entotympanic.
Dried museum specimens tell us little, because you can't distinguish cartilage

and connective tissue adequately.

Cartmill: Do you agree with Starck's identification of a cartilaginous element in the tarsier bulla?

Maier: Dr. Starck was only referring to the enchondral mode of ossification of the middle-ear floor in *Tarsius*, and he suggested that this might indicate the presence of a "true" entotympanic element. I personally doubt that the histogenesis of the bulla in *Tarsius* can prove anything. I would interpret the os bullae as a *Zuwachsknochen* containing a few cartilaginous islets, which even occur sometimes in connection with membrane bones.

CRANIO-DENTAL MORPHOLOGY, TARSIER AFFINITIES, AND PRIMATE SUB-ORDERS

M. CARTMILL and R.F. KAY

Department of Anatomy, Duke University Medical Center, Durham, NC, USA.

Most primatologists acknowledge that the extant primates comprise three monophyletic natural groups: tooth-comb prosimians (lemuriforms and lorisiforms), tarsiers, and anthropoids. Among early primates three groups are also generally recognized: plesiadapoids (archaic primates with relatively large faces and diminutive open orbits), the lemur-like adapids, and a variously named and defined group of "tarsioids" for which we will use the nomen Omomyidae (*sensu* Szalay, 1975b). How are these six groups related phyletically? Only two possibilities currently have defenders. We will refer to them as the "simiolemuriform-plesitarsiiform" (SP) and "haplorhine-strepsirhine" (HS) models, after the subordinal nomina each uses (Fig. 1, A-B). We feel there are flaws in both models, and we wish to point them out and suggest a plausible alternative.

THE S-P MODEL

The SP model was originally formulated by Gingerich (1974), who proposes that primitive plesiadapoids gave rise to adapids and to a more derived plesiadapoid group, the latter in turn giving rise to omomyids (and thence to *Tarsius*). Anthropoids and tooth-comb prosimians are both viewed as adapid descendants. This model is supported by "similarities in dental conformation and middle ear morphology" (Gingerich, 1975a) linking omomyids to plesiadapoids, and by resemblances between some adapids and early anthropoids.
Gingerich (1975a) notes two dental similarities between omomyids and advanced plesiadapoids: enlarged central incisors and loss of the small P1 which adapids retain from a *Purgatorius*-like ancestor (Clemens, 1974). Parallelism is likely here, since the early omomyid *Teilhardina* has small, roughly vertical incisor sockets and may have retained four premolars (Simons, 1960). Incisor enlargement and premolar reduction have occurred in parallel in many mammalian lineages, and their phyletic valence is not great.
Gingerich (1973, 1974, 1975a, 1976) lists four respects in which the bulla of *Plesiadapis* resembles that of the omomyid *Necrolemur*: (1) "Extended tubular auditory meatus." (2) "Tympanic anulus fused into the wall of the bulla...." (3) "... by struts, rather than direct fusion into the lateral wall." (4) "Lateral extension of the ectotympanic to form an external auditory tube." The last item is unfounded; since no sutures are visible in the fossils, the tubular part of the meatus in plesiadapoids or omomyids may have ossified as an extension of the petrosal (like that seen in some *Microcebus*) or of an independent meatal element (as in *Tupaia*). Both plesiadapoids and omomyids resemble adapids in having a topologically primitive, lemur-like tympanic cavity extend-

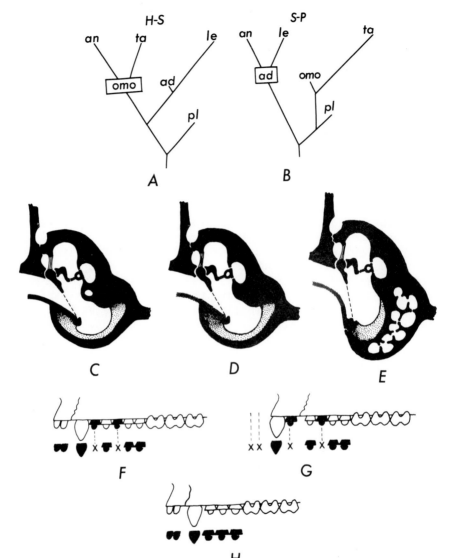

Figure 1. A, "haplorhine-strepsirhine" (H-S) model of primate relationships, contrasted with B, the "simiolemuriform-plesitarsiiform"
(S-P) model. Abbreviations: ad, adapids; an, anthropoids; le,
lemurs (= tooth-comb prosimians); omo, omomyids; pl, plesiadapoids. Boxes indicate taxa of somewhat uncertain phyletic affinities in each model. C-E: diagrammatic coronal sections of the
bulla and tympanic cavity in (C) plesiadapoids and omomyids;
(D) the cheirogaleid Allocebus; and (E) callitrichids (except
Saguinus). Black, bone; uniform stipple, soft tissues; the
hand stipple indicates transverse septa of the bullar floor.
F-G: Diagrammatic representations of Schwartz' hypotheses
concerning tooth loss and homologies in some primate groups;
upper dentitions, viewed from the left. F, the hypothetical
ancestral primate condition, differing from (legend cont.)

Figure 1. (legend cont.) *Gypsonictops only in incisor number; G, the condition in plesiadapoids and tarsioids. H, actual upper dental formula of Tarsius, based on our own work and that of Leche and Greiner. The adult upper dentitions in F-H are shown in place in the upper jaw. The maxillary-premaxillary suture is indicated by a sinuous line. Black symbols represent deciduous teeth, some of which may persist unreplaced in the adult; white symbols represent permanent teeth (molars and replacing antemolars); X's indicate positions where tooth loss has occurred by the developmental failure of deciduous and/or permanent teeth.*

ing laterally below the eardrum. They differ from adapids in exhibiting a derived ossification of the medial end of the meatus (anulus membrane plus part of the extrabullar meatus) and in having shallow transverse septa ("struts") running across the bullar floor medially from the ossified anulus membrane (Fig. 1, C). Similar traits have been derived independently in *Allocebus* and typical callitrichids (Fig. 1, D-E), demonstrating that parallelism is at least possible here.

Gingerich also proposes several resemblances linking some adapids to early anthropoids: (1) small, vertically implanted, spatulate lower incisors; (2) I_2 larger than I_1; (3) upper canine developing a "honing" wear facet against an enlarged anterior lower premolar; (4) fusion of the mandibular symphysis; and (5) a lemur-like ear region with a "free" and "intrabullar" tympanic ring. We doubt that these resemblances represent synapomorphies. Some early omomyids (*Teilhardina, Anaptomorphus, Washakius*) appear to have had small, gently procumbent lower incisors, subequal in size, and this may be primitive for primates of modern aspect. Some adapids have a projecting lower premolar bearing a long, trenchant wear facet produced by the upper canine, and this "canine hone" is admittedly like that seen in cercopithecoids and the supposed primitive catarrhine *Oligopithecus*. However, *Oligopithecus* may not be a catarrhine (Szalay, 1970; Kay, 1977). In most of the undoubted early catarrhines (*Apidium, Parapithecus, Propliopithecus*), the anterior lower premolar has a short, blunt anterior edge and does not project above the occlusal plane. A "canine hone" has been convergently developed in *Lemur fulvus* (Fig. 2).

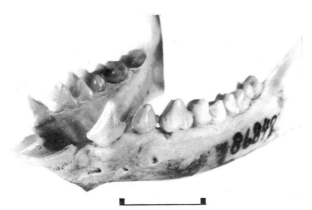

Figure 2. *The lower jaw of Lemur fulvus, U.S. Natl. Mus. 86849, viewed anterolaterally. P₂, the first tooth behind the tooth comb, displays a large wear facet where it wears against the posterior edge of the upper canine. The bar indicates one centimetre.*

Partial or complete fusion of the mandibular symphysis has evolved in many
groups of herbivorous mammals — e.g., *Ailuropoda*, elephants, various ungulates,
and indriids. This parallelism is explicable on biomechanical grounds (Beecher,
1977), and is not a reliable indicator of phyletic affinity. Finally, the
supposedly lemur-like petrosal fragment attributed to the Oligocene catarrhine
Apidium (Gingerich, 1973) is essentially ceboid-like (Cartmill, 1975), but even
if it were not, a lemur-like bulla is primitive for primates in Gingerich's
view (and in ours), and is thus no proof of phyletic affinities between groups
that retain it.

Schwartz (MS; Schwartz and Krishtalka, 1977) offers arguments for the SP
model based on Mc'Kenna's (1975) interpretations of dental homologies in late
Cretaceous placentals. Young individuals of *Gypsonictops* and *Kennalestes* have
five premolars. The P3 is small and inconstant, and is not found in older individuals. McKenna (1975) proposes that this tooth and the small anterior premolar are retained deciduous teeth whose permanent successors have been sup-
pressed (Fig. 1, F). Schwartz (unlike McKenna) thinks that a similar premolar
formula characterized early primates, and persisted among the "plesitarsii-
forms". His argument runs thus: (1) The antepenultimate premolar in *Tarsius*,
omomyids, and plesiadapoids is smaller than its neighbours, and is evidently
in the process of being reduced. (2) In *Tarsius*, and perhaps in the omomyid
Absarokius, this tooth is not replaced. By analogy with *Gypsonictops*, it is
interpreted as a dP3. (3) The antepenultimate premolar in "simiolemuriforms"
is replaced; Schwartz thus regards it as homologous with the P2 of *Gypsonictops*.
(4) In *Tarsius*, the second tooth in front of the supposed dP3 is also unreplaced; therefore, it is homologous with the dP1 of *Gypsonictops*. Adapids
also retain this dP1, but it has been lost (like the dP3) in anthropoids and
tooth-comb prosimians. (5) The most anterior tooth in *Tarsius*' maxilla,
traditionally regarded as a canine, is by the foregoing analysis a P2. This
yields a premolar eruption sequence (P2→P5→P4) like that seen in anthropoids
and strepsirhines, and provides further support for Schwartz' novel dental
homologies. Similar reasoning applies to the lower teeth. (6) The anteriormost upper tooth of omomyids and plesiadapoids, traditionally regarded as an
enlarged incisor, lies directly mesial to the newly-recognized dP1; this
fact and its large size prove it homologous with the canine of anthropoids and
lemurs. On similar grounds, Schwartz identifies the enlarged lower front tooth
of plesiadapoids and omomyids as a canine. *Tarsius* has lost this lower canine;
its antemolar dental formula is thus C^1 dP_1^1 P_2^2 dP_3^3 P_{4-5}^{4-5} (Fig. 1, G).

If this account is sound, plesiadapoids and tarsioids share several synapomorphies, notably loss of all incisors and displacement of canines and premolars
into the premaxilla. We reject Schwartz' account for the following reasons:
1. All the antemolar teeth of *Tarsius* are in fact permanent teeth with deciduous precursors (Fig. 1, H), though the deciduous incisors and antepenultimate deciduous premolar are tiny and may be shed or resorbed before birth
(Leche, 1896; Greiner, 1929). The sectioned *Tarsius bancanus* material which
we examined, and on which we will report more fully elsewhere, clearly retained
small deciduous precursors of the I^1 and P_2. Recognition of the adult tarsier'
antepenultimate premolar as P2 implies that *Tarsius* has a typical primate premolar eruption sequence, and obviates comparisons with *Gypsonictops*.
2. Schwartz' dental homologies imply that "plesitarsioids" differ from other
mammals in having a lower canine which occludes *behind* the upper one. We regard this as unlikely.
3. If Schwartz is right, the premaxillary teeth of *Tarsius* are premolars and
canines. Schwartz apparently takes this to mean that the premaxillary-

maxillary suture has shifted backward relative to the dental lamina of the upper jaw. But the upper dental lamina is not a unitary structure. Its formation is induced by neural-crest cells migrating into the frontonasal eminence (the ancient front end of the vertebrate head) and the maxillary process (derived from the first branchial arch), whose fusion to form a single "upper jaw" (and hence a single dental lamina) is developmentally and phylogenetically secondary. When this fusion fails to occur in fetal mammals, both sets of incisors develop anyway in the isolated premaxilla suspended from the nasal septum. This shows that the premaxillary part of the dental lamina is induced by neural-crest cells which enter the frontonasal process before it fuses with the maxillary process. However, a tooth which develops in one bone may secondarily migrate into the other. In rodents, for instance, the upper incisor's root lies in the maxilla, and so it might be identified as a canine, as *Daubentonia's* upper incisor has been on similar grounds (Tattersall and Schwartz, 1974). But when premaxilla and maxilla fail to fuse in rodents (and presumably in *Daubentonia* too), the upper incisor root grows backward from its original premaxillary locus into the nasal septum, taking the only caudad pathway available (Reed and Snell, 1931). To claim, in the face of this, that the upper incisors of rodents are really canines could only mean that the lamina-inducing cells which migrate into the frontonasal processes of rodents are in some sense homologous with cells which migrate into the maxillary processes of other mammals. Whether or not we would still want to call the resulting teeth homologous is a matter of definition. Yet some such ambiguous and untestable assertion is the only sense we can place on Schwartz's claim that tarsiers lack upper incisors, since published studies of dental ontogeny in *Tarsius* appear to rule out the possibility that the premaxillary teeth of the adult form originally in the maxilla and migrate forward secondarily. We conclude that Schwartz' novel dental homologies are either empirically false or untestable, depending on how we interpret them. They therefore provide no warrant for believing in the SP model.

Yet, it must be admitted that Gingerich and Schwartz have identified several derived features unique to "plesitarsioids", and others which link anthropoids to some adapids. Our rejection of the SP model forces us to dismiss these resemblances as parallelisms. We do so not from mere reactionary perversity, but to avoid a similar dismissal of the traits linking all primates of modern aspect (postorbital bar, enlargement and forward rotation of the orbits, replacement of claws by nails on at least some digits, interorbital narrowing and olfactory reduction), which would have had to appear in parallel if the SP model is correct. The SP model implies that an ancestor something like *Palaechthon*, whose cranial and dental morphology suggests an adaptation resembling that of *Hylomys* (Kay and Cartmill, 1977), gave rise independently to lemur- and tarsier-like prosimian lineages by the beginning of the Eocene. This is not impossible, and perhaps does not alone warrant rejecting the SP model; but it suffices when we take into consideration the many apomorphous features of hard and soft anatomy that link *Tarsius* to the living and fossil Anthropoidea. These features, to which we now turn, must also be explained away as parallelisms if the SP model is adopted.

THE H-S MODEL

If we consider only extant primates, Pocock's (1918) suborders Strepsirhini (tooth-comb prosimians) and Haplorhini (tarsiers and anthropoids) can each be defined by shared derived traits not found in the other: Strepsirhini by the

possession of a tooth comb or some derivation thereof, and Haplorhini by
shared apomorphous features of placentation, carotid circulation, neurology,
narial and olfactory apparatus, and biochemistry. Many of these features are
reviewed by other contributors to this session. No known synapomorphies link
Tarsius (or anthropoids) to any tooth-comb prosimians. The neontological data
therefore indicate that Haplorhini and Strepsirhini (*sensu* Pocock) are monophyletic groups.

Many have tried to apply the haplorhine-strepsirhine division to fossil
primates as well. Szalay (1975a) envisions an initial split between plesiadapoids and primates of modern aspect, with the latter then dividing into Strepsirhini (adapids and their extant strepsirhine descendants) and Haplorhini
(omomyids, anthropoids, and *Tarsius*). Most investigators now subscribe to some
such model (Fig. 1, A).

In assigning omomyids to Haplorhini, Szalay stresses the relatively large
promontory artery and the posteromedial entrance of the internal carotid into
the bulla. However, any mammalian lineage undergoing fore-brain enlargement
might be expected to enlarge the major vessel feeding the anterior branches
of the cerebral arterial circle, and lorisiforms have also shifted the carotid
bullar entrance medially from its primitive posterolateral position. These features in omomyids are thus not a wholly conclusive sign of haplorhine affiniti
Adult *Necrolemur* and *Microchoerus* (and probably *Pseudoloris*) retain an unpneu-
matized apical interorbital septum, like that of *Tarsius* and small anthropoids
(Cartmill, 1975). *Tetonius* shows a similar but smaller septum in this positio
Since the orbits of known omomyids are no larger than those of similar-sized
strepsirhines whose orbital apices are widely separated (Kay and Cartmill,
1977), this feature cannot be dismissed as a convergence produced by the optic
hypertrophy sometimes claimed for *Tetonius* and *Necrolemur* (Simons, 1972; Szalay
1975b). We believe that the persistent fetal septum and the correlated compression of the adult's olfactory fossa (Cave, 1967) are probably haplorhine
synapomorphies, conditioned to some extent by allometry. This and the bullar
features mentioned above warrant the very tentative conclusion that omomyids
are more closely related to extant haplorhines than to other primates.

However, there is no sound evidence that adapids are similarly related to
tooth-comb primates. The traditional arguments for such a relationship rest
on the lemur-like bullae and carotids of adapids. But these are primate symplesiomorphies, of little relevance to phyletic affinities within the order.
Most of the other diagnostic strepsirhine features — naked rhinarium, large
olfactory apparatus, relatively small brain, epitheliochorial placenta — are
also primitive retentions. The extant strepsirhines share certain derived
features, such as the tooth comb, which suggest that they are a monophyletic
group; but these features are not seen in adapids. Gingerich (1975b) notes
that *Adapis parisiensis* has short lower canines whose edges are incorporated
into the incisor series. He suggests that this arrangement may represent a
precursor of the tooth comb. We fail to see how the incorporation of the
canines into a cropping mechanism involving broad, spatulate teeth could be
preadaptive to evolving a tooth comb made up of slender, procumbent lower
teeth that do not occlude with their upper counterparts. Again, other mammalian lineages have developed incisiform lower canines as part of a shearing
mechanism — e.g., cainotheres, pecoran artiodactyls, and *Homo*.

In short, no synapomorphies linking adapids to living strepsirhines have
so far been identified. If the suborder Strepsirhini is defined so as to
include the Adapidae, it is a taxonomic wastebasket for primates of modern
aspect that are not demonstrably haplorhines. Some systematists (e.g., McKenna

1975) have adopted a subordinal division of Primates into Plesiadapiformes, Strepsirhini and Haplorhini, on the supposition that this replaces the old grade boundary between Prosimii and Anthropoidea with a more vertical classification. In fact, this erects two grade boundaries where Simpson (1945) contented himself with one. Since both Strepsirhini and Prosimii are paraphyletic with respect to their sister groups, our choice between them must be dictated by convenience. As Simons (1974) points out, the haplorhine-strepsirhine partition is paleontologically inconvenient, since it would leave many early Prosimii "In an *incertae sedis* position between the two suborders". This is particularly true of the adapids, which are usually regarded as precursors or cladistic sisters of the Madagascar lemurs, but might with equal reason (or lack of it) be regarded as persistently primitive haplorhines. (Such a move would reconcile the SP model to the neontological data supporting the HS model, as Gingerich and Schoeninger [1977] note.) Situations like this underscore the practical necessity of sometimes retaining paraphyletic taxa delimited by grade boundaries.

ALTERNATIVE MODELS AND CLASSIFICATIONS

How are the three haplorhine groups (omomyids, anthropoids, and *Tarsius*) related to each other? Many investigators have posited special affinities between omomyids and *Tarsius,* and Simons (1961) assigns the microchoerine omomyids to the Tarsiidae *sensu stricto*. Most of the features adduced by Simons in support of this reflect the fact that the smaller microchoerines are small animals with gracile jaws and jaw muscles, and thus display slender zygomata, small temporal fossae, capacious bullae, and other adventitious resemblances to tarsiers. If the enlarged lower front tooth of microchoerines is interpreted as an incisor (on the reasonable supposition that they are descendants of some omomyid with enlarged central incisors), they they admittedly share with *Tarsius* a derived loss of I_2. However, the occlusal relationships are different, and the microchoerine arrangement (in which the lower canine occludes with nothing) cannot have given rise to the more primitive condition in *Tarsius* (whose large C_1 retains its occlusal relationships with I^2 and C^1). The tarsier-like fused tibiofibulae assigned to some microchoerines are either dubiously assigned or unfused (Simons, 1961). Other microchoerine and omomyine postcranials are at least as similar to some strepsirhines as to *Tarsius* (Simpson, 1940; Szalay, 1975b).

Derived features shared by *Tarsius* and anthropoids, but not by any omomyids, are more numerous. The most important is the configuration of the middle ear. In *Tarsius* and ceboids, the tympanic cavity proper is small and transversely narrow, but the anterior end of the petrosal is pneumatized by a large diverticulum (trabeculated in ceboids) separated from the main cavity by a transverse septum and communicating with it only via a small foramen adjoining the auditory tube. The internal carotid enters the bulla quite laterally and anteriorly (just medial to the tympanic ring and anterior to the promontorium) and runs almost vertically through the lateral edge of the septum into the braincase. Parallelism is possible here (e.g., due to ventral displacement of the foramen magnum), but the resemblances are detailed, and apparently represent shared derivations from the more primitive condition seen in *Rooneyia* (Szalay, 1975a).

Simons and Russell (1960) argue that the postorbital septum of *Tarsius* is not homologous with that of anthropoids because it is formed largely from the frontal rather than the zygomatic. In fact, the alisphenoid forms much of the postorbital septum in both groups (Hershkovitz, 1974). Expansion of the frontal in *Tarsius* appears to be correlated with postnatal hypertrophy of the eyeball; the anthropoid condition could easily be derived from that seen in a newborn tarsier.

Omomyids are more primitive than extant haplorhines in these and other respects. They retain a subtympanic extension of the tympanic cavity, lost in *Tarsius* and anthropoids. The size difference between the promontory and stapedial arteries is not pronounced in *Necrolemur*, and the tarsier-like disproportion seen in *Rooneyia* thus seems to be a convergence. The large infraorbital foramina of omomyids suggest that they retained well-developed vibrissae and perhaps even a naked rhinarium (Kay and Cartmill, 1977). If the suborder Strepsirhini is defined by shared primitive retentions (as it must be if the adapids are included), then whether we want to call omomyids Haplorhini or not is a matter of taste.

The hypothesis that Anthropoidea and *Tarsius* are sister groups entails a dendrogram like that shown in Fig. 3, in which possible subordinal boundaries are indicated. Considering the adapids' uncertain phyletic position and the resulting necessity of using wastebasket taxa, there is something to be said for retaining a subordinal division of Primates into Prosimii and Anthropoidea, for essentially the reasons advocated by Simons (1974), while recognizing that some prosimians are more closely related to anthropoids than others.

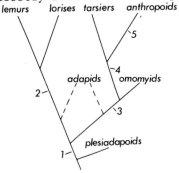

Figure 3. Suggested phylogenetic relationships within the order Primates. Possible subordinal divisions indicated by numerals. 1, division between plesiadapoids and primates of modern aspect; 2, division between Haplorhini and Strepsirhini, using Haplorhini as a wastebasket taxon defined on absence of tooth comb; 3, H-S division based on carotid morphology, using Strepsirhini as a wastebasket taxon; 4, H-S division based on postorbital septum and tympanic cavity; 5, division between Anthropoidea and Prosimii.

ACKNOWLEDGEMENTS

This research was supported by NSF Grant GS-43262 to R.F. Kay, and by NIH Grant 5 K04 HD00083-02 to M. Cartmill. We thank Professor J.P. Lehman of the Institut de Paléontologie, Museum national d'Histoire naturelle, Paris; Dr. A. Petter-Rousseaux of the Laboratoire d'Écologie du Museum, Brunoy, France; Dr. Peter Andrews of the British Museum of Natural History, London; Dr. R.B. Rickards, Sedgwick Museum, Cambridge; and Drs. H.W. Setzer and R.W. Thoringto Jr. of the National Museum of Natural History, Washington, D.C., for allowing us to study specimens in their care. We also thank P.D. Gingerich and A.L. Rosenberger for their helpful comments.

REFERENCES

Beecher, R.M. (1977). *Am. J. Phys. Anthrop.*, 47, 325-336.
Cartmill, M. (1975). *In* "Phylogeny of the Primates", (W.P. Luckett, F.S.
    Szalay, eds), pp. 313-354, Plenum, New York.
Cave, A.J.E. (1967). *Amer. J. Phys. Anthrop.*, 26, 277-288.
Clemens, W.A. (1974). *Science*, 184, 903-905.
Gingerich, P.D. (1973). *Folia primatol.*, 19, 329-337.
Gingerich, P.D. (1974). Cranial anatomy and evolution of early Tertiary
    Plesiadapidae (Mammalia, Primates), Ph.D. thesis, Yale University.
Gingerich, P.D. (1975a). *Nature*, 253, 111-113.
Gingerich, P.D. (1975b). *Contrib. Mus. Paleont. Univ. Mich.*, 24, 163-170.
Gingerich, P.D. (1976). *Univ. Mich. Mus. Paleont. Papers on Paleontology*,
    15, 1-140.
Gingerich, P.D. and Schoeninger, M. (1977). *J. Hum. Evol.*, 6, 483-505.
Greiner, E. (1929). *Z. Anat. Entw.*, 89, 102-122.
Hershkovitz, P. (1974). *Folia primatol.*, 21, 1-35.
Kay, R.F. (1977). *Amer. J. Phys. Anthrop.*, 46, 327-352.
Kay, R.F., and Cartmill, M. (1977). *J. Hum. Evol.*, 6, 19-53.
Leche, W. (1896). "Festschrift zum 70en Geburtstage von Carl Gegenbaur",
    Vol. 3, pp. 127-166, W. Engelmann, Leipzig.
McKenna, M.C. (1975). *In* "Phylogeny of the Primates", (W.P. Luckett and F.S.
    Szalay, eds), pp. 21-46, Plenum, New York.
Pocock, R.I. (1918). *Proc. Zool. Soc., London*, 1918, 19-53.
Reed, S.C. and Snell, G.D. (1931). *Anat. Rec.*, 51, 43-50.
Schwartz, J.H. and Krishtalka, L. (1977). *Ann. Carnegie Mus.*, 46, 55-70.
Simons, E.L. (1960). The phylogeny of the lemuroid and tarsioid primates and
    its relationship to the origin of Hominoidea. Ph.D. thesis, University of
    Oxford.
Simons, E.L. (1961). *Bull. Brit. Mus. (Nat. Hist.) Geol.*, 5, 45-69.
Simons, E.L. (1974). *In* "Prosimian Biology", (R.D. Martin, G.A. Doyle, and
    A.C. Walker, eds), pp. 415-433, Duckworth, London.
Simons, E.L. and Russell, D.E. (1960). *Breviora, Mus. Comp. Zool. (Harvard)*,
    127, 1-14.
Simpson, G.G. (1940). *Bull. Amer. Mus. Nat. Hist.*, 77, 185-212.
Simpson, G.G. (1945). *Bull. Amer. Mus. Nat. Hist.*, 85, 1-350.
Szalay, F.S. (1970). *Nature*, 227, 355-357.
Szalay, F.S. (1975a). *In* "Phylogeny of the Primates", (W.P. Luckett and F.S.
    Szalay, eds), pp. 91-125, Plenum, New York.
Szalay, F.S. (1975b). *In* "Phylogeny of the Primates", (W.P. Luckett and F.S.
    Szalay, eds), pp. 357-404, Plenum, New York.
Tattersall, I. and Schwartz, J.H. (1974). *Anthrop. Papers Amer. Mus. Nat.
    Hist.*, 52, 139-192.

DISCUSSION

Schwartz: Bown has identified *Teilhardina* from North America, and McKenna has now corroborated this identification. Bown's *Teilhardina* definitely has a large front tooth. Isn't there some question about the anterior dentition of the Belgian material?

Kay: I haven't seen Bown's specimens, but I've just come from studying the type material in Belgium, which shows two very small but unmistakable incisor sockets in front of a very large canine, whose cross-sectional area is perhaps 8 or 9 times that of these two incisors. But I would question the number of premolars; the two sockets behind the canine may represent one two-rooted tooth rather than two premolars.

MOLECULAR EVIDENCE ON THE PHYLOGENETIC RELATIONSHIPS OF *TARSIUS*

*M. GOODMAN, **D. HEWETT-EMMETT† and ***J.M. BEARD

*Department of Anatomy, Wayne State University, School of Medicine, 540 E. Canfield Ave., Detroit, Michigan 48201, USA, ** Department of Physiology, Wayne State University, and ***Department of Anthropology, University College London, Gower Street, London WC1E 6BT, UK.

It is clear at this present Congress that questions about the tarsier's phylogenetic affinities continue to fascinate primatologists. Recent investigations of fossils and of the soft as well as hard anatomy of living forms have produced conflicting opinions on the phylogenetic validity of Pocock's (1918) strepsirhine-haplorhine division of the Primates. Szalay (1975) and Luckett (1975), e.g., support it, whereas Gingerich (1975) argues that the Adapidae and later lemuroids and lorisoids are genealogically closer to Anthropoidea than are the tarsioids.

As an attempt to contribute to this debate on the phylogenetic affinities of *Tarsius*, our paper reviews the evidence provided by informational macromolecules, including immunological data on proteins (e.g. Goodman, 1969, 1973; Dene et al., 1976b; Sarich and Cronin, 1976), fragmentary data from interspecies DNA sequence comparisons (Hoyer and Roberts, 1967), and particularly the amino acid sequences of hemoglobin α and β chains gathered in the laboratories of the late Professor Barnicot (Beard et al., 1976) which are to date the only available data on the primary structure of any tarsier protein. As we hope to indicate, the available molecular evidence, while tenuous, appears in balance to justify the phylogenetic validity of the Strepsirhini-Haplorhini division of the Primates.

THE TARSIER MATERIAL

Blood samples and soft tissues were acquired from three *Tarsius syrichta* sent in 1963 to Dr. Dejunge and to one of us (M.G.) by Dr. Galo B. Ocampo, Director of the National Museum of the Philippines. Several years later blood and tissues were acquired from two more *Tarsius syrichta*, which had been brought to the United States from the Philippines by Dr. M.W. Sorenson of the University of Missouri. These materials were the basis for all immunological studies on tarsier proteins carried out in Detroit (Maisel, 1965; Goodman, 1969, 1973; Goodman et al., 1974; Dene et al., 1976, 1976b) and in Berkeley (Cronin and Sarich, 1975; Sarich and Cronin, 1976). DNA isolated from the soft tissues of these tarsiers was also used by Hoyer and Roberts (1967) in their studies. Brain tissue and red blood cells from these same *Tarsius syrichta* were also used in studies of lactate dehydrogenase (LDH) isozymes in primates (Goodman et al., 1969; Koen and Goodman, 1969).

† Present Address: Department of Biochemistry, University of Bristol, Medical School, University Walk, Bristol BS8 1TD, UK.

In the last few years samples of blood were acquired from two *Tarsius bancanus* from Sarawak. One animal was captured by Dr. June Rollinson and the other by Dr. Carsten Niemitz. The former was kept by Dr. Bob Martin in the laboratory of Physical Anthropology of Prof. Nigel Barnicot at University College in London. The blood samples from these animals yielded the hemoglobin upon which the amino acid sequence work was carried out by Beard et al (1976). They were also used to examine selected red cell enzymes and serum proteins by electrophoretic methods (Barnicot and Hewett-Emmett, 1974).

IMMUNOLOGICAL DATA ON PROTEINS

One of the largest bodies of data was obtained from over 6200 immunodiffusion plate comparisons developed with 49 rabbit antisera to protein preparations (in most cases, serum proteins) from 38 species of which 29 were primates (including *Tarsius syrichta*), 6 were tree shrews, 2 were elephant shrews, and one was the flying lemur. These antisera were reacted against antigens from about 125 species representing most genera of Primates and a wide range of mammalian orders (Dene et al., 1976b). The antigen-antibody precipitin results observed in the immunodiffusion plates were converted by computer procedures, described elsewhere (Moore and Goodman, 1968; Goodman and Moore, 1971; Dene et al., 1976), into 38 antigenic distance tables, one for each homologous species (i.e. a species against which antiserum was produced). Each such table was also considered a phylogenetic distance table in that it presented an ordering of species based on increases in the phylogenet distance between homologous and heterologous species. *Tarsius* appeared in 17 of the 38 tables, in its own as the homologous species and in 16 of the other as a heterologous species. A divergence tree of taxa was then constructed from the phylogenetic distances in the 38 tables by the unweighted pair-group method of Sokal and Michener (1958). The groupings in this tree fit exactly that expected for a Strepsirhini-Haplorhini division of Primates and are presented in summary form for the major primate branches in terms of such a classification in Table I.

Tarsioidea (i.e. *Tarsius*) and Anthropoidea diverge from each other by an antigenic distance of 8.90 and, similarly, Lemuriformes and Lorisiformes are comparably separated (antigenic distance, 8.87). In turn the Haplorhini (Tarsioidea plus Anthropoidea) and Strepsirhini (Lemuriformes and Lorisiformes) diverge by an antigenic distance of 10.31. Tree shrews (order Tupaioidea) and flying lemur (order Dermoptera) diverge from Primates in the rabbit immuno diffusion results by antigenic distances of 11.30 and 11.55 respectively, whereas all other non-primate mammalian groups diverge by greater distances ranging from 11.96 for Rodentia to 16.23 for Monotremata (Table V in Dene et a 1976b). . Immunological distances determined by microcomplement fixation tests with rabbit antisera to the purified serum proteins, albumin and transferrin, from a range of primates and other mammalian species (Cronin and Sarich, 1975; Sarich and Cronin, 1976) group Lemuriformes and Lorisiformes together, but do not separate this strepsirhine branch from Anthropoidea by any greater distance than that separating *Tarsius*. Moreover, these microcomplement fixation tests place tupaiids and flying lemur especially close to Primates, so close in fact that their distance from Anthropoidea is no greater than that of either tarsier or strepsirhines. The better resolution of the phylogenetic distances between these taxa in the immunodiffusi data reflects the fact that a number of serum proteins were examined in the immunodiffusion plates, not just albumin and transferrin.

TABLE I

Classification of major taxa of Primates based on immunodiffusion evidence and levels of antigenic divergence at the taxonomic ranks*

	I	II	III	IV	V
I. Order Primates	10.31				
II. Semiorder Strepsirhini		8.87			
III. Suborder Lemuriformes			7.04		
V. Superfamily Lemuroidea					4.15
Cheirogaleoidea					
Daubentonioidea					
III. Suborder Lorisiformes					
V. Superfamily Lorisoidea					4.23
II. Semiorder Haplorhini		8.90			
III. Suborder Tarsioidea					
III. Suborder Anthropoidea			6.52		
IV. Infraorder Platyrrhini					
V. Superfamily Ceboidea					3.05
IV. Infraorder Catarrhini				3.06	
V. Superfamily Cercopithecoidea					1.29**
Hominoidea					2.05

* The numbers shown in the columns under the ranks represent the average antigenic distance between members of the two most divergent sister groups making up the taxa to which the numbers are assigned. These distances were obtained from the divergence tree constructed by the unweighted pair group method using the 38 antigenic distance tables from rabbit antisera results.
** As Cercopithecoidea contains only one family, Cercopithecidae, this number represents the level of antigenic divergence for that family, i.e. the average antigenic distance between its two sister groups, Colobinae and Cercopithecinae.

We are aware of the possibility that certain taxa might have been grouped together in the immunodiffusion divergence tree as a result of plesiomorphic (primitive) antigenic features rather than synapomorphic ones. If the plesiomorphies at the antigenic level had paralleled those at the morphological level, the tree might then have grouped together in one branch all the taxa (Lemuriformes, Lorisiformes, and Tarsiformes) found in Simpson's Prosimii, presuming that plesiomorphic features had actually predominated in these taxa. The fact that such a grouping did not occur suggests that most of the groupings observed in the immunodiffusion data are indeed due to synapomorphic antigenic features more than to any other factor. However, there is one striking exception in the data. Immunodiffusion results obtained with antisera to primate lens proteins (Maisel, 1965) suggest that lorisoids and tarsier have indistinguishable lens proteins. Inasmuch as the lorisoids and tarsier are arboreal nocturnal animals (probably resembling in that regard early Tertiary primates) the apparent identity of their lens protein antigens might well be due to a preponderance of plesiomorphic specificities in these lens antigens. In other words, we are implying that natural selection preserved the lens proteins of the two nocturnal taxa, Lorisiformes and Tarsioidea, at the same state of adaptation which had existed in much earlier primates.

DNA DATA

DNA extracted from the soft tissues of *Tarsius syrichta* was also used in an investigation of the degrees of genetic relatedness among primate taxa (Figure 16 in Hoyer and Roberts, 1967). The fraction of the genomic DNA involved in the interspecies comparisons consisted of families, i.e. different collections of repetitious polynucleotide sequences. The comparisons were carried out from the human standpoint, and the degrees of similarity of the heterologous DNAs to human DNA (percent of matching nucleotide sequences) were chimpanzee 100%, gibbon 94%, rhesus monkey 88%, capuchin monkey 83%, tarsier 65%, lorisoids (galago and slow loris) each 58%, lemur 47%, tree shrew 28%, mouse 21%, hedgehog 19% and chicken 10%. These DNA results are comparable to the immunodiffusion distance results obtained with antisera to human proteins, and indicate that the sister group of Anthropoidea is Tarsioidea.

Of course, this one series of DNA comparisons is too meagre by itself to serve as the basis for phylogenetic conclusions. For that purpose comparisons would have to be carried out from the standpoint of tarsier DNA as well as that of strepsirhine DNAs. Moreover, it is now realized that such comparisons should utilize the fraction of genomic DNA consisting of unique or single-copy sequences because this fraction contains much more genetic information than the repetitious sequences and permits more reliable and discriminating interspecie comparisons (Kohne et al., 1972; Hoyer et al., 1972; Benveniste and Todaro, 1976). Unfortunately, tarsier has not yet been included in comparisons of unique DNA sequences.

MAXIMUM PARSIMONY ANALYSIS OF HEMOGLOBIN SEQUENCES

Amino acid sequence data exist on about 52 α and 60 β hemoglobin chains from 55 vertebrates including 27 primate species. In large stretches of many of these chains only amino acid compositions of peptide fragments had been determined. The sequences proposed for these stretches had been deduced by comparison to the corresponding peptides of related chains with known amino acid sequences. Thirty of the 52 α chains and 31 of the 60 β chains had known sequences — i.e., a majority of their residues had been positioned by actual sequencing procedures. These known sequences were expected to yield the more reliable genealogical trees. However, because of the potential phylogenetic information in the sequences inferred from amino acid compositions and peptide patterns, we constructed genealogical trees by the maximum parsimony method (Moore et al., 1973; Moore, 1976) not only for the 31 more rigorously determined β sequences and the 30 more rigorously determined α sequences, but also for the enlarged collections of 60 β sequences and 52 α sequences.

Our most extensive analysis was carried out with α and β sequences combined in an extended alignment, i.e. an alignment consisting of 287 positions (the 141 of α chains plus the 146 of β chains). There were 24 such combined sequences. They represented a "hybrid" amphibian (newt α plus frog β), chicken, two monotremes, two marsupials, ten primates and eight other eutherian mammals. The criterion for selecting these tetrapods was that in each case a majority of the residues in at least one of the two hemoglobin chain types had been placed by actual sequencing procedures. The ten primate hemoglobins, most of which had been completely sequenced, represented Cercopithecoidea (*Macaca fuscata, Macaca mulatta, Cercocebus atys, Cercopithecus aethiops, Presbytis entellus*), Hominoidea (*Homo sapiens*), Ceboidea (*Ateles geoffroyi, Cebus apella*), Tarsioidea (*Tarsius bancanus*), and Lorisiformes (*Nycticebus coucang*). In the search for the most parsimonious genealogy (that requiring the least number of nucleotide replacements) over 370 alternative branching arrangements for the descent of the 24 tetrapods were examined, testing a wide range of phylogenetic possibilities.* The most parsimonious trees found in this search had a length of 810 nucleotide replacements (NR). There were only two such trees with this minimal length, the one shown in Figure 1 and an alternative in which the branch positions of mangabey (*Cercocebus*) and *Cercopithecus* were interchanged so that *Macaca* was next to *Cercopithecus* rather than next to *Cercocebus*.

As can be noted in Figure 1, tarsier is closer to Anthropoidea than to slow loris (*Nycticebus*), or, conversely slow loris is more anciently separated from the anthropoids than is tarsier. The most parsimonious trees obtained for the separate collections of α and β chain sequences also grouped tarsier with the Anthropoidea. These results support the inclusion of *Tarsius* in the primate subdivision, Haplorhini.

Alternatively joining tarsier first either to the slow loris branch or the ceboid branch or the catarrhine branch rather than to the ancestral stem of the Anthropoidea increased the length of the genealogy by 3 NR (i.e. yielded lengths of 813 NR). Exchanging the position of tarsier and slow loris added 4 NR (i.e. the length of this alternative tree was 814 NR). These increases over the most parsimonious genealogy may seem insignificant for drawing phylogenetic conclusions. It should be noted, however, that the primate region of the genealogy accumulated just 91 NR. Moreover, in the particular region where the branch positions were exchanged (the region of earlier primate ancestors and the descending lineages to slow loris and tarsier) only 38 NR accumulated in the most parsimonious genealogy. This can be noted in Tables II and III which show the nucleotide changes in the ancestral α codons (Table II) and β codons (Table III) of this region. The addition of 3 to 4 NR over the parsimony length by altering the haplorhine position of the tarsier might be considered, therefore, an 8 to 10% increase over the parsimony value. This might not be insignificant. If we accept the phylogenetic validity of the haplorhine grouping of tarsier and Anthropoidea depicted in Figure 1, we can see from Tables II and III that there are 7 haplorhine synapomorphic nucleotide replacements in these sequences linking tarsier and Anthropoidea together(at α 57, GCU→GGU; α 78, AGU→AAU; β 50, AGU→ACU; β 76, AAU→GCU; β 112, AUU→UGU) and 4 anthropoid synapomorphies separating Anthropoidea from Tarsioidea (at α 111, UGU→GCU; β 52, GCU→GAU; β 69, AGU→GGU); that is to say, the haplorhine ancestor diverges more from its precursor the ancestral primate than from its descendant the ancestral anthropoid.

In conclusion, we want to make a plea to the field workers at this 6th International Congress or Primatology to help provide those of us who work with molecules with more material from *Tarsius* and from lorisiform and lemuri-

---

* Hundreds of alternative dendrograms were also examined in the original work (Beard and Goodman, 1976)   when the collection of combined α and β sequences represented 20 lineages (platypus, opossum, mangabey, and tree shrew were not included in them) and again when the collection represented 23 lineages (tree shrew not yet included)

form species. We are reasonably confident that given materials from which
further DNA data and protein sequence data can be obtained, the molecular
approach should produce fairly decisive evidence on the phylogenetic affinities of the tarsier.

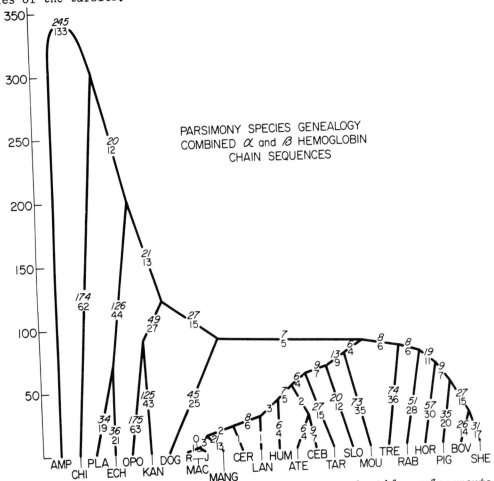

Figure 1.   *Maximum parsimony tree requiring 810 nucleotide replacements
(NR) for 24 taxa on using a combined α and β globin alignment.
Link lengths are the numbers of NR between adjacent ancestral
and descendant sequences; italicized numbers are link lengths
corrected for superimposed replacements by the augmentation
algorithm of Moore and Goodman (Goodman et al., 1974; Moore
et al., 1976).   The ordinate scale in millions of years,
is inferred from fossil evidence on ancestral splitting times
of the taxa represented by the sequences. The taxa are: AMP
(amphibian): newt α (Jones, 1974), frog β (Chauvet and Archer,
1972); CHI (chicken): α (Matsuda et al., 1971), β (Matsuda
et al., 1973g); PLA (platypus): α (Whittaker et al., 1974),
β (Whittaker and Thompson, 1975); ECH (echidna): α (Whittaker et al., 1973), β (Whittaker et al., 1972); OPO (opossum):
α (Stenzel, 1974), β (R.T. Jones, pers. comm.) (legend cont.)*

Figure 1. (legend cont.) *KAN (kangaroo): α (Beard et al., 1971),
β (Air and Thompson, 1969); DOG (dog): α and β (Jones
et al., 1971); R-MAC (rhesus macaque): α and β (Matsuda et
al., 1968); J-MAC (Japanese macaque): α and β (Matsuda et al.,
1973a); MANG (mangabey): α and β (Hewett-Emmett et al., in
press); CER (Cercopithecus): α and β (Matsuda et al., 1973d);
LAN (langur): α and β (Matsuda et al., 1973c); HUM (human): α
and β (Braunitzer et al., 1961); ATE (Ateles): α and β (Matsuda
et al., 1973f); CEB (Cebus): α and β (Matsuda et al., 1973e);
TAR (tarsier): α and β (Beard et al., 1976); SLO (slow loris):
α and β (Matsuda et al., 1973b); MOU (mouse): α (Popp, 1967),
β (Popp, 1973); TRE (tree shrew): α and β (G. Matsuda, pers.
comm.); RAB (rabbit): α (Von Ehrenstein, 1966), β (Best et al.,
1969); HOR (horse): α (Matsuda et al., 1963), β (Smith, 1968;
Dayhoff, 1972); PIG (pig): α (Yamaguchi et al., 1965; Dayhoff,
1972), β (Braunitzer and Kohler, 1966; Dayhoff, 1972); BOV
(bovine): α and β (Schroeder et al., 1967a, b; Dayhoff, 1972);
SHE (sheep): α (Beale, 1967; Dayhoff, 1972), β (Boyer et al.,
1967; Dayhoff, 1972).*

TABLE II

*Unique and synapomorphic α globin codons in descent of slow loris and the
Haplorhine ancestor from the Primate ancestor and of Tarsier and the
Anthropoid ancestor from the Haplorhine ancestor**

Residue No.	15	19	53	57	68	71	78	111	129
Primate Anc	GGU	GGU	GCU	GCU	AAU	GCG	AGU	UGU	CUU
Slow Loris	GAG	AGU	"	"	"	UCG	"	"	"
Haplorhine Anc	"	"	"	GGU	"	"	AAU	"	"
Tarsier	GAU	"	UCU	GGU	ACU	GGG	AAU	"	GUU
Anthropoid Anc	"	"	"	GGU	"	"	AAU	GCU	"

* Codons which are variable among the five sequences and differ from the
primate ancestor are shown. They were reconstructed by the maximum
parsimony method for the genealogy illustrated in Figure 1. Whenever
alternative maximum parsimony solutions existed at a residue position,
the A-solution (as described in Goodman et al., 1974) was utilized in
the tree. A mutation which may appear in a lineage leading either to few
or many contemporary species will choose the few-lineage in the A-solution,
since this decreases the number of times the mutation is counted among
lineages between the most ancestral point of the tree and each contemporary species. The A-solution counteracts more than any other solution
the bias towards grosser underestimation of change on the lineages with
few nodal points.

TABLE III

Unique and synapomorphic β globin codons in descent of slow loris and the Haplorhine ancestor from the Primate ancestor and of Tarsier and the Anthropoid ancestor from the Haplorhine ancestor*

Residue No.	5	6	9	19	21	22	43	50	52
Primate Anc	GCU	GAG	GCU	AAU	GAU	GAG	GAU	AGU	GCU
Slow Loris	GGU	—	UCU	—	—	GAU	GAG	—	UCU
Haplorhine Anc	—	—	—	—	—	—	—	ACU	—
Tarsier	—	GAU	—	GAU	GAG	GAU	—	ACU	—
Anthropoid Anc	—	—	—	—	—	—	—	ACU	GAU

Residue No.	58	69	73	75	76	112	121	128	139
Primate Anc	CCU	AGU	GAU	CUG	AAU	AUU	GAG	GCU	AAU
Slow Loris	—	—	—	—	—	GUU	GAU	UCU	—
Haplorhine Anc	—	—	—	—	GCU	UGU	—	—	—
Tarsier	GCU	AAU	GAG	AUG	GCU	UGU	—	—	ACU
Anthropoid Anc	—	GGU	—	—	GCU	UGU	—	—	—

* Codons which are variable among the five sequences and differ from the primate ancestor are shown. They were reconstructed by the maximum parsimony method for the genealogy illustrated in Figure 1. This tree utilized only A-solution residues.

ACKNOWLEDGEMENTS

    We wish to acknowledge the unfailing support of the late Professor Nigel A. Barnicot who introduced two of us (D.H-E and J.M.B.) to the *Tarsius* problem. We thank Professor Genji Matsuda for providing us before publication with the amino acid sequence of *Tupaia* (tree shrew) α and β hemoglobin chains. We also thank Professor R.T. Jones for providing us before publication with the amino acid sequence of the opossum β hemoglobin chain. The help of Dr. R.D. Martin in securing the blood samples from *Tarsius bancanus* is much appreciated.

REFERENCES

Air, G.M. and Thompson, E.O.P. (1969). *Aust. J. Biol. Sci.*, 22, 1437.
Barnicot, N.A. and Hewett-Emmett, D. (1974). *In* "Prosimian Biology", (R.D. Martin, G.A. Doyle and A.C. Walker, eds), pp. 891-902, Duckworth, London.
Beale, D. (1967). *Biochem. J.*, 103, 129.
Beard, J.M. and Thompson, E.O.P. (1971). *Aust. J. Biol. Sci.*, 24, 765.
Beard, J.M., Barnicot, N.A. and Hewett-Emmett, D. (1976). *Nature*, 259, 338.
Beard, J.M. and Goodman, M. (1976). *In* "Molecular Anthropology: Genes and Proteins in the Evolutionary Ascent of the Primates", (M. Goodman and R.E. Tashian, eds), Plenum, New York.
Benveniste, R.E. and Todaro, G.J. (1976). *Nature*, 261, 101-108.
Best, J.S., Flamm, U. and Braunitzer, G. (1969). *Hoppe-Seyler's Z. Physiol. Chemie*, 350, 563.
Boyer, S.H., Hathaway, P., Pascasio, F., Bordley, J., Orton, C., and Naughton, M.A. (1967). *J. Biol. Chem.*, 242, 2211.
Braunitzer, G., Gehring-Müller, R., Hilschman, N., Hilse, K., Hobom, G., Rudloff, V., and Wittman-Liebold, B. (1961). *Hoppe-Seyler's Z. Physiol. Chem.*, 325, 283.

Braunitzer, G. and Kohler, H. (1966). *Hoppe-Seyler's Z. Physiol. Chemie,* 343, 290.
Chauvet, J-P. and Acher, R. (1972). *Biochem.,* 11, 916.
Cronin, J.E. and Sarich, V.M. (1975). *J. Hum. Evol.,* 4, 357-375.
Dayhoff, M.O. (1972). "Atlas of Protein Sequence and Structure, Vol. 5", National Biomedical Research Foundation, Washington.
Dene, H., Goodman, M., Prychodko, W. and Moore, G.W. (1976). *Folia Primatol.,* 25, 35-61.
Dene, H.T., Goodman, M., and Prychodko, W. (1976b). *In* "Molecular Anthropology: Genes and Proteins in the Evolutionary Ascent of the Primates", (M. Goodman and R.E. Tashian, eds), Plenum, New York.
Gingerich, P.D. (1975). *Contrib. Mus. Paleont., Univ. Michigan,* 24, 163-170.
Goodman, M., Sorenson, M.W., Farris, W., and Poulik, E. (1969). *Am. J. Phys. Anthrop.,* 31, 266.
Goodman, M., Syner, F.N., Stimson, C.W. and Rankin, J.J. (1969). *Brain Research,* 14, 447-459.
Goodman, M. and Moore, G.W. (1971). *Syst. Zool.,* 20, 19-62.
Goodman, M. (1973). *Symp. Zool. Soc. Lond.,* 33, 339.
Goodman, M., Farris, W., Moore, W., Prychodko, W., Poulik, E. and Sorenson, M. (1974). *In* "Prosimian Biology", (R.D. Martin, G.A. Doyle and A.C. Walker, eds), pp. 881-890, Duckworth, London.
Goodman, M., Moore, G.W., Barnabas, J and Matsuda, G. (1974). *J. Mol. Evol.,* 3, 1-48.
Hewett-Emmett, D., Cook, C.N. and Barnicot, N.A. (1976). *In* "Molecular Anthropology: Genes and Proteins in the Evolutionary Ascent of the Primates", (M. Goodman and R.E. Tashian, eds), Plenum, New York.
Hill, W.C.O. (1955). "Primates Comparative Anatomy and Taxonomy, II. Haplorhini: Tarsioidea", University Press, Edinburgh.
Hoyer, B.H. and Roberts, R.B. (1967). *In* "Molecular Genetics, Part II", (H. Taylor, ed), pp. 425-479. Academic Press, New York.
Hoyer, B.H., van de Velde, N.W., Goodman, M., and Roberts, R.B. (1972). *J. Hum. Evol.,* 1, 645-649.
Jones, R.T., Brimhall, B. and Duerst, M. (1971). *Fed. Proc. Fed. Ann. Socs. exp. Biol.,* 30Pt2 of two volumes, abstract 1207.
Koen, A.L., and Goodman, M. (1969). *Biochemical Genetics,* 3, 457-474.
Kohne, D.E., Chiscon, J.A. and Hoyer, B.H. (1972). *J. Hum. Evol.,* 1, 627-644.
Luckett, W.P. (1975). *In* "Phylogeny of the Primates", (W.P. Luckett and F.S. Szalay, eds), pp. 157-182, Plenum, New York.
Maisel, H. (1965). *In* "Protides of the Biological Fluids", (H. Peeters, ed.), pp. 146-148, Elsevier, Amsterdam.
Matsuda, G., Gehring-Mueller, R. and Braunitzer, G. (1963). *Biochem.,* 2, 338-669.
Matsuda, G., Maita, T., Takei, H., Ota, H., Yamaguchi, M., Miyauchi, T. and Migita, M. (1968). *J. Biochem.* (Tokyo), 64, 279.
Matsuda, G., Takei, H., Wu, K.C., and Shiozawa, T. (1971). *Int. J. Peptide Protein Res.,* 3, 173.
Matsuda, G., Maita, T., Ota, H., Araya, A., Nakashima, Y., Ishii, V. and Nakashima, M. (1973a). *Int. J. Peptide Protein Res.,* 5, 405.
Matsuda, G., Maita, T., Watanabe, B., Ota, H., Araya, A., Goodman, M. and Prychodko, W. (1973b). *Int. J. Peptide Protein Res.,* 5, 419.
Matsuda, G., Maita, T., Nakashima, Y., Barnabas, J., Ranjekar, P.K. and Gandhi, N.S. (1973c). *Int. J. Peptide Protein Res.,* 5, 423.
Matsuda, G., Maita, T., Watanabe, B., Araya, A., Morokuma, K., Goodman, M.

and Prychodko, W. (1973d). *Hoppe-Seyler's Z. Physiol. Chem.*, 354, 1153.

Matsuda, G., Maita, T., Watanabe, B., Araya, A., Morokuma, K., Ota, Y., Goodman, M., Barnabas, J., and Prychodko, W. (1973e). *Hoppe-Seyler's Z. Physiol. Chem.*, 354, 1513.

Matsuda, G., Maita, T., Suzuyama, Y., Setoguchi, M., Ota, Y., Araya, A., Goodman, M., Barnabas, J., and Prychodko, W. (1973f). *Hoppe-Seyler's Z. Physiol. Chem.*, 354, 1517.

Matsuda, G., Maita, T., Mizuno, K., and Ota, H. (1973g). *Nature New Biology*, 244, 244.

Moore, G.W. and Goodman, M. (1968). *Bull. Math. Biophys.*, 30, 279-289.

Moore, G.W., Barnabas, J. and Goodman, M. (1973). *J. Theor. Biol.*, 38, 459.

Moore, G.W. (1976)   *In* "Molecular Anthropology: Genes and Proteins in the Evolutionary Ascent of the Primates", (M. Goodman and R.E. Tashian, eds), Plenum, New York.

Moore, G.W., Goodman, M., Callahan, C., Holmquist, R., and Moise, H. (1976). *J. Mol. Biol.*, 105, 15-38.

Pocock, R.I. (1918). *Proc. Zool. Soc. Lond.*, 1918, 19-53.

Popp, R.A. (1967). *J. Mol. Biol.*, 27, 9.

Popp, R.A. (1973). *Biochim. Biophys. Acta*, 303, 52.

Sarich, V.M. and Cronin, J.E. (1976).   *In* "Molecular Anthropology: Genes and Proteins in the Evolutionary Ascent of the Primates", (M. Goodman and R.E. Tashian, eds), Plenum, New York.

Schroeder, W.A., Shelton, J.R., Shelton, J.B., Robberson, B. and Babin, D.R. (1967a). *Archs. Biochem. Biophys.*, 120, 1-14.

Schroeder, W.A., Shelton, J.R., Shelton, J.B., Robberson, B. and Babin, D.R. (1967). *Archs. Biochem. Biophys.*, 120, 124.

Smith, D.B. (1968). *Cand. J. Biochem.*, 46, 825.

Sokal, R.R. and Michener, C.D. (1958). *Univ. Kansas Sci. Bull.*, 38, 1409-1438.

Stenzel, P. (1974). *Nature*, 252, 62-63.

Szalay, F.S. (1975). *In* "Phylogeny of the Primates", (W.P. Luckett and F.S. Szalay, eds), pp. 357-404, Plenum, New York.

Von Ehrenstein, D. (1966). *Cold Spring Harb. Symp. Quant. Biol.*, 31, 705.

Whittaker, R.G., Fisher, W.K. and Thompson, E.O.P. (1972). *Aust. J. Biol. Sci.*, 25, 989.

Whittaker, R.G., Fisher, W.K. and Thompson, E.O.P. (1973). *Aust. J. Biol. Sci.*, 26, 277.

Whittaker, R.G. and Thompson, E.O.P. (1974). *Aust. J. Biol. Sci.*, 27, 591-605.

Whittaker, R.G. and Thompson, E.O.P. (1975). *Aust. J. Biol. Sci.*, 28, 353-365.

Wood Jones, F. (1918). "Arboreal Man", London, Edward Arnold.

Yamaguchi, Y., Horie, H., Matsuo, A., Sasakawa, S. and Satake, K. (1965). *J. Biochem. (Tokyo)*, 58, 186.

DISCUSSION

Cronin (Berkeley): Dr. Sarich and I, using the haemoglobin sequence data provided by Dr. Goodman, have constructed trees associating *Tarsius* with *Nycticebus*, putting *Tarsius* with the anthropoids, and making *Tarsius* an independent branch from the primate ancestry. The summed differences in path lengths between these trees were negligible — on the order of two or three mutations — no matter where we put *Tarsius*. We really need to look at more molecules to sort this problem out.

Hewett-Emmett: Did you keep the mammalian arrangement the same?

Cronin: No, we constructed a different ancestral mammalian sequence. That could alter the interpretation.

Hewett-Emmett: You see, our tree was purely objective. We felt, too, that some of the mammalian arrangements might not be correct, but our tree was the parsimonious one. Once you go in and do something subjective, then you....

Cronin: But the conclusion is similar: no matter what you do with *Tarsius*, you come up with essentially the same minimum path length.

Hewett-Emmett: Yes, but moving *Tarsius* to any neighbouring position on our most parsimonious tree demands three or four extra nucleotide replacements, representing an increase of 8 to 10% in nucleotide-replacement length in the relevant part of the tree. We feel that this increase is probably large enough to be significant.

CLADES VERSUS GRADES IN PRIMATE PHYLOGENY

W.P. LUCKETT* and F.S. SZALAY**

*Department of Anatomy, Creighton University, Omaha, Nebraska 68178, USA
**Department of Anthropology, Hunter College, CUNY, New York, NY 10021,
and Department of Vertebrate Paleontology, American Museum
of Natural History, New York, NY, USA.

INTRODUCTION

Recent years have witnessed an increased interest in the phylogenetic relationships among Primates. This has been stimulated in part by a proliferation of field studies on primate ethology and by a surge of interest in the genetic and molecular biology of primates, with special reference to the molecular evidence of human evolution. During this same period, there has been a renewed interest in the theoretical and operational basis for methods of phylogenetic inference. Much of this interest can be traced to the formalization by Hennig (1950, 1966) of a methodology of phylogenetic reconstruction which adheres rigorously to the use of shared and derived homologous characters for the determination of genealogical relationships, as opposed to shared and primitive similarities.

Evolutionary relationships may be expressed in terms of patristic and cladistic affinities. Patristic relationships are based on the inheritance of both primitive and derived characters from a common ancestor, whereas cladistic relationships are established by the possession of shared and derived characters. The branching sequences of a phylogeny are determined by phylogenetic analysis of as many characters as feasible, in order to distinguish between similarities due to shared ancestral features, shared derived features, convergences, or parallelisms (Hennig, 1950, 1966). In essence, phylogenetic analysis entails the identification of all alternative states of homologous characters, and the subsequent arrangement of these character states in a sequence from most primitive to most derived. The relative primitiveness or derivedness of character states is determined by the distribution of character states in higher categories and by detailed ontogenetic studies (when possible). Character states which are widespread in higher taxa (and in more distantly related taxa) are considered to be primitive retentions of the ancestral condition in that group. Conversely, relatively rare and uniquely acquired character states are generally considered to be derived, particularly when their presence in sister groups can be shown to be the result of common ontogenetic pathways.

RECONSTRUCTION OF PRIMATE PHYLOGENY BY CHARACTER ANALYSIS

The characters evaluated in our study consist primarily of basicranial, reproductive, and embryological data which have been the subject of detailed descriptions and phylogenetic analyses elsewhere (Szalay, 1972, 1975, 1976;

Szalay and Katz, 1973; Luckett, 1974, 1975, 1976). In addition, we have incorporated into our synthesis several features of both "soft" and "hard" anatomy which have been evaluated by other investigators. These include the
nature of the rhinarium (Pocock, 1918; Hill, 1948, 1953), the relationship
between the olfactory bulb and interorbital septum (Cave, 1967; Cartmill,
1972), the occurrence and distribution of a retinal fovea and tapetum lucidum
in the eye (Wolin and Massopust, 1970), the degree and pattern of postorbital
closure (Hershkovitz, 1974a), and the allometric relationship between neonatal
and maternal body weight (Leutenegger, 1973; Martin, 1975).

The presumed primitive and derived states of each character evaluated are
listed in Table I and illustrated in the accompanying cladogram (Fig. 1). In
each case the character state listed is the presumed condition in the morphotype of each taxon. Although intermediate conditions are known for some of
the other morphocline polarities, such intermediate character states are presented only for characters 15-18. The absence of dental characters from the
evaluated data is due to the considerable generic and familial variability in
dental morphology, and the resulting ambiguity in morphotype reconstruction.

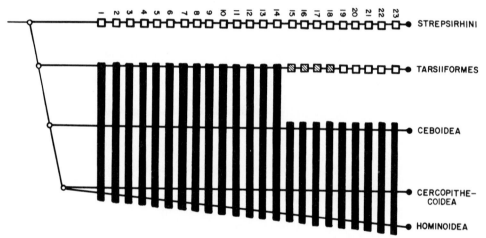

Figure 1. Cladogram of primate higher taxa, derived from phylogenetic
analysis of characters listed in Table I. Horizontal black
bars show derived character states; white boxes indicate
primitive character states.

Our character analysis of various organ systems from both fossil and extant
primates corroborates the hypothesis of an early (Paleocene) dichotomy of
strepsirhine and haplorhine primates. Relationships between Eocene-Recent
strepsirhines and haplorhines and the late Cretaceous-Paleocene paromomyiform
primates remain unclear, due in great part to the scarcity of skulls and postcranials for most known paromomyiforms (for further discussions, see Szalay,
1975). Our analysis indicates that Haplorhini is a monophyletic taxon defined by its common possession of shared and derived homologous characters
(characters 1-14 of Fig. 1). Furthermore, these data corroborate the hypothesis that the common (anthropoid) ancestor of platyrrhines and catarrhines
shared a more recent ancestor with tarsiiforms than with any other primates.
Within the Haplorhini, the morphotype of the Tarsiiformes is distinctly more
primitive than that of the Platyrrhini or Catarrhini; this is consistent with
the hypothesis that the Anthropoidea is derived from a tarsiiform ancestor.

TABLE I

Primitive and Derived Character States of Selected Primate Characters

	Primitive		Derived
1	Naked rhinarium; unfused nasal processes	1	Haired rhinarium; fused nasal processes
2	Noninvasive blastocyst attachment	2	Invasive blastocyst attachment
3	No primordial amniotic cavity	3	Primordial amniotic cavity
4	Transitory choriovitelline placenta	4	No choriovitelline placenta
5	Large vesicular allantois	5	Rudimentary allantois; body stalk
6	Small or reduced promontory branch of internal carotid artery	6	Well developed promontory branch of internal carotid artery
7	Well developed stapedial artery and bony canal	7	Stapedial artery and canal reduced or absent
8	Internal carotid artery primitively enters bulla posterolaterally	8	Internal carotid artery primitively enters bulla posteromedially
9	Interorbital septum marginal or anterior	9	Interorbital septum apical (when present)
10	Olfactory processes pass beneath septum to reach large nasal fossa	10	Olfactory processes pass above septum to reach reduced nasal fossa
11	Retina without area centralis or fovea	11	Retina with area centralis and fovea
12	Tapetum lucidum present	12	Tapetum lucidum absent
13	Ovarian bursa well developed	13	Ovarian bursa reduced
14	Relatively low ratio neonatal wgt./maternal wgt.	14	Relatively high ratio neonatal wgt./maternal wgt.
15	Diffuse epitheliochorial placenta	15	Bidiscoidal, hemochorial placenta
16	Fenestra rotunda ventrally shielded by carotid canal	16	Fenestra rotunda ventrally exposed
17	Orbit and temporal fossa not separated by bony partition	17	Orbit and temporal fossa more completely separated by bony partition formed by malar, alisphenoid, and frontal
18	Ectotympanic is an intrabullar ring	18	Ectotympanic is extrabullar
19	Bullar portion of petrosal and squamosal not pneumatized	19	Bullar portion of petrosal and squamosal pneumatized
20	Blastocyst attachment by para-embryonic pole	20	Blastocyst attachment by embryonic pole
21	Definitive amniogenesis by folding	21	Definitive amniogenesis by cavitation
22	Primary yolk sac only	22	Primary and secondary yolk sacs
23	Bicornuate uterus	23	Simplex uterus

(Note to Table I following page)

Note to Table I: For characters 15-18, the derived character states are
assumed for the morphotype of the Anthropoidea, whereas intermediate character
states (as illustrated by ▨▨▨ on the accompanying cladogram) are indicated for
the tarsiiform morphotype as follows:
  15 Monodiscoidal, hemochorial placenta
  16 Fenetra rotunda partially exposed
  17 Orbit and temporal fossa partially separated by bony partition formed
      by processes of the malar, alisphenoid, and frontal
  18 Extrabullar portion of ectotympanic tubular; intrabullar portion
      variable in size.

A detailed consideration of the nature of the genealogical relationships of
the Platyrrhini and Catarrhini is beyond the scope of the present study; however, our analysis (characters 15-23) provides strong evidence for the monophyly of the Anthropoidea.

PHYLOGENY AND CLASSIFICATION

   Phylogeny may be defined as "the evolutionary history of organisms, to include both cladistic and anagenetic information" (Ashlock, 1974), whereas
classification can be considered as an "essential means of conceptualization,
communication, and storage of information about animals" (Simpson, 1961).
Adherents of both the "evolutionary" (Simpson, 1961; Ashlock, 1974; Mayr,
1974; Ross, 1974) and "cladistic" (Hennig, 1950, 1966; Cracraft, 1974)
schools of classification recognize the importance of character analysis in
phylogenetic inference, but they disagree strongly about the extent to which
cladograms should be translated into classification.
   Methods of phylogenetic inference which include multiple character analyses
of both fossil and extant taxa, coupled with a consideration of the temporal
and paleogeographic distribution of fossils, offer the best approach for the
construction of phylogenetic hypotheses which approximate the actual phylogeny
of a group. A *clade* is of monophyletic origin by definition, whereas evolutionary *grades* reflect a similar level of organization inherited from a common
ancestor, or which may have been reached independently (Simpson, 1961). We
support a classification of primates which emphasizes both cladistic and anagenetic relationships of phylogeny, and which minimizes the utilization of
taxa based on grade similarities.

RELATIONSHIPS OF PRIMATE HIGHER CATEGORIES

   Essentially, three basic patterns of primate subordinal division have been
proposed: (1) Prosimii and Anthropoidea; (2) Lemuroidea, Tarsioidea, and
Anthropoidea; and (3) Strepsirhini and Haplorhini. The basic difference in
each of these schemes is the allocation of the Tarsiiformes. Most taxonomists
would agree that the Tarsiiformes are in many ways "intermediate" between the
Strepsirhini and Anthropoidea, but many of the similarities between Tarsiiformes and the other two taxa have not been phyletically evaluated.
   Simpson's (1945, 1955, 1961) classification of Lemuriformes, Lorisiformes,
and Tarsiiformes as separate infraorders within the Prosimii was based primarily on (1) the relatively primitive *grade* level of organization of these
taxa, when compared to the more advanced Anthropoidea; (2) his assertion
that Eocene Lemuriformes and Tarsiiformes are not clearly distinguishable from

each other (but see Gregory, 1920); and (3) the uncertainty as to whether
Eocene Lemuriformes or Tarsiiformes were closer to the ancestry of the
Anthropoidea. Given his beliefs, Simpson's classification of Primates was
justified at the time. However, the characters evaluated in the present
study clearly indicate that the grouping of Strepsirhini and Tarsiiformes as
the suborder Prosimii represents a grade classification, based primarily on
their retention of numerous primitive primate and eutherian characteristics.
In contrast, Pocock's (1918) bipartite division of the order Primates into
Strepsirhini and Haplorhini marked an early attempt at primate classification
based on cladistic principles. Tarsiiformes and Anthropoidea were clustered
as the Haplorhini, based on their common possession of three shared and derived character states: (1) a haired rhinarium associated with fused medial
and lateral nasal processes; (2) a deciduate hemochorial placenta; and
(3) the presence of a bony postorbital partition. While adopting Pocock's
concept of a Strepsirhini-Haplorhini dichotomy, Hill (1953, 1955) acknowledged the difficulty in further diagnosis of Haplorhini, and his extensive
compilation of data on numerous aspects of strepsirhine and haplorhine biology generally failed to distinguish between phenetic, patristic, and cladistic resemblances. The present analysis provides strong support for a strepsirhine-haplorhine dichotomy during primate phylogeny, and we emphasize the
necessity of defining higher taxonomic categories on the basis of their shared
and derived characters, rather than on the basis of phenetic and patristic
similarities or evolutionary trends.

The contention that postorbital closure may have been acquired independently by parallelism in tarsiiforms, platyrrhines, and catarrhines (Simons and
Russell, 1960; Simons, 1974) is not supported by the examination of numerous
fossil and extant platyrrhines and a large series of *Tarsius* skulls (Hershkovitz, 1974a). Partial to complete postorbital closure is effected in all
haplorhines by the approximation of processes of the malar, frontal, alisphenoid, and maxilla. Differences occur in the degree of postorbital closure
and the extent to which each bony process may contribute to the closure (particularly within platyrrhines), but the absence of comparable postorbital
closure in strepsirhines (and other mammals), including those which exhibit
a considerable degree of orbital frontation, suggests that this is a unique,
shared and derived character of haplorhines.

Although most recent investigators have supported a tarsiiform ancestry of
the Anthropoidea as the hypothesis which is most consistent with comparative
evidence from both fossil and extant primates, Gingerich (1973, 1974, 1975)
has recently presented an hypothesis for the origin of Anthropoidea from a
lemuriform stock. This was based on: (1) the apparent presence of a "free"
ectotympanic ring within the auditory bulla of the primitive catarrhine
Apidium, similar to the condition in lemuriforms; (2) the common occurrence
of small, vertical, and spatulate central incisors in Anthropoidea and Eocene
Lemuriformes, in contrast to the presence of enlarged, procumbent and pointed
central incisors in Tarsiiformes and Plesiadapiformes (=Paromomyiformes); and
(3) the paleogeographic distribution of Eocene Adapidae in both the Old and
New World, providing the probable ancestors for catarrhines and platyrrhines,
respectively. Gingerich (1975) has proposed a new subordinal classification
of Primates to express his concept of primate phylogeny; the Simiiformes
(=Anthropoidea) and Lemuriformes (=Strepsirhini) are grouped together as the
suborder Simiolemuriformes, and the Tarsiiformes and Plesiadapiformes are
clustered as the suborder Plesitarsiiformes.

The occurrence of Eocene Adapidae in North America and Europe is irrelevant to the ancestry of the Anthropoidea, because Eocene omomyid tarsiiforms also occur in both continents (Simons, 1972; Szalay, 1976). In any case, phylogenetic hypotheses should be based primarily on biological rather than paleogeographic evidence.

The fragments of the right petrosal and squamosal of *Apidium* described by Gingerich (1973) provide valuable evidence for the pathway of the intrabullar carotid circulation and the relationship between the ectotympanic and squamosal. The internal carotid artery of *Apidium* entered the petrosal medially and gave rise to an enlarged promontory branch; there was no evidence of a stapedial branch. As noted by Gingerich, all these features are shared with extant Anthropoidea. Gingerich (1973) concluded that "enlargement of the promontory artery and loss of the stapedial, characteristic of all anthropoids, could have occurred with equal probability in lemuroids or tarsioids." This conclusion was based on Gingerich's claim that the diameter of the bony promontory canal was larger than that of the stapedial branch (contra Gregory, 1920) in a single specimen of the adapid *Notharctus*. This observation is at odds with reports on the pattern of carotid circulation in all other specimens of adapids (Stehlin, 1916; Gregory, 1920; Szalay, 1975) and extant lemuriforms (Szalay and Katz, 1973; Szalay, 1975), in which it has been demonstrated that the stapedial branch is as large as, or larger than, the promontory branch. In contrast, the promontory canal is relatively larger than that of the stapedial in all adequately preserved Eocene-Recent tarsiiforms: *Necrolemur*, *Rooneyia*, and *Tarsius* (Gregory, 1920; Simons and Russell, 1960; Szalay, 1975, 1976).

The pattern of relative reduction of the stapedial artery and enlargement of the promontory branch, characteristic of Eocene Tarsiiformes, is an intermediate stage in the loss of the stapedial branch and further enlargement of the promontory artery in all Anthropoidea. An additional shared and derived character state (not discussed by Gingerich, 1973) of all Tarsiiformes, Platyrrhini, and Catarrhini (including *Apidium*) is the medial entry of the internal carotid artery into the bulla (Szalay, 1975, 1976). In contrast, the entry of the carotid into the bulla is posterolateral in Paromomyiformes, Adapidae, Indriidae, Lemurinae (excluding *Lepilemur*), and *Daubentonia*, and this is considered to be the primitive primate condition. Therefore, a consideration of the point of entry of the internal carotid into the bulla in all primates, coupled with an assessment of the relative sizes of the intrabullar stapedial and promontory arteries, contradicts the hypothesis of Gingerich (1973) that the intrabullar carotid pattern of Anthropoidea could have been derived with equal probability from either tarsiiforms or lemuriforms.

Gingerich's (1973, 1974) hypothesis that the Anthropoidea evolved from a lemuriform ancestor rests primarily on his belief that *Apidium* possessed a "free" ectotympanic ring within the bulla, and that the anterior crus of the ectotympanic was not fused to the squamosal. Gingerich (1974) asserted that a "free" ectotympanic ring within the bulla was a primitive condition in primates; if true, such a shared primitive feature between *Apidium* and Lemuriformes would have minimal value in indicating a phyletic relationship between the two taxa. However, the contention that the ectotympanic of *Apidium* was intrabullar is difficult to support, because most of the ectotympanic and the entire bulla is missing. It is equally probable that the ectotympanic ring occupied the lateral margin of the bulla, as it does in the Oligocene catarrhine *Aegyptopithecus* and in platyrrhines. In any case, lack of fusion of the anterior crus of the ectotympanic to the squamosal is not a unique shared feature of *Apidium* and Lemuriformes; Hershkovitz (1974b) has demonstrated such

a relationship in adult *Tarsius* and in many immature and mature cebids. Therefore, it is probable that a ring-like ectotympanic with an unfused anterior
crus was characteristic of the last common ancestor of platyrrhines, catarrhines, and *Tarsius*. Phylogenetic evaluation of the morphology and ontogeny
of the primate bulla has been discussed in more detail by Cartmill (1975) and
Szalay (1975, 1976).

The hypothesis (Gingerich, 1974, 1975) that incisor morphology is of fundamental importance in providing evidence of a plesitarsiiform/simiolemuriform
dichotomy is suspect because of the low phylogenetic valence of this feature
in mammals. Plesiadapiformes and Tarsiiformes were united on the basis of their
common possession of enlarged, pointed, and procumbent incisors (a derived
condition), whereas the Adapidae and Anthropoidea share a pattern of small,
spatulate, and vertical incisors. Incisor hypertrophy has been acquired convergently in numerous lineages of mammals, including multituberculates, dipro-
todont marsupials, tillodonts, rodents, lagomorphs, and ungulates. The claim
that "virtually all of the early genera of primates can be identified as tarsioid or lemuroid on the basis of their anterior dentition alone" (Gingerich,
1974) is not supported by examination of the available evidence. Only a small
proportion of known early Tertiary primates have their incisors preserved, and
it is unlikely that the ancestral tarsiiforms exhibited the incisor specializations of some later omomyids. Admittedly, many anaptomorphines had enlarged
and procumbent incisors (Simons, 1972; Szalay, 1976). However, in a recent
revision of Tarsiiformes (Szalay, 1976), the three most primitive species
(*Teilhardina belgica, Chlororhysis knightensis,* and *Chumashius balchi*), as
judged by the criteria of relative canine size, premolar number, and molar
morphology, have small alveoli for their incisors. Small incisors and large
canines are preserved in *Ekgmowechashala philotau,* and small incisors are
clearly present in *Washakius* and *Dyseolemur*. It is probable that the tarsiiform morphotype included relatively small, vertical, and spatulate incisors,
similar to the morphotype of Lemuriformes and Anthropoidea. In any case, the
sharing of this primitive mammalian condition (i.e., small incisors) by the
latter two taxa does not indicate their close phylogenetic relationship.

THE USE OF SOFT TISSUE CHARACTERS IN PHYLOGENY RECONSTRUCTION

Some paleontologists (Gingerich, 1973, 1974; Simons, 1976) have suggested that the use of soft tissue characters, and placentation in particular,
is of little value in assessing phylogenetic relationships, because these features are not preserved in the fossil record, and therefore it is difficult
to be certain of the primitive, derived, or parallel nature of shared character
states. These authors have presented no discussion of the method used to evaluate the character states of these soft anatomical data, despite the fact that
the majority of investigators using these features have employed careful character analyses. Moreover, in the case of placentation, Gingerich and Simons
have discussed the uncertainty surrounding only a single character, the definitive chorioallantoic placenta, while ignoring the numerous other characters
of the fetal membranes which form the basis for a more extensive analysis and
morphotype reconstruction.

As emphasized elsewhere (Mossman, 1967; Luckett, 1974, 1975), the entire
ontogenetic pattern of the fetal membranes is available for evaluating phylogenetic relationships, not just the morphology of the definitive placenta.
Comparative developmental studies facilitate the recognition of parallel or
convergent evolution of individual characters, such as the convergent evolution

of a hemochorial placenta in some primates, insectivorans, bats, and rodents.
However, the complexity of genetic information involved in fetal membrane
morphogenesis minimizes the possibility of convergence in an entire ontogenetic pattern of shared and derived characters, such as those which characterize
the Haplorhini. The great diversity of soft anatomical, ontogenetic, and
molecular characters available for cladistic analysis serves to offset their
absence in the fossil record.

We reaffirm the basic tenet (e.g., Gregory, 1910; Hennig, 1950, 1966;
Simpson, 1961) that all available characters of organisms, both extant and
fossil, should be utilized in evaluating phylogenetic relationships. In addition, the construction of a phylogram or cladogram based on multiple character
analyses should be an essential prerequisite for constructing or altering a
classification.

REFERENCES

Ashlock, P.D. (1974). *Ann. Rev. Ecol. Syst.*, 5, 81-99.
Cartmill, M. (1972). *In* "The Functional and Evolutionary Biology of Primates",
   (R. Tuttle, ed.), pp. 97-122, Aldine, Chicago.
Cartmill, M. (1975). *In* "Phylogeny of the Primates", (W.P. Luckett and F.S.
   Szalay, eds), pp. 313-354, Plenum, New York.
Cave, A.J.E. (1967). *Am. J. Phys. Anthrop.*, 26, 277-288.
Cracraft, J. (1974). *Syst. Zool.*, 23, 71-90.
Gingerich, P.D. (1973). *Folia Primatol.*, 19, 329-337.
Gingerich, P.D. (1974). "Cranial Anatomy and Evolution of Early Tertiary
   Plesiadapidae (Mammalia, Primates)". Unpublished Ph.D. Thesis, Yale
   University, New Haven.
Gingerich, P.D. (1975). *Contrib. Mus. Paleont., Univ. Mich.*, 24, 135-148.
Gregory, W.K. (1910). *Bull. Amer. Mus. Nat. Hist.*, 27, 1-524.
Gregory, W.K. (1920). *Mem. Amer. Mus. Nat. Hist.*, 3, 49-243.
Hennig, W. (1950). "Grundzuge einer Theorie der phylogenetischen Systematik",
   Deutscher Zentralverlag, Berlin.
Hennig, W. (1966). "Phylogenetic Systematics", University of Illinois Press,
   Urbana.
Hershkovitz, P. (1974a). *Folia Primatol.*, 21, 1-35.
Hershkovitz, P. (1974b). *Folia Primatol.*, 22, 237-242.
Hill, W.C.O. (1948). *Proc. Zool. Soc. Lond.*, 118, 1-35.
Hill, W.C.O. (1953). "Primates. Strepsirhini", Vol. I, Edinburgh University
   Press, Edinburgh.
Hill, W.C.O. (1955). "Primates. Haplorhini: Tarsioidea", Vol. II, Edinburgh
   University Press, Edinburgh.
Leutenegger, W. (1973). *Folia Primatol.*, 20, 280-293.
Luckett, W.P. (1974). *In* "Reproductive Biology of the Primates, Contributions
   to Primatology", (W.P. Luckett, ed.), Vol. III, pp. 142-234, S. Karger,
   Basel.
Luckett, W.P. (1975). *In* "Phylogeny of the Primates", (W.P. Luckett and F.S.
   Szalay, eds), pp. 157-182, Plenum, New York.
Luckett, W.P. (1976). *Folia Primatol.*, 25, 245-276.
Martin, R.D. (1975). *In* "Phylogeny of the Primates", (W.P. Luckett and F.S.
   Szalay, eds), pp. 265-297, Plenum, New York.
Mayr, E. (1974). *Z. Zool. Syst. Evolut.-Forsch.*, 12, 94-128.
Mossman, H.W. (1967). *In* "Fetal Homeostasis", (R.M. Wynn, ed.), Vol. II,
   pp. 13-97, New York Acad. Sci., New York.

Pocock, R.I. (1918). *Proc. Zool. Soc. Lond.*, 1918, 19-53.

Ross, H.H. (1974). "Biological Systematics", Addison-Wesley Publishing Company, Reading.

Simons, E.L. (1972). "Primate Evolution", Macmillan, New York.

Simons, E.L. (1974). *In* "Prosimian Biology", (R.D. Martin, G.A. Doyle, and A.C. Walker, eds), pp.415-433, Duckworth, London.

Simons, E.L. (1976). *In* "Molecular Anthropology", (M. Goodman and R.E. Tashian, eds), pp. 35-62, Plenum, New York.

Simons, E.L. and D.E. Russell (1960). *Brevoria*, 127, 1-14.

Simpson, G.G. (1945). *Bull. Amer. Mus. Nat. Hist.*, 85, 1-350.

Simpson, G.G. (1955). *Bull. Amer. Mus. Nat. Hist.*, 105, 411-442.

Simpson, G.G. (1961). "Principles of Animal Taxonomy", Columbia University Press, New York.

Stehlin, H.G. (1916). *Abh. Schweiz. Pal. Ges.*, 41, 1297-1552.

Szalay, F.S. (1972). *Amer. J. Phys. Anthrop.*, 36, 59-76.

Szalay, F.S. (1975). *In* "Phylogeny of the Primates" (W.P. Luckett and F.S. Szalay, eds), pp. 91-125, Plenum, New York.

Szalay, F.S. (1976). *Bull. Amer. Mus. Nat. Hist.*, 156, 157-450.

Szalay, F.S. and C.C. Katz (1973). *Folia Primatol.*, 19, 88-103.

Wolin, L.R. and L.C. Massopust (1970). *In* "The Primate Brain, Advances in Primatology", (C.R. Noback and W. Montagna, eds), Vol. I, pp. 1-27, Appleton-Century-Crofts, New York.

DISCUSSION

Gingerich (Michigan): In the fossil tarsioids that have very small incisor
alveoli, the teeth simply aren't there. But even in these, the central lower
incisor alveolus is always larger than the lateral one, whereas the reverse
is true in all adapids and anthropoids.

Luckett: No, it's not true in many South American primates.

Kay: Nor in the Belgian specimens of *Teilhardina*, where the I_1 alveolus is
about the same size as the I_2 alveolus.

Martin: As many people have pointed out at this Congress, the tree obtained
depends on which characters you identify as primitive. Two of the characters
in your list illustrate the problems of determining this. How do you know
that relatively small neonatal weights and the presence of tapetum lucidum are
primitive for Primates?

Luckett: The first point is based on looking at outgroups for Primates, particularly Insectivora. The second is based primarily on comparisons within
Primates. I don't know of a good comparative study on the tapetum lucidum in
eutherian mammals, but it does occur in artiodactyls and carnivores. It is
found in all strepsirhines and in no haplorhines — and this is irrespective
of whether the animals are nocturnal or diurnal.

Martin: But the structure and composition in these various groups are markedly
different.

Luckett: Yes. We are just scoring the tapetum as "present" or "absent", and
we haven't utilized the finer details of it.

Ramaswami (Bombay): After restaining Hubrecht's slides and studying them in
great detail, J.P. Hill came to the conclusion that the haemochorial placenta
of *Tarsius* is very peculiar, probably a parallelism *sui generis*, and cannot
be compared with that of monkeys.

Luckett: I agree that there are some unusual features in the placental disc
of *Tarsius*, and I wouldn't propose that it represents the ancestral condition
of Anthropoidea. But the characters of the foetal membranes are more important
for assessing phylogeny than the placental disc itself.

Schwartz: Many of the features of placentation you describe represent what I
would think of as transitory stages of a single process, some of which depend
on the ones that precede them. How can you separate these out as different
characters in your phylogeny reconstruction?

Luckett: In some taxa, certain characters are correlated, whereas in other
taxa they're not. In haplorhines, the body stalk and reduced allantois might
seem functionally correlated with the development of the haemochorial placenta.
But if you look at Dermoptera, for instance, you find a body stalk and haemochorial placenta co-existing with a voluminous allantois, which appears later
in development.

Gingerich: If dermopterans, tarsiers, anthropoids and various other eutherians all have a haemochorial placenta, why do you think it's the derived character state?

Luckett (projecting a slide): In this graph, I've divided all the taxa of eutherians for which we have adequate developmental data into three groups by placenta type. Most of the taxa with an epitheliochorial placenta are condy-larth derivatives, but we also have Strepsirhini and talpid insectivorans in this group. There are also insectivores with endotheliochorial placentae (Soricidae, Tupaiidae), and some with a haemochorial placenta (erinaceids, tenrecids). More eutherian groups have a haemochorial placenta than other types of placenta, and we might regard a haemochorial placenta as primitive if we judged solely by commonality of possession. But you get a different picture if you look at developmental features of the foetal membranes, such as the pattern of amnion formation, the nature of the allantois, and the two features of the yolk sac that are most important in this context. All these features are uniform in the groups that have an epitheliochorial placenta (with the possible exception of one yolk-sac feature in the Talpidae), whereas the endothelio- and haemochorial groups exhibit no agreement or uniformity at all in these features. Because we're dealing with a complex developmental process involving all the germ layers, I would expect agreement in placenta type and all four developmental features of the foetal membranes to occur only as a primitive retention. I conclude that the foetal membrane features associated with the epitheliochorial placenta are primitive; and this is sup-ported by comparisons with outgroups — birds, reptiles, Prototheria, and Metatheria.

Gingerich: What worries me is that you use developmental data to try to decide what's primitive, whereas someone else might take the same data set, use commonality as a criterion, and come up with a different answer. It appears to me that it's very difficult to decide this sort of question without fossil evidence. The other thing that bothers me is that, since the haemo-chorial placenta has evolved many different times in these different groups, there's obviously a lot of convergent evolution going on in these features. In most cases, it's not possible to sort that out. I think that the parsimony assumption of cladistics greatly underestimates the amount of parallelism and convergence that's taken place in the history of mammals.

Luckett: This is the advantage of ontogenetic data. You can distinguish con-vergently evolved haemochorial placentae by the fact that they result from different ontogenetic pathways. As far as commonality is concerned, it should not be the only criterion of phylogenetic analysis if you have others you can use.

THE PROBLEM OF CONVERGENCE AND THE PLACE OF *TARSIUS* IN PRIMATE PHYLOGENY

C.E. OXNARD

Department of Anatomy, The University of Chicago,
1025 East 57th Street, Chicago, Illinois 60637, USA.

INTRODUCTION

In recent years, a number of studies have suggested that tarsiers, monkeys, apes and man form a natural phylogenetic unit from which tree shrews, Lemuriformes and Lorisiformes ought to be excluded. A.J.E. Cave (1973) finds that *Tarsius* and the anthropoids share a pattern of architecture of the nasal fossa which differs radically from that of all other mammals, including the remaining primates. Szalay (1975) concludes that the Tarsiiformes, Platyrrhini and Catarrhini contrast with prosimians in sharing several characteristics of the basicranial region. Studies of primate placentae and foetal membranes have led Luckett (1975) to postulate that the Tarsiiformes and Anthropoidea are "sister groups within the Haplorhini" and therefore cladistically separate from the strepsirhine prosimians. Even immunological data (e.g., those provided by Dene et al., 1975) suggest that the closest affinities of *Tarsius* are with members of the Anthropoidea rather than with the other Prosimii (*sensu* Simpson, 1945; and see also Minkoff, 1974).

A major problem with using the morphology of tarsiers to determine their affinities is that it is so closely associated with their remarkably specialised locomotor pattern. Structures which might reflect an association of tarsiers with anthropoids may be overshadowed by functional adaptations to a saltatory form of locomotion. Evolutionary convergence between *Tarsius* and the similarly specialised Galaginae may also obscure more phylogenetically relevant aspects of its morphology.

An attempt is therefore made here to study the relationships of *Tarsius* using a large battery of data about the overall form and proportions of the body, in the hope that it will prove possible to dissect away (to use an anatomical metaphor) or to partition out (to use a statistical expression) morphological convergence associated with extreme locomotor specialization, and thus provide another view of tarsier affinities.

MATERIALS AND METHODS

The data used in these analyses are those of the late Professor A.H. Schultz and include measurements of 472 non-human primates representing 34 genera. Considerable numbers of specimens are available for most genera. (e.g., *Tupaia, Lemur, Microcebus, Nycticebus, Galago, Tarsius, Leontocebus, Aotus, Cebus, Saimiri, Alouatta, Ateles, Macaca, Cercocebus, Papio, Cercopithecus, Presbytis, Nasalis, Hylobates, Symphalangus, Pongo, Pan, Gorilla* and *Homo*); but in 10 genera (including *Propithecus, Perodicticus, Cacajao, Lagothrix* and

Erythrocebus) there are data from fewer than 4 specimens.

All measurements were made according to the technique defined by Schultz (1929). From these, indices are derived as described by Schultz (1956). The dimensions used in the present study are those of Schultz (1956) less the relative tail length (excluded because of its exceptional variability) but with the addition of relative foot breadth, cephalic index, and a new index describing the ratio of foot length to lower limb length.

The precision of measurement obtained by these techniques is discussed by Schultz (1929). Errors due to unavoidable mensurational inconsistencies are, in general, sufficiently small to render it unlikely that the results of the comparisons made here are affected by them.

The indices provide an overall description of the shape and proportions of the several body regions. So far as overall size is reflected in trunk length, its use as the denominator of some of the indices provides an indication of contrasts emerging from differences in shape rather than of size. But a qualification of the use of the trunk length as defined by Schultz (1929) is necessary because this dimension (suprasternale to symphysion) is influenced by the relative position and form of the suprasternal notch and pubic symphysis, both of which vary in different primate genera. It is also a measure which, even with meticulous technical care, is sensitive to somewhat minor differences in cadaveric orientation. It would have been better if the dimensions could have been reduced to their original measurements with the elimination of trunk height as the denominator in the indices. Nevertheless it has been shown (McArdle, pers. comm.) that replication, even by another investigator, of such an apparently difficult measure is possible.

A second qualification relates to the use of these indices to eliminate effects due to differences in overall size. Such indices can be fully effective only when an isometric relationship exists among the measurements. In practice, relationships are usually allometric so that the best compensation would be some form of regression adjustment. In the present study it has not been possible to apply such methods because of the form in which the data are available. But to make some test of the extent to which factors relating to overall size have been eliminated by expressing each dimension in the form of an index, correlation coefficients have been worked out between the average crown-rump length for each genus and the average value of each index in turn. This test is more fully described in Ashton, Flinn and Oxnard (1975) and the results suggest that it is rather unlikely that size factors not eliminated by the use of indices seriously disturb the final results.

Further tests carried out by Ashton, Flinn and Oxnard (1975), show that there is an increase in the standard deviation of most of the indices when compared with the increasing generic mean values; for this reason all final analyses are based upon logarithmically transformed quantities in which this disturbing feature is found to be absent.

From the transformed data are computed a matrix of squared generalized distances between the genera, coordinates of the means of each genus in a canonical variates analysis, and a minimum spanning tree resulting from examination of the matrix of squared generalized distances (Gower, 1966; Gower and Ross, 1969).

A suite of analyses as outlined in earlier work (Oxnard, 1975a, b) was performed. A *first* analysis is a multivariate compound of the series of overall dimensions of the forelimb alone. The particular indices include relative shoulder breadth, relative upper limb length, brachial index, relative hand length, relative hand breadth, and relative thumb length. A *second* set of

analyses is of the suite of dimensions that describe the overall proportions
of the hindlimb. These include relative hip breadth, relative lower limb
length, crural index, relative foot length, relative foot breadth, and foot
length relative to lower limb length. There are separate multivariate analyses of those indices relating to, *third*, the shape of the trunk (relative
chest circumference, relative shoulder breadth, relative hip breadth and chest
index) and, *fourth*, of the head (relative head size, facial index, relative
face height, relative upper face height, cephalic index, interocular index
and relative ear size). A *final* analysis is of the entire suite of dimensions
including the intermembral index but without, of course, the duplications involved in adding the regional studies together.

On theoretical grounds, it is possible that the pattern of discrimination
in a canonical analysis can, in certain circumstances, be distorted if groups
containing very small numbers of specimens are included (Gower, 1968). But a
series of initial canonical analyses shows that in this investigation the
relative positions of larger groups in the several canonical axes are not appreciably changed whether or not the groups with fewer than four specimens are included. Consequently, then, the final analyses that are presented include all
groups.

RESULTS

The untransformed mean values of individual indices are given in the study
of Ashton, Flinn and Oxnard (1975). Although no individual index makes effective separations among all primate groups, visual assessment of the entire
range of indices seems to suggest that the Hominoidea are divided from the other
Anthropoidea. The Prosimii (*sensu* Simpson, 1945) seem to be discrete from
most of Anthropoidea but merge with some of them; within the Prosimii, there
are obvious major similarities between tarsiers and galagos. But in most dimensions the general scale of these differentiations is quite poor and a univariate analysis of this type does not readily provide good quantitative differentiation.

The multivariate analyses of the various anatomical regions separately
also show that *Tarsius* and *Galago* tend to fall near one another. However, in
each case, the distance between them is not overly close, being 11.5 generalized distance units in the forelimb study, 7.5 units in the hindlimb study, 7.2
units for the trunk study and 4.5 units for the head study respectively. It
certainly cannot be said that the two species are morphologically identical
or even nearly so.

The nature of the separation produced by the multivariate analysis of forelimb dimensions is such that these two genera are outlying at one extremity
of the relatively linear separation of all primates. In the case of the analysis of hindlimb dimensions they are placed so as to lie together in one ray
of a star-shaped arrangement of the other primate genera. These results for
the forelimb and hindlimb have been described in Oxnard (1975a and b).

The nature and degree of separations that appear in the analyses of the
trunk and head differ from those in the limb studies. In the trunk and head
the separations tend to be eclectic, individual genera throughout the primates
being differentiated from one another by one or more of a wide variety of canonical axes. *Galago* and *Tarsius* again tend to fall together as a pair but they
are accompanied by a variety of different other primate genera in each instance.
Thus, although, for example, canonical axis one of the study of the head dimensions places *Galago* and *Tarsius* near one another at approximately -11.0 general-

ized distance units, so also several other primates (*Indri*, *Daubentonia*, *Microcebus* among the Prosimii, *Papio* and *Pongo* among the Anthropoidea) occupy this approximate locus. Canonical axis three of the analysis of trunk dimensions places *Galago* and *Tarsius* at +3.0 units; again, however, this position is shared with *Lichanotus* [*Avahi*] and *Propithecus* (but not *Indri*) among the Prosimii and *Erythrocebus* and *Nasalis* among the Anthropoidea.

It is in the combination of data from multivariate analysis of all bodily proportions, however, that the best separations are achieved. Thus Figure 1 (upper frame) provides a view of the minimum spanning tree that obtains from all the primate genera. This tree is drawn within the set of directional arrangements derived from consideration of the first three canonical axes (those producing the greatest discriminations). The pictorial convention adopted in this figure utilises ligands drawn in perspective fashion to show connections lying in front of or behind the plane of the diagram. Because, however, this minimum spanning tree exists within space of more than three dimensions, the only distances that are correct within the diagram are those links shown as solid lines. Dotted lines indicate regions where the minimum links are so similar that alternate arrangements could have easily obtained, given slightly different expressions of biometrical variation.

Bearing these factors in mind, it can be seen that the primates are linked in a long chain that is curled around itself like a figure "six". At one end of the chain are hominoids, next lie Old World monkeys, next New World monkeys, and in the curled head of the "six" lie various prosimians. Emanating from this linear skeleton are a series of side chains.

Figure 1 (lower frame) shows the minimum spanning tree as it exists if we straighten out the linear skeleton and adopt certain conventions (regularly used with this mode of display: e.g., Ashton, Flinn, Oxnard and Spence, 1976) to demonstrate the side chains. It is thus evident that the side chains comprise, in each case, sets of genera the same as those on the main skeleton to which they are linked. Thus hominoids on a side chain link with hominoids on the main skeleton, Old World monkeys with Old World monkeys, New World monkeys with New World monkeys, and, of course, prosimians with prosimians. The separations appear, therefore, to be of phylogenetic import except for the contiguity of *Tarsius* and *Galago*. These two genera do not appear to be well separated as might be thought befits their separate infraordinal status within the Prosimii. They lie together on a side chain emanating from a part of the main skeleton occupied by *Microcebus*.

DISCUSSION

The discussion of the placement of all primates in the fore- and hindlimb analyses has already been presented in Oxnard (1975a, b). Those arrangements seem to have biological relevance to some aspects of the function of the fore- and hindlimbs respectively.

Thus the forelimb analysis appears to grade the primates so that those with more highly mobile, more tension-bearing limbs lie nearer one end of a broad band and those with less highly mobile, more compression bearing limbs lie near the other end of this band-like spectrum. *Galago* and *Tarsius* fit this interpretation. Their limbs are not overly mobile and presumably bear heavy compressive stresses relative to their size in landing after long leaps; they lie appropriately near the end of the spectrum which they share, distantly, with other genera (e.g., *Papio* and *Erythrocebus*) whose forelimbs bear weight by compression (although not because of great leaping but because they are so highly terrestrial).

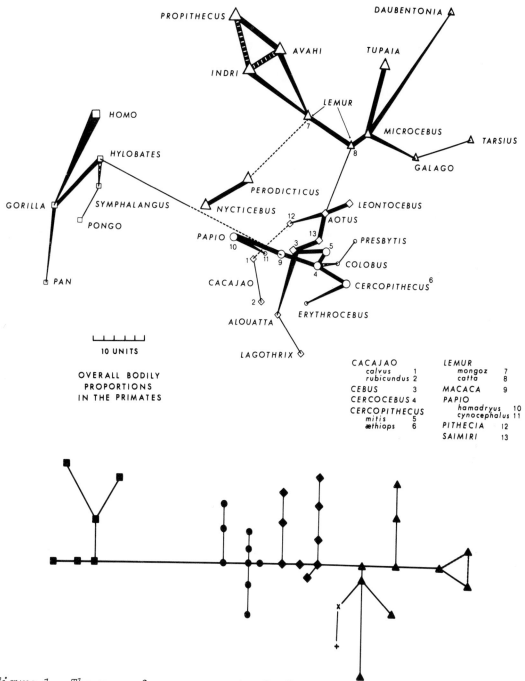

Figure 1. The upper frame represents the "C"-shaped minimum spanning tree of generalized distance connections among the various primate genera as explained in the text. The marker equals ten standard deviation units. The lower frame represents this tree in diagrammatic form as explained in the text. Squares=hominoids, circles= Old World monkeys, diamonds=New World monkeys, and triangles= prosimians. The symbols x and + represent Galago and Tarsius respectively.

In the case of the hindlimb analysis, the various genera are separated into a star-shaped arrangement in which generalized quadrupedal species (irrespective of taxonomic group) tend to lie in the centre of the star and those genera that have more extreme functional patterns within the hindlimbs (e.g., slow-climbing: pottos; hindlimb-hanging: uakaris) lie, separately, in outlying rays of the star. *Tarsius* and *Galago*, both extreme leaping forms, share such a ray linked to the main body of quadrupedal species through the genus *Microcebus* which is lying intermediately.

These arrangements seem, then, to be evidence of functional convergence between the structures of the fore- and hindlimbs of *Galago* and *Tarsius* whatever other information they may contain.

As described in the results the correspondences between *Galago* and *Tarsius* in the analysis of the trunk and head, and their variable relationships with some of the other primates are clear-cut but do not seem to lend themselves readily to any biological explanation. Conjunctions of genera such as *Tarsius, Galago, Indri, Daubentonia, Microcebus, Papio* and *Pongo* (axis one of the study of head dimensions) or *Tarsius, Galago, Lichanotus* [*Avahi*], *Propithecus, Papio* and *Nasalis* (axis three of the study of trunk dimensions) are truly unlikely. Although it is possible that these arrangements have some biological relevance that is not obvious, detailed examination of rank orders of genera in individual canonical axes and of contributions of individual dimensions to those canonical rank orders provide no easy solutions. And in all these separate analyses, even the proximity of *Galago* and *Tarsius* is not as total as may appear from scanning early canonical axes alone. In each analysis, separately sufficient discrimination is built up in statistically significant higher axes that there are major distinctions between these two genera.

How different are the separations obtained in the overall analysis of the total set of dimensions. As foreshadowed in Oxnard (1975) but here much more clearly evident, the primates are separated on the basis of overall morphology in a manner that concurs very considerably with the current taxonomic classification. All major taxonomic arrangements among the primates are adhered to with one exception: the close conjunction of *Tarsius* and *Galago*.

However, further examination of the nature of the network of generalized distance linkages surrounding *Galago* and *Tarsius* provides unexpected information. If we look beyond the minimum link between *Tarsius* and *Galago* and examine other neighbouring links, a different picture is provided. *Galago* truly "belongs" morphometrically with *Microcebus, Lemur catta* and *Lemur mongoz*. But *Tarsius* does not possess near minimum links with any other prosimian genus; its next nearest neighbours are *Leontocebus, Aotus,* and *Saimiri*, among the New World monkeys and *Cercopithecus* and *Cercocebus*, from the Old World (Fig. 2). It is then possible that the minimum link between *Tarsius* and *Galago* results from convergent features of the data that are so heavy that they overshadow the taxonomic evidence in the data? Is it possible that removing this particular link dissects out convergence and allows another view of the morphological associations of *Tarsius* and *Galago*? We cannot truly know the answers to these questions; but the picture presented is not unconvincing. The morphological affinities of *Tarsius*, once we allow for convergence, may truly be with man, apes and monkeys (Anthropoidea); those of *Galago* are clearly with lemuroids and lorisoids.

It might be thought that overlooking the minimum link between *Tarsius* and *Galago* in this manner is special pleading for a phylogenetic placement of *Tarsius* with Anthropoidea. It must therefore be pointed out that nowhere else in this analysis of the overall proportions of the primates does removal of a

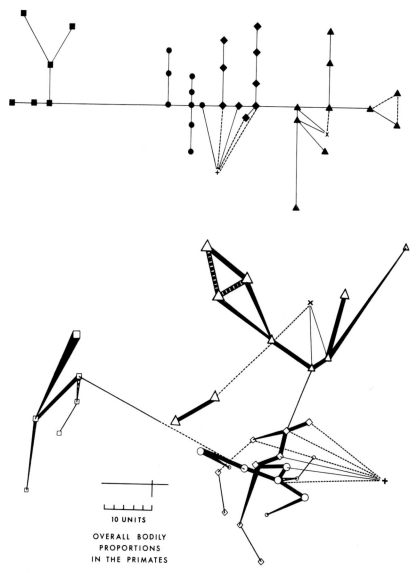

Figure 2. *The upper frame gives a diagrammatic view of the minimum span-*
ning tree after the link between Galago and Tarsius has been
severed. The lower frame recreates the full "C"-shaped mini-
mum spanning tree but again with the link between Tarsius and
Galago severed. Conventions as in Figure 1.

particular minimum link make any special change to the morphological relationships of the genera. Removal, for instance, of the link between *Pongo* and
Symphalangus still leaves a next close link between *Pongo* and *Hylobates*; removal of the link between *Homo* and *Gorilla* still leaves as man's nearest neigh-

bour the genus *Pan*. And it should also be noted that overlooking the minimum
link between *Tarsius* and *Galago* does not provide this unique result when
applied to the various sub-analyses relating to each individual anatomical
region. For both the forelimb and hindlimb analyses, for example, the next
nearest neighbour of *Tarsius* is another prosimian, *Microcebus*, a morphological
relationship presumably speaking to the functional similarity between the incipient leaping of *Microcebus*, the developed leaping of *Galago* and the extreme
leaping of *Tarsius*.

A second point that is of interest for this discussion relates to the more
usual ways of making phylogenetic assessments for these species. The kinds of
evidences upon which we most depend for assessments are features such as patterns of tooth cusps, sutural arrangements in orbits, middle ear architectures,
and so on. The choice by Professor Cave of characteristics of the nasal fossa
or by Professor Szalay of the architecture of the basicranium are exactly such
attempts to isolate architectural arrangements that might relate less to habi-
tus and perhaps, therefore, more to heritage. Such a weighting of biological
characteristics of this type is a usual practice even though the bases upon
which it may be done may vary from one investigator to another. In the present study, cutting the link between *Tarsius* and *Galago* and relying subsequently upon the picture provided by the next nearest minimum links with other
species may be the equivalent of making such a weighting. However, in this
case we are not weighting the individual characters themselves but rather certain features of their multivariate compound.

Given the hindsight provided by this study, can we go back to the original
dimensions and define among them those that appear to provide the information
about the similarity with *Galago* (convergence) and those that seem to be speaking to the likeness to monkeys (phylogenetic proximity)? The answer appears
to be that, in this case, we cannot. The contributions of the original dimensions to the multivariate separations are such that almost all dimensions contribute to almost all separations. Each original datum pertains to both convergent and phylogenetic information. Presumably, therefore, the way in which
the original dimensions contain information associated with these two biological phenomena: convergence and phylogenetic proximity, is such as to lie, not
within the separations of species by individual dimensions, but within the complex patterns of correlations among the dimensions.

This discussion leads in turn to the final point. For although a form of
weighting has been carried out in this study in order to see more deeply into
the nature of the relationships of *Tarsius* and *Galago* with other primates, no
such weighting has been necessary for the great bulk of primate species. Presumably this is because, although there are many anatomically localised parallels and convergences among the other primates, summation of information from
many anatomical regions overpowers the individual adaptive contents from the
separate regions and provides an overall picture relating to phylogeny. Or
is it that mere summation of adaptive information provides, directly, a phylogenetically related result?

There can be little doubt that variations in the form and proportions of
primates are associated with variations in function. In this sense these features differ from many of the characters traditionally used in primate taxonomic assessments; the more traditional characters, although usually from
some overall region with obvious adaptive significance, often themselves display arrangements within that adaptive feature that do not appear to be adaptive. But then I suppose, even for these traditional features, it is not that
we know that their arrangements are not of functional significance; rather is
it that we do not know if they are.

ACKNOWLEDGEMENTS

I am indebted to the late Professor A.H. Schultz for permission to use his data and to Professor Lord Zuckerman, Professor E.H. Ashton, and Mr. T.F. Spence whose collaborations have formed the background of this work. Thanks are also due to Professors James A. Hopson and Leonard Radinsky whose discussion of and comments on the text have been most valuable. The investigations are supported by NSF grant no. GS 30508.

REFERENCES

Ashton, E.H., Flinn, R.M. and Oxnard, C.E. (1975). *J. Zool.*, 175, 73-105.
Ashton, E.H., Flinn, R.M., Oxnard, C.E. and Spence, T.F. (1976). *J. Zool.*, 179, 515-556.
Cave, A.J.E. (1973). *Biol. J. Linnean Soc.*, 5, 377-387.
Dene, H., Goodman, M., Prychodki, W. and Moore, G.W. (1976). *Folia primatol.*, 25, 35-61.
Gower, J.C. (1966). *Biometrika*, 53, 588-590.
Gower, J.C. (1968). *Biometrika*, 55, 582-585.
Gower, J.C. and Ross, G.J.S. (1969). *Appl. Statist.*, 18, 54-64.
Luckett, W.P. (1974). *Contrib. Primatol.*, 3, 142-234.
Minkoff, E.C. (1974). *Amer. Nat.*, 108, 519-532.
Oxnard, C.E. (1975a). "Uniqueness and Diversity in Human Evolution: morphometric studies of australopithecines", The University of Chicago Press, Chicago.
Oxnard, C.E. (1975b). *Proc. Symp. 5th Congr. Int. Primatol. Soc.*, 269-286.
Schultz, A.H. (1929). *Contrib. Embryol.*, 20, 213-257.
Schultz, A.H. (1956). *Primatologia*, 1, 886-964.
Simpson, G.G. (1945). *Bull. Amer. Mus. Nat. Hist.*, 85, 1-350.
Szalay, F.S. (1975). *In* "Primate Functional Morphology and Evolution, (R.H. Tuttle, ed.), pp. 3-22, Mouton, The Hague.

DISCUSSION

Martin. Can you exclude the possibility that removing *Galago* from the comparison simply leads to emphasis of primitive primate resemblances between *Tarsius* and various monkeys, which tell us nothing about evolutionary relationships?

Oxnard. I don't know any way in which I can answer that. The picture I've arrived at merely shows that when you look at all the data, you get a pattern of resemblances which fit in reasonably well with phylogeny throughout the primates, except for putting *Tarsius* very close to *Galago*. When you cut that link, which is the one thing that's grossly wrong with the picture, *Tarsius* turns out to be nowhere near the prosimians. But I'm not prepared to answer the question you asked, because that requires subjective judgement. To me, the interesting question is this: why should I get the result I did without having to use the words "primitive" and "derived" in presenting the data?

PHYLOGENY RECONSTRUCTION AND THE PHYLOGENETIC POSITION OF *TARSIUS*

P.D. GINGERICH

Museum of Paleontology, University of Michigan, Ann Arbor, Michigan 48109, USA.

Tarsiers — "These animals are astonishingly deliberate and stupid-appearing in behavior, so much so that it seems a miracle that they can survive." D. Dwight Davis, 1962

Tarsius, whatever its aptitude, has managed to survive, and its phylogenetic relationships, classification, and evolutionary significance remain among the most interesting unsolved problems in primatology. First classified as a lemur, when that term included all prosimians, then separated in a group distinct from the other primates, *Tarsius* is now sometimes ranked with anthropoids. Thus the tarsier has achieved in a century of classificatory revision what most "lemurs" failed in 50 million years of evolution!

The phylogenetic position of *Tarsius* is fundamental to any consideration of either its classification or its evolutionary significance, and the phylogenetic relationships of the genus will receive most attention in the following discussion. First, it is necessary to outline and justify the methodology of phylogeny reconstruction advocated here (not because it is a new methodology, but because it has rarely been stated explicitly and it has recently fallen out of favour with a vocal majority of systematists). Second, it is necessary to review briefly the fossil record of primates, with emphasis on the early Tarsiiformes. Finally, a comment will be added concerning both the classification and the evolutionary significance of *Tarsius*.

It should be noted at the outset that it is impossible in a paper of this length to deal specifically with individual criticisms of my earlier conclusions regarding the relationships of *Tarsius* and the origin of anthropoid primates, such as those recently put forward by Szalay (1976) and others. These will be discussed at length in a monograph on the evolution of Eocene Adapidae now in preparation. Suffice it to say that I think most disagreement is due to differences between stratophenetic and "cladistic" approaches to phylogeny reconstruction, and this is the problem I wish to discuss at greatest length here.

PHYLOGENY RECONSTRUCTION

A phylogeny is generally understood to be the history of the various lines of evolution within a group of organisms. The very concept of *history* implies a time dimension, and only historical data, records of past times, can be used to reconstruct history with any confidence. This is as true of the study of animal phylogeny as it is of the study of 'prehistory' or the history of Victorian England. The historical data of animal phylogeny, fossils, are the objects of research of a branch of science, paleontology, which is uniquely suited to the study and reconstruction of phylogeny. If there is an "undesirable" characteristic of paleontological methodology, it is the following: when historical data are inadequate to permit determination of the lines of

evolution within a given group, paleontology yields no phylogeny. When the
fossil record is inadequate, sound paleontological methodology has often been
abandoned in favour of analytical algorithms that do yield answers, whether
the available raw data are adequate to solve the particular problem or not.
Thus, it is little wonder that recently proposed phylogenies based chiefly on
the comparative anatomy of living primates have proven so unstable and so controversial.

The essence of phylogeny reconstruction based on paleontology ("strato-
phenetics" see Gingerich 1976b) has been described and applied, explicitly
or implicitly, many times (see Simpson, 1961, for example) but I am not aware
that anyone else has stated the method quite so simply as it is presented
here. Three steps are involved: (1) data organization, (2) linking, and
(3) testing.

Data organization, simply stated, involves arranging the available fossil
specimens in chronological order (i.e., stratigraphical order). This is conveniently done in a diagram, where morphological attributes of the oldest
fossils are recorded at the bottom and attributes of successively more recent
ones are recorded in sequence, with the youngest known (Recent, if there are
any living members of the group) being plotted at the top.

Phenetic linking joins the different species in any one time interval to
the most similar species samples in adjacent intervals. Ideally, the criterion
of acceptable linking would be near identity of population samples in adjacent
intervals, and this is sometimes possible (Gingerich, 1976a, b). In other
cases gaps in the fossil record introduce varying levels of uncertainty, depending on the size of the gap and the morphological distinctness of the two
species populations being linked. The pattern of phenetic linking derived
from this step itself constitutes the principal phylogenetic hypothesis.

Critical testing requires that a high level of uncertainty be placed on a
phylogeny lacking a dense and continuous fossil record. Once the pattern of
stratophenetic linking is worked out, it is possible to go back over it and
study the evolution of individual morphological characters. Links based
largely on characters seen to be retentions of primitive morphology (as between the Adapidae and living lemurs, see below) are somewhat weaker than
links based on newly acquired characteristics. However, the crucial test
falsifying a phylogenetic hypothesis constructed stratophenetically is usually
the discovery of new fossil evidence that cannot be accommodated into the pattern of phenetic linking previously advanced. It is a positive characteristic
of phylogenies constructed stratophenetically that they do tend to be relatively stable.

In view of the wide use of "cladistics" to reconstruct phylogenetic history, some specific criticisms will here be directed toward that approach to
phylogeny. To put this into perspective, the approach outlined above and advocated here can be summarized as follows:
1. Fossils are collected from different stratigraphic intervals.
2. These fossil taxa are linked together based on their overall similarity
and stratigraphic proximity.
3. If a fairly dense and continuous pattern of linking is found, the whole
pattern is accepted as the probable phylogeny of the group under study.
Several additional steps are usually taken to establish evolutionary patterns
and erect a classification:
4. Individual characters are studied to trace their change through time,
giving some idea of the relative importance of character divergence, convergence, parallelism, and other patterns in evolution.

5. Individual lineages are grouped into clades at various levels, based on common ancestry. Morphological characters held in common by species within each clade are used to diagnose various clades from each other, and verbal classifications are constructed from this.

The end result is an understanding of the phylogenetic history of the group being studied, better knowledge of how evolution works, and finally a verbal classification that can be used to organize species and discuss them in groups at whatever level is preferred.

In contrast to this approach, cladistics requires *a priori* assumptions about the phylogenetic history of the group under study and about the evolutionary patterns of individual characters. Cladistics attempts to provide conclusions comparable to those of steps 4 and 5 above, without any independent means of constructing the phylogeny from which the conclusions must be derived. This problem arises because cladistics is basically a method of classification, and not a method of phylogeny reconstruction. Clades themselves are parts of a phylogeny, not something with an independent existence that can be used to construct a phylogeny. In other words, clades have no existence until the desired phylogeny is already constructed.

Circularity is manifest at every stage of cladistic analysis as it is currently being used to reconstruct phylogenies. As presented by numerous authors, cladistic analysis involves:

1. Identification of the alternative states of homologous characters.

Comment: However, homologous characters are characters that can be traced back to the same feature in a common ancestor — which can only be done *after* the phylogenetic history of the group is known.

2. Arrangement of alternative states into a "morphocline" for each character.

Comment: The only justification for a bipolar morphocline is an operational one — this simplistic, one way, primitive-to-derived ordering is *assumed* in order to make the subsequent analysis manageable. Examples of evolutionary radiations including both evolutionary reversals and multipolar character radiations have been documented in the fossil record (see Gingerich, 1976b) and are undoubtedly both common and important in evolution.

3. Assignment of "polarity" to the morphocline, i.e. one end of the morphocline is identified as primitive, the other as derived.

Comment: For every rule used to assign polarity to a morphocline there is an equally valid converse: a widely distributed character state (such as the presence of hair in mammals) is assumed to be primitive, whereas it may only be the result of a secondary radiation within the group (nails on the terminal phalanges of most primates are an example of a widely distributed, but probably derived character state, claws being primitive or sometimes probably secondarily derived), illustrating that widely distributed character states are not necessarily primitive. Character states that appear early in ontogeny are assumed to be primitive because ontogeny usually recapitulates phylogeny, but neoteny is a well known and important developmental process leading to the converse.

4. Species sharing large numbers of "derived" character states are clustered together in a cladogram.

Comment: Even when two living species are known to possess derived character states, it is often not possible to demonstrate that these are shared because of common inheritance — adaptation, unfortunately for the comparative anatomist, has been very effective in moulding similar morphological patterns independently, whether convergently or in parallel.

5. The resulting cladogram is used to infer phylogenetic relationships.

Comment: Cladograms represent an attempt to summarize the most parsimonious possible classification of animals according to the distribution of morphological character states — they contain no historical information beyond that assumed in their construction.
To repeat a conclusion of the preceding paragraph, cladistics is a method of classification and not a method of phylogeny reconstruction.

FOSSIL RECORD OF TARSIIFORMES

Turning to the fossil evidence, it is interesting to note that among the most respected evidence discussed at the 1918 Zoological Society symposium on the relationships of *Tarsius* was the apparent dental similarity of the Oligocene primate *Parapithecus* to the tarsier. "*Parapithecus* retains sufficient of the primitive traits to establish the truth of the Tarsioid ancestry of the Apes, ..." (Elliot-Smith, 1919). However, the most striking similarity of *Parapithecus* to *Tarsius* has since been shown to be an artifact of breakage (Simons, 1972): the mandible of *Parapithecus* appeared to have a V-shaped mandible, a mobile mandibular symphysis, and retain but a single pair of incisors only because the symphyseal region and the central incisors were broken away before the left and right rami were found and reassembled (see Fig. 1). *Parapithecus* is still considered a very primitive anthropoid, but its morphology, as now known, supports the stratophenetic hypothesis that higher primates originated from Eocene Adapoidea, leaving the tarsier in a much different phylogenetic position (see below).

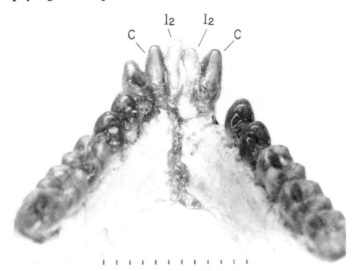

Figure 1.   *Type and only mandible of the primitive anthropoid primate*
Parapithecus fraasi from the Fayum Oligocene of Egypt. Note
breakage between left and right rami causing the specimen to
falsely resemble tarsioid primates by making it appear to have
had a V-shaped mandibular arch, unfused symphysis, and only one
pair of lower incisors (the left and right lateral incisors).
Photograph of specimen in Stuttgart Natural History Museum,
Ludwigsburg; scale is in mm.

It is an unfortunate accident of biogeography and paleontological discovery that no fossils are known that closely resemble *Tarsius* from strata younger than 35 million years before present. In the Eocene, two such subfamilies of Tarsiiformes are known, the Omomyinae (recently reviewed by Szalay, 1976) and the Microchoerinae (reviewed by Simons, 1961, and more recently by Sudre). Some members of both of these subfamilies are known from relatively complete skulls and dentitions, and there is general agreement that both groups are tarsiiform — stratophenetically they link more closely to the living tarsier than to any other mammalian group. Close resemblances in dentition (see Figure 2), and cranial and basicranial structure are sufficient to constitute a strong link between Eocene Omomyinae-Microchoerinae and *Tarsius*, in spite of a 35 my gap in the fossil record. Furthermore, Eocene tarsiiform primates link closely stratophenetically with Paleocene plesiadapiform primates. Their close stratigraphic proximity, similar dentition (especially the enlarged lower central incisors), and similar auditory region (especially the tubular ectotympanic) being the strongest evidence favouring this close linking (Gingerich, 1976b).

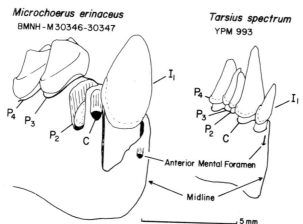

Figure 2. *Comparison of the herbivorous late Eocene microchoerine Micro-choerus, with the insectivorous Recent Tarsius. Note especially the unfused mandibular symphysis and large pointed central incisors found in both — these are important characteristics by which both differ from Eocene adapids and primitive anthropoids. (Figure from Gingerich, 1976b.)*

How do the remaining living primates relate to this broad plesiadapiform-tarsiiform evolutionary pathway? Living lemuriform primates are, like *Tarsius*, separated from their closest possible ancestors in the fossil record by a nearly complete gap of some 35 my. The only Eocene primates possibly ancestral to the living Lemuriformes are the Adapidae, which share with modern lemurs certain cranial features thought to be primitive for primates (annular ectotympanic) and lack other important derived features of living Lemuriformes (such as the tooth comb; but see Gingerich, 1975).

It was mentioned above that the anterior dentition of *Parapithecus*, as now known, supports an adapoid rather than tarsioid origin for higher primates. This linking between primitive anthropoids and Eocene adapoids is one of the strongest in all of primate phylogeny. The Oligocene anthropoids from the Fayum can hardly be distinguished from late Eocene adapids in any morphological characters yet known, i.e., in molar structure, in their anterior dentitions, or in middle ear morphology. *Amphipithecus* is a late Eocene primate of particular importance to this discussion because authorities disagree about whether it is anthropoid or adapoid in morphology — further substantiating the broad linking of these two groups.

The general pattern of stratophenetic linking in primates can be diagrammed as follows, where solid lines indicate relatively strong linkings and dashed lines indicate relatively weaker linkings:

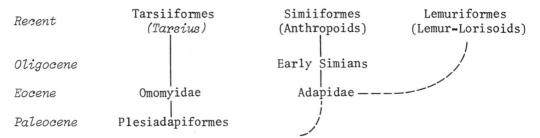

This phylogenetic pattern is discussed in more detail in Gingerich and Schoeninger (1976). It should be noted that cladistic testing of the above pattern shows that many of the similarities shared by recent Lemuriformes and Eocene Adapidae are characteristics (such as the free annular ectotympanic) thought to be primitive in primates, thus weakening this link between the two groups.

CLASSIFICATION AND EVOLUTIONARY SIGNIFICANCE OF *TARSIUS*

Several different classifications could be proposed for the pattern of phylogenetic relationships diagrammed above. Classification of Tarsiiformes and Plesiadapiformes together in one suborder, and Simiiformes and Lemuriformes together in another seems to be the most natural classification since both halves of this dichotomy appear to be strictly monophyletic. If it is found that in fact modern Lemuriformes were not derived from the Adapidae, then adapids should probably be ranked as primitive Simiiformes, and the Lemuriformes retained as a third suborder. Inclusion of Tarsiiformes and Simiiformes as a single suborder, with Plesiadapiformes and/or Lemuriformes excluded, makes the resulting "Haplorhini" polyphyletic or paraphyletic, depending on which group or groups are excluded.

To conclude, it appears that *Tarsius* should properly be regarded as a side branch derived from the earliest and most primitive primates, rather than a close relative of early anthropoids. The few distinctive "derived" characters of soft anatomy shared by *Tarsius* and anthropoids may have evolved independentl or they may prove in fact to be primitive characters of primates. Without a fo sil record to document successive stages in the evolution of a given anatomical character, it is very difficult (perhaps impossible) to use the character to infer relationships within a group of mammals.

ACKNOWLEDGEMENTS

Participation in this symposium was made possible by grants from the International Primatological Society — Wenner-Gren Foundation, and the Turner Fund, Department of Geology and Mineralogy, University of Michigan. I thank Dr. J. Fleagle for reading an early draft of the manuscript, and Mrs. Gladys Newton for typing it.

REFERENCES

Davis, D.D. (1962). *Bull. Singapore natn. Mus.*, 31, 1-129.
Elliot Smith, G. (1919). *Proc. Zool. Soc. Lond.*, 1919, 465-475.
Gingerich, P.D. (1975). *In* "Lemur Biology", (I. Tattersall and R. Sussman, eds), pp. 65-80, Plenum, New York.
Gingerich, P.D. (1976a). *Am. J. Sci.*, 276, 1-28.
Gingerich, P.D. (1976b). *Univ. Mich. Pap. Paleont.*, 15, 1-140.
Gingerich, P.D. and Schoeninger, M.J. (1976). "The fossil record and primate phylogeny", *J. Human Evolution*, 6, 483-505.
Simons, E.L. (1961). *Bull. Br. Mus. nat. Hist.*, Geol. 5, 43-65.
Simons, E.L. (1961). "Primate Evolution", Macmillan Co., New York.
Simpson, G.G. (1961). "Principles of Animal Taxonomy", Columbia University Press, New York.
Szalay, F.S. (1976). *Bull. Am. Mus. nat. Hist.*, 156, 157-450.

DISCUSSION

Kay: It seems to me that most of the features that we use to recognize extant primates are shared by adapids and omomyids: namely, orbits that are frontated and somewhat enlarged, small interorbital breadths, small infraorbital foramina nails rather than claws, and so on. If plesiadapoids were ancestral to tarsioids but not to adapids, then all of those characters would have had to be evolved in parallel in the two Eocene groups. Why do you think that's more probable than evolving enlarged incisors in parallel?

Gingerich: Throughout the Eocene, we can separate tarsioids very clearly from adapids. But in the late Paleocene and early Eocene we get forms like *Berruvius* and *Navajovius* (and even *Plesiadapis* at one time) being confused with omomyids; whereas, to my knowledge, there are no forms intermediate enough to confuse them between plesiadapiform primates and adapids.

Kay: *Plesiadapis rex,* which I believe was mistaken for an omomyid, is known from one isolated tooth. I believe *Berruvius* is only known from two lower teeth. I don't think anyone today would make the mistake of assigning species of *Anemorhysis* or *Teilhardina* to the plesiadapoids. Where we have adequate fossils, it seems to me that the distinctions between plesiadapoids and omomyids are quite great.

Gingerich: You haven't mentioned the tubular ectotympanic, which we see in *Tarsius,* in the Eocene tarsioids, and then again in the plesiadapiform primates. We don't have a lot of evidence to go on, but what we have seems to tie the omomyids back into the plesiadapiform stock.

Luckett: You list a free ectotympanic ring as a primitive primate character. But the ring isn't free in the earliest primates, the plesiadapoids. How does that fit in with your stratophenetic method?

Gingerich: Microsyopids seem to be an early branch of the main plesiadapiform stock. The microsyopid ectotympanic isn't known, but it clearly wasn't tubular like that of the other plesiadapiform primates, and it may have been a free ring. It is known to have been free and intrabullar in leptictid insectivores, which I think are very closely related to microsyopids.

SECTION V

MOLECULAR AND CHROMOSOMAL EVOLUTION
Chairman and Section Editor: A.E. Romero-Herrera (Detroit)

BLOOD GROUPS OF NONHUMAN PRIMATES: NEW CONCEPTS, THEIR APPLICATIONS AND CONCLUSIONS OF FORTY YEARS OF RESEARCH*

J. MOOR-JANKOWSKI, W.W. SOCHA and A.S. WIENER

The Primate Blood Group Reference Laboratory and World Health Organization Collaborating Centre for Haematology of Primate Animals, at the Laboratory for Experimental Medicine and Surgery in Primates (LEMSIP) of the New York University Medical Center, New York, N.Y., USA.

Until the early 1960s much more was known about the blood groups of mice, cattle, chickens and some other domestic and feral animals than about non-human primates, as can be seen by perusing the comprehensive volume "Blood Groups of Infrahuman Species" (Cohen, 1962). A complete review (Franks, 1962) of all publications on blood groups of primate animals listed only 66 articles for the period 1911 to 1962, 16 of them by Alexander S. Wiener. Aside from the paucity of the then available publications, the reported findings were mostly on single animals, and results of several of the papers were disproved by subsequent studies.

To understand this lack of information as recent as 15 years ago, it should be kept in mind that it applied at that time to the majority of data on non-human primates, with the notable exception of ethological observations and of investigations on dead animal material.

The senior author of this paper (A.S.W.) recalls the difficulties in obtaining blood from primate animals 15-20 years ago: considerable preambles were necessary to secure the permission, and laborious preparations had to be made in a zoological garden where the procedure could only be performed in the early dawn hours to bar any witnesses or interference with daily routines. The animals were caught and held down by keepers obviously untrained in handling primate animals for medical experimentation, and only a few samples could be obtained at each of the carefully prearranged visits.

The only serological technique available at that time was extremely laborious (Landsteiner & Miller, 1925) and allowed solely to determine what we later defined as the "human-type" blood groups. The reagents were prepared by treating human red cells with blood typing sera and then eluting the antibody from the red cell surface; the so eluted antisera were free of nonspecific hetero-agglutins which cause false positive reactions (Socha et al., 1972), but were of low titer and unstable. The results obtained were often unreliable and new batches had to be prepared for each experiment and used while still fresh.

The bulk of information presently available on blood groups of nonhuman primates has been accumulated by our Primate Blood Group Reference Laboratory created at the New York University School of Medicine in 1962. In 1974 it was appointed as the WHO Collaborating Centre for Haematology of Primate Animals. The work at the Laboratory has resulted in more than 150 publications des-

* Aided in part by U.S. Public Health Service grant GM 12074 and Contract RR-4-2184.

cribing the new methodology, concepts and findings which will be summarized in this paper.

METHODOLOGY

In 1963 we developed a new, rapid and dependable method of testing primate animal red cells (Wiener & Moor-Jankowski, 1963). It consists of absorbing the standard human blood typing reagents with chimpanzee group O red cells. This absorption removes nonspecific heteroagglutinins which are present in all human sera and are directed against red cells of chimpanzees and of most other simian species. If not absorbed, such heteroagglutinins mask the differences due to blood group antigens. The absorption procedure has no effect on the blood group specific titer of the antisera. The standard human reagents used are high-titered, so that the antisera remain potent after the absorption. They can be diluted should they still contain some heteroagglutinins directed against primate species other than chimpanzees; such remaining heteroagglutinins are usually of a titer no greater than 1:4 and are thus easily rendered ineffective by low grade dilution.

In addition to the use of suitably absorbed and diluted human antisera, we have introduced other specific reagents for primate animal blood grouping such as lectins, prepared from plant extracts. Some lectins were already used previously in tests on human blood (Gold & Balding, 1975), while others were introduced by us in the course of the primate animal studies (Moon & Wiener, 1974; Wiener & Moon, 1975; Wiener, Moor-Jankowski & Gordon, 1969). Other new testing reagents were prepared from snail extracts (Wiener, Brain & Gordon, 1969) and using antisera from fish, immunized with human saliva (Wiener et al., 1968).

Because of the considerable number of publications reporting false results, we have always insisted that *testing of primate animal red cells has to conform with the now well-defined standardized methods for typing of human red cells for clinical and forensic purposes*. In the blood typing at our Primate Blood Group Reference Laboratory we use the same techniques as in our work for blood banks and in forensic tests for the Chief Medical Examiner of New York City, and for Federal agencies. The only difference consists of the above-described absorption and dilution to render the reagents specific for nonhuman primates.

Blood typing of simians in our laboratory includes all the methods used in routines of human blood typing, namely, saline agglutination, antiglobulin and enzyme-treated red cells, and inhibition tests on saliva to detect the presence or absence of A-B-H and Lewis blood group substances. These methods were described in detail elsewhere (Socha et al., 1972; Erskine, 1973) and have consistently produced sharply defined reproducible results.

More recently, some new investigators have attempted to apply to nonhuman primates methods which have been used for some other animals, e.g., mice, in which blood grouping is done by adding dextran to the reagents in order to maximize agglutination reactions (Hirose & Balner, 1969; LaSalle, 1968), because, in mice, unlike simians, the blood groups reactions cannot be elicited by methods used for human red cells. Another group of investigators (Duggleby & Stone, 1971) unable to produce sufficient amounts of typing sera, has resorted to the routine use of reagents diluted to minimal titers so that centrifugation is required to strengthen otherwise imperceptible reactions of the weak reagents. One of the workers who used the dextran method and also the group using the highly diluted reagents and centrifugation worked in our

Laboratory to attempt to reproduce their results. Their tests, however, proved to be poorly reproducible to an extent that would be unacceptable for typing of human blood.

MODERN CONCEPTS IN PRIMATE BLOOD GROUPING

Blood grouping of apes and monkeys for medical experimentation may directly involve the human patient, as in organ transplantation (Cortesini, 1970) or cross-circulation of patients in hepatic coma (Goldsmith & Moor-Jankowski, 1971); therefore, blood typing standards must equal those used for human blood grouping.

The first phase of our large-scale investigation consisted of testing numerous primates from as many species as possible, using suitably prepared human blood typing reagents. A sufficient amount of information became available in 1965 to make possible the general conclusion that some of the reagents which detected blood group polymorphisms in man detected them also in some, but not all, species of apes and Old World and New World monkeys, but not in prosimians (Moor-Jankowski & Wiener, 1972). We defined these polymorphisms as the *human-type* blood groups of simians. The continuation of serological and genetic analysis of these human-type blood groups has made us realize that they are the *homologues* of the blood groups of man (Moor-Jankowski, Wiener, Socha, Gordon & Kaczera, 1973; Wiener et al., 1972; Wiener et al., 1974) in the same sense that the wing of a bird is a homologue of a human arm.

In our tests for these homologues we were limited to the blood group systems A-B-O, M-N, Rh-Hr and I-i to which highly specific potent reagents are available.

Already at the start of the first phase of our investigations in 1962, however, we believed that in addition to the homologues of the human blood groups, the nonhuman primates will also possess their "own" blood groups which would be detected by antisera prepared against their own red cells. In early 1964, isoimmune sera produced in chimpanzees detected the first of what we have later called the *simian-type* blood groups of nonhuman primates (Moor-Jankowski, 1964). A virtual avalanche of new findings on simian-type blood groups in chimpanzees, gibbons, baboons, rhesus monkeys and crab-eating macaques followed (for review see Moor-Jankowski & Wiener, 1972).

The simian-type reagents were shown to react also with primate species other than those in which they were raised, e.g., chimpanzee (*P. troglodytes*) isoantibodies react with red cells of other apes (Moor-Jankowski, Wiener, Socha, Gordon & Kaczera, 1973); baboon isoantibodies can be used for typing red cells of macaques (Moor-Jankowski, Wiener, Socha, Gordon & Davis, 1973). The reagents were produced either by isoimmunization, i.e., injections of red cells from an animal of the same species, or by what we have called cross-immunization, namely, injections of red cells of primates of another species, including man. The flurry of these newly obtained information is still only partially classified and brought into a cogent system. It is already evident, however, that the simian-types may be considered *analogues* of the human blood groups.

The latest phase of our experimentation includes to a considerable extent the exploration of cross-reaction of simian-type antisera among several species of primates (Moor-Jankowski, Wiener, Socha, Gordon & Davis, 1973; Moor-Jankowski et al., 1974; Wiener et al., 1971). The cross-reactions may prove helpful in tracing phylogenetic relationships among primate species, including man.

REFERENCES

Cohen, C. (ed.), (1962). "Blood Groups in Infrahuman Species", *Ann. N.Y. Acad. Sci.*, 97, 1-328.

Cortesini, R., Casciani, C. and Cuccharia, G. (1970). *In* "Infections and Immunosuppression in Subhuman Primates", (Balner and Beveridge, eds), pp. 239-248, Munksgaard, Copenhagen.

Duggleby, C.R. and Stone, W.H. (1971). *Vox Sang.*, 20, 109-123.

Erskine, A.G. (1973). "Principles and practice of blood grouping", pp. 307-315, C.V. Mosby, St. Louis.

Franks, D. (1962). *Symp. Zool. Soc.*, *Lond.*, 10, 221.

Gold, E.R. and Balding, P. (1975). "Receptor-specific proteins; plant and animal lectins", Amer. Elsevier Pub- Co., Inc., New York.

Goldsmith, E.I. and Moor-Jankowski, J. (eds.), (1971). *In* "Medical Primatology 1970", pp. 52-89, Karger, Basel and New York.

Hirose, Y. and Balner, H. (1969). *Blood*, 34, 661-681.

Landsteiner, K. and Miller, C.P., Jr. (1925). *J. Exp. Med.*, 42, 663.

La Salle, M. (1968). *In* "Proc. 2nd int. Cong. Primatol., Atlanta, Ga.", vol. 3 (H.D. Hofer, ed.), pp. 120-128, Karger, Basel.

Moon, G.J. and Wiener, A.S. (1974). *Vox Sang.*, 26, 167-170.

Moor-Jankowski, J. and Wiener, A.S. (1972). *In* "Pathology of Primates", Part I, (R.N.T-W. Fiennes, ed.), pp. 270-317, Karger, Basel.

Moor-Jankowski, J., Wiener, A.S. and Rogers, C.M. (1964). *Science*, 145, 1441-1443.

Moor-Jankowski, J., Wiener, A.S. and Socha, W.W. (1974). *Folia primatol.*, 22, 59-71.

Moor-Jankowski, J., Wiener, A.S., Socha, W.W., Gordon, E.B. and Davis, J.H. (1973). *J. Med. Primatol.*, 2, 71-84.

Moor-Jankowski, J., Wiener, A.S., Socha, W.W., Gordon, E.B. and Kaczera, Z. (1973). *Folia primatol.*, 19, 339-360.

Socha, W.W., Wiener, A.S. and Moor-Jankowski, J. (1972). *Transpl. Proc.*, 4, 107-111.

Wiener, A.S., Brain, P. and Gordon, E.B. (1969). *Haematologia (Budapest)*, 3, 9.

Wiener, A.S., Chuba, J.V., Gordon, E.B. and Kuhns, W.J. (1968). *Transfusion*, 8, 226-234.

Wiener, A.S., Gordon, E.B., Moor-Jankowski, J. and Socha, W.W. (1972). *Haematologia (Budapest)*, 6, 419-432.

Wiener, A.S. and Moon, G.J. (1975). *Haematologia (Budapest)*, 9, 235-241.

Wiener, A.S. and Moor-Jankowski, J. (1963). *Science*, 142, 67-69.

Wiener, A.S., Moor-Jankowski, J. and Gordon, E.B. (1969). *Int. Arch. Appl. Immunol.*, 36, 582-591.

Wiener, A.S., Socha, W.W. and Gordon, E.B. (1971). *Haematologia (Budapest)*, 5, 227-240.

Wiener, A.S., Socha, W.W. and Moor-Jankowski, J. (1974). *Haematologia (Budapest)*, 8, 195-216.

KARYOLOGY OF THE GENUS *CERCOPITHECUS*

B. CHIARELLI

Institute of Anthropology, University of Turin, Italy.

In his synthetic approach to the evolutionary history of Prosimians presented at this meeting, R. Martin utilized my early analysis on Prosimian chromosomes (Chiarelli, 1972) in which the main feature taken into consideration was the number of arms of the chromosomes (the Fundamental Number of Matthey).

However, from 1972 up to now a considerable improvement in the interpretation of chromosomal changes both in phylogenetic and taxonomic studies occurred. This new approach is mainly due to the fact that the fundamental structure of the chromosomes is made up by a folded filament of DNA surrounded by protinaceous matrix (Du Praw, 1970) and by the fact that this filament probably interconnects the different chromosomes of the karyotype (Chiarelli, 1976).

This new interpretation of the chromosome structure is the result of both the new banding techniques and a way to properly interpret this important new artifact of the chromosome preparations (Chiarelli, 1973).

This new concept of chromosome ultrastructure produces an obvious reinterpretation of the traditional concept of chromosomal mutations as centric fusion and fission, as due to simple change of polarity in the chromosome attachment to the nuclear membrane.

Apart from these, moreover, two new additional concepts have to be considered for gross chromosomal variations in the study of phylogeny and classification. They are: introgressive hybridization and genome intrusion (chromosome insertion).

I will not go into details to present these two newly developed concepts as they will be published shortly (Chiarelli, 1977). However, with such a preamble I can now present a brief synthesis of a possible reinterpretation of my 1968 data on *Cercopithecus* chromosomes, utilizing the mechanisms of chromosomal insertion on which I worked in the last few months.

The genus *Cercopithecus* includes a dozen species which have a large number of sub-species and geographical populations. In spite of the remarkable external morphological variability, they show a notable anatomical-structural uniformity.

The phylogenetic history of this genus is uncertain and it is not known whether this genus has an ancient or recent origin (Buettner-Janusch, 1963), but the most interesting problem lies in its karyological heterogeneity. The chromosome morphology and a regular increase of nuclear DNA in relation to the chromosome number has been measured (Chiarelli, 1967).

The most simple hypothesis to explain this karyological heterogeneity is a gradual reduction in the chromosome number due to centric fusion. In order to support this hypothesis it is necessary to assume that the ancestral karyotype, common to all the living species, was higher than 72.

TABLE I

Chromosome numbers and morphology in different species of Cercopithecus

| Species | 2n | Chromosome morphology | | | X | Y |
		M	A	Mark		
C. nigris	48	22	-	1	S	A
C. patas	54	18	7	1	S	A
C. talapoin	54	19	6	1	S	A
C. diana	58	21	6	1	S	A
C. i'hoesti	58-60	21	6-7	1	S	A
C. neglectus	58-62	22-23	5-6	1	S	A
C. nigroviridis	60	18	10	1	S	A
C. aethiops	60	17	11	1	S	A
C. cephus	66	23	8	1	S	A
C. mona	66-88	23	8-9	1	S	A
C. nictitans	66-70	23	8-10	1	S	A
C. mitis	72	24	10	1	S	A

TABLE II

Total haploid chromosomal length

Cercopithecus 6	No. of mitosis measured	Average	lit. 0.05
2n=54	20	94.5±14	±7
2n=60	20	101.6±24	±11
2n=66	20	112.1±13	±6
2n=72	20	125.1±20	±9

Several different considerations, however, lead us to think that this hypothesis is unlikely (Chiarelli, 1968). Another possible explanation of this karyological variability is the proposed circular interconnected chromosome model and its consequence. This model provides for simple genomic addition and change in chromosome morphology by changing the valve of the point of attachment of the telomere and centromere at the nuclear membrane. Therefore it explains both the increase of the chromosome material and the different changes in the centromeric position in the chromosomes.

This hypothesis is not alternative, but only complementary and integrative to the traditional chromosome mutations.

The use of this model in the karyological researches on Primates could moreover explain the karyological heterogeneity of other groups as that of Hylobatidae (2n=44-50-52).

On another occasion I shall present data of introgressive hybridization and its interaction with early human evolution and the differentiation of apes.

REFERENCES

Ardito, G. and Mortelmans, J. (1975). *J. Hum. Evol.*, 4, 377-381.
Buettner-Janisch, J. (1963). *In* "Evolutionary and Genetic Biology of Primates", vol. I., Academic Press, New York.
Chiarelli, B. (1967). *Experientia*, 23, 672.
Chiarelli, B. (1968). *Cytologia*, 33, 1-16.
Chiarelli, B. (1972). *J. Hum. Evol.*, 1, 61-64.
Chiarelli, B. (1973). *J. Hum. Evol.*, 2, 337-340.
Chiarelli, B. (1974). *In* "Prosimian Biology", (G.A. Doyle and A.C. Walker, eds), pp. 871-880, Duckworth, London.
Chiarelli, B. (1976). *Nucleus*, 19, 71-74.
Chiarelli, B. and Vaccarino, C. (1964). *Atti Ass. Gen. Ital.*, 9, 273-282.
Du Praw, E.J. (1970). "DNA and chromosomes", Holt, Rinehart, New York.

NUCLEAR CYTOCHEMISTRY AS AN APPROACH TO PRIMATE TAXONOMY AND EVOLUTION

M.G. MANFREDI-ROMANINI and G.F. DE STEFANO

Institute of Histology, Embryology and Anthropology, Pavia, Italy.

The nuclear DNA content of *Tupaia glis* (Feulgen positive material after 60' of hydrochloric acid hydrolysis) has been calculated to be approximately 30% lower than the average DNA content in mammals. This finding has been interpreted as an indication that the *Tupaiidae* are a highly differentiated group, and have a lower degree of evolutional plasticity than either Primates or Insectivores (De Stefano, 1972). This hypothesis seems to be supported by the high number of acrocentric chromosomes observed in this genus, which is an indication of "primitivity" (Bender and Chu, 1963). The previous hypothesis is also supported by the low DNA content found in *Chiroptera*, which appears to be related to their evolutional stability (Manfredi-Romanini and Capanna, 1972; Bachmann 1972).

On the other hand, it has been recently claimed that the speciation process might imply a change in the amount of heterochromatin rather than a decrease in the overall genome size (Stock, 1973, 1975; Manfredi-Romanini et al., 1976). Obviously, in this context, the term heterochromatin indicates the DNA fraction strongly bound to nuclear basic proteins showing little or no genetic activity (Kiefer et al., 1972; Mittermayer et al., 1971). It may therefore be important to establish whether the low DNA content in *Tupaia* indicates a real decrease in the genome size as compared to Primates and Insectivores, or merely represents a corresponding increase in the heterochromatic DNA fraction, within the framework of the hypothesis suggested above.

The Feulgen reaction, after different times of hydrolysis, is a sensitive tool for an *in situ* study of the qualitative aspects of DNA in the nucleus (as shown, for example, by a higher or lower degree of heterochromatinization). The kinetics curves thus obtained have allowed at least two DNA fractions to be distinguished in the interphasic nucleus. These fractions exhibit a different resistance to hydrolysis, the most acid-resistant fraction represents the firmly protein-tied DNA (in the meaning specified above).

By means of a scanning microdensitometer (Vickers M86; ob. 100X; A.N. 1.25; oc. 10X; 560 ± 5 nm), we have analyzed the kinetics of the Feulgen reaction on the peripheral lymphocytes of female *Tupaia glis, Erinaceus europaeus, Macaca irus* and man at different times of hydrolysis. The resultant curves, together with those of *Rhinolophus ferrumequinum* and *Cercopithecus aethiops*, will be published elsewhere. A comparison of the latter two species is particularly interesting since *R. ferrumequinum* belongs to *Chiroptera*, a group of mammals by now regarded as evolutionarily stable (Simpson, 1945), while *C. aethiops* is thought to be in a phase of active evolution (Formenti, 1975; Chiarelli, 1973).

In all the species studied, the optimal response to the Feulgen reaction was obtained after 70' of hydrolysis. Under these conditions, the data

relative to *T. glis*, *E. europaeus* and man were in agreement with those previously obtained by one of us (De Stefano, 1972). In particular, *T. glis* exhibited
a value which was statistically lower than the one observed either in man (t =
4.75; P ≤ 0.01) or in *E. europaeus* (t = 5.37; P ≤ 0.01), which are practically
identical. On the other hand, the value of *M. irus* was found to be statistically lower than the value of either *E. europaeus* (t = 3.20; P ≤ 0.01) or man
(t = 2.69; P ≤ 0.05).

Beyond this point of maximal response (70' of hydrolysis), the Feulgen
reaction curves show a similar and steady decrease in the species examined,
regardless of their DNA content. In none of these species was any acid-resistant fraction observed which could be responsible for the masking of the genome
size value, and therefore related to the heterochromatinization phenomena. This
would have been considered a useful clue for evaluating the evolutionary condition of the individual species. In fact, the acid-resistant fraction is absent in both the evolutionally stable species (*T. glis* and particularly *R.
ferrumequinum*) and those believed to be in a stage of active evolution (*Cercopithecus*).

In agreement with other experiments, the curve relative to *H. sapiens* shows
a different trend: There was a second maximum at 160' and the curve remained
fairly high up to 360' of hydrolysis. A second peak at 160' was also observed
in the curve relative to *Erinaceus*. However, it is premature to make any conclusions in the absence of similar data from other species of insectivores.

On the whole, the curves with a maximum at 70' followed by a gradual decrease seem to be a feature of a particular cell type (Bernocchi et al., 1975,
1976; De Stefano et al., 1975) rather than a characteristic of a given species. On the other hand, these data seem to confirm the taxonomic significance
of the differences in the nuclear DNA content, as already reported by one of
us (Manfredi-Romanini, 1973). Such differences could be attributable to the
acid-labile DNA fraction.

REFERENCES

Andersson, G.K.A., Kjellstrand, T.T. (1971). *Histochemie*, 27, 165-172.
Andersson, G.K.A., Kjellstrand, T.T. (1975). *Histochemistry*, 43, 123-130.
Bachmann, K., Harrington, B.A., Craig, J.P. (1972). *Chromosoma*, 37, 405-416.
Bender, M.A., Chu, E.H.Y. (1963). *In* "Evolutionary and Genetic Biology of
   Primates, I", (J. Buettner Janusch, ed.), pp. 261-310, Academic Press,
   New York.
Bernocchi, G., De Stefano, G.F. (1975). XIII Congr. Soc. It. Istoch., Perugia
Bernocchi, G., De Stefano, G.F., Porcelli, F., Redi, C.A., Manfredi-Romanini,
   M.G. (1976). (in press)
Chiarelli, B. (1973). "Evolution of the Primates", pp. 161-186, Academic
   Press, New York.
De Stefano, G.F. (1972). *Bull. Zool.*, 39, XLI Conv. UZI, Trieste, 3-6 October
De Stefano, G.F., Formenti, D., Bernocchi, G. (1975). XIII Congr. Soc. It.
   Istoch., Perugia, 26-28 May.
Formenti, D. (1975). *Genetica*, 45, 307-313.
Kiefer, R., Kiefer, G., Sandritter, W. (1972). *Histochemie*, 30, 150-155.
Manfredi-Romanini, M.G., Capanna, E. (1972). 4th Int. Congress of Cytochem.
   Histochem., Kyoto, 213-214.
Manfredi-Romanini, M.G. (1973). *In* "Cytotaxonomy and Vertebrate Evolution",
   (B. Chiarelli and E. Capanna, eds), pp. 39-81, Academic Press, New York.
Manfredi-Romanini, M.G., Pellicciari, C., Bolchi, F., Capanna, E. (1975).
   Mammalia, 39, 675-683.

Manfredi-Romanini, M.G., Redi, C.A. (1976). 5th Int. Congress of Histochemis-
    try and Cytochemistry, Bucarest.
Manfredi-Romanini, M.G., Redi, C.A. (1976). (in press)
Mittermayer, G., Madreiter, H., Lederer, B., Sandritter, W. (1971). *Beitr.
    Path.*, 143, 157-171.
Sandritter, W., Jobst, K., Rakow, L., Bosselman, K. (1965). *Histochemie,* 4,
    420-437.
Simpson, G.G. (1945). *Bull. Am. Mus. Nat. Hist.*, 85, 1-350.
Stock, A.D. (1975). *Cytogenet. Cell Genet.,* 14, 34-41.
Stock. A.D., Hsu, T.C. (1973). *Chromosoma,* 43, 211-224.

SEROPRIMATOLOGY: SEROLOGICAL REACTIONS AS A TAXONOMIC TOOL*

J. MOOR-JANKOWSKI, W.W. SOCHA and A.S. WIENER

The Primate Blood Group Reference Laboratory and World Health Organization Collaborating Centre for Haematology of Primate Animals, at the Laboratory for Experimental Medicine and Surgery in Primates (LEMSIP) of the New York University Medical Center, New York, N.Y., USA.

Starting with an early finding in baboons (Moor-Jankowski et al., 1964a), we demonstrated more than a decade ago that species or other population groups of nonhuman primates are characterized by statistically significant differences in distribution of their human-type and simian-type blood groups (Moor-Jankowski & Wiener, 1967; Moor-Jankowski et al., 1966). That this is true in man was shown as early as 1919 (Hirszfeld & Hirszfeld, 1919) and has been expanded into the discipline of seroanthropology (Bernard & Ruffie, 1966; Mourant, 1954). In primate animals, the differences in blood group distribution enabled us to distinguish among population groups in chimpanzees (Moor-Jankowski & Wiener, 1967, 1971; Moor-Jankowski et al., 1966), in baboons (Moor-Jankowski et al., 1964a; Wiener et al., 1970) and in gibbons (Moor-Jankowski & Wiener, 1967, 1972; Moor-Jankowski et al., 1964b).

In view of the above findings, the taxonomic value of blood groups begins to be known to primatologists, although it is still less recognized than that of the morphological characteristics or even of enzymes and blood proteins.

Much less known among taxonomists is the method of recognizing differences among population groups by what is called by immunologists the "species-specific immune reaction" (Landsteiner, 1945). We speak of such a reaction when immune sera are found to react with antigen(s) present in all animals of one population, but absent in all animals of other populations. ("All animals" are understood to mean a statistically significant number of individuals of the population group under study.) Past experience reaching back to the early 1900s (Nuttall, 1904; Tchistovitch, 1899; Uhlenhuth, in Landsteiner, 1945) showed that such a population group corresponded to the taxonomic entity of a species, hence the *immunological term* of "species-specific" reactions.

This term was first used at the turn of the century to define differences among serum proteins from different species. The differences were demonstrated with the aid of precipitating antisera prepared by immunizing rabbits with sera from other animals (Nuttall, 1904; Tchistovitch, 1899). However, precipitating antisera made in rabbits failed to distinguish clearly between sera from closely related species, such as chimpanzee and man, due to the similarities between human and chimpanzee proteins. A finer serological analysis to distinguish between closely related species was then developed (Uhlenhuth, in Landsteiner, 1945) by cross-immunization between such closely related species,

---

* Aided in part by U.S. Public Health Service grant GM 12074 and Contract RR-4-2184.

e.g., by injecting rabbits with hare serum, resulting in production of anti-sera that precipitated sera from all hares but not from rabbits. This pheno-menon of higher specificity of antisera produced in closely related species has been characterized by Landsteiner (1945) as "immunological perspective", i.e., the closer the donor species is to that of the immunized animal, the more specific is the immune reaction.

It has been known for more than a century (Landois, 1875) that when serum and red cells of animals of different species are mixed, the red cells are hemolyzed or agglutinated.* To investigate this phenomenon further, Land-steiner and Miller (1925) immunized rabbits with red cells of man and apes. Similarly to the previously described tests for serum protein differentiation in closely related species, the anti-red cell sera reacted strongly not only with red cells of the *homologous* species, i.e., the species used as donor for the immunization, but also with the red cells from the other, *heterologous,* primate species studied. However, the agglutinating technique used in red cell studies is more sensitive than the precipitating technique used in the serum protein studies. This allows removal of antibodies reactive with the heterologous species by absorption with red cells of that heterologous species, leaving behind antibodies specific for the red cells of the homologous species. Such absorbed antisera make it possible to differentiate between red cells of closely related primate species, e.g., man and chimpanzee, on the basis of their species-specific serological reactions. (Sera of rabbits immunized with chimpanzee red cells will initially react with red cells not only of the homologous chimpanzee, but also of the heterologous man. However, absorption of the antiserum with the heterologous human red cells will leave behind anti-bodies reactive specifically for the homologous chimpanzee.)

Landsteiner (1945) was mainly interested in blood type differences within the species, i.e., blood group polymorphism, and, therefore, did not pursue the problem of primate species specificity by the next logical step, that of cross-immunization (cf. above-mentioned work on serum proteins (Uhlenhuth, in Landsteiner, 1945)). Such studies of red cell specificities in primates were not carried out until our experiments on gibbons in which two different populations of gibbons, *Hylobates lar lar* and *Hylobates lar pileatus* were cross-immunized with each other's red cells (Moor-Jankowski et al., 1964b). When this experiment was performed in 1964, it was not generally recognized that there was more than a "racial" or "strain" difference between these two gibbon populations. Actually, at the time of our immunization experiment, we were not aware of any taxonomic differences between the immunized animals and our aim was to produce isoimmune antibodies for the study of blood groups polymorphism in *Hylobates lar,* thought by us to be a well-defined homogeneous species. To our surprise, the immunized gibbons produced antibodies against red cells which reacted reciprocally between the two groups of gibbons. Only then did we find that one of the groups could be defined as *H. lar lar* and the other as *H. lar pileatus*. Sera of *H. l. lar* immunized with red cells of a *H. l. pileatus* animal agglutinated red cells of all *H. l. pileatus* tested and of none of the *H. l. lar,* while vice versa, serum of *H. l. pileatus* immunized with red cells of a *H. l. lar* animal agglutinated red cells of all *H. l. lar* tested, but none of the *H. l. pileatus.*

---

* However, this does not always occur between closely related species, e.g., while human sera regularly clump or lyse red cells of all chimpanzees, sera from nonimmunized chimpanzees usually do not clump human red cells (Wiener & Moor-Jankowski, 1972).

These reactions, thus, represented a classic example of immunological species-specific reactions and indicated to us that the two groups of gibbons belonged, immunologically, to two different species. This conclusion, based on species-specific reactions was confirmed definitively six years later by the work of Groves (1970) who used the classic geographic and morphologic criteria.

More recently, these studies have been expanded and demonstrated species-specific differences between yellow and olive Kenya baboons on the one hand, and hamadryas and chacma baboons on the other (Moor-Jankowski et al., 1973), while the geladas represent a third immunologically separate group (Moor-Jankowski et al., 1974).

REFERENCES

Bernard, J. and Ruffié, J. (1966). "Hématologie géographique, Masson, Paris.
Groves, C.P. (1970). *Symp. zool. Soc. Lond.*, 25, 127-134.
Hirszfeld, L. and Hirszfeld, H. (1919). *Lancet*, 2, 675.
Landois, L. (1875). "Zur lehre von der bluttransfusion", Leipzig.
Landsteiner, K. (1945). "The specificity of serological reactions", reprinted 1962, Dover, New York.
Landsteiner, K. and Miller, C.P. (1925). *Science*, 61, 402.
Moor-Jankowski, J. and Wiener, A.S. (1967). *In* "Progress in Primatology", (Schneider and Kuhn, eds), pp. 373-381, Fischer, Stuttgart.
Moor-Jankowski, J. and Wiener, A.S. (1971). *In* "Medical Primatology 1970", (Goldsmith and Moor-Jankowski, eds), pp. 232-244, Karger, Basel.
Moor-Jankowski, J. and Wiener, A.S. (1972). *In* "Pathology of Simian Primates", (Fiennes, ed.), Part I, pp. 270-317, Karger, Basel.
Moor-Jankowski, J., Wiener, A.S., and Gordon, E.B. (1964a). *Transfusion*, 4, 92-100.
Moor-Jankowski, J., Wiener, A.S., and Gordon, E.B. (1964b). *Transfusion*, 5, 235-239.
Moor-Jankowski, J., Wiener, A.S., Kratochvil, C.J. and Fineg, J. (1966). *Science*, 152, 219.
Moor-Jankowski, J., Wiener, A.S., Socha, W.W. and Gordon, E.B. (1973). *J. med. Primatol.*, 2, 71-84.
Moor-Jankowski, J., Wiener, A.S., Socha, W.W. (1974). *Folia primatol.*, 22, 59-71.
Mourant, A.E. (1954). "The distribution of the human blood groups", Blackwell, Oxford.
Nuttall, G.H.F. (1904). "Blood immunity and blood relationship", Cambridge University Press, London.
Tchistovitch, Th. (1899). *Ann. Inst. Pasteur*, 13, 406-425.
Uhlenhuth, P. Cited in Landsteiner (1945), p. 68.
Wiener, A.S. and Moor-Jankowski, J. (1972). *In* "Primates in Medicine", (Goldsmith and Moor-Jankowski, eds), vol. 6, pp. 115-144, Karger, Basel.
Wiener, A.S., Socha, W.W., Moor-Jankowski, J. and Gordon, E.B. (1970). *Amer. J. phys. Anthrop.*, 33, 433.

SERUM PROTEINS IN THE CEBIDAE

T. SCHWEGLER

Anthropologisches Institut der Universität Zürich, Switzerland.

INTRODUCTION

During the last year we have carefully examined blood samples of several species of South American primates (Cebidae). The aims of our research programme were: (1) determination of monomorphisms and polymorphisms in the given serum proteins (by electrophoresis); (2) formal genetics of structural variants (by analysis of family data); (3) utilisation of the serum proteins in taxonomic questions (by interspecific comparison). The first two points will be dealt with in this article. The third and more complex one will be discussed in a separate essay.

MATERIAL AND METHODS

The proteins were examined by: (1) high voltage agarose gel electrophoresis and immunoelectrophoresis after Teisberg (1970) for the albumin (Al), antitrypsin (At), group specific component (Gc), third component of the complement factor (C3), and the transferrin (Tf); (2) Agar gel immunoelectrophoresis after Hirschfeld (1955) for the group specific component (Gc) and the macroglobulin (Mg).
We mainly worked with the combined agarose gel immunoelectrophoresis because nearly all of our proteins could be made visible by that method.
We will refer to the species of *Aotus trivirgatus* (11 specimens), *Ateles paniscus* (5 sp.), *Ateles belzebuth* (2 sp.) and *Saimiri sciureus* (79 sp.). They were examined with reference to the mobility and the number of bands of the already mentioned proteins.

RESULTS

As an example of monomorphism there is *Saimiri sciureus*, the squirrel monkey; the Al, C3 and Tf bands show in a sample of nearly 80 specimens one and the same phenotype.
If we turn to *Aotus trivirgatus*, the owl monkey, we notice a polymorphism in several proteins (Fig. 1).
. Albumin. The Al shows in the family B a fast moving component in the male and a slow moving one in the female. The offspring shows the intermediate phenotype. This is in good accordance with the Mendelian laws. We mixed the sera of the adults and got the same result as in the child. Hence, we have homozygous parents and a heterozygous offspring, as expected. Comparable results were found in the other family A.

2. Antitrypsin. The At bands show a slow mobility in the male and a fast one in the female. In the young animal we find the heterozygous phenotype. This is comparable to the mixed serum.
3. Group specific component. The Gc is represented by one band in each individual. All of them are moving with the same mobility, i.e. they are monomorphic.
4. Macroglobulin. The Mg shows different moving bands in the parents. The offspring stands between them.
5. Third component of the complement factor. The C3 exhibits four different moving patterns. In our *Aotus* family B we have a heterozygous father and a homozygous mother. The child is a heterozygote again.
6. Transferrin. Last we examined the Tf which shows in one individual two bands. This is most probably a heterozygous form.

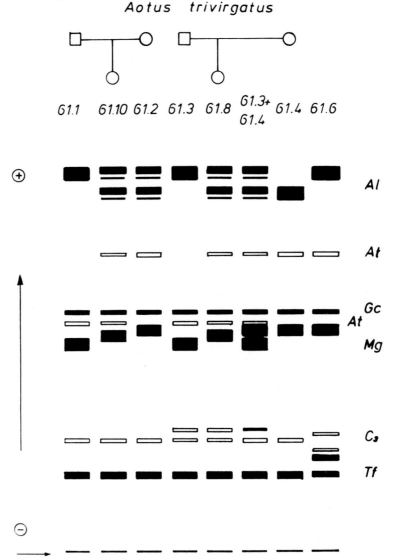

Figure 1. Electropherogramme of Aotus trivirgatus. Family A left, B right

s a summary of the data of the formal genetic analysis in *Aotus* we can state
hat there exists a polymorphism in five of six serum proteins. The structural
ariants of each protein seem to be genetically determined by at least two
odominant alleles.

Another example of polymorphism was found in *Ateles paniscus* and *Ateles*
elzebuth. In the pherogramme of the spider monkey we notice in the area of
he group specific component five distinct phenotypes (Fig. 2).

In the family the father has the two most extreme variants, the mother shows
wo fast ones, whereas the child has got the slow band of each of its parents.

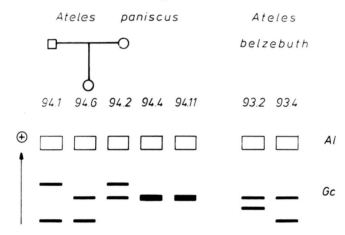

Figure 2. Gc-variants in Ateles.

JMMARY

The following statements are to be made:

. In the Cebidae we examined six different serum proteins. All of them show
polymorphism.

. Our family data support the assumption that the structural variants recog-
ized in the albumin, the antitrypsin, the group specific component, the third
omponent of the complement factor and in the transferrin are due to codominant
lleles.

EFERENCES

irschfeld, J. (1959). *Acta path. microbiol. scand.*, 47, 160-172.
cheffrahn, W. and Glaser, D. (1977, in press). *J. Hum. Evol.*
eisberg, P. (1970). *Vox. Sang.*, 19, 47-56.

SERUM PROTEINS IN THE CALLITRICHIDAE

W. SCHEFFRAHN

Anthropologisches Institut der Universität Zürich, Switzerland.

INTRODUCTION

Our blood research programme has the following aims:
. Testing a protein in respect to monomorphism or polymorphism
. Studying the mode of inheritance of intraspecific protein variants
. Consideration of the taxonomic value of blood proteins.
This paper presents as examples of these aims some blood protein data of the South American primate family, the Callitrichidae. The species involved and the number of individuals are given in Table I. The proteins albumin (Al), group specific component (Gc), α_1-antitrypsin (At), haptoglobin (Hp), β_1C-globulin (C3) and transferrin (Tf) have been investigated by means of the combined agarose immuno-electrophoresis (Teisberg, 1970), agar electrophoresis (Hirschfeld, 1959) and starch gel electrophoresis (Scheffrahn et al., 1974).

TABLE I

Protein systems, species and number of individuals
(+ polymorphism, - monomorphism)

Species	n	Al	At	Gc	Hp	C3	Tf
Saguinus nigricollis	15	-	-	-	-	+	-
Saguinus fuscicollis	12	-	+	-	-	+	-
Saguinus oedipus	23	-	-	-	-	-	-
Saguinus midas	19	-	-	-	-	-	+
Saguinus leucopus	1	-	-	-	-	-	-
Callithrix jacchus	115	-	+	-	-	+	+
Callimico goeldii	2	-	-	-	-	-	-

MONOMORPHISM, POLYMORPHISM

In reference to the first purpose of our studies, Table I summarizes the data.
In all species mentioned above there exists a monomorphism in the albumin (Al) and group specific component (Gc). α_1-antitrypsin (At) is polymorphic in

Callithrix jacchus and *Saguinus fuscicollis,* β_1C-globulin (C3) in *Callithrix jacchus, Saguinus fuscicollis* and *Saguinus nigricollis.*

Haptoglobins

A single haptoglobin phenotype is found in all species, which is comparal to the human Hp 1-1. In Fig. 1 the human Hp phenotype 2-1 is taken as a ref erence pattern; on the level of the human fraction 1 a similar component i: observable in *Callithrix jacchus.* One of our cotton tops (*Saguinus oedipus*) shows a so-called hypohaptoglobinemia (Fig. 2) which closely resembles the

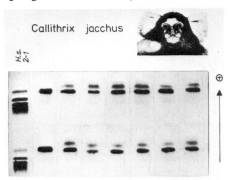

Figure 1. Hp phenotypes in Callithrix jacchus. Hp 2-1 of Homo is given as reference sample.

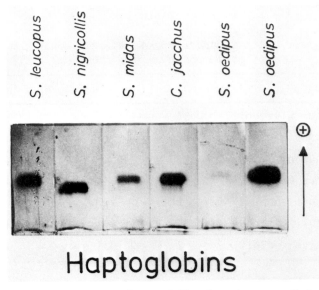

Figure 2. Interspecific comparison of Hp phenotypes. Note the hypo-haptoglobinemia (ahaptoglobinemia) in Saguinus oedipus.

human cases of hypohaptoglobinemia or ahaptoglobinemia respectively. Elect phoresis has been undertaken many times in a borate and a phosphate buffer

ystem, always with the same result. Weak reactions were recognized three
ays after the electrophoretic analysis in this *Saguinus oedipus* individual.
his example of a hypohaptoglobinemia (or ahaptoglobinemia) does not lead
onvincingly to the conclusion that a second allele is existent at the Hp-
ocus besides the normal one. In comparison to the human findings, hypo-
aptoglobinemia and ahaptoglobinemia may have many causes, not only genetic
nes.

ransferrins

   The agarose pherogrammes of *Callithrix jacchus* and *Saguinus midas* show
ariations in an area which has been localized by immuno-diffusion as the
ransferrin region. There are at least 7 distinct phenotypes in *Callithrix
acchus* (Fig. 3) and 3 phenotypes in *Saguinus midas* (Fig. 4). Notice that
in contrast to *Callithrix jacchus*) the homozygotes in *Saguinus midas* show
 two-banded pattern (main bands), while the heterozygotes are characterized
y a four-banded pattern in this species.

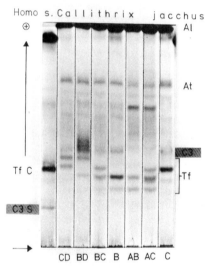

*Figure 3. Localization of the Tf bands by the immuno-diffusion method
and identification of the 7 Tf phenotypes in Callithrix
jacchus.*

)RMAL GENETICS

   The mode of inheritance of these Tf phenotypes has been clarified with the
elp of family data. Also, in comparison to the genetics of the transferrins
n catarrhine primate species (*Macaca mulatta, Papio doguera, Papio ursinus,
an troglodytes, Pan paniscus* and *Homo sapiens*) it seems reasonable to assume
aat the Tf phenotypes are determined by at least 4 autosomal co-dominant
lleles in *Callithrix jacchus*, by at least 2 autosomal co-dominant alleles in
aguinus midas.

XONOMIC VALUE OF BLOOD PROTEINS

   The comparison of electrophoretic banding patterns between different species
lways involves some difficulties. Electrophoresis can only detect similarities

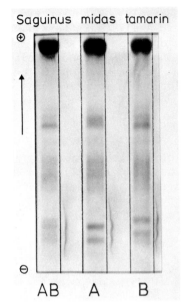

Figure 4. *Localization of the Tf bands by the immuno-diffusion method and identification of the 3 Tf phenotypes in Saguinus midas.*

of a given protein in different species. Electrophoretic similarities may deal with the identical protein or protein variant, but at present we must accept it as unproven until the amino acid composition of similar variants in different species is known. Yet we need just such homologous similarities for purposes of taxonomy. Nevertheless, it may be of some interest to look at the electrophoretic differences of a given protein in some species. In general the interspecific electrophoretic comparison in the Callitrichidae reveals Hp bands which have a similar electrophoretic mobility. *Saguinus nigricollis* is the only exception to this rule, as this species yields a slower-moving Hp fraction(cf. Fig. 2).

In the species studied, β_1C-globulin takes a different position in the agarose gel electrophoretic pherogramme. Namely, in nearly all platyrrhine species (*Cebus capucinus* seems to be the only exception with a cathodal β_1C-globulin position very close to the transferrins) β_1C-globulin is to be found anodal to the transferrins, while in the catarrhine primates so far studied (*Homo sapiens, Papio hamadryas, Papio anubis, Cercopithecus aethiops, Macaca mulatta, Macaca arctoides, Macaca irus* and *Macaca nemestrina*) this protein is generally positioned cathodal to the transferrins. Platyrrhine and catarrhine primates, therefore, can be clearly differentiated on the basis of the β_1C-globulin position on the agarose gel electrophoretic pherogramme.

REFERENCES

Hirschfeld, J. (1959). *Acta path. microbiol. scand.,* 47, 160-172.
Scheffrahn, W., Lipp, H.P. and Mahler, M. (1974). *Archiv für Genetik,* 47, 96-104.
Teisberg, P. (1970). *Vox. Sang.,* 19, 47-56.

ISOZYMES AND PLASMA PROTEINS IN EIGHT TROOPS OF
GOLDEN MANTLED HOWLING MONKEYS (*ALOUATTA PALLIATA*)

LINDA A. MALMGREN and A.H. BRUSH

*Biological Sciences Group, University of Connecticut, Storrs,
Connecticut 06268, USA.*

INTRODUCTION

This investigation involved behavioural observations and starch and polyacrylamide gel electrophoresis of blood samples collected from natural populations of mantled howling monkeys, *Alouatta palliata* (Gray 1849) to determine the effect of population structure and social behaviour on the nature of
protein polymorphisms within and among troops. In particular, data was used
to estimate the effective migration rate among troops.

MATERIALS AND METHODS

Techniques developed for capture and marking of low forest monkeys enabled
the capture of one hundred and thirty-two howling monkeys from 17 troops during
1972-6 on Finca "La Pacifica", 7 km NW of Canas, Guanacaste Province, Costa
Rica (Scott, Scott, and Malmgren, 1976). All animals were marked and measured
and blood was taken from one hundred and seventy of them. Most samples came
from eight troops. In addition, random samples were taken from troops in other
areas of the ranch, from two other sites in Costa Rica and from Panama. The
latter samples were taken to provide information concerning the nature of
protein variability over a broader geographic range.
Plasma proteins were separated by disc polyacrylamide gel electrophoresis.
Red cell samples were analysed by horizontal starch gel electrophoresis followed by specific staining for 15 enzyme systems. Enzymes considered primarily
to reflect genetic rather than environmentally or physiologically indiced polymorphisms were chosen for the study.

RESULTS AND DISCUSSION

Little variation and only two polymorphisms were detected among the systems
assayed in the first 110 samples. MDH and leucine-amino-peptidase were polymorphic. The additional 60 samples also were screened for M-MDH variants, and
combined results concerning allele frequency data for M-MDH are described here.
There were three phenotypes of M-MDH in the populations studied, similar to
those described for man (Davidson and Cortner, 1967) and for Japanese macaques
(Shotake and Nozawa, 1974). Heterozygotes show three main bands: one with the
same mobility as the main band of the common M-MDH, an intermediate band which
is most intense, and a third band nearer the origin. So, M-MDH was considered
to be a dimer controlled by a codominant allelic system (Shotake and Nozawa,
1974). There was no evident association of alleles at the M-MDH locus with age
or sex of the animals.

The average frequency of "wild-type" in the eight troops studied was 0.88. The eight troops combined were in Hardy-Weinberg equilibrium, but at a low level of significance: $X^2 = 6.438$; $P = 0.011$. However, consideration of values within troops gave chi-square values of from 0.025-0.244 (P between 0.9 and 1.0), with the exception of two troops ($X^2 = 11.0$ and 6.93) which are exceptional in other ways and will be discussed later. The variance in allele frequencies among troops was $V_p = 0.04$. There was no significant correlation between geographic proximity and allelic frequency within troops.

Migration rates were estimated from allele frequencies using Wright's island model. This model involves several assumptions. First, populations must be internally panmictic genetic islands. Our calculations show that this is so for six of the eight troops studied. Another, potentially questionable assumption dictates equal likelihood of genetic exchange between population units. From observations of Carpenter (1965) and others, and from our observations, it seems that the most probable vehicle of gene exchange in howlers is the peripheral male. If this is so, troops on the ranch may indeed have equal chance for gene exchange. Peripherals often travel long distances, the ranch is only 1,330 hectares and almost all territories on the ranch are connected in some way by forest strips or short terrestrial walks.

Effective migration rates were calculated from the formula:

$$V = \frac{(1-m)^2}{2N_e - 2(N_e-1)(1-m)^2}$$

Where V is the weighted variance in allele frequency, and N_e is the effective population size, m is the migration rate. The average effective population size was calculated from census data. When Crow and Kimura's (1970) equation

($N_e = \dfrac{4N_m N_f}{N_m + N_f}$) was used, the average effective population size was 9.73

for the eight troops. Nozawa's (1972) equation to calculate N_e takes into account male dominance within troops of Japanese monkeys. As *A. palliata* also shows similar dominance behaviour in troops, we used Nozawa's N_e for a second, more conservative estimate. Here, $N_e = 5.93$. Using these derived figures, the effective migration rate was $m = 0.01$ with Crow and Kimura's N_e and $m = 0.026$ with Nozawa's.

The number of genetically effective migrants per generation (N_g) was calculated next: $N_g = N_e m$. For the two estimates of N_e, there were 0.154 or 0.1 genetically effective migrants per generation.

If a few further assumptions are made, the average number of migrants among troops per year can be calculated. Long term capture-recapture data indicate that howlers probably do not mature until 4-5 years of age and may not breed until 6 or 7. Taking the extreme values we got a minimum range of 0.4 - 0.7 and a maximum range of 0.62 - 1.08 migrants among troops per year. If these values are correct, some degree of behavioural isolation between troops is indicated.

The distribution of variants among troops reinforced our behavioural data. First, troop 3 contains a homozygote for the mutant allele, but no heterozygotes. A parsimonious explanation for this is that this troop contained an individual originally from another troop. Troop 3 and its neighbouring troop (4) were the only ones not in Hardy-Weinberg equilibrium and were those with the most active shuffling of males observed during my total of 9 months in the field.

REFERENCES

Crow, J.F. and Kimura, M. (1970). *In* "An Introduction to Population Genetic Theory", p. 103, Harper and Row, New York, Evanston, and London.
Davidson, R.G. and Cortner, J.A. (1967). *Nature*, 215, 761-762.
Gray, J.E. (1845). *Am. Mag. Nat. Hist. Ser.* 1 (16), 217-221.
Nozawa, K. (1972). *Primates*, 13 (4), 381-393.
Scott, N.J., Scott, A.F., and Malmgren, L.A. (1976). *Primates*, 17 (4), (in press).
Shotake, T. and Nozawa, K. (1974). *Primates*, 15 (2-3), 219-226.

PRIMATE HIGHER TAXA: THE MOLECULAR VIEW

J.E. CRONIN and V.M. SARICH

Departments of Anthropology and Biochemistry, University of California, Berkeley, California 94720, USA.

The picture of interordinal relationships among the placental mammals lacks a good deal in the way of resolution. The molecular contribution to the clarification of that picture has been limited to immunological comparisons of their serum albumins and the sequencing of a limited number of globins. The albumin data strongly suggest that all extant placental mammal lineages fall into two clearly delineated clades, and the globin data are consistent with that picture. The less diverse clade contains the following monophyletic units: Perissodactyla, Lagomorpha, Cetacea, Suiformes, Ruminantia, Hyracoidea, Sirenia, and Proboscidea. The Carnivora clearly fall into the other, and this change, along with the association of the lagomorphs with the ungulates rather than with the rodents, are the only substantive modifications of the classic division of the placentals into Unguiculata and Ferungulata, plus Glires (rodents and lagomorphs) and Mutica (cetaceans)(Simpson, 1945), or lepticids and paleoryctids (Lillegraven, 1969).

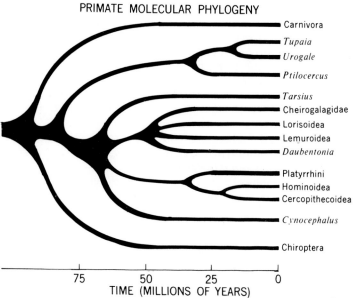

Figure 1.   *Our current view as to the phylogenetic relationships among the taxa indicated based on immunological comparisons of the appropriate albumins and transferrins and on the available globin sequence data. The other related nonprimate lineages have been omitted for the sake of clarity, but are mentioned in the text.*

The monophyletic clade which includes the unquestioned primates (*Tarsius*, Anthropoidea, Lemuriformes-Lorisiformes) also contains the so-called "flying lemur" (*Galeopithecus* or *Cynocephalus*) and the tupaiids (Fig. 1). The mammalian groups most closely associated with this clade include the insectivores, bats, carnivores, and "edentates", with the rodents probably representing a somewhat earlier diverging lineage.

Only the albumin immunological data bear directly on the questions of interordinal relationships within the two groups generally and the "primate" clade specifically, as the transferrins evolve too rapidly to provide useful interordinal cross reactions. The albumin data quite reliably (i.e., with excellent internal consistency) show an early primate adaptive radiation leading to four extant lineages (*Tarsius*, *Cynocephalus*, Anthropoidea, Lemuriformes plus Lorisiformes) where no pair can be reliably associated to the exclusion of the other two. The tupaiids represent more of a problem, as the amount of change *measured* along the *Tupaia* albumin lineage is not independent of the antiserum used to effect the measurement. This is true whether or not the antisera had been prepared against primate albumins, and the empirical observation can be made to the effect that from some perspectives the albumin of *Tupaia* looks rather conservative, and from others almost as changed as those of the Anthropoidea. Thus one cannot with any degree of confidence and precision place the tupaiid divergence along the ancestral primate lineage using only the albumin data. The transferrin of *Tupaia* is also clearly that of a primate, and our analysis of the hemoglobin sequence data places the *Tupaia* divergence approximately half-way between those of *Canis* and *Tarsius* along that lineage leading to the primates.

Recently Dr. Adrian Friday of Cambridge made available to us a sample of *Ptilocercus* serum which has allowed the first protein placement of that genus. The summed albumin plus transferrin immunological distances from *Tupaia* are: *Urogale*, 90; *Ptilocercus*, 131; and any nontupaiid is at least 260 units away

Some disagreement has arisen concerning the molecular placement of *Tarsius* We emphasize that cladistic analyses (that is, taking rates of change into account) of the albumin and transferrin immunological data do *not* allow any specific association of *Tarsius* and the Anthropoidea to form the clade Haplorrhini. The suggestion of Goodman and Moore (in Beard, Barnicot, and Hewett Emmett, 1976) that the hemoglobin sequence data support such a clade is not necessarily the most economical interpretation of those data, and our analysis of the sequences involved indicates that, in fact, the most probable order of divergences along the nonungulate clade is *Mus*, *Canis*, *Tupaia*, *Tarsius*, *Nycticebus-Lemur*, and intra-anthropoid.

Our primary recent emphasis has been on the lemuriform-lorisiform clade, using, in the main, serum samples generously provided by Dr. Yves Rumpler. We find that those taxa fall into four major lineages (Lorisiformes, cheirogalines, lemuroids, and *Daubentonia*) which diverged from one another at about the same time, though it is possible that the aye-aye line diverged at a slightly earlier point. The important observation is that the Malagasy lemurs do not form a monophyletic grouping, a fact made more palatable by the recent evidence that Madagascar was separated from Africa by the subsidence of the Mozambique Channel in the Eocene or early Oligocene (Walker, pers. comm.). Thus the loris-lemur adaptive radiation could well have taken place in Africa-Madagascar prior to that subsidence, with the lemurs on Madagascar then simply being the relict survivors of that radiation. The intra-lorisiform radiation, on the other hand, is a much more recent event which most probably began sometime in the early Miocene. We emphasize the following: (1) There is no specia relationship between the cheirogalines and galagos; (2) Indriids are at least as closely related to the *Lemur-Hapalemur-Varecia* line as is *Lepilemur;* (3) *Hapalemur* is more closely related to *Lemur* than is *Varecia*;   (4) *Hapalemur simus* and *H. griseus* are clearly part of the same monophyletic clade;   and (5) The *Lemur fulvus-coronatum-macaco* complex derives from a very recent radiation within the last 2MY or so with *mongoz* diverging perhaps 4-5My ago, and *catta* and *Hapalemur* representing two lineages which diverged from that leading to the above *Lemur*, and from each other, on the order of 20MY ago.

In closing, we wish to draw attention to certain features and implications of our figure. When all available data are taken into account, there emerges a rather clear-cut sequence of divergence events. Going back in time from the anthropoid-prosimian radiation, these are: tupaiids, most non-ungulate placentals, rodents, ungulates, and finally marsupials. That we can still see these internodal lineages so clearly today must be taken to indicate that a significant number of derived features (at both the molecular and morphological levels) must have developed along each of them. Such developments take time, and their existence bears strongly on the nature of the time scale to be used in viewing mammalian evolution. It is unnecessary to push the marsupial-placental divergence much beyond 120MY, where, in any case, the molecular data place it; and it is highly unlikely that the beginnings of the nontupaiid primate adaptive radiation could be other than late Cretaceous (65-70MY). It is thus quite in order to contemplate the adaptive radiation of the placental mammals as encompassing at least the latter half of the Cretaceous (back to about 100MY), and it is certainly untenable to continue the classic latest Cretaceous-early Paleocene placement of that radiation. It should also be noted, of course, that the complexity of the cladistic framework involved places serious constraints on the degree of "give" in the dating of any nodes within it. This observation applies to the primates as well, and should be kept in mind when questions as to the dating of divergence events among the primates arise.

REFERENCES

Beard, J.M., Barnicot, N.A. and Hewett-Emmett, D. (1976). *Nature*, 259, 338-340.

Cronin, J.E. and Sarich, V.M. (1975). *J. Hum. Evol.*, 4, 357-375.

Lillegraven, J.A. (1969). *University of Kansas Paleontological Contributions*, 50.

Lillegraven, J.A. (1974). *Annual Review of Ecology and Systematics*, 5, 263-283.

Sarich, V.M. and Cronin, J.E. (1976). *In* "Molecular Anthropology", (M. Goodman and R.E. Tashian, eds), Plenum, New York, pp. 139-168.

Simpson, G.G. (1945). *Bull. Amer. Mus. Nat. Hist.*, 85.

ATYPICAL EVOLUTION OF PAPIONINE α-HAEMOGLOBINS AND INDICATION
THAT *CERCOCEBUS* MAY NOT BE A MONOPHYLETIC GENUS

D. HEWETT-EMMETT* and C.N. COOK

*Department of Anthropology, University College London,
Gower Street, London, UK.*

The tribe Papionini comprises macaques (*Macaca*), mangabeys (*Cercocebus*),
mandrills and drills (*Mandrillus*), baboons (*Papio*) and the gelada baboon
(*Theropithecus*). They all possess a 42 chromosome karyotype and their haemoglobins are proving of interest both to primatologists and to students of molecular evolution. The latter aspect has been emphasised elsewhere (Hewett-
Emmett et al., 1976a, 1976b) but a brief review is essential to exploring the
primatological implications.

BACKGROUND

Haemoglobin is a mixed (heterotrophic) tetramer comprised, in the mammalian
adult, of α and β chains ($\alpha_2\beta_2$). Early primate studies (Hill et al., 1963;
Barnicot et al., 1965) involved digesting the tetramers with trypsin and mapping the resultant tryptic peptides in 2 dimensions (electrophoresis plus
chromatography). Surprisingly (for the idea of "molecular clocks" was introduced about then; Zuckerkandl and Pauling, 1965), the maps or fingerprints
of *Macaca* and *Papio* differed more than the fingerprints of *Macaca* and *Homo*.
Later *Cercocebus* (*albigena*) was shown to be almost identical to *Papio* (Barnicot and Wade, 1970).
  It was concluded that the *Papio* and *Cercocebus albigena* globin chains had
incorporated a large number of charge-altering amino acid substitutions whereas, *if* a comparable number of substitutions had accrued in the *Macaca* and *Homo*
lineages, they were of a highly conservative nature and undetected by finger-
printing. Sequence data has shown that the *few* differences between *Macaca
mulatta* and *Homo* are indeed conservative (Matsuda, 1976) whereas the large
number between the non-macaque papionines and *Homo* (or *Macaca*) are quite radi-
cal (Hewett-Emmett et al., 1976b). This has implications with respect to the
neutral mutation theory which predicts the random-walk incorporation of "non-
harmful" mutations and hence, within a Poisson scatter, uniformity of both
rate and type of substitution in all descendent lineages. There are complicating factors: non-allelic α-chain genes (i.e. gene duplicates) have been
noted in several *Macaca* species (reviewed by Barnicot, 1969; Nute, 1974), in
some *Cercocebus* species (Barnicot and Hewett-Emmett, 1972; Hewett-Emmett et
al., 1976b) and *Mandrillus* (Buettner-Janusch et al., 1969; Barnicot, 1969 and
unpublished). In the case of *Macaca fascicularis* and *Cercocebus atys*, it is
probable that some individuals possess 3 non-allelic α-chain genes. It is
well recognised that gene duplication provides a "short circuit" for evolving

* Present address: Department of Biochemistry, University of Bristol, Medical
School, University Walk, Bristol BS8 1TD, UK.

new functions; the duplicate is free from the editing role of negative (sta-
bilising) selection (Ohno, 1972). Consequently bursts of evolution occur until
a new function evolves, as documented in the globin family (Goodman et al.,
1975).

Catarrhine α and β Hb Phylogenies

The maximum parsimony approach tests many alternative trees and selects
that tree which requires the fewest mutations in order to accommodate the extant sequences (Goodman et al., 1975). Space is limited here and the reader
is referred to figures 3-6 in Hewett-Emmett et al., 1976b, for the Catarrhine
haemoglobin trees which form the basis of this discussion. This reference,
Dayhoff (1972), Matsuda (1976) and Sullivan et al. (1976) are good sources for
primate haemoglobin sequences.

Substitution rates

It is not intended to get into discussion here about the date at which various primate ancestors existed. In general terms, the catarrhine β-chains
have accepted mutations at a rate rather slower than the β-chains overall, calculated by Fitch and Langley (1976) to be 0.60 DNA mutations/10^9 yrs/nucleotide
position. The distribution of mutations in the different lineages is not inconsistent with a Poisson distribution. The overall α-chain rate (Fitch and
Langley, 1976) is 0.53 DNA mutations/10^9 yrs/ nucleotide position. Again,
the catarrhine α-chains, with the notable exception of the non-macaque papionines, have accepted mutations at a rather lower rate, particularly the *Homo*,
Macaca and *Cercopithecus* lineages. The non-macaque papionines, however, have
rates varying from 1.9 to 4.5 DNA mutations/ 10^9 yr/nucleotide position. These
rates approach the "silent" mutation rate (e.g. Leu→ Leu) calculated from globin
m-RNA sequences to be 5.8 DNA mutations/10^9 yr/nucleotide position (Fitch and
Langley, 1976). There has certainly been one, and probably two, α-chain gene
duplications in the non-macaque papionines and it might be argued that slack-
ening of negative selection at the duplicate locus has allowed mutations to be
incorporated at a rate close to the "silent" rate. However, if this were so,
the substitutions should also be random in nature and they are not.

Charge-altering substitutions

All 17 substitutions in the non-macaque papionines are charge-altering*
whereas only one of 19 catarrhine β-chain substitutions (β52 Asp → Ala in
Theropithecus and *Papio*) is charge-altering. However, the α-chain charge
changes balance each other closely, so that overall charge is essentially conserved. If a protein reflecting the composition of the genetic code is allowed to mutate randomly, only 38% of the substitutions are charge-altering.
Analysis of eutherian globin phylogenies shows that this figure is closely
approximated in both α and β-chains. The non-macaque papionines differ significantly from eutherian globins as a whole, and the conclusion must be
drawn that elimination of negative selection *alone* cannot explain both the

---

* Of the 20 amino acids, Glu and Asp are acidic and Lys and Arg are basic.
At physiological pH, His is partially charged and in this analysis is regarded as basic and equivalent to Lys and Arg. The other 15 are neutral.

rapid substitution rate *and* the fact that each substitution represents an alteration of charge. A candidate for positive (Darwinian) selection has surely been found.

Taxonomic implications

The single most important finding is the indication that the arboreal mangabey, *Cercocebus albigena* may belong with *Theropithecus* and *Papio* and not with the terrestrial mangabeys. At α57, *C. albigena*, *Theropithecus* and *Papio* have Lys whereas *C. atys* and all other catarrhines have Gly. At α56 *C. albigena* and *Papio* have Asx (most probably Asn) whereas *Theropithecus*, *C. atys* and all other catarrhines have Lys. A back mutation (Asn → Asp) at α47 also links the three species together (Hewett-Emmett, 1973; Hewett-Emmett et al., 1976b Sullivan et al., 1976). No β-chain data is available, although fingerprint evidence does hint that *C. albigena* may share the charge-altering substitution (β52 Asp → Ala) with *Papio* and *Theropithecus* (Barnicot and Wade, 1970; Barnicot et al., 1965; Hewett-Emmett et al., 1976b; Sullivan et al., 1976b) whereas *C. atys* does not (Cook and Barnicot, unpublished). Recently microcomplement fixation studies of antisera raised again mangabey transferrins and serum albumins have led to similar conclusions, namely that the genus *Cercocebus* is not monophyletic and the mangabeys have a dual origin (Cronin and Sarich, 1976). It is hoped to pursue studies on the β-chain of *C. albigena* to add to this growing body of data. Cronin and Sarich (1976) rightly pointed out that a DNA hybridisation study, of the type subsequently published by Beneveniste and Todaro (1976), might prove further compelling evidence.   Jacob and Tappen (1957) can little have expected that the study of mangabey blood proteins would lead to the realisation that the genus *Cercocebus* is not monophyletic when they first described "abnormal haemoglobins" in mangabeys.

ACKNOWLEDGEMENTS

Readers are referred to Hewett-Emmett et al (1976b) for recognition of those who supplied blood to the late Professor N.A. Barnicot, and who discussed various aspects of Old World monkey haemoglobin evolution with us. C.N.C. was supported by a S.R.C. (London) grant to Prof. Barnicot. D. H-E. was supported by University College London during the course of the experimental work, some of which formed part of his Ph.D. thesis.

REFERENCES

Barnicot, N.A. (1969). *Sci. Progr.*, 57, 459-493.
Barnicot, N.A. and Hewett-Emmett, D. (1972). *Folia Primatol.*, 17, 442-457.
Barnicot, N.A. and Wade, P.T. (1970). *In* "Old World Monkeys", (J.R. Napier and P.H. Napier, eds), pp. 227-260, Academic Press, London and New York.
Barnicot, N.A., Jolly, C.J., Huehns, E.R. and Dance, N. (1965). *In* "The Baboon in Medical Research", (H. Vagtborg, ed.), Vol. I, pp. 323-338, Univ. of Texas Press, Austin.
Beneveniste, R.E. and Todaro, G.J. (1976). *Nature,* 261, 101-108.
Buettner-Janusch, V., Buettner-Janusch, J. and Mason, G.A. (1969). *Int. J. Biochem.*, I, 322-326.
Cronin, J.E. and Sarich, V.M. (1976). *Nature,* 260, 700-702.
Dayhoff, M.O. (1972). "Atlas of Protein Sequence and Structure", Vol. 5, National Biomed. Res. Foundation, Silver Spring, Maryland.

Fitch, W.M. and Langley, C.H. (1976). *In* "Molecular Anthropology", (M. Goodman and R.E. Tashian, eds), pp. 197-219, Plenum Press, New York.

Goodman, M., Moore, G.W. and Matsuda, G. (1975). *Nature*, 253, 603-608.

Hewett-Emmett, D. (1973). Ph.D. thesis, University of London.

Hewett-Emmett, D., Cook, C.N. and Goodman, M. (1976a). *Fed. Proc.*, 35, 1605 (Abstract #1256).

Hewett-Emmett, D., Cook, C.N. and Barnicot, N.A. (1976b). *In* "Molecular Anthropology", (M. Goodman and R.E. Tashian, eds), pp. 257-275, Plenum Press, New York.

Hill, R.L., Buettner-Janusch, J. and Buettner-Janusch, V. (1963). *Proc. Nat. Acad. Sci. USA.*, 50, 885-893.

Jacob, G.F. and Tappen, N.C. (1957). *Nature*, 180, 241-242.

Matsuda, G. (1976). *In* "Molecular Anthropology", (M. Goodman and R.E. Tashian, eds), pp. 223-237, Plenum Press, New York.

Nute, P.E. (1974). *Ann. N.Y. Acad. Sci.*, 241, 39-60.

Ohno, S. (1972). *J. Hum. Evol.*, I, 651-662.

Sullivan, B., Bonaventura, J., Bonaventura, C. and Nute, P.E. (1976). *In* "Molecular Anthropology", (M. Goodman and R.E. Tashian, eds), pp. 277-288, Plenum Press, New York.

Zuckerkandl, E. and Pauling, L. (1965). *In* "Evolving Genes and Proteins", (V. Bryson and H.J. Vogel, eds), pp. 97-166, Academic Press, New York.

CONCLUDING REMARKS

A. E. ROMERO-HERRERA

*Department of Anatomy, Wayne State University,
Detroit, Michigan 48201, USA.*

Ascertaining the phylogenetic relationships between the various taxa and the evolutionary mechanisms which are operative in establishing those relationships is indeed a task fraught with frustrations. Comparative anatomy and paleontological studies have allowed the construction of a basic phylogenetic tree which supplies information on ancestral lines and times of divergence. However, much ambiguity surrounds portions of such a tree. The development of the field of molecular evolution has helped to resolve some of these ambiguities as well as redefine other areas and will continue to do so as more macromolecules are investigated. The application of the study of blood groups to problems of primate relationships, pioneered by Dr. Wiener, has developed over several decades and, strengthened by the introduction of new techniques still holds promise for the future.

It has become apparent that the mere use of chromosome numbers and morphology has not provided any indices which correlate with the relationships derived from other methods. As has been stressed, the organization of the DNA throughout the chromosomes, the mechanisms of structural mutations, and identification of mutant chromosomes are not well enough understood at this time to allow for a systematic classification of species. In addition to the structural mutations (e.g., simple deletion, symmetrical translocation, pericentric inversion, centric fusion and so on), two new concepts are complicating the picture, namely, introgressive hybridization and genome intrusion. Just as new theories of DNA distribution throughout the karyotype are allowing for plausible explanations of the increase and decrease in the 2n number in members of the same genus, likewise ameans of monitoring chromosome evolution could provide new insights into the speciation process. While it is readily appreciated that the geographical isolation of one population from another of the same species may result in the emergence of a new species, it may be equally important that even without geographical isolation separate species could develop through the mechaniam of chromosomal mutations.

Drs. Manfredi-Romanini and De Stefano have raised the question of the role played by the total nuclear DNA quantity and distribution in determining the evolutionary plasticity of a species. An important consideration along this line and that of the speciation process is the significance of heterochromatin and its quantity relative to the euchromatin. Utilization of the Feulgen reaction with varying times of acid hydrolysis of the total nuclear DNA have allowed for the construction of kinetics curves. There are important differences between the curves of some of the species investigated. It is hoped that further studies in this field will enhance our knowledge of the phylogenetic relationships and distances.

Drs. Moor-Jankowski, Socha and Wiener have drawn our attention to the recognition of serological "races" within primate species; some of these "races"
have been recognised subsequently as being distinct species on other criteria.

Drs. Schwegler and Scheffrahn's investigation of the serum proteins of
Cebidae and Callitrichidae reveal that polymorphisms are commonplace. These
results are in contrast to the findings of Drs. Malmgren and Brush who found
in fifteen enzyme systems only 2 instances of polymorphism among several troops
of howling monkeys. It could be speculated that this lack of polymorphism
is the result of genetic drift due to the mechanism of the bottle-neck effect
by which the number of individuals is substantially reduced as the result of
an environmental catastrophe. Thus as Dr. Malmgren mentioned in this meeting,
a recent outbreak of yellow fever, which could have greatly lowered the population of howling monkeys living in this region, could well have been responsible for the lack of polymorphism.

A better understanding of primate phylogeny during the last years has been
provided by the information gained through immunological and protein sequencing
studies. Most of the information derived from the comparative anatomy and the
fossil record has been substantiated by the biochemical evidence. However, the
phylogenetic position of *Tarsius* is still debatable. On the one hand, the
immunological studies of Drs. Cronin and Sarich are placing *Tarsius* as the earliest branch of the primate monoclade. Furthermore, according to these authors
the hemoglobin sequence supports this placement of *Tarsius*. On the other hand,
information from Dr. Goodman's laboratory, using the computerized parsimony
technique, places *Tarsius* as diverging from the anthropoid stem.

To clarify this apparent disagreement it will be necessary to gather more
sequence information on *Tarsius* since the available data is rather scant. To
one who is familiar with the parsimony technique (by hand or computer) it is
evident that a good deal of uncertainty is attached to it since several parsimonious solutions can be derived from the same set of data. A good example of
this is the contested position of *Tupaia*. As proven in our own laboratory using
myoglobin information, several parsimonious positions for *Tupaia* can be found.
It is salutory to remember at this stage that the most parsimonious solution
is not always providing the truest reconstruction of past events. It is only
due to the present lack of a better approach that we assume proteins have always utilized the shortest pathways during the course of their evolution. These
shortcomings may be overcome when information from different proteins becomes
available.

Of very relevant interest are the findings of Drs. Hewett-Emmett and Cook
on the taxonomic relationships involving the terrestrial and arboreal mangabeys. Traditionally these two groups of mangabeys were classified as having
a monophyletic origin; however, the hemoglobin information, besides some other
biochemical evidence, leads to the placement of the arboreal mangabeys closer
to *Theropithecus* and *Papio* than the terrestrial mangabeys. In such instances
of clarifying phylogenetic relationships the biochemical evidence has proven
to be most helpful.

SECTION VI

METHODS OF PHYLOGENETIC INFERENCE: ROUND-TABLE DISCUSSION
Chairman and Section Editor: M. Cartmill (North Carolina)

METHODS OF PHYLOGENETIC INFERENCE

M. CARTMILL

*Department of Anatomy, Duke University Medical Centre,
Durham, North Carolina, 27710 USA.*

INTRODUCTION

   Most evolutionary biologists, whether Hennigian cladists or Simpsonian
evolutionary systematists of the old school, would agree that Hennig and his
followers have contributed greatly to the question of phylogenetic reconstruction by explicitly systematising a set of ideas that most of us have been
assuming tacitly for a long time. The most important of these is the distinction between primitive (or *plesiomorphous*) traits and derived (or *apomorphous*) traits. This distinction underlies the fundamental premise used by
most students of phylogeny: namely, that phyletic relationships must be
assessed on the basis of shared derived traits (synapomorphies) rather than
shared primitive retentions (symplesiomorphies). For example: crocodiles
look more like lizards than like birds. But the few traits which crocodiles
share with birds but not with lizards are apomorphous, and outweigh the host
of reptilian symplesiomorphies retained in crocodiles and lizards but lost in
birds. We conclude that crocodiles are nearer akin to birds than to lizards.
A purely phenetic approach to the construction of phylogenetic trees, in
which we just count up all resemblances of whatever sort, would obscure this
relationship.
   Even this much would be questioned by some people. But granting this,
there still remain several further questions which have to be answered in
some fashion by anyone trying to reconstruct the phylogeny of some group of
animals. I wish to raise three such questions as a starting-point for our
discussion.
   1. *What counts as a trait*? We encounter this question whenever we deal
with allometric differences. For instance: *Indri* is larger than *Microcebus*,
and it has a greater proportion of leaves in its diet. *Indri* shows folivorous specialisations of its gut and dentition; these specialisations are
lacking in *Microcebus*. Is this four differences between *Indri* and *Microcebus*,
or two, or one? Suppose two animals differ in two features of unknown functional significance. What gives us the right to count them as two features,
rather than one? If they are truly functional, they may both turn out to be
aspects of a single specialisation, like the enlarged incisors and elongated
third finger of *Daubentonia*. If they are genuinely functionless, they may be
pleiotropic effects of genes being selected for in some other context. Primary
biochemical data avoid problems of this sort, but we face such problems con-
tinually in dealing with traits further removed from the genome.
   2. *How do we tell which traits are primitive*? This is of paramount importance for any theory of phylogenetic reconstruction which dismisses shared

primitive retentions as irrelevant. Some assert that palaeontological data
are indispensable for determining which traits or character states are primitive; others regard them as no more significant than neontological data,
and rely instead on various ways of making comparisons with taxa outside the
group under study.

3. *How do we weight traits*? In almost any group of real animals, we find
what might be called *crossing synapomorphies*; some subgroup of the group
under study will share some obviously derived traits uniquely with another
subgroup, and other such traits with yet another subgroup. Among Palaeogene
primates, for instance, the tarsioids are linked to early anthropoids by
derived features of the facial skeleton and carotid arteries, and to the
plesiadapoids by derived features of the bulla and dentition. Hence the
current debate over tarsier affinities. In a situation like this, at least
one of the sets of shared derived features must be the product of parallel
evolution. How do we determine which one? Should we devalue traits that
have an obvious functional significance and thus seem more likely to have
evolved in parallel — in effect, placing greatest significance on traits we
don't understand yet? Should we weight more heavily traits produced by the
addition of novel ontogenetic processes, less likely to have been developed
in parallel, and devalue traits that result from the retention of foetal or
juvenile morphology? Should we devalue traits that offer serious zoogeographic problems? Are there privileged sorts of data — e.g., do biochemical dat
provide an escape from the problem of assessing the likelihood of convergence?
All of us can think of difficulties with these and other criteria of privi-
lege; but we do in fact make judgements of this sort. It seems intuitively
obvious, for instance, that banding patterns of chromosomes ought to be
weighted more heavily than banding patterns on agouti-type hairs in assessing
phyletic affinities. What principles warrant this sort of judgement, and can
they be systematized and made explicit?

DISCUSSION

Kay (North Carolina): There should perhaps be a fourth question. Once we've
agreed on what our traits are, which ones are derived, and how those are to
be weighted, how do we proceed from there to the actual reconstruction of the
phylogenetic tree? Is maximum parsimony necessary and sufficient, and if so,
how should we define parsimony?

Cartmill: Can you list some of the alternatives?

Kay: The procedure that I think should be used runs as follows: (1) Set up
all the topologically non-equivalent trees possible for the groups under
consideration; (2) take each character which has been determined to be important, and superimpose it on each of the possible phylogenies; (3) count
the parallelisms required for each phylogeny with that character; (4) weight
the characters, and (5) determine which phylogeny involves the least amount of
parallelism after weighting.

Martin (London): This problem has already been tackled in biochemical tree-
building. One of the first obstacles you encounter is that if you have more
than a few species the number of possible trees becomes astronomical, and
nobody can afford to pay for the computer time needed to assess all of them.
The biochemists therefore have to have some way of eliminating the more

obviously unsuitable trees. A first step would be simply to take only those
hypothetical trees that other people have already suggested, because these
represent only a small sample of the total number of possibilities.

Cartmill: But I was struck, in listening to Dr. Joysey's presentation at this
Congress, by the fact that wildly different trees (including some that seemed
obviously preposterous) turn out to be about equally likely if your assumptions
about what's primitive and what's derived in each case are tailored to be most
parsimonious for the tree in question - a fact I found both interesting and
depressing.

Joysey (Cambridge): Of course, with biochemical data, one can't actually
determine what's primitive and what's derived. As Bob Martin referred to computer techniques just now, it seems appropriate to mention that the cladograms
discussed in my presentation were all hand-worked by Alex Romero-Herrera. He
was prepared to sweat it out and his performance was just as good as that of
the computer programmes available to us.

Wood (London): We haven't yet mentioned the question of temporal polarity.
People get very worried about the age of the KBS tuff when they're trying to
fit the Koobi Fora material into hominid phylogeny, for instance. I wonder
whether it makes any difference.

Tattersall (New York): Phylogenetic hypotheses can be raised at a number of
different levels of complexity, not all of which incorporate temporal information. Three principal levels can be distinguished. First, you can group organisms into nested sets on the basis of the sharing of derived characteristics,
without indicating the type of evolutionary relationship involved. This is
what's usually called the *cladogram*. At this stage, you're just saying what
is most closely related to what. The *tree* is the second stage, where you're
making a more complex type of hypothesis - adding stratigraphic information
(if it's available) and specifying the type of evolutionary relationship
involved, whether that of ancestor to descendant or that of two sister taxa
derived from a common ancestor. Here you are already getting into speculation,
since there is to my knowledge no reliable way of determining which kind of
relationship is involved. And from the tree you go to a third, still more
complex level, sometimes called the *scenario*, where you're adding to the tree
information on adaptation, ecology, zoogeography, and so forth. Many trees
can be derived from the same cladogram, and for any tree you can devise a large
number of scenarios. As you progress to higher-order hypotheses, you get
further and further away from the basic data, which is after all morphological. Antiquity is no guarantee of primitiveness, although the two may be
correlated on the average.

Cartmill: But the ecological, functional and zoogeographical hypotheses expressed in the scenario may be involved at the level of building a cladogram,
since they influence our judgements about the likelihood of parallel evolution
of various traits.

Tattersall: The ideal is to work out the parallelisms by generating and testing hypotheses of relatedness - not specifying the type of relationship - at
the level of the cladogram on the basis of the morphology. If you can't, it's
probably because the morphological data aren't adequate. I don't think zoo-

geography ever tells you anything about relationships. I don't think time
necessarily tells us anything; certainly not at the most basic level of what
is most closely related to what.

Thorington (Washington): When you talk about settling it at the level of the
cladogram, are you referring to analysis of the character in sister groups,
et cetera?

Tattersall: Yes; you try to determine what's primitive and what's derived
and to nest taxa together on the basis of a minimum amount of parallelism. At
the level of the cladogram, by the way, you needn't claim to be dealing with
species; you can just be dealing with aggregations of material. When you
start talking about species, particularly where fossils are concerned, you're
imputing characteristics to the material that are just not directly observable.

Kay: How would you deal with the weighting of traits? Would you regard
traits that have an obvious functional significance as more or less heavily
weighted than traits of unknown funciton?

Tattersall: There are primitive and derived traits, and this is all you're
worrying about. Once you've decided what counts as a trait, which is where
the question of function enters in, you ignore function.   I would weight
traits only in terms of regarding them as primitive or derived at the appropriate level of the system you're constructing. I don't otherwise know what
weighting means in this context. Would you say that some trait is 58% derived,
or 20% more primitive than something else? If a trait looks likely to be a
parallelism in terms of the overall mass of traits you're discussing, then it
simply has no weight.

Joysey: Would you ignore the one character which is inconsistent and go on
the weight of evidence pointing in the other direction, even though you're
aware that twenty correlated characters are functionally related and should
be thought of as a single character?

Tattersall: This is what we should be discussing under the rubric of what
counts as a trait, which has to be resolved before you begin constructing the
cladogram.

Cartmill: Here's the sort of situation where I personally begin worrying
about weighting traits: suppose we have three taxa, A, B and C. We find
that all three have some specialisations - apomorphies - in common, which
suggests that they form a monophyletic group. A and B share 4 specialisations
not found in C; B and C share 5 not found in A; and A and C share 6 not
found in B. Now, do we simply say that because A and C share the largest
number of apomorphies, they form a monophyletic subgroup which is the cladistic
sister of B? Or do we look at other aspects of these derived traits besides
mere numbers?

Martin: One thing most of us rely on in weighting traits is their complexity:
that is, some subjective estimate of how much genetic information is necessary
to maintain the trait in question. Characters that appear to be complex in
this sense deserve to be weighted more heavily than relatively simple ones.

Thorington: This amounts to a probabilistic assessment of homology; you
would assume that shared specialisations that are more complex are corres-
pondingly less likely to have been derived in parallel. I'd agree with that.
But I think that the whole problem Matt Cartmill's posing here is just the
logical result of the Hennigian assumption that cladistic events are always
dichotomous. If we look at the zoogeographic evidence, it appears that changes
in climate and forest distribution, particularly during the Pleistocene, have
been a very potent force in primate evolution by breaking up large forests
into smaller isolated units. By this means, a single primate population could
split into several isolated daughter populations simultaneously. I suspect
that a lot of South American primate evolution has, for this reason, taken
the form of what Simpson called "bushes" rather than dichotomously branching
trees. It is therefore possible that all three of your groups - A, B and C -
originated at the same time; and if so, the problem you're posing disappears.

Cartmill: I agree that cladistic events aren't necessarily dichotomous. What
I had in mind here was not speciation, but relationships at higher levels of
the Linnaean hierarchy - for example, the relationships between plesiadapoids,
tarsiers, and higher primates. No one would assert, I think, that these
three developed simultaneously from a single parent species: two of them must
therefore be more closely related to each other than to the third.

Joysey: But the strict Hennigian - and I'm not one - would insist that your
problem should only be answered at the species level.

Maier (Frankfurt): I was quite surprised to find that American investigators
draw a clear-cut distinction between the terms *phylogeny* and *evolution*. Your
concept of phylogeny and the way you deal with it seems to me to be quite
typological. I would suggest that phylogeny and evolution can't be separated.
Weighting characters, or even distinguishing between derived and primitive
characters, always involves questions of function. Phylogenetic reconstruction requires conscious consideration of the functional meanings of traits.
A character should be defined, not in terms of its primitive or derived
status, but in terms of the extent to which form and function are correlated -
as a form-function complex. Only a form which is understood functionally is
a trait. Phylogeny can't be reconstructed in a purely abstract and inductive
fashion.

Cartmill: Many of us would agree with most of that. Pat Luckett, for example,
has relied extensively on questions of function in trying to sort out derived
and primitive features of primate placentation. In making decisions of this
sort, especially when we lack fossil data, we often need to try to determine
the functional significance of each of the alternative character states and
to postulate a (preferably testable) selective regime that would account for
the transition from one to the other. For instance, in attempting to decide
whether the ancestral tooth-comb prosimians had a lemur-like or loris-like
bulla, one not only considers the fact that all the early Eocene primates have
a lemur-like intrabullar eardrum, one also asks under what circumstances it
might be advantageous to transform one type into the other; and this gets
one into considerations of acoustics, fluid mechanics, and so on.

Gingerich (Michigan): If we could build up phylogeny from the fossil record,
then we could actually see that in fact some particular lineage went from a

free ectotympanic ring to the ring being fused into the wall of the bulla, or whatever. But as long as we go on manipulating these characters to make some parsimonious arrangement that pleases us, we will never really know what happened. It's not sound procedure to labour over functional or ecological explanations for changes that are merely hypothetical to begin with.

Tattersall: Carrying your logic through requires the assumption that every fossil you find is an ancestor; because if a fossil just represents a collateral branch, then it has the same evidential status as a living form. If you want to assert that paleontology gives us an information source which is categorically superior to neontological data, you're compelled to assert that none of the fossils are collaterals, and that extinction has never occurred.

Gingerich: But fossil evidence does have a special status, because of the time dimension.

Tattersall: Yes, you have a time dimension; but in order to conclude that the differences you observe between fossils at different stratigraphic levels represent evolutionary change, you have to assume an ancestor-descendant relation. I think this is a dubious assumption to make about every fossil we find.

Tobias (Johannesburg): But is that assumption really inherent in Phil Gingerich's argument?

Tattersall: I think it is. As I understand it, Phil doesn't believe in the comparative approach for reconstructing phylogeny, precisely because the comparative approach doesn't show you what actually happened; you don't have evidence of the intermediates. Phil, correct me if I'm wrong.

Gingerich: The comparative approach is fundamentally different from mine, and reflects fundamentally different interests. I'm interested in ancestors. In the case of *Pelycodus*, for instance, we now have a very dense and continuous fossil record which enables us to trace actual ancestor-descendant relationships between the various species that were originally described as discrete entities. Obviously, many parts of the fossil record still contain extensive gaps, where we have to rely on inference. When a certain group displays some uniform morphology - e.g., incisor morphology in the Adapidae - I think we can take that as reflecting the way that group was from its inception. When we trace anthropoids back through time, and find they have the same thing, I think we can say something about the group that was ancestral to anthropoids. All that matters is that the morphology seen in the Eocene adapids is similar to what we find in Oligocene anthropoids. It doesn't matter whether it's primitive or derived with respect to the outgroup, because we can "demonstrate" that it's either one, depending on what outgroup comparisons we make. In most cases, and the primates are no exception, we aren't sure what to take as an outgroup. If you take tree shrews, you get one conclusion; if you take dermopterans, you get another; if you take Eocene leptictids, you get a third. You can manipulate what's primitive and what's derived by selecting the outgroup you take as a comparison.

Martin: What you're saying is that if we have a very reasonable hypothesis about relationships based on living forms, but we can't find any fossil evidence to support it, then we should disregard it.

Gingerich: Take, for example, the division of the primates into haplorhines and strepsirhines. The Eocene and Oligocene primates just don't fit this dichotomy. A phylogeny is basically a history; and a history based entirely on recent occurrences, which receives no support from any of the historical documents, is not the sort of thing that I find very convincing.

Tobias: There are obviously different ways of probing phylogeny. Fossils are one legitimate and valid way to do so, and the comparative method is another. If the phylogeny we reconstruct by the latter approach is at strong variance with one reconstructed from the actual hard evidence of the fossils, it doesn't mean that one whole field is invalidated. It means we've got to go back and see where one or the other reconstruction has gone off the rails. Maybe it's a matter of a simple observational error, or a mistaken inference, or a problem in determining what a trait is. A finding of this kind challenges us to harmonise the two reconstructions. When this sort of thing happens, it means there are still important problems left to solve - and that's reassuring.

SECTION VII

ECOLOGY AND DISTRIBUTION OF SOUTH-EAST ASIAN PRIMATES
Chairman and Section Editor: J.R. Mackinnon (Cambridge)

ECOLOGY AND DISTRIBUTION OF SOUTH-EAST ASIAN PRIMATES: INTRODUCTION

J. MACKINNON

Sub-department of Veterinary Anatomy, University of Cambridge, UK.

The Malesian archipelago has long been recognised as one of the most complex and fascinating zoogeographic regions in the world. No one who has worked on South-east Asian primates, from whatever original viewpoint, can fail to have become intrigued by the great diversity of species and their erratic distributions. Surely if we could explain how and why such multiplicity of forms arose and achieved their present geographical distributions we would learn a great deal about the processes of Evolution in general and primate evolution in particular as well as clarifying the significance of many aspects of pelage colour, markings and the form of species vocalisations.

The present session was planned as a discussion to follow four brief papers. The introductory paper by Wendell Wilson described what criteria should be used to distinguish species and plot their distribution in the wild. Two closely related papers by Warren Brockelman and Paul Gittins described species boundaries and limited gene exchange between allopatric forms of gibbon. The final paper consisted of a neat climatic deterioration model proposed by Douglas Brandon-Jones to explain the anomalies of Asian colobine monkeys.

VARIATION AMONG PRIMATE POPULATIONS IN SUMATRA

W.L. WILSON

Regional Primate Research Center, University of Washington, Seattle, Washington 98195, USA.

There are 16 primate species on Sumatra and offshore islands; the 12 described display differing degrees and types of geographic and intratroop variation. Social system as defined in terms of reproductive group composition and intertroop spacing mechanism seems to be the most conservative character, particularly among hylobatine genera. Pelage variation is slight in *Macaca*, *Presbytis cristata* and *P. thomasi*, and is absent in *Hylobates klossii* and *Symphalangus*. *Presbytis melalophis* and *P. femoralis* display geographic variations in coat colour which correspond to subspecific designations. *Hylobates agilis* and *H. lar* show intratroop variation in some parts of their distribution while in other locations only one colour morph is present. For all species, pattern of coloration below the head is generally more conservative than the pelage colour itself. Several explanations of the observed variation are discussed. Little geographic variation in morphology is positively correlated with widely ranging habits, male transfer between groups, ability to swim, terrestriality, and occupancy of mountainous or riverine habitats. Considerable geographic variation is produced by genetic isolation and independent metachromic processes; isolation can be a result of obvious geographic barriers to travel such as water to non-swimmers or extensive volcanic activity; or it can be a function of preference for habitats which are or were discontinuously distributed.

We have described, and attempted to interpret, several types of variation observed among species of Sumatran primates. Exposure to essentially identical geological events has influenced different species in different ways. Some of this could be traced to different times and paths of dispersal and the stages of evolutionary development they achieved prior to encounters with particular ecological situations. We have not dealt with this aspect of the problem here, nor have we attempted to describe hypothetical dispersal patterns of Southeast Asian primates. However, some conclusions and hypotheses may be summarized at this point.

One might expect the greatest geographic variation and speciation in the most sedentary species. This factor alone cannot account for our data; present evidence does not indicate marked differences in mobility among *Presbytis* and hylobatines, yet they show wide ranges of types and degrees of variation.

Macaques show the least geographic variation over a wide distribution. They are the least sedentary and have the largest home ranges. They are also good swimmers; *M. fascicularis* in particular would only be isolated by marine barriers of considerable breadth. The terrestrial habits of macaques permit travel over gaps in forested areas. It is suspected that male emigration from the natal troop to a new troop is common and would be an avenue for gene flow. Presumably, distances travelled by such males are greater than those travelled by emigrés (or "evictees") from troops of other species, but there is no concrete evidence for this. It is probably easier for a macaque to join a new troop or at least to engage in clandestine copulation than it is for a langur.

Contrasts in habitat preference account for much of the difference between *P. cristata* and members of the *P. aygula-melalophos* group. Habitat preference as an explanatory mechanism relies on the pattern of distribution of the favored type of habitat. If the habitat is or was distributed in "pockets", or "refuges" genetic isolation and subsequent subspecific variation might result. Mountainous or riverine habitats provide avenues for gene flow. The boundary between swamp forest and dry-land forest may be, because of habitat preference, as effective a barrier as is a large river to a non-swimming species.

Once there is relaxation of selection for concealing agouti pelage, metachromic saturation and bleaching may produce a variety of non-agouti pelages. Some species, e.g., the isolated macaques, the siamang, *P. cristata*, and some gibbons, may have remained a saturate black or blackish due to stabilizing selection, possibly as a function of positive assortative mating. The *P. aygula-melalophos* group has evolved a variety of colour combinations, and colour pattern is far more conservative than actual colouration. It is hypothesized that such patterning combined with silent freezing behaviour provides camouflage in an arboreal habitat. Facial patterning and ornamentation may be linked to camouflage, but may also be a result of ethological selection for enhanced discrimination of intraspecific communicative gestures.

Local variation in morphology may facilitate individual identification. This seems to be a reasonable explanation for the relatively greater amount of intertroop variation in macaques, which normally form larger troops than do members of the other genera discussed. Variation in some gibbon populations is poorly explained on this basis since other populations show great homogeneity. The advantages of recognition of siamang by gibbons were discussed as a possible reason for the existence of polymorphic, non-black gibbon populations.

In summary, little geographic variation is positively correlated with wide-ranging habits, male transfer between groups, ability to swim, terrestriality and preference for mountainous or riverine habitats. Much geographic variation is produced by genetic isolation and independent metachromic processes; isolation can be a result of obvious geographic barriers to travel such as water to non-swimmers or extensive volcanic activity; or it can be a function of selection of habitats which are/were distributed in pockets or refuges. The relative strength and universality of these correlates may be tested by examining other primate populations showing similar patterns of variation. For example, South American marmosets could be compared with such less variable genera as *Saimiri*; or the variable *Cercopithecus* species might be compared with other African primates showing more homogeneity. Careful examination of types of variation within a particular geographic area and comparison with other regions and kinds of fauna is contributing to our understanding of the processes of evolution.

Finally, it is all too obvious that undisturbed tropical rainforests are shrinking rapidly, making it increasingly urgent that natural populations be studied while they still exist. Investigators fortunate enough to be able to study these populations may collect various kinds of data not directly related to their own research goals, but of great interest to scientists in different fields.

A fuller description of the behavioural and morphological variation of the Sumatran primates will be published elsewhere (Wilson and Wilson, 1976).

REFERENCE

Wilson, C. and Wilson, W.L. (1976). Behavioral and Morphological Variation
    among Primate Populations in Sumatra. *Yearbook of Phys. Anthrop.*, 20.

PRELIMINARY REPORT ON RELATIONS BETWEEN THE GIBBONS
HYLOBATES LAR AND *H. PILEATUS* IN THAILAND

W.Y. BROCKELMAN

Faculty of Science, Mahidol University,
Rama VI Road, Bangkok, Thailand.

Hylobates lar and *H. pileatus* occur in different regions of Thailand, but a zone of overlap occurs in Khao Yai National Park about 120 km northeast of Bangkok. This area is in the headwaters of the Takhong River, a tributary of the Mun River which flows eastward into the Mekong, which suggests that these rivers have served as geographical isolating barriers. However, very little forest is left near these rivers; the few gibbon groups remaining support the idea. The gibbon groups in the zone of overlap are being mapped and studied to determine the extent of genetic isolation between the species, the amount and nature of behavioural interaction between them, and the importance of behaviour, especially vocal, as a reproductive isolating mechanism. The species have distinct pelage colouration and vocalizations. Nevertheless, there is evidence for limited hybridization. Several mixed-species groups have been found, and some individuals with atypical pelage and vocalizations which may be hybrids. The evidence thus far suggests that territorial vocalizations may function as partial isolating mechanisms and are under a high degree of genetic control. *H. lar* and *H. pileatus* still deserve specific rank despite limited interbreeding.

INTRODUCTION

Most of the nine recognizable species of Hylobatids are allopatric, separated by rivers or straits (Groves, 1972). One of the few known cases of interspecific contact lies in the relatively secure area of Khao Yai National Park in Thailand, where an area of sympatry between the white-handed gibbon (*Hylobates lar*) and pileated gibbon (*H. pileatus*) occurs (Marshall et al., 1972). This area is being studied by J.T. Marshall, Jr., and myself, and although work is far from complete, some interesting patterns have turned up which merit reporting at this time. First, I will clarify the geographical separation of the species, and second, I will discuss relations between the two species in the zone of sympatry.

Hylobates pileatus, described by Gray (1861), was regarded as conspecific with *H. lar* by Kloss (1916), Pocock (1927), Napier and Napier (1967), Fooden (1969) and others. This confusion persisted so long because of ambiguity in the description of early authors, a lack of field studies (especially on *H. pileatus*) and the rather complicated and still puzzling colour dimorphisms in these species, in which *H. lar* is asexually, and *H. pileatus* sexually, dimorphic (Fooden, 1969). It was not until further field work was carried out and an attempt to apply the biological species concept was made that the separation of *H. pileatus* from *H. lar* could again be justified. Marshall et al. (1972)

reported that the taxa occurred in their rather distinct forms in an area of sympatry and described the rather marked differences in their territorial vocalizations. Both species are sexually divocal, and the call of either sex can be used to distinguish the species in the forest.

Here I present recently obtained field evidence that genetic separation between the species is not complete, and that a limited amount of hybridization between the species is occurring.

ZOOGEOGRAPHY

The pileated gibbon occurs in southeastern Thailand and Cambodia, presumably in all suitable forest west of the Mekong River. The important northern and northwestern boundaries of its range, however, where it gives way to *H. lar*, have not been adequately investigated and are imprecisely indicated on range maps (Marshall et al., 1972; Groves, 1972). The northern limit of this species is now the northern slopes of the Dong Rek Range which parallels the Mun River. Gibbons have been recently heard in Sisaket Province near the Cambodian border (P. Enderlein and J.F. Maxwell, personal communication). Pileated gibbons may have reached the Mun River to the north, which flows eastward into the Mekong, in preagricultural times, but the Mun River valley is now completely deforested and relatively arid. It is not known to what extent gibbons occurred in the drier lowland forest types which have now been virtually eliminated from Thailand.

In Khao Yai Park, where both species are still abundant, the area of sympatry is around where the Takhong River enters a canyon and flows northward out of the park. This river is a tributary of the Mun, and it was suspected that the Takhong-Mun River served as a barrier between these two species' distributions. Nearly complete deforestation in this region, especially in the range of pileated gibbon, makes it urgent to test this theory as quickly as possible (Brockelman, 1975). Brief surveys for gibbons are being carried out in nearly all remaining patches of forest. A range of northwest-extending mountains reaching 700 to 800 m in elevation north of Khao Yai Park contains nearly the only patches of forest remaining in the region. This is crossed by the northeastward-flowing Takhong and Phra Phloeng Rivers which are dammed at these points to form reservoirs. Pileated gibbons (male and female in duet) have been heard between the reservoirs, providing the most northwestern record for this species. Other records farther south include the Sakaerat Biosphere Reserve, where pileated gibbons were last heard in 1966; the northeastern corner of Khao Yai Park; and at Lam Phyathan Ranger Station in the southeastern corner of the park, where a group was recently heard. Pileated gibbons have also been heard in the range east of Khao Yai Park near Khao Lamang (N.K. Kobayashi, personal communication).

The closest *H. lar* record to the Takhong River is on the mountain Khao Phrik about 3 km from the river, where a male and female were heard together in May, 1976. There is no other gibbon habitat left close to the river on the north side, and the 5 square km of forest remaining on Khao Phrik will be gone in a few more years. The nearest *H. lar* habitat of significant extent outside of western Khao Yai Park is in the Petchabun Mountains about 200 km to the north, where Groves (1972) has cited several records for *H. lar carpenteri*.

It thus appears that the Takhong River is the only barrier which could have separated the species. This river is only about 30 m wide at the reservoir, and less farther west toward Khao Yai.

In Khao Yai Park, the zone of sympatry extends from the Takhong River canyon south to the moutains Khao Rom and Khao Khieo, an area about 9 km across. The mountains rise over 1200 m in elevation, too high to support the diverse tree flora required by gibbons. Another zone of sympatry also exists on the south slopes, but the area has not been explored well yet. South of Khao Yai Park lie rice fields which extend to the Gulf, and we will never know what separated the species in this region; perhaps it was the Bang Pakong River which drains into the northeastern corner of the Gulf. Possibly there was a broad zone of interspecific contact in the upper reaches of the river.

EVIDENCE FOR HYBRIDIZATION

There are two types of evidence for hybridization: the presence of some mixed-species groups, and a few individuals of odd or intermediate pelage and vocal patterns.

The mixed-species groups number at least four so far. All involve a female *lar* mating with male *pileatus*. In one case, a female *lar* is a member of a group with an adult male and adult female *pileatus* and offspring. In two others, only two adults are involved, with no young, and in the last, the pair has an infant about 1 yr old. Both black and buff *lar* females are involved in interspecific pairing. Mixed groups are in the minority, comprising only about 6% of the groups in the area.

Four individuals have so far been discovered which are atypical in either pelage or voice, and at least three are atypical in both respects (the fourth has been heard, but not seen, so far). Some odd vocalizations have been noted as early as 1968 by J.T. Marshall, but the discovery of the mixed groups in 1972, 1975 and 1976 has re-raised the whole issue of hybridization. These odd vocalizations have been consistently atypical — individual calls are rather stable in pattern even over several years, and are a very useful aid in individual recognition. These individuals will be described in detail, and the full implications discussed, in a future report. At this point the following statements appear justified.
1) Most individuals in the zone of sympatry are clearly one species or the other and quite typical; there is no swamping of the gene pool.
2) Most individuals apparently prefer to mate with their own species, and therefore a partial behavioural isolating mechanism must be present. For this reason and (1) above, no taxonimic revision appears justified.
3) The two sexes may not have an equal tendency to form interspecific pair bonds.
4) The territorial vocalizations are likely to be under a high degree of genetic control, and it is hypothesized that they are important in species recognition (i.e., help serve as isolating mechanisms).
Future research will concentrate on the role of territorial vocalizations in intraspecific vs. interspecific communication, and on the fate of the mixed-species groups and atypical individuals.

ACKNOWLEDGEMENTS

I thank P. Suwanakorn, Chief, National Park Division for his generous co-operation, also W. Yarnpirat and W. Na Nakorn for their hospitality and help. J.T. Marshall has been generous with his own data and notes, and he and P. Rand have provided much useful discussion and criticism.

REFERENCES

Brockelman, W.Y. (1975). *Nat. Hist. Bull. Siam. Soc.*, 26, 133-157.
Fooden, J. (1969). *Evolution*, 23, 627-644.
Gray, J.E. (1861). *Proc. Zool. Soc. London*, 1861, 135-140.
Groves, C.P. (1972). *In* "Gibbon and Siamang", (D.M. Rumbaugh, ed.), Vol. 1, pp. 1-89, S. Karger, Basel.
Kloss, C.B. (1916). *Proc. Zool. Soc. London*, 1916, 27-75.
Marshall, J.T., Ross, B.A. and Chantarojvong, S. (1972). *J. Mammal.*, 53, 479-486.
Napier, J.R. and Napier, P.H. (1967). "Handbook of Living Primates", Academic Press, New York and London.
Pocock, R.I. (1927). *Proc. Zool. Soc. London*, 1927, 719-741.

THE SPECIES RANGE OF THE GIBBON *HYLOBATES AGILIS*

S.P. GITTINS

Sub-department of Veterinary Anatomy, University of Cambridge, Tennis Court Road, Cambridge, CB2 1QS, UK.

In a recent survey in Peninsular Malaysia and Sumatra (1974-6), the boundary of the species range of the gibbon *Hylobates agilis* was located more precisely (Fig. 1) than was previously known (e.g. Fooden, 1969; Chivers, 1972). Gibbons were rarely found over 1000 metres, and high mountain chains form effective barriers to dispersal. Rivers were found to act as boundaries, separating *H. agilis* from its neighbour *H. lar*. In several places the two species were in auditory contact across the rivers. The boundaries were not always the widest rivers, and a river could still be the boundary upstream where it was narrow enough to be crossed.

Why then are particular rivers barriers to dispersal? It could be just chance: the two species of gibbon arriving from opposite directions and meeting there, the river acting as a partial barrier until the other species has occupied the opposite bank and made expansion there difficult. Another possibility is that the rivers are not in fact the barriers, but that they fall close to an ecological break in the habitat. For example, the 'Kra ecotone' described by Whitmore (1976) where lowland evergreen rainforest is replaced by semi-evergreen rainforest, lies very close to the northern boundary of the species range of *H. agilis*.

One area was found near the head waters of the Muda River where the river barrier breaks down and allows *H. agilis* and *H. lar* to come into contact. Here on the shores of the lake created by the Muda dam (6°08'N, 100°53'E) in 1968, three groups have been found containing individuals that have crossed this barrier. One group consists of a *lar* male which has crossed to the *agilis* side and formed a group with an *agilis* female. There is a juvenile in this group with *agilis* colouring and markings giving the typical *agilis* call. Here the *agilis* characters seem to be dominant. There is also a very young infant in this group. The second group consists of an *agilis* male which has crossed to the *lar* side of the barrier and formed a group with a *lar* female. There is a juvenile in this group with *agilis* colouring, but giving the *lar* call. Here there seems to be a mixing of the different characters. There is another individual in this group which was not seen properly, but which gave female calls. Auditory contact was made with a third group which gave *agilis* calls on the *lar* shore. This could be a mixed group or an *agilis* group which has crossed the lake.

The data presented show that when *agilis* and *lar* come together in the wild, group formation can occur, and interbreeding takes place. There seems to be no effective "behavioural isolating mechanism" despite the fact that calls of each species are distinct and play an important part in group formation (Mac-Kinnon and MacKinnon, 1977). Although there is a limited amount of inter-

breeding in the wild, this is only in a narrow zone, and geographic isolation
of the two species is almost complete. This together with the fact that there
are consistent differences in coat colour markings and calls, it seems best to
leave *H. agilis* and *H. lar* as distinct, but closely related species.

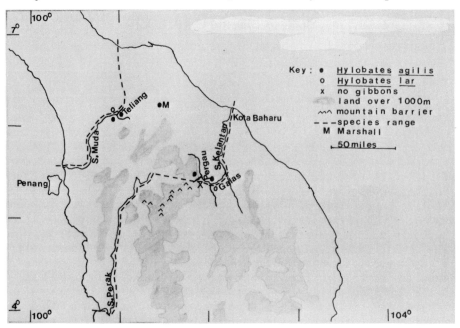

Figure 1a. Gibbon distribution in Peninsular Malaysia and Thailand.

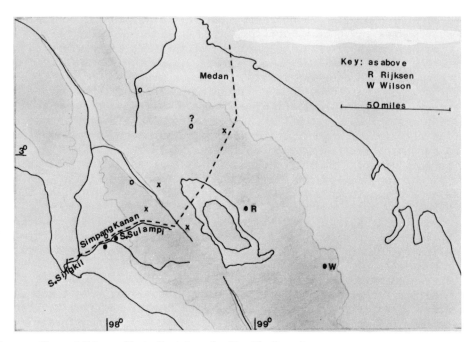

Figure 1b. Gibbon distribution in North Sumatra.

ACKNOWLEDGEMENTS

The survey was facilitated by information from D. Chivers, J. Marshall, W. and C. Wilson, J. MacKinnon, M. Borner and H. Rijksen. I would like to thank all Government Departments of Malaysia and Indonesia who helped in this project. The study was financed by the Science Research Council, U.K., The Royal Society, The New York Zoological Society and the Bartle Frere Fund.

REFERENCES

Chivers, D.J. (1972). *In* "Gibbon and Siamang", (D. Rumbaugh, ed.), Vol. 1, Karger, Basel.
Fooden, J. (1969). *Evolution,* 23, 627-644.
MacKinnon, J. and MacKinnon, K. (1977, in press). *Primates,* 3, 18.
Whitmore, T.C. (1975). "Tropical Rain Forests of the Far East", Oxford University Press.

THE EVOLUTION OF RECENT ASIAN COLOBINAE

D. BRANDON-JONES

*Unit of Anatomy with special relation to Dentistry, Anatomy Department,
Guy's Hospital Medical School, London SE1 9RT, UK.*

The present zoogeographic distribution of Asian colobines is rife with anomalies.

The restricted ranges of *Pygathrix*, *Rhinopithecus* and *Presbytis francoisi* in south-eastern China and eastern Indo-China are inconsistent with the existence of apparently comparable tropical forest habitats in immediately adjacent areas.

Two distinctive but related forms, *Nasalis* and *Simias*, both apparently highly adapted for an open forest habitat, are indigenous to Borneo and the Mentawai Islands respectively. These islands are predominantly covered with primary forest and are separated by Sumatra which has no *Nasalis* or *Simias*-like form.

One of the three cranial morphotypes found in the colobines currently classified as *Presbytis* shows a split in its geographic distribution between the southern Indian subcontinent and eastern Asia. Cranial characters link *Presbytis francoisi* with this group. Its tegumentary characters, geographic distribution and ecology strongly suggest that *Presbytis francoisi* is a relict assemblage of taxa: the central remnant of a population of this morphotype once continuous from southern India to Java. Comparable disjunctions occur in the distributions of other fauna and flora (Blandford, 1901).

In south-east Asia the tripartite geographic distribution of the grey races of *Presbytis* (sensu stricto) shows a close correlation with that of the chromatically monomorphic subspecies of *Hylobates* (Fooden, 1969). The faunal boundary of *Presbytis* and *Hylobates* taxa in north-western Sumatra is well-documented. This boundary closely corresponds to the limit of *Pongo* distribution in Sumatra. The three zones of grey *Presbytis* distribution can reasonably be regarded as relicts of a single population, rather than representing areas of convergent evolution (Medway, 1965). This interpretation is supported by the distribution of other mammals such as the black and red races of *Callosciurus prevostii* (Chasen and Kloss, 1925) and the genus *Dremomys* whose range in south-east Asia (Moore and Tate, 1965) is restricted to that area of north-east Borneo postulated as the region of *Presbytis* relict distribution.

Cranial, dental (Colyer, 1936), skeletal (Washburn, 1944), neonatal (Tilson, 1976) and adult pelage characters as well as its pattern of vocalization (Wilson and Wilson, 1975) clearly ally *Presbytis potenziani* from the Mentawai Islands with the *Presbytis* rather than with the *Trachypithecus* group although its pelage character affinities with *Presbytis francoisi* are interesting and significant. The geographic distribution and taxonomic affinities of the *Presbytis* group can best be explained by postulating that *Presbytis potenziani*,

restricted to the Mentawai Islands by the elimination of suitable habitats in
its former range, was the progenitor of all living members of the group. By
analogy, *Hylobates klossii* must have been the progenitor of most, if not all,
extant *Hylobates* (sensu stricto). Similarly, *Macaca pagensis* of the Mentawai
Islands could well have been the progenitor of *Macaca nemestrina*.

The west Javan *Trachypithecus* is the most probable candidate as the nucleus
for recent evolution of the east Asian branch of this group. The Ceylonese
Trachypithecus radiation has stemmed from *Presbytis johnii*, with the *Semnopithecus*
group most probably arising in southern Sri Lanka.

Recent Asian colobine population dispersion has been accompanied by succes-
sional changes in tegumentary colouration comparable with those demonstrated
by Hershkovitz (1968) in South American primates. The constancy of this chro-
matic series in unrelated populations living in dissimilar environments suggests
it has no adaptive significance. It is postulated that the geographic distribution of the different colour phases may indicate successive stages of population dispersion. Since minor tegumentary and cranio-facial differences appear
to effectively inhibit interbreeding between populations of colobines under
natural conditions, it is possible that changes in these characters constitute
the principal means of speciation in this group.

No unequivocal case of speciation in reproductive isolation has been found
in this study. Conversely, there appear to be at least three instances where
Asian colobine populations have undergone prolonged and widespread isolation
without significant morphological differentiation. In at least one case an
ancestral population has apparently been displaced by the species descended
from it. Sea barriers are found to have played an unexpectedly minor part in
limiting recent colobine dispersion. 'Rafting', over the short distances involved, must have been more common than has been assumed.

These results carry the wider implication that the distinctive biogeography
of the Oriental Region can be better explained as resulting from climatic change
rather than by the presence of topographic barriers to faunal and floral dispersion. To explain such widespread disruption of both faunal and floral distributions, it is necessary to postulate at least one period in which the climate changed significantly. The distribution of the relict zones corresponds
more closely to that of the current winter, rather than summer, rainfall. Since
these areas are characterised by high altitude or maritime isolation, a period
of prolonged aridity, with its ineluctable effect on forests, would seem the
most plausible explanation. The zoogeographical anomalies of south-east Asia
can best be explained by postulating two such arid periods.

It follows that correlation of the faunal and floral distributions of an
area makes it possible to draw conclusions on the diet and ecology of animals
as yet imperfectly studied in the field: for instance, the unequal distribution
of the *Presbytis*, *Trachypithecus* and *Semnopithecus* groups reflects their different ecological specialisations. It is suggested that such differences in
ecology may be of value in assessing the taxonomic status of a natural assemblage of species, a question of particular relevance to the conspicuous im-
balance in current Asian colobine taxonomy.

REFERENCES

Blanford, W.T. (1901). *Phil. Trans. Roy. Soc. Ser. B.*, 194, 335-436.
Chasen, F.N. and Kloss, C.B. (1925). *J. Malay Brch. R. Asiat. Soc.*, 3, 97-99.
Colyer, J.F. (1936). "Variations and Diseases of the Teeth of Animals", Bale,
    Sons and Danielsson, London.

Fooden, J. (1969). *Evolution*, 23, 627-644.
Hershkovitz, P. (1968). *Evolution*, 22, 556-575.
Medway, Lord (1965). *Fedn. Mus. J. (n.s.)*, 9, 95-101.
Moore, J.C. and G.H.H. Tate,(1965). *Fieldiana, Zool.*, 48, 1-351.
Tilson, R.L. (1976). *J. Mammal.*, 57, 766-769.
Washburn, S.L. (1944). *J. Mammal.*, 25, 289-294.
Wilson, W.L. and Wilson, C.C. (1975). *In* "Contemporary Primatology", (S.
    Kondo, M. Kawai and A. Ehara, eds), pp. 459-463, Karger, Basel.

SUMMARY OF DISCUSSION

J. MACKINNON

Sub-department of Veterinary Anatomy, University of Cambridge, UK.

The general phenomenon of transatlantic speakers' inability to contain themselves within allocated time limits totally precluded any discussion from sche-
duled programme but fortunately we were able to find extra time for a smaller
gathering to make some useful and stimulating discussion on the problems of
interpreting the South-east Asian primate puzzle.

Deforestation has denied us the opportunity of seeing what species inhabit
some key areas of the region and consideration was given as to how much reliance could be given to early records by Chasen, Fooden etc. to fill in some
of these gaps. Where earlier collectors have secured well labelled specimens
prior to deforestation these locality records may be legitimately used but it
was generally agreed that little reliance could be put on mere sightings described by early collectors whose knowledge of species field-characteristics
was often poor.

The discussion moved on to consider the Brandon-Jones model for explaining
colobine monkey distributions. There was general agreement that the colobine
picture and indeed the gibbon picture also looks like the adjunction and overlap of species' ranges from previously smaller isolated distributions. The
main argument lay in the causes of such isolations. Other models to explain
gibbon distributions, such as that proposed by Chivers (1977), rely primarily
on changing sea levels to cause the isolation that has led to divergence and
speciation. We know that such changes in sea level did occur in the Pleistocene
and have a fair idea of the dates and land shapes involved. Not only is there
no physical evidence for climatic deteriorations postulated by Brandon-Jones,
but the data we have on pollen analysis and the survival of such enormous faunal
and floral diversity with typical rainforest structure and physiognomy indicate
that extensive rain-forest survived the Pleistocene intact. In Africa by contrast climatic deteriorations undoubtedly occurred during the Pleistocene and
rain-forest, where present today, shows atypical physiognomy and comparatively
impoverished diversity.

Brandon-Jones argued that climatic deteriorations of only twenty years
would be sufficient to explain his faunal contractions and that such short
deteriorations would not be registered in the pollen record nor since plant
seeds could survive these periods would there be much evidence of floral im-
poverishment.

Whilst Brandon-Jones' model might well be the simplest to explain the colobine distributions, no such model of climatic or geological changes can approach
the full historical picture unless it also fits with the distribution of all
other floral and faunal groups in the area. The discovery of one squirrel
species whose distribution of forms accords with the Brandon-Jones model seems

weak support in view of the thousands of other species that could have been considered. The model is too simple to explain adequately the more complex picture of gibbon distributions nor does it satisfactorily explain the problem of the proboscis monkey's confinement to Borneo although this confinement forms one of the cornerstones of his argument. The proboscis monkey occurs in large numbers throughout almost all the riverine and coastal forest of Borneo and is a good swimmer as well as being anatomically adapted for locomotion through open forest. Even in the most severe climatic deteriorations we would expect this monkey to be one of the most capable of adapting to open conditions and certainly during any period of low sea level that we would expect to accompany climatic deterioration, the coastlines of exposed Sundaland must still have sported swampy forest ideal for the spread of proboscis monkeys. Neither climatic deteriorations nor changes in sea level adequately explain why the proboscis monkey is absent from Sumatra.

It was felt that much more data on the distribution of fauna and flora, better pollen analyses from Pleistocene deposits and better geological dating of recent changes in land shape and sea levels are needed before models of zoogeographic events in south-east Asia can be given greater confidence. Such models should consider the effects of those climatic and sea level changes that can be documented before invoking events for which there is no physical evidence. Premature speculation is only useful when it makes new predictions that can be checked or corroborated in the field.

The participants continued on equally shaky ground to discuss the adaptive significance of gibbon and leaf monkey coat colour and distinctive markings. Brandon-Jones pointed out that relict or ancestral stocks, e.g. *S. syndactylus*, *H. klossi*, *M. nigra*, *P. potenziani*, *P. cristata*, *P. francoisi* and *Callosciurus prevostii* etc. are black in colour whilst forms that appear to have undergone more recent speciation are more agouti in colour, often with interesting and distinctive patterns. It was suggested that black was in fact an extremely conspicuous colour in a tropical forest but why such a colour should be retained by surviving relict stocks but abandoned by more recent forms was not explained.

Gibbon coat colour shows interesting geographical variation. The northern forms show marked sexual dichromatism and the *lar* gibbons in Thailand are also dichromatic though not sex-linked. On the south-west of the old Sunda shelf we find both *lar* and *agilis* with variable colours from buff to dark brown but on the eastern side of the Sunda shelf *muelleri* and *moloch* (and perhaps Kalimantan *agilis*) show much grayer pelage. These changes in colour fit better with geographical distribution than with phylogeny and strongly suggest that the colours are adapted to suit subtle climatic differences.

No one could suggest why the eastern gibbons should be grey but it was suggested that northern dichromatism might be related to the drier conditions where more trees in the forests are deciduous and where canopy height is not so great. Gibbons in such forests may have better visibility than in the dense rain-forests further south permitting them greater use of visual signals of individual recognition. It must certainly be of advantage for a gibbon to be able to identify the sex and group of any neighbour it may encounter if it is to take appropriate action efficiently. A neighbouring female is far less of a threat to a resident male than is another male and vice versa (see Tenaza, 1975). Speedy recognition is mutually important for gibbons to decide whether to ignore, attack, call at, display at or flee from each other. Thus it is very probable that *concolor*, *pileatus* and *hoolock* separately evolved sexual dichromatism in parallel and that *lar*, which may be a more recent inhabitant of the northern forests, has evolved a compromise dichromatism but not yet sex-linked.

Whilst there is no real evidence of greater visibility in northern forests we do know that gibbon densities in these forests are lower than in the south (Chivers, 1977) and we can infer that territorial behaviour and individual recognition consequently have to operate across wider distances.

Alternatively or additionally Gittins suggested that dichromatism may give the gibbon family greater camouflage from potential predators. Individual shapes are broken up by the disruptive colour patterns and repetition of pattern is avoided by each animal in the group being marked differently. If visibility in the canopy of northern forests is indeed greater, this additional defence could be important enough to have been evolved independently by different types of gibbon.

Gittins asked why *lar* gibbons should have white hands like those of *pileatus* although neither were otherwise similar in song or colour. Was this parallel convergence, coincidence, shared ancestry or what? It was agreed that white hands were effective long range gibbon recognition signals (particularly during brachiation) except on pale phase individuals so they might have been evolved independently but a more interesting possibility is that these are an example of territorial mimicry.

The *lar* gibbon appears, for several reasons, to be the most recent to speciate. Their expansion in distribution may well have been northwards (Chivers, 1977), and at the expense of *pileatus* with whom *lar* still shows a poorly defined boundary with some overlap (Brockelman, this vol.). Such displacement would only be possible if *lar* was ecologically better adapted to the moister southern forests as well as territorially exclusive with *pileatus* which is certainly the case in Khao Yai Park (Marshall et al., 1972). Thus the species vocal and visual display must not only be sufficiently specific for the individual to attract a mate of the right species but also sufficiently general to be recognised as a territorial threat by another gibbon species. It is possible, therefore, to interpret the *lar* white hands as having been evolved so that *pileatus* would not fail to recognise *lar* as a gibbon. Even the wailing great call of *lar* can be interpreted as mimicking the great call of *agilis* with whose distribution *lar* is intermittent in the south (Gittins, this vol.) and not necessarily reflective of a close phylogenetic relationship between these species as many authors have inferred. Wilson even mentioned that *H. agilis* showed a very high proportion of dark almost black forms in parts of its distribution in Sumatra where there were no siamangs. Where *agilis* distribution overlapped with that of the black siamang there were no black gibbons. This may be the opposite of mimicry in that the gibbons whose ranges do overlap with those of siamangs cannot afford to be mistaken for siamangs territorially.

Such problems could be clarified by comparison with some of the literature on pelage and song characteristics of sympatric congeneric bird species but much could also be learned by more closely directed fieldwork on the primates concerned. Hopefully all the participants that have contributed to this discussion will have some new ideas to take back with them into shady forests or equally shady museum vaults.

REFERENCES

Chivers, D.J. (1977). *In* "Conservation of Non-human primates" (Prince Rainier and G. Bourne, eds), Academic Press, New York.

Marshall, J.T., Ross, B.A. and Chantarojvong, S. (1972). *J. Mammal.*, 53, 479-486.

Tenaza, R.R. (1975). *Folia primat.*, 24 (1), 60-80.

SECTION VIII

HOMINID EVOLUTION
*Chairmen and Section Editors: M.H. Day (London)
and P.V. Tobias (Johannesburg)*

INTRODUCTION

M.H. DAY

St. Thomas's Hospital Medical School, London, UK.

The Hominid Evolution session of the Sixth Congress of the International Primatological Society was an outstanding occasion in that all of the invited speakers, (Bilsborough, Wood, Tobias, Stringer and Preuschoft) gave excellent papers. These contributions were followed by a series of shorter papers, also of high quality, covering a wide range of subjects. The international nature of the occasion was emphasised by the large number of countries that were represented including Great Britain, South Africa, the United States of America, West Germany, Japan, France, Holland and India. Czechoslovakia should also be included here although Jelinek was unable to travel at the last minute.

The session was opened by Bilsborough who tackled some basic taxonomic questions in hominid studies as they relate to morphological differentiation among the australopithecines, and cranial evolution within the species *Homo erectus*. His discussion centred on the mosaic evolutionary process and was illustrated by the multivariate analysis of certain cranial and gnathic complexes in *A. africanus*, *H. erectus* and *H. sapiens*. Bilsborough was followed by Wood who considered the classification and phylogeny of East African hominids using a comparative anatomical approach for cranial and dental material that highlighted some of the very real differences that exist in the interpretation of the rapidly enlarging sample of early hominid remains from this area. Tobias continued the session with a contribution that centred specifically upon an important current problem, the place of *A. africanus* in hominid evolution. His paper was of a special interest since he chose this occasion on which to announce new finds from his excavations at Sterkfontein, fossils recovered only days before the conference began, and a fitting reward for ten years' continuous work at this important site. Stringer then spoke on the possible relationships between the Neanderthalers and modern man, a topic that he had investigated by means of size and shape analysis and by univariate plots of cranial measurements.

The last of the invited papers was given by Preuschoft on his investigations into the biomechanics of bipedality, and this included a general review of the evolution of this form of locomotion.

The remainder of the day was devoted to proffered papers on a wide range of topics including ecology, behaviour, Neanderthal man, *Ramapithecus*, comparative studies and experimental work; all of the papers given in the session are reproduced and arranged here so that papers of similar subject are adjacent. From this arrangement it is apparent that four themes were of prime current interest to students of hominid evolution at this time.

The first of these was the problem of taxonomy both in the genus *Australopithecus* and in the genus *Homo* illustrated by fossil material from both South

Africa and East Africa (Bilsborough, Wood, Tobias); the taxonomic approach
was continued in relation to *Homo erectus*, Neanderthal man and *Homo sapiens*
(Bilsborough, Stringer, Trinkaus and Jelinek). The second theme concerned
biomechanical problems as they relate to bipedality and upright posture (Preuschoft, Ishida et al., Kimura et al., and Sakka) while the third theme dealt
with comparative studies (Cramer and Zihlman, Sirianni, Prasad). Finally the
fourth theme was that of ecology and behaviour (Rijksen, Kortlandt).

This diversity of approach is in itself an interesting commentary on the
present state of hominid studies since we can see both old problems being
tackled by new methods, and new problems being investigated by traditional
methods. There would seem to be, from this conference, a healthy determination
on the part of primatologists in the hominid field to make use of the methods
that are most appropriate to the problem in hand and to resist the temptation
to pursue techniques for their own sakes.

All of the contributions provoked questions and discussion from an interested and active audience; indeed it was their participation, coupled with the
excellence of the papers, that produced a most stimulating and valuable symposium.

SOME ASPECTS OF MOSAIC EVOLUTION IN HOMINIDS

A. BILSBOROUGH

Department of Physical Anthropology, Pembroke Street, Cambridge, UK.

INTRODUCTION

The concept of mosaic or 'palimpsest' evolution — that is, that different structures, responding to varying selection pressures, evolve at differential rates at differing times — is now well established in palaeoanthropology, and indeed, is central to a number of recent studies (e.g. Robinson, 1972; Holloway, 1973; R.E.F. Leakey, 1974; McHenry, 1975). Moreover, recognition of the phenomenon of mosaicism is crucial for making phyletic and taxonomic judgements; it is necessary, for example, to distinguish between 'characters of common inheritance' and 'characters of independent acquisition', and between cladistic and patristic aspects of morphological affinity according to the principles adumbrated by for example Clark (1962), Cain and Harrison (1960), Harrison and Weiner (1963), Hennig (1966) and others. Aspects of such studies relating to primate systematics are dealt with elsewhere in this volume; here it is necessary only to note that failure to draw these distinctions almost invariably results in confusion, a point well illustrated by the needlessly prolonged controversy over the phyletic status of *Australopithecus*.

Despite the conceptual advances noted above, there are still relatively few data upon which to base estimates of the pattern and tempo of morphological change within individual lineages and hominids are no exception to this generalisation. Several early studies (e.g. Haldane, 1949; Boné, 1962; Campbell, 1963; Bilsborough, 1969) attempted to provide information upon rates of change in cranial capacity and dental dimensions within the Hominidae, but they have been superseded by developments in chronology and the recovery of more extensive fossil material. More recently Bilsborough (1973) provided data on neurocranial proportions, whilst McHenry (1975) has considered mosaic evolution in the post cranial skeleton. This paper summarises information upon certain aspects of cranial evolution within the Hominidae, and suggests ways in which the analysis of mosaicism may aid phyletic studies.

THE ANALYSIS OF MOSAIC EVOLUTION

As an aid to investigating patterns of morphological change in hominids, I have divided the skull into a number of functional regions and attempted to describe the 'total morphological pattern' (Clark, 1962) of each by appropriate dimensions. These regions, which are shown in Figure 1, are as follows (numbers of characters in brackets): upper face (16); upper jaw (13); lower jaw (16); cheek region and masticatory musculature (12); articular region (8); balance (14); basicranium (12); cranial vault (16). In addition

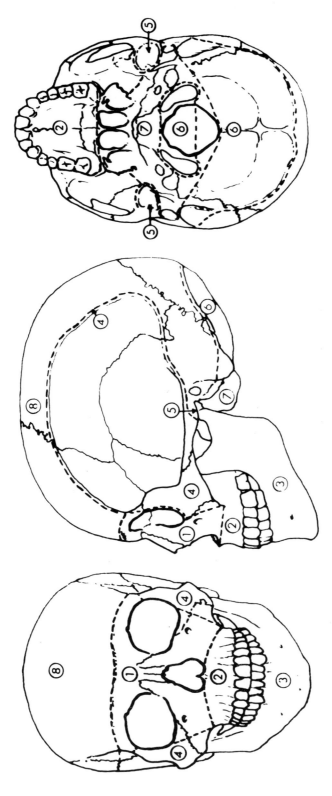

Figure 1. Front, profile and basal views of modern human skull to show functional complexes used in this study: (1) upper face; (2) upper jaw; (3) lower jaw; (4) cheek and masticatory complex (zygomatic region and extent of masticatory muscles); (5) articular region; (6) balance (extent of nuchal musculature, orientation of occipital condyles and relative proportions of pre- and post-condylar sections of head lever); (7) basicranium; (8) cranial vault. Characters were also selected from each of the above to provide a set measuring overall skull dimensions.

characters from each of these complexes were selected and combined to provide a ninth set measuring overall skull dimensions, so as to establish a reference standard against which to compare changes in each region. It is then possible to investigate changes in individual characters within a particular complex or, by deriving some multivariate statistic based upon all the characters for each complex, changes in different regions may be compared, thereby giving a quantitative basis to the concept of mosaic evolution.

Most previous studies of this kind have expressed change in *darwins* (Haldane, 1949), a measure of proportional change in a single character. However, since most long term evolutionary change results from the action of natural selection upon the phenotypic variation present within populations, it seemsdesirable in phyletic studies to relate between-population differences to within-group variability. This approach similarly results in a dimensionless measure of change, thereby permitting comparisons between organisms and structures of differing sizes, and has the advantage of being more meaningful in terms of evolutionary processes.

Such a measure may be derived by relating differences between samples within a lineage to the average variation exhibited by the lineage at any one time, as measured by the individual sample variances (Lerman, 1965). These are then pooled to obtain a weighted mean variance for the lineage as a whole, and the resulting standard deviation used as a common unit of evolutionary change within the lineage. This approach will reveal whether characters are evolving slowly or rapidly in terms of their within-group variation, irrespective of the magnitude of the change in 'raw' metrical terms.

The multivariate extension of this is to use some appropriate distance statistic, e.g. D — the square root of the Generalised Distance — to express change in a character complex. D^2 relates between-group differences in several characters to a common within-group dispersion matrix, after rejecting redundant information due to character inter-correlation. In the examples given below, because of the very restricted samples of Tertiary/early Quaternary Hominidae available for analysis, the within-group matrix is based upon the variation exhibited by late Pleistocene and recent human samples. The distances were obtained by a Q-technique (Gower, 1966) which indicates which of the original characters most influence the separation; further information and examples are given in Bilsborough, 1971, 1973, 1976).

When expressed on a chronological scale both univariate and multivariate statistics may be read as a measure of the rate of evolution of a character or structure, and change in differing structures compared by means of the corresponding D values. The technique allows for intercorrelation of characters within each complex but not between complexes, and since the various regions form part of an integrated structure, the skull, they will inevitably be correlated and the D distances will reflect this.

Correspondence in D values may denote any of several causal relationships: approximately equal distances may reflect the interdependence of the complexes which are, in effect, part of a larger functional unit (e.g. upper and lower jaws or those aspects of facial morphology which are mechanical concomitants of increased jaw size). Alternatively, equivalent changes in two complexes may result if both are determined in similar manner by changes in a third component e.g. overall size; certain aspects of pongid neurocranial anatomy seem to be determined in this manner. Finally, changes in one region may correlate with changes in another because both are determined by the pleiotropic effects of common genes. These may be selected because of the functional importance of only one of the complexes, the other having little or no adaptive significance, although such an explanation is unlikely to obtain if functional criteria have been used in delimiting the complexes.

   The fact that groups vary in their separation on the basis of different complexes is information of obvious biological importance, since it shows which complexes are under strong directional selection and which are not. This should prove especially valuable when one compares successive populations within a single lineage, since it provides evidence from which, in conjunction with palaeoecological (and, for hominids, archaeological) data, inferences may be drawn concerning the selective factors operating on past populations. Differential separation is also of importance in the investigation of adaptive differences in contemporary groups: if otherwise similar forms differ markedly in one or two complexes, such complexes are likely to reflect directly ecological contrasts between the groups e.g. the otherwise cranially similar chimpanzee and gorilla show marked dissimilarity in gnathic, especially mandibular, characters that appear to be directly related to the dietary differences between these species.

   The remainder of this article summarises aspects of mosaic evolution in hominid populations providing examples of the evolutionary phenomena noted above. These are: morphological differentiation among australopithecines (species diversity at a single time horizon) and cranial evolution within Middle Pleistocene *Homo erectus* (phyletic change within a chronospecies).

CRANIAL DIFFERENTIATION WITHIN *AUSTRALOPITHECUS*

   Figures 2 and 3 summarise data on cranial differences for the complexes shown in Fig. 1 between robust australopithecines (East African *A. boisei* and South African *A. robustus*) and *A. africanus* (South Africa); a sample of modern *H.s. sapiens* is also included as a reference standard.

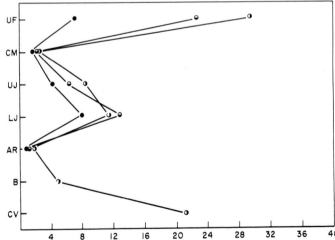

Figure 2. Patterns of morphological mosaicism in Plio-Pleistocene hominids. Vertical axis represents individual cranial complexes. UF = upper face; CM = cheek region and masticatory musculature; UJ = upper jaw; LJ = lower jaw; AR = articular region; B = balance; CV = cranial vault.
   ●——● *East African robust australopithecines – South African robust australopithecines.*
   ○——○ *A. africanus – South African robust australopithecines.*
   ◑——◑ *A. africanus – East African robust australopithecines.*
Horizontal axis in distance units. For further explanation see te:

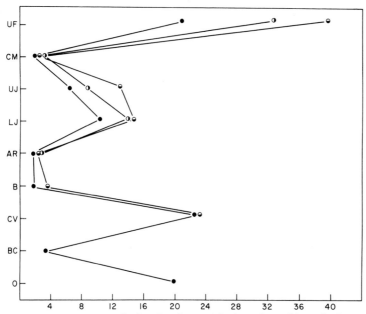

Figure 3.   *Patterns of morphological mosaicism in Plio-Pleistocene hominids.*
Vertical axis represents individual cranial complexes.   UF =
upper face;   CM = cheek region and masticatory musculature;
UJ = upper jaw;   LJ = lower jaw;   AR = articular region;
B = balance;   CV = cranial vault;   BC = basicranium;   O = over-
all.
●———●*A. africanus - modern H. sapiens.*
◐———◐*South African robust australopithecines - modern H. sapiens.*
◉———◉*East African robust australopithecines - modern H. sapiens.*

As expected the two robust samples lie closer to one another than they do
to the other groups; their separation is most marked in face and jaws, reflecting the considerable size variation in these regions. The cheek and
articular complexes — both morphologically conservative within Hominidae —
show much less separation. Both robust groups are well separated from *A.
africanus*, particularly in the upper face, jaws and cranial vault, partly
due to size differences, but principally a function of the distinctive
splanchnocranial morphology of the robust forms. Separation is most marked
in the upper face reflecting fundamental differences in facial architecture
between the groups, largely a consequence of the specialised nature of this
region in robust australopithecines. This is shown by the handle-bar shaped
supra-orbital torus with prominent glabellar region shelving into a triangular
planum and tapering laterally, the size and proportions of the orbits and the
dish-shaped profile with little or no prognathism in the nasal and mid-facial
regions. The gnathic region is similarly characteristic: reduced anterior
dentition and massive cheek teeth, minimal alveolar prognathism, a mandible
with a deep broad corpus, strong symphysial buttressing, high ramus and a
shallow sigmoid notch. Contrasts in the jaws and dentition between gracile
and robust australopithecines are reflected in the morphology of the cheek
region and proportions of the masticatory musculature where the separation,
while less striking than that of face and jaws, is still almost as great as
that between *A. africanus* and modern man. Contrasts in articular dimensions

are less marked, but still impressive in view of the relatively invariant nature of this region within the Hominidae; separation is more marked in the balancing complex reflecting both overall size differences, and the differing proportions of splanchnocranium and neurocranium in the two taxa.

The analysis of cranial vault proportions reveals striking differences between the australopithecine species; the robust frontal region is lower and generally less developed, with a marked post-orbital constriction, whilst the biparietal arch is longer, lower and wider, and the occipital not as curved as in *A. africanus*. In addition, the relationship between neurocranium and facial skeleton differs in the two forms, the robust braincase being hafted to the face at a much lower level than in *A. africanus* (Tobias, 1967). This may be partially due to the overall size difference between the forms — a similar phenomenon is seen in *Pan* and *Gorilla* — but it results also from contrasts in the cranial proportions of the two species which do not appear to be simply size dependent.

The essential similarity of the two robust samples in cranio-facial proportions, and their distinctive morphology compared with other hominids, are seen by their separation from *A. africanus* and *H.s. sapiens*, where they show the same relative separation from the two reference groups. By contrast, *A. africanus* is much closer to *H.s. sapiens*, and to other hominid samples generally (not included here), than is either robust group. This proximity compared with the robust forms is predictably more marked in facial and gnathic complexes than in the neurocranium, and reflects the generally smaller, less specialised dental proportions and weaker jaws of *A. africanus*, together with a basic similarity in facial architecture to all members of the genus *Homo*. Basicranial differences are few, due to the homogeneity of this region in all Hominidae, but neurocranial contrasts are more evident in the vault due to the small size of this region in *A. africanus*, especially the unexpanded frontal and parietal regions.

Discussion

The marked contrasts revealed by discriminatory analysis of the australopithecine skull provide a quantitative assessment of the differences noted by other workers who have examined the material (e.g. Robinson, 1954, 1968, 1972; Tobias, 1967). Moreover, the picture obtained by analysing — both separately and conjointly — the morphological pattern of the various complexes within the *A. robustus* skull is of a hominid showing considerable functional specialisation of the splanchnocranium which has also affected neurocranial proportions. This is most plausibly explained as adaptation to exploit a dietary niche significantly different from that of other Plio-Pleistocene Hominidae, although its precise nature is controversial (e.g. Robinson, 1963, 1972; Tobias, 1967; Jolly, 1970). The multivariate pattern revealed by *A. africanus* is similarly consonant with views on its phyletic status — a small-sized, small-brained hominid, whose facial, dental and gnathic regions are sufficiently like those of Middle and Upper Pleistocene Hominidae to make it the possible ancestor of the genus *Homo*.

Whatever their precise adaptive significance may be, the morphological differences between robust and gracile australopithecines appear sufficiently distinctive to indicate genetic discontinuity. Having established this, the taxonomic level at which such differences are indicated remains largely a matter of individual taste. Robinson (1968, 1972), on the basis of the inferred adaptive significance of these and other morphological features, advocated generic division of the group, retaining *Paranthropus* for the robust australopithecines, whilst sinking *Australopithecus (sensu stricto)* within *Homo*. It might be thought that support for generic distinction (although not necessarily for inclusion of *A. africanus* within *Homo*) is provided by the wide separation of the australopithecine groups in multivariate space, particularly in the analyses of facial and gnathic complexes. This is not necessarily the case. Quite apart from the consideration that taxonomic categories may be required to reflect desiderata other than morphological affinity, the present study is a phenetic one, and in the making of phyletic judgements it is necessary to weight the characters, since some contrasts include a size-dependent component. When this is done the differences are somewhat less obtrusive, whilst the similarities between the taxa in certain cranial and postcranial features suggest close phyletic relationship, and separation at the specific level appears most appropriate (Clark, 1964; Mayr, 1963; Tobias, 1967).

Equally striking is the essential similarity of morphological pattern revealed by the multivariate analyses of the robust samples. This suggests that schemes distinguishing two robust species *A. robustus* (South Africa) and *A. boisei* (East Africa) may require revision, and that the two groups are perhaps most appropriately accommodated within a single polytypic species. Many of the characters cited as differentiating *A. boisei* from *A. robustus* (Tobias, 1967) are size-dependent differences and the formal definitions do not encompass any spectrum of individual and/or polytypic variability, whilst some of the morphological criteria (e.g. the proportions of the face) used to distinguish the two groups appear to have been eroded by recent discoveries. In the preamble to the formal definition of *A. boisei*, Tobias (1967, p. 233) is clearly aware that the taxon may require revision; no doubt detailed description of the East Rudolf australopithecines will clarify this point.

Attempts to reconstruct the factors determining australopithecine differentiation include the 'dietary hypothesis' (Robinson, 1963, 1968, 1972) and the 'seed-eating hypothesis' (Jolly, 1970). Jolly's model, in particular, is more plausible than many alternative theories which posit the early development of tools/weapons and hunting propensities among early Hominidae, but both theories appear to involve difficulties. In view of the wide dietary flexibility of extant hominoids (and indeed of most terrestrial primates), it is perhaps unlikely that hominid craniodental features were adaptations to one food-stuff, be it mammalian flesh or cereal grains.

Moreover, both Jolly and Robinson agree (although for different reasons) in viewing the robust morphology as generally characteristic of basal Hominidae, albeit in Jolly's case with the superimposition of long term seed-eating adaptations. This does not accord with the available chronological data, whilst Simons (1964, 1968) has emphasised that *Ramapithecus* is almost indistinguishable in dental and gnathic features from *A. africanus* (not *robustus*), and is retained as a separate genus only because of its temporal and spatial separation from the South African fossils.

On the other hand, Tobias (1967) has argued that *A. robustus* represents a specialised lineage which evolved from early or middle Pliocene hominid populations whose resemblances were to *A. africanus* rather than to the known robust australopithecines. This view accords with current evidence of the chronology of australopithecine populations and is morphologically economical; many of the contrasts between *A. africanus* and *A. robustus* result from the greater body size of the latter, whilst others reflect the specialisations of the dentition, face and jaws which also affect neurocranial shape.

The results and conclusions summarised above accord with those expressed

by most workers on the skull and dentition of *Australopithecus (sensu lato)*
and on the postcranial skeleton (e.g. Napier, 1964; Robinson, 1972; McHenry,
1975); they are however in marked contrast to those obtained from a recent
multivariate study of certain postcranial elements (Oxnard, 1975). Oxnard
has studied the scapula, clavicle, innominate and talus, and considered in
less detail phalanges, humeri and femora. He considers that their morphologies suggest locomotor activity distinct from that of modern man, and also
unlike those of other living hominoids. On the basis of such a unique locomotor mode and the absence of unequivocal evidence of striding bipedalism, he
concludes that 'it is rather unlikely that any of the australopithecines ...
can have any direct phylogenetic link with the genus *Homo*' (Oxnard, 1975, p.
122). However, the hominid affinities of the australopithecines are based
principally upon the cranial and dental evidence, not upon the postcranial
skeleton, and it is not to be expected that Villafranchian Hominidae would
necessarily possess a locomotor skeleton like that of the extant representatives of the family. Moreover, Oxnard's decision not to distinguish between
australopithecine taxa in his analysis but to lump both gracile and robust
into a single sample obscures the potential discrimination possible, and
undermines the phyletic inferences based upon his study.

EVOLUTIONARY CHANGE WITHIN *H. ERECTUS*

Figures 4-7 summarise dental and cranial changes in the Middle Pleistocene
chronospecies *A. erectus*, and compare them with *A. africanus*, early (Holstein)
H. sapiens (represented by the Steinheim and Swanscombe crania) and a sample
of modern *H. sapiens*. The *H. erectus* material is divided into two groups -
an earlier sample based upon specimens from Java, Olduvai and Lantian, assigned
a mean age of 1,000,000 years, and a later population from Choukoutien dated
at 300,000 years. Dental dimensions (mesio-distal and bucco-lingual diameters)
for each cheek tooth are expressed in mean standard deviation units for the
lineage overall, so as to obtain a measure of average within-group variation
to which between-group changes may be related. These data are discussed in
detail elsewhere (Bilsborough, 1976), and only salient features will be summarised here.

The cheek tooth data reveal the markedly mosaic nature of hominid dental
reduction. Premolar dimensions are relatively stable throughout the earlier
Quaternary, undergoing substantial and rapid reduction during the later Middle/
Upper Pleistocene. The molar teeth undergo reduction much earlier, and there
is significant diminution between *A. africanus* and early *H. erectus*, particularly in view of the increased body size between these forms. Substantial
molar reduction continues throughout the Middle Pleistocene, so that later
H. erectus populations have molar dimensions well within the range of some
modern populations, and post-*erectus* changes are relatively slight. Moreover,
within the molar teeth a definite gradient is discernible, in accord with
Butler's Field theory (1941): the posterior molars undergo an earlier, more
rapid and more sustained reduction than those further forward which are posited
to be closer to the centre of the molarisation field, so that M_3, which is already much smaller in early *H. erectus* than in *A. africanus*, continues to reduce during the Middle Pleistocene, the rate of diminution trailing off in the
later Quaternary. First molar reduction is initiated later and occurs more
slowly; there is little change between *A. africanus* and early *H. erectus*, a
distinct reduction between the two *H. erectus* samples, and a more rapid diminution during the Upper Pleistocene.

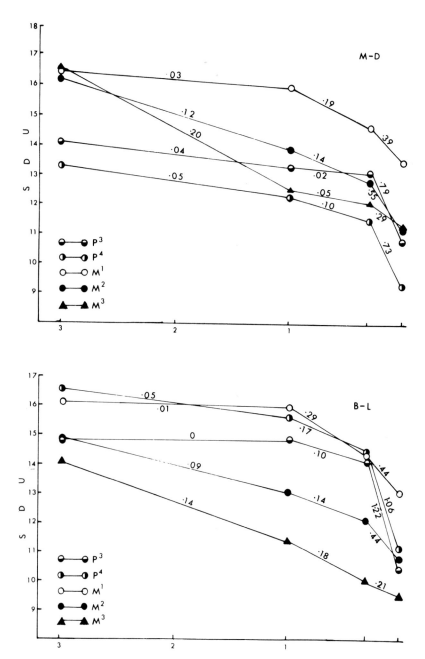

Figure 4. Standardised dimensions and rates of change in mesio-distal
(M-D) and bucco-lingual (B-L) diameters of hominid maxillary
teeth. Vertical scale represents mean tooth dimensions expressed
in average standard deviation units (SDU) calculated over Plio-
Pleistocene Hominidae (excluding A. robustus). Horizontal axis
represents time. The four hominid groups represented are (from
left to right): A. africanus (3 million years); early H. erectus (1 million years); late H. erectus (3 million years);
modern H. sapiens. Figures represent rates of change in mean
standard deviation units per hundred thousand years. See text.

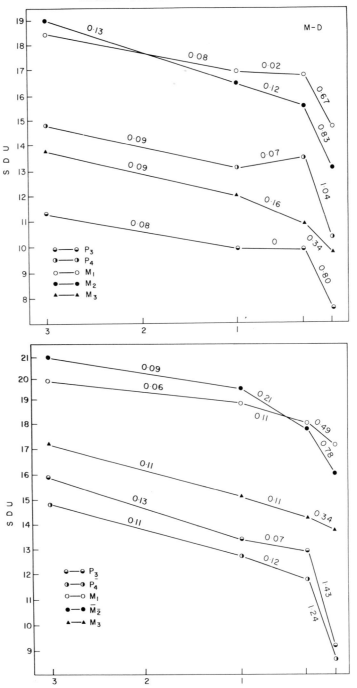

Figure 5. *Standardised dimensions and rates of change in mesio-distal (M-D) and bucco-lingual (B-L) diameters of hominid mandibular teeth. Symbols as for Fig. 4.*

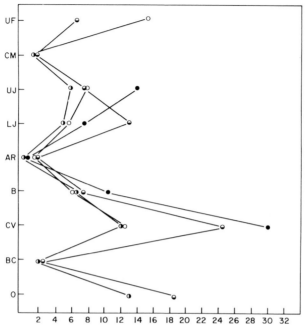

Figure 6. Patterns of mosaic evolution in cranial complexes of Plio-Pleistocene Hominidae. For abbreviations see Fig. 3.

●———● *A. africanus – early H. erectus*
○———○ *early H. erectus – late H. erectus*
◕———◕ *A. africanus – late H. erectus*
◑———◑ *late H. erectus – modern H. sapiens*

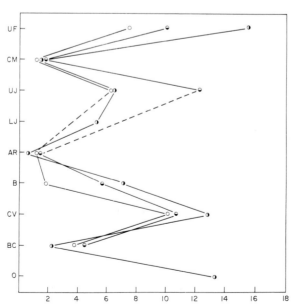

Figure 7. Patterns of mosaic evolution in cranial complexes of Middle and Upper Pleistocene Hominidae. For abbreviations see Fig. 3.

○———○ *late H. erectus – early (Holstein) H. sapiens*
◑———◑ *late H. erectus – modern H. sapiens*
◕———◕ *early (Holstein) H. sapiens – modern H. sapiens*

These dental changes have influenced cranial, especially gnathic, morphology in *H. erectus*. Despite increased body size the jaws of early *H. erectus* are already smaller than those of *A. africanus*, and diminution continues throughout the Middle Pleistocene, so that late *H. erectus* specimens have substantially reduced maxillary and mandibular regions. Other complexes associated with the masticatory apparatus also diminish throughout the chronospecies: the zygomatic and masseteric regions are reduced, whilst the relatively invariant articular region undergoes major contraction during the period. Reduction of the jaws shortens the anterior cranial segment (the load arm of the head lever), whilst concurrent neurocranial expansion is extending the power arm, with a consequent reduction of the nuchal musculature in later *H. erectus* specimens. Within the taxon there are substantial changes in neurocranial dimensions; overall capacity is increased, and later *H. erectus* specimens have a higher more expanded frontal, the parietal arcs are longer and broader, whilst the occipital reveals the 'rolling up' (kyphosis) of the braincase.

Analysis of the cranial complexes thus reveals evidence of substantial morphological change within the taxon *H. erectus*, presumably the result of directional selection pressures. These changes are of such a magnitude that in my opinion the usual distinction into subspecies (*H. erectus erectus* and *H. erectus pekinensis*) is insufficient to reflect the fundamental morphological contrasts evident within the chronospecies. This point is underlined by comparing the separation of the two *H. erectus* groups relative to one another and to early *H. sapiens*. In every case the Choukoutien material is closer to early *H. sapiens* than it is to the early *H. erectus* sample, and in some complexes (maxillary, zygomatic, articular and basicranial regions) the late *H. erectus* group is actually closer to modern man than is the early *H. sapiens* material. It is tempting to conclude that, but for the relatively trivial characteristic of neurocranial shape, the Choukoutien material *is* sapient, and that it is only the phyletic weighting traditionally accorded to braincase characteristics which determines the current position of the taxonomic interface between *H. erectus* and *H. sapiens*. In this situation the conventional hierarchic, Linnean system of nomenclature obscures rather than illuminates the patterns of evolutionary change.

Figure 8 reveals further the contrasting pattern of human evolution by comparing the individual segments of Plio-Pleistocene hominid evolution, and expressing change in each complex relative to change in the morphologically most conservative one, the articular region. I have selected articular characteristics rather than overall skull dimensions as the reference unit, so as to include the relatively incomplete early *H. erectus* material in the comparisons.

It is clear that during the earliest segment (*A. africanus* - early *H. erectus*) neurocranial evolution is dominating the separation, and cranial vault changes are much greater than those within any other complex. However, changes in gnathic dimensions and the balancing complex, whilst less than in the braincase, are still substantial, and all regions are evolving rapidly relative to articular dimensions. This pattern continues throughout the Middle Pleistocene (early-late *H. erectus*) although the relative magnitude of the changes in each complex is less than in the previous segment owing to the major changes in articular dimensions which occur during this period.

Consideration of evolutionary changes for the Lower/Middle Pleistocene as a whole (*A. africanus* - late *H. erectus*) provides more complete data which confirm the above; it is clear that neurocranial characters were evolving much more rapidly than those of other complexes during this period, and their separation exceeds by a substantial margin that of skull dimensions overall.

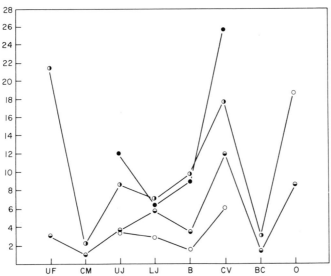

*Figure 8. Patterns of mosaic evolution in cranial complexes of Plio-
Pleistocene Hominidae. Change in each complex expressed in
terms of change in articular dimensions, the morphologically
most conservative complex. For further explanation see text.*

●——● *A. africanus – early H. erectus*
○——○ *early H. erectus – late H. erectus*
◕——◕ *A. africanus – late H. erectus*
◑——◑ *late H. erectus – modern H. sapiens*

However, the later Middle/Upper Pleistocene segment (late *H. erectus*/
modern *H. sapiens*) reveals a sharply contrasting pattern: neurocranial
changes, whilst marked, are less than those of overall skull dimensions and
no longer dominate the analysis. Of the other cranial regions, separation in
mandibular parameters is lower than previously, that in maxillary dimensions
little changed, but that of the face increases spectacularly to exceed overall skull dimensions and is paralleled by co-respondingly obtrusive changes
in the balancing complex. Clearly neurocranial evolution predominated during
the earlier phases of hominid phylogeny and were largely accomplished by the
end of the Middle Pleistocene, whereas facial and gnathic changes, although
initiated during the Villafranchian, did not gain momentum until the Middle
Pleistocene and were completed only during the later Quaternary.

Discussion

The possible selective factors underlying these morphological changes are
discussed in detail elsewhere (Bilsborough, 1976). In summary they probably
reflect, at least in part, responses to a shift in the dietary niche of hominids during the late Pliocene/Lower Pleistocene and the selection pressures
associated therewith. Such a transition, although rapid in evolutionary
terms, would be a long, drawn-out process in relation to the time scale appropriate to hominid studies, occupying much of the Villafranchian. It is to be
expected that behavioural adaptation, facilitated by the gradual development

of material culture, would precede morphological response. It is also likely
that hominid communities exhibited considerable diversity in their subsistence
patterns and technological capabilities during this period, but with the overwhelming proportion of the diet obtained from vegetable sources. In this respect I would regard Robinson's (1968, 1972) reconstruction of *A. africanus* as
an omnivorous hominid, with a morphology already determined by the selective
requirements of a predatory niche, as unrealistic. 'Forager-scavenger'
would probably describe the food-getting activities of these hominids better
than 'hunter-gatherer'.

The shift, although gradual, was none the less real, and it is
evident from East African sites such as Olduvai and Rudolf (Turkana) that
the hominid communities there, even allowing for the dietary bias inherent in
the archaeological record, were effectively omnivorous in a manner which cannot be demonstrated for australopithecine populations. Behavioural patterns
during this phase, as revealed by the archaeological evidence, are discussed
by Isaac (1976). Such adaptive shifts often correspond to phases of extremely
rapid morphological change (Simpson, 1953); the diversity apparent in the early
East African hominids may well reflect this phenomenon.

Once entered, consolidation and radiation within the new adaptive zone was
relatively rapid, and is reflected in the extended geographical range of
hominids and their increased numbers, as well as by the tempo of morphological change. It surely cannot be fortuitous that the Middle Pleistocene provides evidence of the expansion of hominid occupation sites and that towards
the end of that period detailed and specifically human behaviour patterns are
recorded from sites such as Torralba-Ambrona (Howell, 1966), Vértesszöllös
(Kretzoi and Vértes, 1965) and Choukoutien, all well into higher latitudes.
Such geographical expansion would itself increase diversity by acting as a
stimulus to polytypism which, through gene flow, would have an accelerating
effect upon phyletic change, since novel advantageous characters, arising in
one population, could be rapidly disseminated throughout the entire species
range. The variation in cranial shape in later Middle Pleistocene populations discussed above, and reflected in the conventional *H. erectus/H. sapiens*
boundary, is perhaps an example of this, where an initially polytypic morphological variant became established across the whole species.

This interpretation of late Tertiary/Quaternary hominid adaptation and
evolution may be expressed in terms of Simpson's (1953) general model relating
evolutionary rates to adaptive thresholds. *Australopithecus africanus* would
provide an example of a species occupying a *relatively* stable adaptive zone,
with perhaps some gracile populations entering a prospective adaptive phase.
It is, however, unlikely that *Australopithecus* could be categorized as
bradytelic (slowly evolving) as defined by Simpson; more probably it represents a *horotelic* lineage (exhibiting somewhat more rapid, or average, evolutionary tempo). The East African hominids from Rudolf, Olduvai, and other
sites between 1.5 and 2.5 million years ago document the adaptive shift,
necessarily of short duration, and represent very rapidly evolving (*tachytelic*)
populations. Middle Pleistocene *Homo erectus* populations would then document
the post-adaptive phase, characterised by accelerating evolutionary change
towards the upper limits of horotely.

Detailed analysis of the East African skeletal material representing local
populations from well dated time horizons, provides the most likely possibility
of establishing the validity or otherwise of this interpretation of hominid
evolution.

CONCLUSIONS

The above examples illustrate the markedly mosaic nature of hominid cranial evolution, and the extent to which different regions of the skull vary in the degree and tempo of their morphological change. This would not be especially remarkable except that traditional phyletic analysis weights a small selection of characters heavily at the expense of others, and tends to emphasise differences, a tendency reinforced by the conventional Linnean nomenclature. Analyses such as those above are phenetic, indicating similarities as well as differences and thus provide additional information.

I do not mean to imply thereby that they are superior to, or should replace the usual morphological analyses and comparisons. The two approaches are complementary, being designed to provide different kinds of information. As noted above, taxonomic categories are not necessarily required to reflect overall similarity, but they may well obscure certain patterns of resemblance. The use of biometrical techniques incorporating some measure of overall resemblance may therefore yield useful additional information (particularly in the case of a lineage segmented into chronospecies), which helps illuminate evolutionary processes in the Hominidae, and the selective factors determining them.

REFERENCES

Bilsborough, A. (1969). *Nature, Lond.,* 223, 146-149.
Bilsborough, A. (1971). *Man,* 6, 473-485.
Bilsborough, A. (1973). *J. hum. Evol.,* 2, 387-404.
Bilsborough, A. (1976). *J. hum. Evol.,* 5, 423-439.
Boné,E. (1962). *In* "Bibliotheca Primatologica", (H. Hofer, A. Schultz and D. Stark, eds), Karger, Basle and New York.
Butler, P.M. (1941). *Am. J. Sci.,* 239, 421-450.
Cain, A.J. and Harrison, G.A. (1960). *Proc. zool. Soc. Lond.,* 135, 1-31.
Campbell, B.G. (1963). *In* "Classification and Human Evolution", (S.L. Washburn, ed.), pp. 50-74, (Viking Ed. Publ. Anthrop. 37), Aldine, Chicago.
Clark, W.E. Le Gros (1962). "The Antecedents of Man", 2nd ed., Edinburgh University Press, Edinburgh.
Clark, W.E. Le Gros (1962). "The Fossil Evidence for Human Evolution", 2nd ed. Chicago University Press, Chicago.
Gower, J.C. (1966). *Biometrika,* 53, 588-590.
Haldane, J.B.S. (1949). *Evolution,* Lancaster, Pa., 3, 51-61.
Harrison, G.A. and Weiner, J.S. (1963). *In* "Classification and Human Evolution", (S.L. Washburn, ed.), pp. 74-84, Aldine, Chicago.
Hennig, W. (1966). "Phylogenetic Systematics", (D.W. Davis and R. Zanger, transl.), University of Illinois Press, Urbana.
Holloway, R.L. (1973). *J. hum. Evol.,* 2, 449-459.
Howell, F.C. (1966). *In* "Recent Studies in Palaeoanthropology", (J.D. Clark and F.C. Howell, eds), pp. 83-201, *Am. Anthrop. Special Publ.*
Isaac, G.L. (1976). *In* "Human Origins: Louis Leakey and the East African Evidence", (G.L. Isaac and E.R. McCown, eds), pp. 483-514, Benjamin, Menlo Park.
Jolly, C.J. (1970). *Man,* 5, 5-26.
Kretzoi, M. and Vertes, L. (1965). *Curr. Anthrop.,* 6, 74-87.
Leakey, R.E.F. (1974). *Nature, Lond.,* 248, 653-656.
Lerman, A. (1965). *Evolution,* Lancaster, Pa., 19, 16-25.

McHenry, H. (1975). *Science, N.Y.,* 190, 425-431.

Mayr, E. (1963) *In* "Classification and Human Evolution", (S.L. Washburd, ed.), pp. 332-346, Aldine, Chicago.

Napier, J. (1964). *Arch. Biol. Liège* (suppl.), 75, 673-708.

Oxnard, C. (1975). "Uniqueness and Diversity in Human Evolution: Morphometric Studies of Australopithecines", University of Chicago Press, Chicago and London.

Robinson, J.T. (1954). *Am. J. phys. Anthrop.,* 13, 429-446.

Robinson, J.T. (1963). *In* "African Ecology and the Human Evolution", (F.C. Howell and F. Bourlière, eds), pp. 385-416, Aldine, Chicago.

Robinson, J.T. (1968). *In* "Evolution and Hominisation", 2nd ed., (G. Kurth, ed.), pp. 150-175, Fischer, Stuttgart.

Robinson, J.T. (1972). "Early Hominid Posture and Locomotion", University of Chicago Press, Chicago.

Simons, E.L. (1964). *Proc. nat. Acad. Sci. USA,* 51, 528-535.

Simons, E.L. (1968). *S. Afr. J. Sci.,* 64, 92-112.

Simpson, G.G. (1953). "The Major Features of Evolution", Columbia University Press, New York.

Tobias, P.V. (1967). "The Cranium and maxillary dentition of *Australopithecus (Zinjanthropus) boisei,* Olduvai Gorge", Vol. 2, (L.S.B. Leakey, ed.), Cambridge University Press, Cambridge.

NOTE ADDED IN PROOF

The age of 300,000 years assigned here to the Choukoutien *H. erectus* sample was based upon the work of Huang (1960), Kurten and Vasari (1961), and Kahlke and Chow (1961) which suggested a date equivalent to an early phase of the Holstein interglacial complex of Europe. Recent evidence, in particular the indication of a palaeomagnetic reversal in the Choukoutien stratigraphic column (Woo, pers. comm. 1977) suggests that this age may be too young. If so, the evolution rates given for dental dimensions in Figs. 4 and 5 will need to be revised; those in the segment early-late *erectus* will be too low, those in the segment late *erectus*-modern *sapiens* too high. However the patterns of change of *one structure relative to another,* which is the essence of mosaic evolution, will be unaltered by any revised age determination.

ADDITIONAL REFERENCES

Huang, W.P. (1960). *Vertebr. palasiat.,* 1, 45-48.

Kahlke, H.D. and Chow, B.S. (1961). *Vertebr. palasiat.,* 3, 234-240.

Kurten, B. and Vasari, Y. (1960). *Comment. Biol. Helsingf.,* 23, 1-10.

CLASSIFICATION AND PHYLOGENY OF EAST AFRICAN HOMINIDS

B.A. WOOD

Department of Anatomy, The Middlesex Hospital Medical School, Cleveland Street, London W1P 6DB, UK.

INTRODUCTION

The last decade has seen an unprecedented increase in the available sample of fossil hominid remains. Invaluable progress is being made at the known cave sites in South Africa (Brain, 1970, 1973; Tobias and Hughes, 1969; Tobias, 1975); not only are new hominid fossils being discovered but our understanding of the stratigraphy and taphonomy of these cave deposits is also increasing (Brain, in press). Nonetheless, the most dramatic advances have been made in East Africa. Sites that ten years ago were only partly known or even totally unexplored have yielded remarkable collections of fossil hominids. The Omo (Howell, 1976, for a review), Koobi Fora (formerly East Rudolf) (Leakey, R.E. and Isaac, 1976), Hadar (Johanson and Taieb, 1976) and Laetolil (Leakey, M.D. et al., 1976) have contributed approximately five hundred hominid specimens some of which are partial skeletons and fine, nearly complete skulls. It has been a formidable task merely to keep pace with the discoveries let alone to have considered how this new evidence affects hypotheses about the systematics of the Pliocene and Lower Pleistocene stages of hominid evolution.

During the same decade there has been an increasing awareness of the need to integrate anatomical, archaeological and palaeoecological data in order to investigate the behaviour of early hominids. Behaviour here includes locomotion, diet, habitat preference, technological ability and all the other conceptual subdivisions that scientists use to study early hominid adaptation. It must be recognized that these investigations are based on inference, on data that are one stage removed from the morphology of the fossils which is the primary information we have available. These inferential data are properly included in a broad biological study of hominids but they should be excluded from discussions about classification and phylogeny.

CLASSIFICATION

A. *Theoretical issues*

Classification starts with the process of the ordering of animals into groups (Simpson, 1961). In this particular case the task is to decide whether the patterns of variation are sufficiently well defined within the samples of fossil hominids for them to have been drawn from distinct populations. Time is not relevant to this decision, the samples may just as well be synchronic or allochronic. The level of distinction is arbitrary, but the aim is to recognize distinctions which are equivalent to species differences or above. In

practice, the problem of what constitutes a significant distinction is commonly a vexing issue in hominid and other palaeontology. Sexual dimorphism and
allometric relationships are well recognized as sources of intra-group variation and the problem of how to identify and then remove their effects to
focus on inter-group differences is an active field of study (Mosimann, 1970;
Corruccini, 1972; Gould, 1975; Wood, 1976a and b).

     As if the task of classification was not already difficult enough, there is
the additional problem of how much assumed or actual phylogenetic information
should be used when deciding about the groupings. Should one draw cladograms,
phenograms or dendrograms? At the risk of misinterpreting their positions
Hennig (1966, 1975) and Mayr (1974) best represent this conflict of opinion.
Hennig would say that only by previously establishing groups can one "determine
the chronological sequence of speciation events in lineages of phylogenetic
descent" (Hennig, 1975, p. 244). Mayr and others contend that a satisfactory
classification is not possible without injecting into it what is known about
the relevant evolutionary history. Those who hold these viewpoints have been
unhappily labelled the 'cladists' and the 'evolutionary taxonomists' respectively. The debate is a continuing one and is often conducted with near patri-
otic fervour (Nelson, 1974; Rosen, 1974).

     This author's view is that classification should be as objective as pos-
sible and though not all the criteria can be expressed metrically it is in many
ways a statistical problem (Walker, 1976). The subsequent and separate process
of linking the groups is the opportunity to give some evolutionary polarity
to a classification scheme and is the time when the origin of the characters
shared between the groups becomes important. Preliminary attempts to look at
hominid phylogeny in a more rigid cladistic fashion have been made (Eldredge
and Tattersall, 1975; Szalay, 1975). The significance of whether a character
shared by two groups is exclusive to them or is something they and many other
groups have inherited from a common ancestor is discussed in general by Hennig
(1966) and in relation to primate systematics in particular by Martin (1968).
Unchanged characters in groups which are monophyletic are called *plesiomorphous*
characters developed subsequently and restricted to a subset of the groups are
called *apomorphous*. For any subdivision of the fossil groups to be 'real' it
must be based on synapomorphy; the sharing of characters that are common to
the whole group is no indication of any closer relationship than that which
binds all group members. The terminology is admittedly cumbersome but to substitute terms such as 'derived' to describe characters invites misunderstand-
ing. Decisions about whether characters are apomorphous are subjective and
often difficult, but if they are not made and a purely phenetic approach is
adopted the value of the resulting scheme is greatly reduced. A list of apomorphous features is equivalent to a differential diagnosis of a taxon but
initially, at least, such a list should be regarded as a means of communicating ideas and encouraging discussion rather than constituting a formal and
limiting taxonomic definition (Walker, 1976).

     A further practical problem is how to deal with different anatomical regions
and avoid having the post-cranial skeleton of one type of hominid ultimately
associated in a phylogenetic scheme with the cranial remains of another! This
problem may well be solved by the discovery of many more specimens where limb
bone and cranial remains are associated. A similar problem also applies to
upper and lower jaw remains but, although the usefulness of teeth as a taxonomic indicator has been doubted, the size and proportions of preserved teeth
in jaws do allow dental remains to be matched. Nonetheless it is a sensible
strategy to base any morphological groupings on the more complete specimens.

This review will be restricted to the cranial evidence.

B. *Fossil evidence: Koobi Fora*

1. Cranial remains   TABLE I

Cranial Remains

Koobi Fora	Olduvai
a) KNM-ER 406, 733	a) O.H.5.
b) KNM-ER 1470, 1590, 3732	b) O.H.9
c) KNM-ER 407, 732	c) O.H.7, 13, 16, 24
d) KNM-ER 1813	
e) KNM-ER 1805	
f) KNM-ER 3733	

Up to thirty-two specimens had been recovered. Ten specimens are complete enough to attempt a provisional classification.
 (a) KNM-ER 406, 733. These two specimens, fragmentary though one of them is, share the same distinguishing features. KNM-ER 406 (Leakey, R.E.F., 1970; Leakey, R.E.F. et al., 1971) is a nearly complete adult cranium lacking the roots and crowns of most of the anterior teeth and the crowns of all the posterior teeth. It combines a relatively small neurocranium with extreme pneumatization of the cranial base and massive jaw, facial and ectocranial structures. The glabella is rounded and projects in front of the supraorbital torus. The torus and glabella form the base of a frontal trigone; there are no post-glabellar or post-toral sulci. The piriform aperture lies in a hollow and the nasal bones taper inferiorly. KNM-ER 733 (Leakey, R.E.F., 1971; Leakey, R.E.F. and Walker, 1973) is made up of ten fragments of an adult skull which match closely the corresponding parts of the more complete cranium. In addition the right side of the mandibular body is preserved and it and a maxillary fragment each bear a single large molar tooth.
 (b) KNM-ER 1470, 1590, 3732. KNM-ER 1470 (Leakey, R.E.F., 1973b; Day, et al., 1975) is an adult cranium without tooth crowns which is damaged in the malar regions and at the base. It combines a large neurocranium and relatively large palate and shows less bony buttressing of the upper jaw and none of the excessive ectocranial cresting that are seen in KNM-ER 406. The cranial capacity has been determined by Holloway from a latex mould and is between 770 and 775 cm^3. This compares with a capacity of 510 cm^3 for KNM-ER 406, estimated by the formula of MacKinnon, et al. (1956) working from cranial measurements (Holloway, 1973). The neurocranial shape is more elongated than in the smaller brained skull nonetheless, despite the presence of parietal eminences, the greatest width of the skull is across the supramastoid crests. The glabella lies in a slight hollow between the medial ends of the superciliary parts of the supra-orbital torus. There is no post-glabellar sulcus but there are slight hollows posterior to the middle third of each supraorbital torus. The nasal margins are everted and the nasal bones themselves project forward and widen as they approach the piriform aperture. The malar region is robust but is badly broken; the roots of the zygomatic processes take off at the level

of M^1. The preserved anterior palatal margin is flat and broad.

KNM-ER 1590 (Leakey, R.E.F., 1973b) is an immature specimen consisting of most of two parietal bones and the germs and exposed crowns of eleven upper teeth. The parietals are similar in shape to, but larger than, those of KNM-ER 1470. Both permanent canine germs are preserved and have relatively large cusps and the central incisor crown is also very large.

KNM-ER 3732 (Leakey, R.E.F., 1976a) is a calotte fragment. It also resemble KNM-ER 1470 but it is a little smaller in overall dimensions.

An obvious way to explain the differences between these cranial remains and those in group (a) on the basis of intra-specific variation would be to invoke sexual dimorphism. The problem is to decide which of the two groups are the males. There are no records of primates in which ectocranial structures predominate in females rather than males (Wood, 1976a). Huxley (1932) and Freedman (1957, 1962) showed in baboons, and Wood (1976a) confirmed in a wider range of primates that jaw size is positively allometric when related to the neurocranium so that males have proportionately larger jaws, and therefore jaw muscles, than females. Ectocranial structures point unambiguously to crania like KNM-ER 406 being males yet many measurements are smaller than those in KNM-ER 1470. Cranial capacity relationships in the majority of primate species are such that male average values generally exceed those of the female. The capacity of KNM-ER 406 is 66% of KNM-ER 1470, a most unlikely male/female relationship: in a recent study (Wood, 1976b) the least sexual dimorphism found in cranial capacity was seen in *Pan*, but even so the smallest male value was still 82% of the largest female value. Many external skull dimensions are larger than those in KNM-ER 1470; therefore it is unlikely that sexual dimorphism accounts for the differences between these two groups of crania. Gross differences in shape of the mastoid, frontal and zygomatic regions provide ample evidence for recognising that these two groups of specimens were sampled from populations which were morphologically quite distinct.

(c) KNM-ER 407 and 732. KNM-ER 407 (Leakey, R.E.F., 1970) is an adult calvaria which was extensively damaged, but skilful cleaning and restoration by R.J. Clarke have produced a valuable specimen. KNM-ER 732 (Leakey, R.E.F., 1971; Leakey, R.E.F., et al., 1972) is an adult cranial fragment with parts of the right side of the base, face and vault and part of the left side of the vault preserved. The neurocranium of both specimens is globular and KNM-ER 407 has a tightly curved occipital profile. By reconstructing the region of the posterior cranial fossa Holloway (1973) has estimated the cranial capacity of KNM-ER 732 as 506 cm^3. Neither specimen has a sagittal crest but the temporal lines are well marked. The base of both skulls is well pneumatized and the bulging supramastoid crests are the widest points on the skull in both specimens; the mastoids project further laterally in KNM-ER 407 than in KNM-ER 732. The mandibular fossae are wide and extend laterally well beyond the sides of the cranial vault. The root of the zygomatic process of the temporal bone is broad and flares laterally. The malar region in KNM-ER 732 is broad and slightly hollowed; there is no central facial prominence and the nasal bones are widest superiorly. The zygomatic process of the maxilla arises between P^3 and P^4. There is evidence of considerable post-cranial constriction in both specimens. In KNM-ER 732 the glabella is prominent and with the medial ends of the supra-orbital torus forms the anterior edge of a frontal trigone; there are no signs of a post-toral sulcus. Although both specimens are incomplete, enough in common survives to indicate that they can be considered together. Initial taxonomic attributions suggested otherwise (Leakey, R.E.F., 1970) but later discussions have acknowledged their similarity (Leakey, R.E.F., 1976b).

The smaller size and lack of sagittal cresting of the two specimens formed
the basis for the proposal (Leakey, R.E.F., 1970) that they should be considered female sexual dimorphs of the specimens in group (a). KNM-ER 732
is sufficiently complete on one side to enable an approximate reconstruction
of the whole vault to be made by mirror imaging. Dimensions taken from the
reconstruction, imperfect though they are, and measurement of the cranial
base of KNM-ER 407 show percentage differences from KNM-ER 406 which closely
match average sexual differences in a selection of primates (Wood, 1976b). This
fact, taken together with the similarity of temporal, upper facial and malar
morphology, makes it very likely that groups (a) and (c) should be combined
for taxonomic purposes.

   (d) KNM-ER 1813 (Leakey, R.E.F., 1974). This is a nearly complete adult
cranium. Apart from I^1, good examples of each type of upper tooth crown are
preserved. The neurocranium is rounded and, apart from small nuchal crests and
slight crest formation at the anterior extent of the temporal lines, there are
no other ectocranial features. Preliminary estimates of cranial capacity indicate a value between 500 and 550 cm^3. The widest part of the skull is near
the base but the mastoids themselves are small and end inferiorly in a point.
The zygomatic processes of the temporal bones are flared but the mandibular
fossae lie with approximately half their width within the vertical plane of
the lateral skull wall. The zygomatic process of the maxilla arises at the
level of M^1 and is hollowed anteriorly before it leads on to the projecting
nasal margins. The medial ends of the rounded and arched supra-orbital torus
lie just anterior to glabella and posterior to both torus and glabella is a
marked sulcus which separates them from the sloping frontal squama. The nasal
bones widen inferiorly and the palate has a rounded anterior border.

   The alternatives for any classification of this specimen are that it may
properly belong with the combined groups (a) and (c), or that it should be
included in group (b) or, finally, that it samples a population so far unre-
presented. The anatomy of the mastoid, malar and frontal region is sufficiently
distinct for it to be unlikely that KNM-ER 1813 should be assigned to groups
(a) and (c). When compared to specimens in group (b) there are obvious differences in overall size and not a few shape differences. Cranial capacity
estimates (estimating KNM-ER 1813 as 525 cm^3) show that the smaller skull has
a volume which is 68% of the volume of KNM-ER 1470; this compares with average
sex differences in mean cranial capacity for primates of between 80 and 90%.
Skull dimensions show a similar discrepancy. The shape of the cranial vault
is broadly similar but the smaller skull lacks the parietal eminences seen in
KNM-ER 1470. There are both size and shape differences between the facial
regions of the two skulls. The brow ridges are more prominent and the face
more projecting in KNM-ER 1813, contrasting with the flatter, broader face
and palate of KNM-ER 1470. However they share some features, a relatively
receding glabella, nasal bones which broaden inferiorly and everted nasal
margins.

   The teeth of KNM-ER 1590 allow some dental comparisons. All preserved teeth
of KNM-ER 1813 are smaller than the germs of KNM-ER 1590. Both I^1 and $\underline{C}$ show
size discrepancies of 50% and, while sexual dimorphisms of this order are recorded for primate canines, no known modern primate shows incisor sexual dimorphism greater than 85% (Wood, 1976a). The differences in molar crown area
between the two fossil dentitions are much less: they average out at 70%. The
crown area of the upper canine is not functionally equivalent to molar crown
area but comparison of the two does highlight the extraordinarily large canines
of KNM-ER 1590. In KNM-ER 1813 canine area is 47% of M^1 crown area, in KNM-ER

1590 the figure is 73%. Strong positive allometry in canine growth is the
reason for disproportionately large canines in some primates. The relatively
small body size difference between males and females in *Homo* means that this
positive allometry had no significant effect on the C/M$\underline{^1}$ ratio in males and
females in a collection of *Homo* teeth, but in *Gorilla* the positive allometry
produced a change in C/M$\underline{^1}$ proportion from 84% in the females to 156% in the
males (Wood, 1976b). Data taken from Drennan (1929) show that dentitions as
different in size as those of the Bushmen and Australians do not change their
canine molar tooth proportions. Thus, despite the shape and cusp morphology
of the teeth being similar, and even taking into account the errors inherent
in basing comparisons on single dental arcades, the overall size differences
and the differences in tooth proportion between the dentitions of KNM-ER 1813
and KNM-ER 1590 are profound. If the dentitions were sampled from a single
population, the pattern of dental variability would be unlike that of any
living primate populations that have been studied.

(e) KNM-ER 1805 (Leakey, R.E.F., 1974). This adult skull lacks the supraorbital region and much of the base; the mandible is badly cracked, only the
crowns of the right M_1 and M_2 remain but most of the maxillary teeth are preserved.

The neurocranium is elongated and made up of moderately thick vault bones.
Cranial capacity is estimated to be between 575 and 600 cm^3. There are posteriorly situated parasagittal crests which connect with nuchal crests. The
greatest skull width is across the nuchal crests, the mastoid processes themselves are relatively small. The nasal aperture and the interorbital region
are wide and the whole face short. The zygomatic processes of the maxillae
take off at the level of M$\underline{^1}$ to form what obviously were quite broad malar
regions. The central incisor was obviously a large tooth but the I^2 is small
and peg-like and a small diastema separates it from the worn canine crowns.
Worn as the upper canines are, they nonetheless are relatively large teeth.
While they represent only 50% of the crown area of the M$\underline{^1}$ the canines have an
exposed crown area equal to that of the anterior premolars (a similar relationship is found in the teeth also of KNM-ER 1590). The effects of age make the
mandibular morphology difficult to assess.

Despite the presence of parasagittal crests many other contrasting features
of the skull KNM-ER 1805 make it highly improbable that it should be included
with groups (a) and (c). The wide face and large canine are reminiscent of
group (b). However, in overall size it compares better with KNM-ER 1813 and
as facial width is subject to greater sexual dimorphism than facial length
(Giles, 1970; Wood, 1976b) it is possible, especially as large canines and
cresting are masculine features, that KNM-ER 1805 is a male sexual dimorph of
the population represented by KNM-ER 1813.

(f) KNM-ER 3733 (Leakey, R.E.F. and Walker, 1976). This cranium is complete
except for all the anterior tooth crowns, some posterior teeth and parts of the
malar region. The neurocranium is elongated with an angled occiput. Though
the greatest breadth is low down on the skull the mastoids are not excessively
pneumatized. A prominent supra-orbital torus, a deep sulcus dividing it and a
sharply rising frontal squama and a wide nose with prominent nasal bones characterize the facial skeleton. Only a very preliminary description is available
and it would be premature to make detailed comparisons. Nonetheless this skull
displays features which, though they may have been noted singly in descriptions
of other specimens, have never been associated. Its particular size and shape
suggest that it was sampled from a population not represented in groups (a) to
(e).

2. Mandibular remains   TABLE II

Mandibular Remains

Koobi Fora	Omo
a) KNM-ER 729, 818, 1469, 1806	a) L18-18(1967), L75-14a (1969), L427-7, L222 (1973-2744)
b) KNM-ER 992	
c) KNM-ER 1801, 1802	b) L57-41(1968), L7A-125, L74A-21
d) KNM-ER 1482	c) L860-2

Up to 1976 forty-two mandibles varying from virtually complete specimens to fragments have been recovered from Koobi Fora. It is a feature of this material, particularly the larger fossils, that pre-fossilization expansion and cracking is very common and the crowns of the teeth have often been lost, thereby reducing their potential value as specimens. R.E.F. Leakey (1972) suggested that the sample of mandibles could be subdivided by a combination of a measure of the cross-sectional area of the mandibular corpus and the relative proportions of anterior and posterior tooth crown areas into a group of usually large, robust mandibles on the one hand and smaller mandibles, but with relatively large anterior teeth, on the other. A further analysis (Wood, 1976c) confirmed the separation into two major groups. Subsequent finds have confirmed the basic trends but suggest that more than the two groups originally proposed may be represented in the sample.

(a) KNM-ER 729, 818, 1469, 1806. These specimens show in various ways the typical morphology of the group of large, robust mandibles. The features of this group are a large and robust mandibular body with a rounded base. The lateral swelling which gives rise to the anterior border of the ascending ramus extends anteriorly often to between P_4 and M_1. The ascending ramus is tall and well buttressed. The inner aspect of the symphysis is usually buttressed by two equally massive transverse tori; the outer aspect of the symphysis is usually rounded but vertically orientated. The molar teeth are relatively massive with molarized premolars; the posterior premolar crown is larger than that of the anterior. The anterior teeth are relatively and absolutely reduced in size.

A number of adult mandibles do not conform to the set of preliminary criteria given for group (a). They are KNM-ER 730, 992, 1482, 1501, 1502, 1801 and 1802. These specimens tend to have smaller mandibular bodies and are less robust at the level of M_1 than mandibles in group (a) but there is an overlap between the two groups.

(b) KNM-ER 992 (Leakey, R.E.F., 1972; Leakey, R.E.F. and Wood, 1973) is the most complete specimen in this group. The mandibular body is less massive than those in group (a) and it lies at the lower end of their range of robusticity. The ascending ramus is short and relatively lightly constructed. The canines have a spatulate shaped crown and their crown dimensions enclose an area which is 62% of the M_1 crown area, to give an idea of their relative size. The crowns of the anterior premolars are set obliquely in the tooth row and they are dominated by the buccal cusp. Their dimensions exceed those of the posterior premolars which have subequal lingual and buccal cusps. The teeth are arranged in a parabolic shaped arcade. Three other specimens, an aged adult mandible KNM-ER 730 (Leakey, R.E.F., 1971; Day and Leakey, R.E.F., 1973), an abraded right half of an adult mandibular body KNM-ER 1501, and part of a sub-adult mandible KNM-ER 1502 (Leakey, R.E.F., 1973a; Leakey, R.E.F. and Wood, 1974a) have either small mandibular bodies or small molar teeth, or both - the mean M_1 crown area of this small sample being just over 50% of the M_1 crown area of KNM-ER 729 which is by no means the largest mandible in group (a).

It has been proposed that the differences between mandibles such as KNM-ER 992 and the large robust sample are the result of intra-specific variation of which body size sexual dimorphism is the major component (Brace, et al., 1972). However the pattern of percentage differences between individual teeth in KNM-ER 992 and in dentitions such as KNM-ER 729 makes this most unlikely. The canine dimensions are the same whereas molar crown area of KNM-ER 992 is only 50% of that of the larger specimen. With very few exceptions canine dimensions are the most sexually dimorphic variables in extant primate species (Wood, 1976a) and in no known case does molar size dimorphism exceed it.

(c) KNM-ER 1801 and 1802 (Leakey, R.E.F., 1974) can be conveniently dis-cussed together. KNM-ER 1802 is the more complete and has relatively unworn teeth. The mandibular body is strongly built but has an everted base and its external surface is marked by sulci which are unusual in the robust mandibles of group (a). The symphysis shows none of the excessive buttressing by trans-verse tori seen in group (a) mandibles. The anterior premolar crowns lie obliquely across the tooth row and are asymmetrical with the buccal cusp pre-dominating; the posterior premolar crowns in contrast have subequal main cusps with cusps developed on the distal enamel shelf. The dental arcade is shaped so that the molar rows are nearly parallel. The teeth of KNM-ER 1801 are too worn to show details of cusp morphology, but the root of the anterior pre-molar runs obliquely across the mandibular body, matching the orientation of the crown in KNM-ER 1802. The same combination of an everted base and obliquely orientated premolar roots is seen also in another specimen, KNM-ER 1483. Their overall size and the size and proportions of the teeth suggest that they form a subdivision, morphologically distinct from the specimens which include KNM-Er 992.

(d) Since its initial description mandible KNM-ER 1482 has been singled out as a distinctive specimen (Leakey, R.E.F., 1973a; Leakey, R.E.F. and Wood, 1974b; Groves and Mazák, 1975; Howell and Isaac, 1976). The post-incisive planum, obvious transverse tori and small anterior dentition are features seen in group (a), whereas its relatively small size and obliquely orientated an-terior premolar are distinguishing features, as is the V-shaped dental arcade. It may prove to belong to a separate morphotype, or improved understanding of the variation in the whole mandibular sample may lead it to be included in one of the other categories.

Juvenile mandibles have not been discussed in detail but of two fine speci-mens, one, KNM-ER 1477, can with confidence be included in group (a) whereas KNM-ER 820 most probably belongs with group (b).

3. Teeth

Given the difficulties of sorting crania with nearly complete dentitions and the paucity of any firm criteria for sorting the 'non-robust' group of mandibles it is not surprising that the allocation of isolated teeth is often problemati-cal. Maxillary teeth, particularly molars, on preliminary inspection, show a consistent morphology whatever their overall size or the size of the anterior teeth associated with them. Mandibular teeth offer more hope of providing distinguishing features and the deciduous dentition has been used in the past as a taxonomic indicator (Robinson, 1956). The molarization of the posterior premolar and more symmetrical anterior premolar teeth have already been referred to in connection with the identification of group (a) mandibles. Mandibular molar crown morphology in this group shows a combination of characters which may form the basis of a reliable set of sorting criteria. The molars tend to be low crowned relative to their large size (this may be an allometric phenomenon). They tend to have bulging lingual and buccal surfaces which are continuous with the occlusal surface. This leads to a low cusp profile with the peaks of the cusps near the centre of the tooth. Extra distal cusps are common but additional lingual cusps are rare. Detailed studies of tooth shape, crown profiles and estimates of relative cusp areas may either substantiate or invalidate these tentative proposals. In many cases the sheer size of the teeth is the best indicator of their affinity.

C. *Olduvai*

The catalogue numbers of Olduvai Gorge hominids now go up to O.H.54, and though some new numbers are due to reorganization of the catalogue there has been a steady accumulation of specimens from this important site. Apart from O.H.5 none of the other specimens has been published in detail and discussion of their classification is necessarily limited.

1. Cranial remains (see Table I)

(a) O.H.5 (Leakey, L.S.B., 1959; Tobias, 1967) is a virtually complete sub-adult cranium. Its neurocranium is dwarfed by massive malar and facial buttressing and the small anterior dentition combined with large molars and molarized premolars complete the same suite of characters which distinguished the crania of group (a) at Koobi Fora. An endocranial capacity of 530 cm^3 (Tobias, 1963) confirms the small size of the neurocranium.

(b) O.H.9 (Leakey, L.S.B., 1961) is an adult calvaria. The angled occiput, massive supraorbital torus, recessed glabella, large neurocranium with a cranial capacity estimate of around 1000 cm^3. (Tobias, 1971; Holloway, 1973) all contrast with O.H.5. O.H.12 (Leakey, L.S.B. and Leakey, M.D., 1964) is an incomplete cranium that has been compared morphologically with O.H.9 (Tobias, 1965a; Leakey, M.D., 1971), despite the fact that the endocranial volume estimated from a reconstruction is only 727 cm^3 (Holloway, 1973).

(c) O.H.7, 13, 16, 24. These specimens are skulls or cranial remains varying in their degree of completeness.

The only remains of the cranial vault of O.H.7 are parts of two thin parietal bones (Leakey, L.S.B., 1961). Despite their thinness the parietals are quite large and they exceed in size those in the reconstructed cranium of O.H.13 and the parietals of O.H.24. Some discussion surrounds the acceptability of the partial endocast method of cranial capacity estimation but the value of 687 cm^3 estimated in this way by Tobias (1971) has found acceptance by an acknowledged expert (Holloway, 1973, 1975). O.H.13 includes much of the vault of the skull and a palate containing the roots and crowns of molar and premolar teeth. The skull bones are thin, like those of O.H.7, and they are from a cranium with a rounded occiput. As in O.H.7 there are no signs of ectocranial crests. The mandibular fossae lie beneath the neurocranium and there is little or no pneumatization of the temporal bone. Cranial capacity estimates of 639 cm^3 from the partial endocast method (Tobias, 1971) and

650 cm^3 from a latex mould of a reconstruction (Holloway, 1973) are comparable. The preserved maxillary tooth crowns are relatively small and are approximately 50% of the crown area of equivalent teeth of O.H.5.

O.H.16 (Leakey, L.S.B. and Leakey, M.D., 1964; Leakey, M.D., 1969) is the shattered remains of a skull apparently broken up by cattle. The vault remains have been set into plaster but a reliable reconstruction is difficult to achieve because many of the fragments are isolated. Nonetheless, a rounded skull vault with a slightly angulated occiput, arched supraorbital torus and post-toral sulci are fairly certain features of the original skull form. The endocranial capacity estimate is only as reliable as the reconstruction, but Holloway (1973) accepts Tobias's estimate of 633 cm^3 (1971) as unlikely to be grossly inaccurate. The dental proportions are very different from those of O.H.5 for, despite O.H.1 having molar crown areas only approximately 70% of those of O.H.5, the anterior teeth of O.H.16 are significantly larger.

O.H.24 (Leakey, M.D., 1969) was found crushed and embedded in calcareous matrix but skilful restoration has produced a valuable specimen (Leakey, M.D. et al., 1971). The reconstructed skull has a rounded bulging occiput, supra-orbital torus and post-toral sulci and only moderate post-orbital constriction. The endocranial capacity is estimated at 590 cm^3 (Holloway, 1973). The teeth are not as megadont as those of O.H.7 or O.H.16 but there are no gross differences in cusp morphology between the three specimens.

2. Mandibular remains

Apart from specimens such as O.H.22, O.H.23 (Leakey, M.D., 1969) and O.H.37, most mandibular remains are parts of skulls such as O.H.7 and O.H.13 (although in the case of the latter specimen the mandible body is almost totally destroyed). None of the specimens has very large molar teeth and a feature common to all those where anterior teeth are preserved is that there is no sign of the marked reduction seen in the maxillary teeth of O.H.5. The premolar teeth are narrow buccolingually, especially in O.H.7, and the crowns of the anterior premolars show a dominant buccal cusp in an asymmetric crown. Molar size tends to decrease posteriorly in at least two of the specimens, O.H.13 and O.H.16.

3. Teeth

Collections of isolated teeth have been found e.g. O.H.30 and O.H.39 but they are as yet unpublished.

Although group (c) of the crania and their associated mandibular remains have been discussed in a single group to reflect their published taxonomic attribution, this is by no means a universally held view. Robinson (1965) suggested that O.H.13 was distinct from O.H.7. Leakey, M.D., Clarke and Leakey, L.S.B. in 1971, when they described the cranium of O.H.24, commented that there were significant differences between these specimens, particularly in the morphology of the occipital region. Brace (1972) and Brace, et al (1972) also consluded that O.H.7 and O.H.13 belonged in different taxa. More recent consideration of this material has dwelt on the possibility of sub-dividing the material in the same way with O.H.7 and O.H.16 allied on the one hand and O.H.13 and O.H.24 on the other (Leakey, R.E.F., 1974, 1976b). The possibility and implications of more than one morphotype being represented in this sample will be discussed under the heading of phylogeny.

D. Omo

The most recent and intensive period of exploration of the sediments associated with the Omo river began in 1967. Since the initial reports published
by the French and American expeditions (Arambourg and Coppens, 1967; Howell,
1968) a steady stream of specimens have been discovered and reported (Coppens,
1970a, b, 1971, 1973a, b; Howell, 1969; Howell and Wood, 1974). Useful
reviews of the collection have been published also by Howell and Coppens
(1973, 1976) and Howell et al. (1974). The vast majority of the well over
200 specimens are isolated teeth (Bishop, 1976), a fact which is probably related to the predominantly fluviatile sedimentary environment (de Heinzelin,
Haesaerts and Howell, 1976).

1. Cranial remains

Small fragments of skull vault of an adult, L.P.-996, and of a sub-adult,
L345(A)-11, are notable but the most complete cranium is a juvenile specimen
comprising the occipital and parietal bones, L338(Y)-6. Juvenile remains are
notoriously difficult to assess, but an estimated adult cranial capacity of
440 cm^3 (Howell and Coppens, 1976) and the strong development of temporal lines
in such a young individual suggest that the adult form had relatively large
jaws compared to the neurocranium.

2. Mandibular remains (see Table II)

A recently published assessment of the fossil hominids from the Omo
(Howell and Coppens, 1976) places them in three groups.
   (a) L18-18 (1967) (Arambourg and Coppens, 1967; Coppens, 1970b), though
originally ascribed to a new genus *Paraustralopithecus aethiopicus*, has been
likened to another adult specimen L75-14a (1969) (Coppens, 1971). Both have
quite low mandibular bodies but the former specimen has an anteriorly narrowing arcade, whereas published photographs of L75-14a (1969) indicate more
parallel sided dental rows. L18-18 (1967) shows quite prominent transverse
tori with an obvious post-incisive planum; both premolar teeth show talonid
development. Two immature specimens L427-7 and L222 (1973-2744) are also
included in this group.
   (b) L57-41 (1968) (Coppens, 1970b) and L7A-125 and L74A-21 (Howell, 1969)
have been grouped together. These specimens all have tall but robust mandibular bodies and, though the crown of L57-41 (1968) is broken, the posterior
premolars are large and molarized. The symphysis in all three specimens is
well buttressed by transverse tori. The incisors of specimens L57-41 (1968)
and L7A-125 are small but the preserved canines in two of the three mandibles
have base areas of 75 mm^2 and 85 mm^2.
   (c) L860-2 has been placed in a group of its own. The vertical internal
symphyseal border and the everted base are presumably features which led
Howell and Coppens (1976) to compare it with group (c) of the Koobi Fora
collection.

3. Teeth

Isolated dental remains have been assigned to one of several categories
known from the Omo sites, with a confidence which contrasts with the more cau-
tious approach to the classification of isolated teeth from Koobi Fora (Leakey,

R.E.F., 1974). This may be either because the teeth at the Omo are more distinctive or because the preponderance of isolated teeth in the Omo collections has generated, in those that study them, a more perspicacious attitude towards the assessment of dental remains.

(a) A group of teeth has been recognized from the Usno and Shungura Formations which are distinguished by their canine morphology, relatively broad lower premolars without excessive talonid development and relatively small and elongated lower molar teeth.

(b) A second group of teeth is distinguished by the cusp morphology and large size of the molar crowns, and molarization of the posterior lower premolar crown.

(c) A very few specimens showed sufficiently narrowed and elongated anterior lower premolars to suggest that a third and distinct premolar morphology is represented.

Howell and Coppens (1973), in a study of the deciduous dentitions, came to the conclusion that the teeth were being sampled from more than a single population. This conclusion was based largely on dm_2 crown morphology.

E. *Hadar*

Since its first formal field season in 1973 the International Afar Expedition, led by Johanson, Taieb and Coppens, has accumulated a most important sample of hominid remains. Discoveries in 1973 and 1974 have been supplemented by a large, but as yet unpublished, collection in 1975.

A preliminary assessment (Johanson and Taieb, 1976) of the material suggests that the maxillary and mandibular remains can be divided into two groups and that a temporal fragment, AL166-9, samples a third morphotype. The temporal fragment is heavily pneumatized with a large bulbous mastoid process and thus resembles this area of the crania of groups (a) and (c) from Koobi Fora and O.H.5 from Olduvai.

(a) One group of jaw fragments is typified by AL-288-1, the associated remains of a partial skeleton. There is a cranium associated with this specimen which apparently has thin vault bones and little sign of ectocranial structures. The incisor and canine part of the alveolar process of the mandible is very restricted and the anterior premolars show a dominant buccal cusp. The dental arcade is V-shaped.

(b) The second group of remains, AL199-1, 200-1, 266-1 and 277-1, vary in overall size but dental proportions and tooth morphology are similar in both large and small forms. The central incisors ofthe bigger maxillary fragment are large and the canine base area is nearly as great as the anterior premolar crown area in both specimens. The two mandibular remains have worn teeth with moderately elongated anterior premolars.

F. *Laetolil*

The first hominids from the exposures of the Laetolil (Vogel River) Beds were found in 1939 (Kohl-Larsen, 1943). The largest fragment, a piece of right maxilla, contained both premolars (Weinert, 1950). These were small teeth with two main cusps and no sign of molarization.

Exploration by Mary Leakey in 1974 and 1975 produced a further fourteen hominid specimens, all the remains being teeth and jaws or isolated teeth. Despite some size disparity all the remains are morphologically consistent and may well have been sampled from a single population (Leakey, M.D. et al., 1976)

Relatively large canines, asymmetric anterior lower premolars and moderate sized molar teeth characterize the collection.

G. *Other sites*

Several fossils in East Africa have yielded either single fossils or small collections.

1. Lothagam.

Part of the right side of a hominid mandible, KNM-ER 329, with the crown of M_1 only was found in 1967 (Patterson, et al., 1970). The fragmentary nature of the specimen does not allow any really confident assignation to a morphotype.

2. Natron

A complete adult mandible was found in 1964 in sediments exposed by the Peninj River (Leakey, L.S.B. and Leakey, M.D., 1964). The massive, robust mandibular body, high broad ascending rami, small anterior teeth, molarized premolars and large molar teeth correspond to a reasonably distinct morphotype well represented in the Koobi Fora fossil collection.

3. Baringo Basin

Single molar teeth from the Ngorora Formation (Bishop and Chapman, 1970) and the Lukeino Formation (Pickford, 1975) can at this stage be classified only as hominoid, probably hominid.

(a) Chemeron. Most of a right temporal bone was found in 1965 in sediments of the Chemeron Formation. The bone was less pneumatized and bulbous than the equivalent region in O.H.5 and no generic or specific attribution was made in the initial report (Martyn and Tobias, 1967).

(b) Chesowanja. A cranium, KNM-CH-1, comprising most of the right facial skeleton, including the tooth crowns of $\underline{C}$ to $M^{\underline{3}}$, part of the right frontal region and parts of the base of the skull were found in 1970 (Carney, et al., 1971). The form of the facial skeleton, the zygomatic processes of the maxilla and the size and proportions of the teeth are all suggestive of morphotypes well represented at Koobi Fora and Olduvai; the crown area of the worn canine is only 50% that of the anterior premolar crown. However the lack of marked post-orbital constriction and the more steeply rising frontal squama suggest some deviations from the usual morphological pattern of this morphotype. A lower molar tooth fragment, KNM-CH-302, was found in 1974.

4. Ndutu

A cranium was found during excavations at Lake Ndutu in 1973. Preliminary reports suggest that it shows a mixture of *Homo erectus* and *Homo sapiens* features; for the moment it has been assigned to *Homo erectus*, subspecies as yet undetermined (Clarke, 1976).

TAXONOMY

A. *Preamble*

In 1963 Simpson made a very clear and useful statement about taxonomy and taxonomic procedure (Simpson, 1963). He proposed a process whereby name sets were given at each level of classification. His basic, N_1, name sets were the reference numbers given to specimens e.g. KNM-ER 729 and O.H.9. The next level of classification he applied to populations, and the designations (a), (b) etc. used within the collections from each fossil site reviewed in this paper correspond to Simpson's N_2 name sets. The process he recommended continued with the recognition of these populations as formal units in a hierarchic classification, thereby each population becomes a taxon. The N_3 name sets for taxa are Linnaean or Neo-Linnaean and close similarity between two populations can be recognized by linking them under an 'umbrella' taxonomic label; the degree of similarity will determine at what level in the hierarchy they are united.

Historically the important hominid fossils from East Africa, with the exception of the 1939 finds from Laetolil, all postdate the formal classification of the South African hominids. Thus inevitably East African fossil hominid populations are most often compared with formal taxa of South African hominids.

Most assessments of the South African collections place the material in one of two taxa (Robinson, 1954)(excluding specimens from Swartkrans referred to *Telanthropus* or *Homo* and the material from the West Pit of the Sterkfontein excavation). Some workers, though subscribing to the two taxa theory, admit that similarities between them suggest that gene exchange may have been possible (Tobias, 1969). Finally there are those who hold that there is insufficient morphological evidence for more than one taxon in the South African sample (Brace, 1972; Wolpoff, 1973). The results of the initial proliferation of genus and species names have been subsumed into two major South African taxa, *Australopithecus africanus* (Dart, 1925) and *Australopithecus robustus* (Broom, 1938). An alternative nomenclature is that of Robinson (1965) who suggested that the former taxon be included in the genus *Homo* and that the latter should retain its original generic name, *Paranthropus*. Whatever nomenclature is employed the hypodigms are identical in both schemes. Uncertainty has been expressed about the 'lineage' of the type specimen of *Australopithecus africanus* (Tobias, 1973a, 1976) but, notwithstanding, most parameters of this taxon are usually derived from the collections of fossil hominids from Sterkfontein and Makapansgat; parameters for *Australopithecus robustus* usually relate to the fossil collections from Swartkrans and Kromdraai.

B. *East African Taxa*

1. *Australopithecus boisei*

Apart from the reference of Garusi I from Laetolil to *Meganthropus africanus* (Weinert, 1950), the first major early hominid find that could be compared with the South African taxa was O.H.5. Leakey, L.S.B. (1959) attributed the cranium to a new genus and species of the sub-family Australopithecinae, *Zinjanthropus boisei*, though Robinson (1960) contended that generic distinction from *Paranthropus* was not justified. Leakey, L.S.B., et al. (1964) sank *Zinjanthropus* as a genus and reduced it to a sub-genus of *Australopithecus (sensu lato)*, thus *Australopithecus (Zinjanthropos) boisei*. Tobias (1967) continued the process of rationalization and sank the sub-genus *Zinjanthropus* and included *boisei*, along with *africanus* and *robustus*, as three species of the genus *Australopithecus (sensu lato)* as defined by Le Gros Clark (1964). The mandible from Natron was also included in the hypodigm of *Australopithecus boisei* (Leakey, L.S.B. and Leakey, M.D., 1964).

In the first major taxonomic pronouncement on a wide range of specimens from the Omo (Howell, 1969) material was referred to *Australopithecus boisei* or compared jointly with *Australopithecus boisei* and *Australopithecus robustus*. Subsequently cranial remains, and groups (b) of the mandibular and dental fossils from the Omo have either been assigned to, or compared with, *Australopithecus boisei*.

The most complete cranium in group (a) of the Koobi Fora hominids, KNM-ER 406, was assigned to *Australopithecus boisei* (Leakey, R.E.F., 1970). Subsequent specimens, such as the cranium KNM-ER 732 and mandible KNM-ER 729 were referred to *Australopithecus* sp. indet. which was also the level of taxonomic attribution followed in a series of more detailed anatomical descriptions. Robinson (1972a) pointed out the possible confusion caused by using the term australopithecine to describe characters such as disproportionately small anterior teeth which are diagnostic features of neither *Australopithecus (sensu lato or sensu stricto)* but are characteristic of only the *boisei* and *robustus* species groups. A cranium from Chesowanja has been assigned to *Australopithecus boisei* (Carney, et al., 1971) and a temporal fragment from Hadar has been compared with that of *Australopithecus boisei* (Johanson and Taieb, 1976).

Thus at Olduvai, Omo and Koobi Fora there is abundant evidence of the 'robust' or 'large' form of australopithecine with the typical cranial, gnathic and dental features which have been well summarized by others (Tobias, 1967; Robinson, 1972b). Sexual dimorphism is marked in this taxon as demonstrated by the range of cranial remains from Koobi Fora. There is widespread acknowledgement that the 'robust' australopithecine species in East and South Africa resemble each other more than either resemble synchronic 'gracile' populations sampled in the two areas and there is a proposal that this should be recognized taxonomically by combining the two 'robust' taxa in a single superspecies (Tobias, 1973b and 1976).

2. *Homo habilis* and *Homo* sp. indet.

Leakey, L.S.B., et al. (1964) proposed that the sample of hominids from the lowest levels at Olduvai Gorge should be subdivided and proposed the nomen *Homo habilis* for fossils they considered were distinct from *Australopithecus boisei*. Their assessment of the significance of the morphological differences between the Olduvai fossils and *Australopithecus africanus* did not go unchallenged (Robinson, 1965, 1966) and since that time others have taken the view that the differences between *Homo habilis* and *Australopithecus africanus* are no greater than would be expected from intra-specific variation (Brace, 1972; Brace, et al., 1972).

The occurrence of *Homo* sp. indet. at Koobi Fora was hinted at (Leakey, R.E.F., 1970, 1971) and specimens began to be listed according to generic attribution (Leakey, R.E.F., 1972), but no formal specific attributions were ever made. Attribution at the generic level was all that was attempted for the cranium KNM-ER 1470. Inclusion of this specimen within the taxa *Homo erectus* or *Homo habilis* was discussed, but rejected pending more detailed studies (Leakey, R.E.F., 1973b). Nonetheless a year later it was suggested that a mandible KNM-ER 1802 and a cranium KNM-ER 1590 were conspecific with

KNM-ER 1470 and that as a group they resembled *Homo habilis* (Leakey, R.E.F.,
1974), but in doing so Leakey questioned the homogeneity of the material attributed to that taxon.

Teeth resembling those of *Homo habilis* have been identified at the Omo
(Howell, 1968; Coppens, 1971; Howell and Coppens, 1976) and O.H.7 and KNM-
ER 1802 have been cited as comparisons for specimens referred to *Homo* from Hadar
(Johanson and Taieb, 1976) and Laetolil (Leakey, M.D. et al., 1976).

3. *Homo erectus*

The cranium O.H.9 from Olduvai was originally assigned to *Homo* (Leakey,
L.S.B., 1961) but it is interesting to note that it was considered to resemble
Steinheim, Broken Hill and Saldanha, all cranial remains which are held to
belong to *Homo sapiens* (Campbell, 1963). Apart from a hasty attribution to a
new species (Heberer, 1963), O.H.9 has been with general agreement assigned to
Homo erectus (Tobias, 1965a; Leakey, L.S.B., 1966). Two other specimens from
Olduvai, O.H.12, a cranium and O.H.22, a mandible, have been either attributed
to or compared with *Homo erectus*. Although O.H.13 was designated as a para-
type of *Homo habilis* (Leakey, L.S.B. et al., 1964), Robinson (1965) and Brace
(1972) have both expressed the opinion that it should be included within *Homo
erectus*.

At Koobi Fora mandibular remains, e.g. KNM-ER 992 and 730, have been compared with, but not attributed to, *Homo erectus* (Wood, 1976c). A cranium
found in 1975, KNM-ER 3733, has been unequivocally assigned to *Homo erectus*
(Leakey, R.E.F. and Walker, 1976) and resemblances between *Homo erectus*
and small cranial fragments from Koobi Fora and Omo have been noted (Leakey,
1973a; Howell and Coppens, 1976). A recently described cranium from Ndutu
(Clarke, 1976) has been assigned to *Homo erectus*, although its relatively advanced cranial vault morphology contrasts with Asian variants.

4. *Australopithecus africanus*

The debate about whether *Australopithecus africanus* populations were being
sampled in East African fossil collections can be divided into two successive
stages. The first concerned the problem of whether *Australopithecus africanus*
or *Homo* was represented in the collections; latterly the discussion has turned
on the problem of whether *Australopithecus africanus* and *Homo* are being sampled

Robinson (1965, 1966) put forward persuasive arguments that the *Homo habilis*
taxon was not valid and that the material assigned to it by L.S.B. Leakey et al.
(1964) was not sufficiently distinct from *Australopithecus africanus*. This proposal was vigorously debated (Tobias, 1966a) and steadfastly defended (Tobias,
1965b, 1966b, 1967, 1971). Robinson (1965) proposed a taxonomic device whereby
Australopithecus (sensu stricto) should be transferred to *Homo* but this solution has never received widespread approval.

The second stage of the debate was heralded by the proposal that isolated
teeth from the Omo region were best compared with those of *Australopithecus
africanus* (Howell, 1969). This was followed by the categorization of the mandible from Lothagam as *Australopithecus* c.f. *africanus* (Patterson, Behrensmeyer and Sill, 1970). The hominid remains recovered from Koobi Fora initiall
made it seem 'unlikely' (Leakey, R.E.F., 1971) and provided 'no clear evidence'
(Leakey, R.E.F., 1972) that a second species of *Australopithecus* was being
sampled in East Africa. New evidence, in particular a fine cranium KNM-ER
1813, prompted a revision of this view. Because the new cranium was recovered

from a horizon that most probably significantly post-dated those that yielded the larger brained crania, KNM-ER 1470 and KNM-ER 1590, it was proposed that the question of the possibility of an East African population of *Australopithecus africanus* be reopened (Leakey, R.E.F., 1974). Meanwhile the proposal had been made that 'gracile' specimens, such as the mandible KNM-ER 992 from Koobi Fora, were equivalent to *Australopithecus africanus* in South Africa (Robinson, 1972a).

R.E.F. Leakey (1974) also had posed the problem that the 'gracile' specimens from Olduvai may have been drawn from more than one taxon. This possibility had already been raised by Robinson (1965) and L.S.B. Leakey (1966).

In an interesting paper L.S.B. Leakey had suggested that O.H.16 was a 'protopithecanthropine' and was synchronic with, but differed from, *Homo habilis* specimens such as O.H.13. The possibility of 'taxonomic variation' within the gracile specimens at Olduvai, aligning O.H.24 with O.H.13 and O.H.7 with O.H. 16, was discussed again several years later (Leakey, M.D., et al., 1971).

A preliminary report on the Hadar remains (Johanson and Taieb, 1976) suggested that a population is being sampled at this site which shows affinities with *Australopithecus africanus*.

5. *Paraustralopithecus aethiopicus*

A mandible, L18-18 (1967), found in 1967 at the Omo was placed in a new genus and species (Arambourg and Coppens, 1967). In a more recent review it is redesignated *Australopithecus c.f. africanus* (Howell and Coppens, 1976). Attention has been drawn to similarities between this mandible from the Omo and a distinctive mandible from Koobi Fora, KNM-ER 1482 (Leakey, R.E.F., 1974).

6. *Homo ergaster*

A recent publication has suggested the creation of a new species of *Homo*, with KNM-ER 992 as the holotype, to include a series of hominid specimens from Koobi Fora (Groves and Mazàk, 1975). Until more detailed studies of this material are undertaken and a more wide ranging comparative sample is employed such a definitive taxonomic statement is unwise and premature.

7. An alternative to the apparent taxonomic diversity described above is summarized in a recent proposal to revise the diagnosis of *Australopithecus* (Wolpoff and Lovejoy, 1975). The new diagnosis is so wide that such a taxon would apparently include specimens such as O.H.5, KNM-ER 406, KNM-ER 1470 and KNM-ER 1813; in fact it would include all the previously proposed East African hominid taxa except *Homo erectus*.

PHYLOGENY

This review of the fossil collections and the taxonomic survey have deliberately been presented with little reference to relative and absolute dates. In maintaining this independence of time, and by relying solely on morphology, the taxa presented are most obviously subdivided into the 'robust' australopithecine taxon *Australopithecus boisei*, and another group consisting of all the remaining taxa. When dating information is added it is apparent that examples of *Australopithecus boisei* are known from sediments which range from less than 3 million to between 1 and 1.5 million years old (Tobias, 1976). Although comprehensive cranial evidence does not span this whole time period

the mandibular and dental evidence is distinctive throughout and has led to
recognition of this group as a distinct taxon in recent reviews of fossil
collections (Leakey, R.E.F., 1974; Howell and Coppens, 1976; Tobias, 1976;
Wood, 1976c) but for a contrary view see Wolpoff (1973).

At three East African sites which sample hominids from a wide time range,
'non-robust' hominid populations are found synchronically with *Australopithecus boisei*. In contrast to the conformity within the 'robust' lineage the
'non-robust' hominids display a wide range of variation. Calibrating of
these 'non-robust' hominids, and using their stratigraphic relationships within
collecting areas does help to interpret some of this variation. Though crania
referred to *Homo* from the lower levels at Koobi Fora, such as KNM-ER 1470 and
1590, show apomorphous features such as an enlarged neurocranium and details
of the nasal and glabellar architecture, they also share symplesiomorphous
features with the 'robust' lineage such as the greatest width across the base
of the skull, broad malar region, relatively orthognathic face and canines
reduced when compared to the apes. At a higher stratigraphic level at Koobi
Fora another cranium, KNM-ER 3733, has been found which, though differing from
KNM-ER 1470 and 1590 does so only because it expresses to a more marked degree
hominine synapomorphous features seen in the early crania.

At Koobi Fora, this basically simple phylogenetic scheme is apparently com-
promised by specimens such as KNM-ER 1813 which appear to show characteristic
features of both the 'robust' lineage e.g. cranial volume, and of the *Homo*
lineage e.g. skull shape and dental size and proportions. There are two possible explanations for these findings. Either the range of variation in the
Homo lineage is greater than first thought or a third lineage is being sampled
along with the other two lineages. Such a third lineage could have been derived from a single speciation event in hominid evolution, or it could have
shared a more recent common ancestry with either the 'robust' australopithecines or with the *Homo* lineage. In summary, if specimens like KNM-ER 1813
represent a third lineage, should they be regarded as a small brained *Homo* or
a small toothed *Australopithecus*? Without more details about cranial capacity
estimates and an adequate comparative sample detailed quantitative comparisons
are premature. Much will depend on the analysis of the origin of morphological features that are common to some specimens and not others; if these are
symplesiomorphous to hominids as a whole their value as taxonomic indicators
will be lessened.

The collections of hominids between about 3 and 3.75 million years old from
Laetolil and Hadar (Johanson and Taieb, 1976; Leakey, M.D., et al., 1976) provide important pointers to the ancestral hominid morphotype. Many of the
trends in the dental remains from these two early sites can be matched in
specimens taken from the East African *Homo* lineage as well as those belonging
to *Australopithecus africanus* from South Africa. If features such as the
orientation and unequal lower cusps of the anterior premolars are plesiomorphous ancestral hominid characters, then their presence in a wide range of
'gracile' hominid samples is to be expected. Whatever the nomenclatural de-
vices given to the various 'gracile' populations, they need to be distinguished
from the 'robust' australopithecines which have developed a whole suite of apomorphous cranial, gnathic and dental features of their own.

If specimens such as KNM-ER 1813 do represent a third hominid lineage in
East Africa during the Plio-Pleistocene, as others have pointed out, it calls
into question many assumptions that have been made about the 'gracile' hominids
from Olduvai Gorge. Much of the present uncertainty hinges around our inability
to decide whether at least part of the *Homo* sample in East Africa and *Australo-*

pithecus africanus are in the same clade, or in an equivalent grade, or whether, in addition to symplesiomorphous similarities, they show sufficient significant distinguishing features to justify their placement in separate taxa. L.S.B. Leakey, et al (1964) considered that sufficient characters did exist to distinguish the 'gracile' specimens in the Olduvai collection from *Australopithecus africanus*. In the author's opinion similar distinguishing features are also evident in specimens from Koobi Fora such as KNM-ER 992 and 820. Thus, if there is taxonomic variation within East African 'gracile' hominids it is by no means certain that it is due to the presence of *Australopithecus africanus*.

ACKNOWLEDGEMENTS

The author would like to thank Richard Leakey for the opportunity to study and report on the fossil hominid collection from Koobi Fora. Research that has been incorporated into this review was supported by the Natural Environments Research Council, Royal Society, Central Research Fund of the University of London and the Boise Fund.

REFERENCES

Arambourg, C. and Coppens, Y. (1967). *C.r. hebd. Séanc. Acad. Sci., Paris,* 265, 589-590.
Bishop, W.W. (1976). *In* "Earliest man and environments in the Lake Rudolf basin", (Y. Coppens, F.C. Howell, G.L. Isaac and R.E. Leakey, eds), pp. 585-589, University of Chicago Press, Chicago and London.
Bishop, W.W. and Chapman, G.R. (1970). *Nature, Lond.,* 226, 914-918.
Brace, C.L. (1972). *Yearb. phys. Anthrop.,* 16, 31-49.
Brace, C.L., Mahler, P.E. and Rosen, R.B. (1972). *Yearb. phys. Anthrop.,* 16, 50-68.
Brain, C.K. (1970). *Nature, Lond.,* 225, 1112-1119.
Brain, C.K. (1973). *Jl. S. Afr. biol. Soc.,* 13, 7-22.
Brain, C.K. (in press). *In* "Early hominids of Africa", (C. Jolly, ed.), Duckworth, London.
Campbell, B. (1963). *In* "Classification and human evolution", (S.L. Washburn, ed.), pp. 50-74, Aldine, Chicago.
Carney, J., Hill, A., Miller, J.A. and Walker, A. (1971). *Nature, Lond.,* 230, 509-514.
Clarke, R.J. (1976). *Nature, Lond.,* 262, 485-487.
Clark, W.E. Le Gros (1964). "The fossil evidence for human evolution", University of Chicago Press, Chicago and London (2nd ed.).
Coppens, Y. (1970a). *C.r. hebd. Séanc. Acad. Sci., Paris,* 271, 1968-1971.
Coppens, Y. (1970b). *C.r. hebd. Séanc. Acad. Sci., Paris,* 271, 2286-2289.
Coppens, Y. (1971). *C.r. hebd. Séanc. Acad. Sci., Paris,* 272, 36-39.
Coppens, Y. (1973a). *C.r. hebd. Séanc. Acad. Sci., Paris,* 276, 1823-1826.
Coppens, Y. (1973b). *C.r. hebd. Séanc. Acad. Sci., Paris,* 276, 1981-1984.
Corruccini, R.S. (1972). *Syst. Zool.,* 21, 375-383.
de Heinzelin, J., Haesaerts, P. and Howell, F.C. (1976). *In* "Earliest man and environments in the Lake Rudolf basin", (Y. Coppens, F.C. Howell, G.L. Isaac and R.E.F. Leakey, eds), pp. 24-49, University of Chicago Press, Chicago and London.
Day, M.H. and Leakey, R.E.F. (1973). *Am. J. phys. Anthrop.,* 39, 341-354.
Day, M.H. and Leakey, R.E.F., Walker, A.C. and Wood, B.A. (1975). *Am. J. phys. Anthrop.,* 42, 461-476.

Drennan, M.R. (1929). *Ann. S. Afr. Mus.*, 24, 61-87.
Eldredge, N. and Tattersall, I. (1975). *In* "Approaches to primate paleobiology",
    (F.S. Szalay, ed.), pp. 218-242, Karger, Basel.
Freedman, L. (1957). *Ann. Transv. Mus.*, 23, 122-259.
Freedman, L. (1962). *Growth*, 26, 117-128.
Giles, E. (1970). *In* "Personal identification in mass disasters", pp. 99-
    109, Smithsonian, Washington.
Gould, S.J. (1975). *In* "Approaches to primate paleobiology", (F.S. Szalay,
    ed.), pp. 244-292, Karger, Basel.
Groves, C.P. and Mazák, V. (1975). *Cas. Miner. Geol.*, 20, 225-247.
Heberer, G. (1963). *Z. Morph. Anthrop.*, 53, 171-177.
Hennig, W. (1966). "Phylogenetic systematics", University Illinois Press,
    Chicago.
Hennig, W. (1975). *Syst. Zool.*, 24, 244-256.
Holloway, R.L. (1973). *Nature, Lond.*, 243, 97-99.
Holloway, R.L. (1975). *In* "Primate functional morphology and evolution",
    (R. Tuttle, ed.), pp. 391-415, Mouton, The Hague and Paris.
Howell, F.C. (1968). *Nature, Lond.*, 219, 567-572.
Howell, F.C. (1969). *Nature, Lond.*, 223, 1234-1239.
Howell, F.C. (1976). *In* "Human Origins", (G.L. Isaac and E.R. McCown, eds),
    pp. 227-268, Benjamin, Menlo Park and Reading.
Howell, F.C. and Coppens, Y. (1973). *J. Hum. Evol.*, 2, 461-472.
Howell, F.C. and Coppens, Y. (1976). *In* "Earliest man and environments in the
    Lake Rudolf basin", (Y. Coppens, F.C. Howell, G.L. Isaac and R.E. Leakey,
    eds), pp. 522-532, University of Chicago Press, Chicago and London.
Howell, F.C., Coppens, Y. and de Heinzelin, J. (1974). *Am. J. phys. Anthrop.*,
    40, 1-16.
Howell, F.C. and Isaac, G.L. (1976). *In* "Earliest man and environments in the
    Lake Rudolf basin", (Y. Coppens, F.C. Howell, G.L. Isaac and R.E. Leakey,
    eds), pp. 471-475, University of Chicago Press, Chicago and London.
Howell, F.C. and Wood, B.A. (1974). *Nature, Lond.*, 249, 174-176.
Huxley, J.S. (1932). "Problems of relative growth", Methuen, London.
Johanson, D.C. and Taieb, M. (1976). *Nature, Lond.*, 260, 293-297.
Kohl-Larsen, L. (1943). *Auf den Spuren des Vormenschen*, 2, 379-381.
    Stuttgart, Strecker und Schroder Verlag.
Leakey, L.S.B. (1959). *Nature, Lond.*, 184, 491-493.
Leakey, L.S.B. (1961). *Nature, Lond.*, 189, 649-650.
Leakey, L.S.B. (1966). *Nature, Lond.*, 209, 1279-1281.
Leakey, L.S.B. and Leakey, M.D. (1964). *Nature, Lond.*, 202, 5-7.
Leakey, L.S.B., Tobias, P.V. and Napier, J.R. (1964). *Nature, Lond.*, 202, 7-9.
Leakey, M.D. (1969). *Nature, Lond.*, 223, 754-756.
Leakey, M.D. (1971). Olduvai Gorge, 3. "Excavations in Beds I and II, 1960-
    1963", Cambridge University Press, Cambridge.
Leakey, M.D., Clarke, R.J. and Leakey, L.S.B. (1971). *Nature, Lond.*, 232,
    308-312.
Leakey, M.D., Hay, R.L., Curtis, G.H., Drake, R.E., Jackes, M.K. and White,
    T.D. (1976). *Nature, Lond.*, 262, 460-466.
Leakey, R.E.F. (1970). *Nature, Lond.*, 226, 223-224.
Leakey, R.E.F. (1971). *Nature, Lond.*, 231, 241-245.
Leakey, R.E.F. (1972). *Nature, Lond.*, 237, 264-269.
Leakey, R.E.F. (1973a). *Nature, Lond.*, 242, 170-173.
Leakey, R.E.F. (1973b). *Nature, Lond.*, 242, 447-450.
Leakey, R.E.F. (1974). *Nature, Lond.*, 248, 653-656.

Leakey, R.E.F. (1976a). *Nature, Lond.*, 261, 574-576.
Leakey, R.E.F. (1976b). *In* "Earliest man and environments in the Lake Rudolf basin", (Y. Coppens, F.C. Howell, G.L. Isaac and R.E. Leakey, eds), pp. 476-483, University of Chicago Press, Chicago and London.
Leakey, R.E.F. and Isaac, G.L. (1976). *In* "Human Origins", (G.L. Isaac and E.R. McCown, eds), pp. 307-332, Benjamin, Menlo Park and Reading.
Leakey, R.E.F., Mungai, J.M. and Walker, A.C. (1971). *Am. J. phys. Anthrop.*, 35, 175-186.
Leakey, R.E.F., Mungai, J.M. and Walker, A.C. (1972). *Am. J. phys. Anthrop.*, 36, 235-251.
Leakey, R.E.F. and Walker, A.C. (1973). *Am. J. phys. Anthrop.*, 39, 341-354.
Leakey, R.E.F. and Walker, A.C. (1976). *Nature, Lond.*, 261, 572-574.
Leakey, R.E.F. and Wood, B.A. (1973). *Am. J. phys. Anthrop.*, 39, 335-368.
Leakey, R.E.F. and Wood, B.A. (1974a). *Am. J. phys. Anthrop.*, 41, 237-244.
Leakey, R.E.F. and Wood, B.A. (1974b). *Am. J. phys. Anthrop.*, 41, 245-250.
Mackinnon, I.L., Kennedy, J.A. and Davies, T.V. (1956). *Am. J. Roentg.*, 76, 303-310.
Martin, R.D. (1968). *Man*, n.s. 3, 377-401.
Martyn, J. and Tobias, P.V. (1967). *Nature, Lond.*, 215, 476-479.
Mayr, E. (1974). *Z. Zool. Syst & Evolutions forsch.*, 12, 95-128.
Mosimann, J.E. (1970). *J. Am. statist. Ass.*, 65, 930-945.
Nelson, G. (1974). *Syst. Zool.*, 23, 452-458.
Patterson, B., Behrensmeyer, A.K. and Sill, W.D. (1970). *Nature, Lond.*, 226, 918-921.
Pickford, M. (1975). *Nature, Lond.*, 256, 279-284.
Robinson, J.T. (1954). *Am. J. phys. Anthrop.*, 12, 181-200.
Robinson, J.T. (1956). *Trans. Mus. Mem. No. 9*, 1-179.
Robinson, J.T. (1960). *Nature, Lond.*, 186, 456-458.
Robinson, J.T. (1965). *Nature, Lond.*, 205, 121-124.
Robinson, J.T. (1966). *Nature, Lond.*, 209, 957-960.
Robinson, J.T. (1972a). *Nature, Lond.*, 240, 239-240.
Robinson, J.T. (1972b). "Early hominid posture and locomotion", University of Chicago Press, Chicago and London.
Rosen, D.E. (1974). *Syst. Zool.*, 23, 446-451.
Simpson, G.G. (1961). "Principles of animal taxonomy", Columbia University Press, New York and London.
Simpson, G.G. (1963). *In* "Classification and human evolution", (S.L. Washburn, ed.), pp. 1-31, Aldine, Chicago.
Szalay, F.S. (1975). *Man*, n.s. 10, 420-429.
Tobias, P.V. (1963). *Nature, Lond.*, 197, 743-746.
Tobias, P.V. (1965a). *Curr. Anthropol.* 6, 391-399, 406-411.
Tobias, P.V. (1965b). *Science, N.Y.*, 149, 918.
Tobias, P.V. (1966a). *Nature, Lond.*, 209, 953-957.
Tobias, P.V. (1966b). *Curr. Anthropol.*, 7, 579-580.
Tobias, P.V. (1967). Olduvai Gorge, 2. "The cranium and maxillary dentition of *Australopithecus (Zinjanthropus) boisei*", Cambridge University Press, Cambridge.
Tobias, P.V. (1969). *Yearb. phys. Anthrop.*, 15, 24-30.
Tobias, P.V. (1971). "The brain in hominid evolution", Columbia University Press, New York and London.
Tobias, P.V. (1973a). *Nature, Lond.*, 246, 79-83.
Tobias, P.V. (1973b). *In* "L'Origine dell'Uomo", pp. 63-85, Acad. Naz. dei Lincei, Rome.
Tobias, P.V. (1975). *Optima*, 25, 24-35.

Tobias, P.V. (1976). *In* "Human Origins", (G.L. Isaac and E.R. McCown, eds),
    pp. 377-422, W.A. Benjamin, Menlo Park.
Tobias, P.V. and Hughes, A.R. (1969). *S. Afr. Archaeol. Bull.*, 24, 158-169.
Walker, A. (1976). *In* "Earliest man and environments in the Lake Rudolf
    basin", (Y. Coppens, F.C. Howell, G.L. Isaac and R.E. Leakey, eds), pp.
    484-489, University of Chicago Press, Chicago and London.
Weinert, H. (1950). *Z. Morph. Anthrop.*, 42, 138-148.
Wolpoff, M.H. (1973). *Yearb. phys. Anthrop.*, 17, 113-139.
Wolpoff, M.H. and Lovejoy, C.O. (1975). *J. Hum. Evol.*, 4, 275-276.
Wood, B.A. (1976a). "An analysis of sexual dimorphism in primates", Universit
    Microfilms, Michigan and London.
Wood, B.A. (1976b). *J. Zool. Lond.*, 180, 15-34.
Wood, B.A. (1976c). *In* "Earliest man and environments in the Lake Rudolf
    Basin", (Y. Coppens, F.C. Howell, G.L. Isaac and R.E. Leakey, eds), pp.
    490-506, University of Chicago Press, Chicago and London.

THE PLACE OF *AUSTRALOPITHECUS AFRICANUS* IN HOMINID EVOLUTION

P.V. TOBIAS

University of the Witwatersrand, Johannesburg, South Africa

Since World War II, Africa has seen a marked acceleration in the tempo of discovery of fossils assigned by most competent students of palaeoanthropology to the family Hominidae. Indeed, not merely new fossils but new sites and fossiliferous areas have been found with increasing rapidity. It is sobering to reflect that by 1939, at the outbreak of that War, early fossil hominids had been found only at Taung in the Cape Province, Sterkfontein and Kromdraai in the Transvaal and Ngarusi in Tanganyika, if we include L. Kohl-Larsen's discovery of hominid remnants in 1939, though they were not published until 1942. Even the Olduvai Gorge had, at that stage, yielded only a few non-descript calvarial fragments to Mary Leakey in 1935 (Olduvai hominid 2) - apart from the much more recent skeleton of 'Oldoway man' that Hans Reck had uncovered in 1913.

By 1949, a decade later, two more South African sites had been added to the list, namely Makapansgat and Swartkrans; by 1959, Olduvai was added more effectively to the list with the first of the major finds, that of the type specimen of *Australopithecus boisei* (Olduvai hominid 5). Another ten years brings us to 1969 and the tally of localities has been swelled by Peninj in Tanzania, Chemeron, Kanapoi, Lothagam, Koobi Fora and Ileret in Kenya, Omo in Ethiopia and Yayo (Koro Toro) in Chad. Seven years of the next decade have passed, revealing Chesowanja in Kenya and Hadar-Afar in Ethiopia as further sites of early hominid fossils. The growth in the number of fossils has been stupendous, indeed virtually exponential!

So frenzied has been the quest to create new research reputations by the wresting of Africa's antique fossils from its soil that discovery has out-stripped description, description has surpassed rigorous evaluation, while behind evaluation there lags systematization of the ever-growing number of fossils and fossil taxa. One is confronted with a now bewildering array of published and unpublished, evaluated and formless, classified and anonymous fossils. Small wonder that the new worker in the field of palaeoanthropology finds himself often unable to see the wood for the trees. Indeed, he hovers between the extremes of two dire perils: either he is in danger of seeing a new species lurking behind every tree; or else he is obsessed by preconceived ideas about the misdeeds of the taxonomic splitters, about sexual dimorphism, variability, orthogenetic heresies, evolutionary strategies and scenarios, con-flicting and sometimes faintly teleological functional interpretations and many other fads and fantasies with few facts - to such an extent that he is apt to retreat into a lumper's fool's paradise, in which he may convince himself (if not many of his colleagues) that he sees all hominid fossils, ancient and modern, as variations on the theme of a single genus, admitting perhaps to the merest suggestion that more than one species of hominid has ever existed. These are the Scylla and Charybdis between which subtle, skilful and scrupulous sailing is necessary!

Some classificatory problems

A new genus and species were created by Dart (1925) to accommodate the Taung child. In that pioneering paper, he gave as the distinguishing characteristics of *A. africanus* the following summary:
"...it represents a fossil group distinctly advanced beyond living anthropoids in those two dominantly human characters of facial and dental recession on one hand, and improved quality of the brain on the other. Unlike Pithecanthropus, it does not represent an ape-like man, a caricature of precocious hominid failure, but a creature well advanced beyond modern anthropoids in just those characters, facial and cerebral, which are to be anticipated in an extinct link between man and his simian ancestor. At the same time, it is equally evident that a creature with anthropoid brain capacity, and lacking the distinctive, localised temporal expansions which appear to be concomitant with and necessary to articulate man, is no true man." (Dart, 1925)
In the 51 years that have elapsed since that time, it is probably correct to say, as Wolpoff and Lovejoy (1975) have done in their introduction to the most recently proposed re-definition of *Australopithecus*, that:
"In spite of the fact that the Taung individual was a juvenile, with the first permanent molar having just erupted into occlusion, Dart's diagnosis has proven to be a farsighted one. Subsequent diagnoses by Broom and Schepers (1946), Robinson (1954), Le Gros Clark (1964), and Tobias (1967) have broadened the diagnosis without substantially altering any of the original criteria." (Wolpoff and Lovejoy, 1975)
The redefinitions have generally sharpened the criteria originally recognized by Dart and, further, have added to them, as new material made more and more aspects of the bodily structure of *Australopithecus* available for study.
A complicating feature has been the addition of at least two further kinds of early hominid to the available material, namely the group referred to loosely as the robust australopithecines and known since the discovery of the Kromdraai specimen in 1938, and secondly the early members of the genus *Homo*, presaged by the finding of what was originally called 'Telanthropus capensis' by Broom and Robinson (1949) and formalised systematically by Leakey et al.(1964) with their recognition of the species *Homo habilis*. Some of the alterations in the definition of *Australopithecus* have been proposed by those who consider the robust ape-men to belong to the same genus: a wider definition to accommodate the South African robust species, *A. robustus* from Kromdraai and Swartkrans, along with *A. africanus*, was thus proposed by Le Gros Clark (1955, 1964). When my study of Olduvai hominid 5, the hyper-robust ape-man of East Africa, led me to classify it, too, in *Australopithecus*, it was necessary to broaden the definition slightly further to include this species, *A. boisei* (Tobias, 1967, 1968a). This classification of all the small-brained 'ape-men' or 'australopithecines' into a single genus, *Australopithecus*, still seems to enjoy a consensus of support today, to judge by the published papers and books.
Another viewpoint retains the robust australopithecines in a separate genus *Paranthropus*, leaving the genus *Australopithecus* to hold only the 'small' or 'gracile' australopithecines (Robinson, 1954). On this view, a somewhat narrower or unexpanded definition of *Australopithecus* sufficed and it was a mono-

typic genus with but a single species, *A. africanus.*

Later, Leakey et al. (1964) recognised the larger-brained, probably stone tool-making, little hominid of Olduvai as representing a new species which they named *Homo habilis.* They regarded it as distinct from *A. africanus* and accordingly made no attempt further to enlarge the compass of the definition of *Australopithecus.* For them, *Australopithecus* referred to the small-brained early hominids, while *H. habilis* connoted the earliest of the species of larger-brained hominids.

Robinson (1965) could not accept the proposal that *H. habilis* represented a distinct taxon, while accepting our claim that the specimens to which the name was applied were links between *A. africanus* on the one hand and later forms of *Homo* (*Homo erectus*) on the other. Thus, he wrote,

"If the interpretation here is correct, then clearly no new species name is needed: the situation is simply one common in palaeontology when specimens are found which link two already existing taxa. Creating a new taxon here is no solution; the two taxa between which the new one falls are already so similar that insufficient morphological distance exists between them to justify the insertion of another species. As is well recognized, conventional Linnean taxonomy is not suited to dealing with a problem such as this...." (Robinson, 1965, p. 123)

Since he recognized that the *habilis* fossils were, as Leakey et al. had claimed, intermediate, Robinson was therefore confronted with the need to resolve the taxonomic question. He suggested that it would be reasonable to extend the genus *Homo* to include not only the new Olduvai material (as Leakey and his colleagues had done) but also the material then regarded as belonging to *A. africanus* from Taung, Sterkfontein and Makapansgat. It was thus directly consequent upon our claim that the *habilis* fossils belonged to *Homo* that Robinson proposed to resolve the taxonomic and phylogenetic problems this presented, by lumping *A. africanus* into *Homo.* In his 1965 paper, he suggested that there be recognised only two species of *Homo*, namely *H. transvaalensis* (to include the material formerly assigned to *A. africanus* and to *H. habilis*) and *H. sapiens.* Later, after a somewhat intricate argument on whether the name *africanus* was available, in terms of the International Code of Zoological Nomenclature, Robinson (1972a) decided that it was indeed available and that it enjoyed priority over *transvaalensis:* he then proposed to use the term *H. africanus* as referring to the earlier species of *Homo* (including *A. africanus* and *H. habilis*). It should be noted that for Robinson, *Australopithecus* was a monotypic genus; hence, for him, the lumping of the species *A. africanus* into *Homo* meant the disappearance of the genus *Australopithecus* from the nomenclature: it was a genus that had lost its only species, hence its disappearance.

However, it should be clearly appreciated that a considerable number of workers in the field considered that *Australopithecus* was not monospecific, but included at least two species, *A. africanus* and *A. robustus*, and for many, a third, namely *A. boisei.* Now it was only one of these species that Robinson proposed to lump into *Homo* - and this was *A. africanus.* At first glance, then, it might be thought that, even if one were disposed to accept the proposal to lump *A. africanus* into *Homo*, the genus *Australopithecus* could remain as comprising *A. robustus* and *A. boisei.* But Article 42 (b) of the I.C.Z.N. states quite clearly, "Each taxon of the genus-group is objectively defined only by reference to its type-species." In other words, remove the type-species of a genus and the genus is no longer considered to be objectively defined. However, the I.C.Z.N. in Article 42 (c) refers to certain

biological arrays known as "Collective groups" that is, "an assemblage of
identifiable species of which the generic positions are uncertain; treated
as a genus-group for taxonomic convenience" (Glossary, I.C.Z.N., 1961). It
would seem from this definition of a collective group that it is hand-woven
to meet the difficulties of the group of australopithecine species! For
Article 42 (c) adds that "collective groups require no type-species". Hence,
since collective groups are to be treated as generic names in the meaning of
the Code, it follows that the name *Australopithecus* (which has historical
priority) could survive as the name for a collective group, even if it were
agreed that *A. africanus*, the type-species, be removed to the genus *Homo*.

Current Classifications of the Hominidae

    While Robinson uses the terms *Homo* and *Paranthropus* for the two genera he
recognizes, others use *Homo* and *Australopithecus*. The latter pair of terms,
it should be clear, is used in at least four different ways:
(1) *Australopithecus* – comprising *A. africanus*, *A. robustus* and *A. boisei*
                          (c.f. Le Gros Clark, 1964; Tobias, 1967, 1968a)
     Homo   – comprising *H. habilis*, *H. erectus* and *H. sapiens*
(2) *Australopithecus* – comprising *A. africanus*, *A. habilis*, *A. robustus*
                          and *A. boisei* (c.f. Pilbeam, 1972; Simons, 1972;
                          Wolpoff and Lovejoy, 1975)
     Homo   – comprising *H. erectus* and *H. sapiens*
(3) *Australopithecus* – comprising *A. africanus* (including *A.a. robustus* and
                          A.a. habilis) and *A. boisei* (c.f. Campbell, 1972)
     Homo   – comprising *H. erectus* and *H. sapiens*
    As Robinson (1972b) has pointed out, the way in which R.E.F. Leakey (1971,
1972) uses the term *Australopithecus* corresponds to Robinson's use of *Paranth*
ropus, that is, Leakey seems to envisage *Australopithecus* as a robust form,
whilst relegating much of what others would call the gracile australopithecine to *Homo*. This provides a fourth possible usage:
(4) *Australopithecus* – comprising *A. boisei* and perhaps *A. robustus* and some
                          A. africanus
     Homo   – comprising an equivalent of some   *A. africanus*,
                          as well as *H. habilis*, *H. erectus* and *H. sapiens*
                          (c.f. R.E.F. Leakey, 1971, 1972)
    As Walker (1976) has declared, "It seems that except for Robinson there is
a general consensus that most, if not all, the South African fossil hominids
belong in the genus *Australopithecus*."
    Robinson's classificatory system, however, retains the generic name *Paran-*
thropus. Until very recently, as I pointed out in 1967 and 1973 (Tobias,
1967, 1973) and Walker in 1976, Robinson's view has had little support and
has been virtually unique among modern classificatory systems, in abandoning
Australopithecus and recognizing only *Paranthropus* and *Homo*. Lately, two
other workers, Groves and Mazak, have proposed another system based on the
latter two genera. However, their system and Robinson's are not consistent.
Thus, in contrast with Robinson's system, in which *H. habilis* is lumped into
H. africanus, Groves and Mazak recognize *H. habilis* as "a good species by any
standards" (1975, p. 242). They have gone further and, on the basis of two
hemi-mandibles, KNM-ER 992, from Ileret east of Lake Turkana, and of a number
of other jaws and a few referred parts, they have erected a new species, *H.*
ergaster, which they feel differs in dental and gnathic features from *H.*
habilis as well as from *H. africanus*. Thus, their version of *H. africanus*

differs from that of Robinson:
(1) *Paranthropus* - comprising *P. robustus* (including *P. robustus boisei*)
     Homo   - comprising *H. africanus*(including *H. habilis*)
                       and *H. sapiens* (including *H.s. erectus*) (c.f. Robinson, 1972a)
(2) *Paranthropus* - (comprising *P. robustus*)
     Homo   - comprising *H. africanus*, *H. habilis*, *H. ergaster*
                       and, presumably, *H. erectus* and *H. sapiens* (c.f. Groves and Mazák, 1975)

While these six nomenclatural variants certainly do not ring the changes on the published views, they all agree in one important respect, namely that the early hominids may best be viewed as two genera (Chart 1). (I am leaving out of consideration here such earlier claimants for hominid status as *Ramapithecus*, *Graecopithecus* and *Gigantopithecus*.) It seems that few workers in this field today find it helpful or convenient - or in accord with their idea of reality - to think of the available fossils as sorting themselves into three genera.

CHART 1

Two main approaches to Plio-Pleistocene hominid classification

2 Genera: *Australopithecus* and *Homo* (at least 4 versions)

2 Genera: *Paranthropus*   and *Homo* (at least 2 versions)

The problem of genera not confined to Palaeoanthropology

A further point deserves to be stressed at this point: these 6 variants exist, not because some workers are confused, or are abusing the I.C.Z.N., or have failed to understand it or interpret it correctly; they exist because of varying judgments on the part of the scientists in this field. Judgments vary on such matters as how much weight to lay on a particular morphological trait or complex of traits, what is sometimes called phyletic valence (a vague concept that I have contested elsewhere - Tobias, 1966 - and which Groves and Mazák, 1975, seem likewise to distrust). Judgments vary, too, on the relative values of phenetic and phylogenetic classifications, on what Blackwelder (1964) calls an Omnispective Classification, and on the role and value of numerical taxonomy. Adjudicators differ also on the degree to which classification should be based on Hennig's principles (1965, 1966) - the study of the distribution of derived or *apomorphous* features as distinct from primitive or common ancestral or *plesiomorphous* traits.

These and many other theoretical and procedural problems preoccupy many palaeontologists and are largely responsible for the marked variations in their classificatory systems. I have been at pains to stress this point because one so frequently hears ill-informed comments about the lack of familiarity of palaeoanthropologists with zoological principles and with the I.C.Z.N. - as though such problems were the preserve of workers in hominid palaeontology! Far from it; every branch of palaeontology is beset with this kind of problem, and it is the judgmental variations, rather than ignorance of, or indifference to, the Code of Zoological Nomenclature, that is responsible for the overwhelming majority of the divergences of opinion. Consider the words of Thomas W. Amsden, introducing the symposium of the North American Paleontological Convention on 'The Genus: a Basic Concept in Paleontology':

"It is probably safe to assume that all paleontologists regardless of
their persuasion have used the generic category in one way or another,
and some have been known to propose a new genus now and again (for some of
us, it might even be said to be again and again). Even so, the question,
'What is a genus?' is as elusive today as in the time of Linnaeus."
(Amsden, 1970, p. 156)
    Sweet and Bergström (1970), working on those fascinating extinct marine
organisms, the Conodonts, state,
    "As long as there are two generic concepts for conodonts, and thus two
    specific concepts, at least some of the Linnean binomens employed will
    have two meanings.' (op.cit., p. 170)
    In the botanical area, Beck (1970) says that the establishment of a valid
conception of a fossil plant genus is extremely difficult and he adds,
    "...I believe that we may never be able to achieve a uniformly acceptable
    means of generic delimitation in paleobotany. Even among living plants,
    the genus is essentially an arbitrary category, based in those cases where
    variation is continuous, on the judgment of experienced systematists.
    Where there is discontinuity by virtue of past extinction, one still must
    decide whether the group delimited by gaps should be called a genus or
    assigned to some other taxonomic category." (op.cit., p. 183)
This is what Cooper (1970) writes of brachiopods:
    "Brachiopod genera are now appearing at a rapid rate. Since the appear-
    ance of the American Treatise (1965) more than 300 genera have been pro-
    posed indicating that new genera are appearing at a rate of about 75 a
    year.... For years I have been recording and tracking down the names
    applied to brachiopods. My count is well over 2600 genera and I regard
    about 2300 of these as useful names..." (op.cit., p. 196)
    And after all that, he is forced to admit, "It seems clear from the above
remarks that no general rule can be made relating to generic characters"
(op.cit., p. 240).
    Talking of the genus concept in Bryozoa, Boardman and Cheetham of the Smith
sonian Institution and Cook of the British Museum (Natural History), in a
joint paper (1970), have this to say:
    "... all supraspecific categories, including the genus, lack...direct
    relationship to gene pools. Thus genera and other supraspecific taxa are
    necessarily more broadly based on phenetic properties and as a result can
    be more variable in content; that is, each taxon of any given categorical
    rank above species may include varying combinations of several evolutionary
    lineages or segments of lineages depending on the phenetic criteria used
    in taxonomic grouping." (op.cit., p. 294)
    I have taken my examples from such diverse areas to illustrate the point
that the kind of problems palaeo-anthropologists are grappling with are not
peculiar to them, but are encountered right across the board of living things
And in case you may think that refuge is to be found in numerical taxonomy and
that all problems will disappear, as though by the wave of a wand, ponder for
a moment what Rowell (1970) has to say on numerical taxonomy and the genus
concept:
    "The category Genus, like any other category, is an artefact, but a useful
    one in the taxonomic hierarchy that we commonly employ. It is however ex-
    ceedingly difficult to define in any nonarbitrary manner.
    "Numerical methods, although useful in other aspects of taxonomy, have
    to date not contributed much to the concept of categories. Indeed, it
    would seem beyond their capability to do so. Methods with fundamentally

different philosophical approaches ... have been developed, that in their different ways enable one to explicitly determine relative ranks of taxa within a study. Taxa are assigned ranks dependent on the amount of divergence, or the related degree of similarity that they display. However, how much homogeneity related taxa must display to be considered as of generic rank or what is the critical divergence for a genus can only be determined empirically, if at all. These numerical approaches cannot answer the question of how big, how diverse should a taxon be to be considered of generic rank."
That is the verdict of a numerical taxonomist.

The scope of the genus, Australopithecus

From the foregoing, it emerges clearly that the genus, *Australopithecus*, means different things to different people. Tobias's formal definitions proposed in 1967 and slightly modified by him in 1968 (Tobias, 1967, 1968a), included three species within the genus, namely *A. africanus*, *A. robustus* and *A. boisei*. The views of Clark Howell and of Coppens, though not presented formally, may be inferred from several writings to be similar.* Campbell's approach is very similar, except that he lumps *A. robustus* into *A. africanus*. R.E.F. Leakey's views are nearly the same, except that it appears he would like to remove some (? most) fossils from the hypodigm of *A. africanus* into *Homo*. Pilbeam, Simons, Wolpoff and Lovejoy recognize a genus *Australopithecus*, nearly as defined by Tobias, save for the addition of some or all of the hypodigm of *H. habilis* which they would subsume in *Australopithecus*.
Thus, a large number of active palaeoanthropologists recognize the existence of a single extinct genus, *Australopithecus*, that embraces the groups colloquially referred to as the gracile and robust australopithecines.

Hominid Classification and Phylogeny

While having a phenetic basis, all of the six classificatory systems reviewed here are, to a greater or lesser degree, phyletic. Hence, the diversity of classificatory systems reflects a variety of views on hominid phylogeny. Thus, Pilbeam (1972), in his figure 73 (op.cit., p. 144), shows four alternative phylogenies for *Australopithecus* species, while Edey in *The Missing Link* (1973) portrays a spectrum of six different views of the hominid family tree, attributed respectively to B.G. Campbell, W.E. Le Gros Clark, J. Napier, P.V. Tobias, C.L. Brace and L.S.B. Leakey. These varying approaches will not be reviewed here; my point in mentioning them is to stress that, in the present state of play, the available fossils are amenable to more than one systematic and phylogenetic interpretation, even at the hands of serious, competent and well-informed scholars.

---

* Howell (1975, and footnote in Howell and Coppens, 1976) would, it seems, go a little further and recognize the robust australopithecines of Swartkrans as belonging to the species, *A. crassidens*, while those from Kromdraai would remain as representatives of the species, *A. robustus*. *A. boisei* of East Africa, on this terminology, is closer to *A. crassidens* than to *A. robustus*.

The hypodigm of A. africanus

Though conflicting views may exist on the nature and scope of the genus *Australopithecus*, there is much less room for doubt when it comes to its type species, *A. africanus*. The species was created by Dart in 1925. Its type specimen is the Taung skull which remains, to this day, one of the completest early hominid skulls yet found. The hypodigm comprises a very large sample of fossils from Sterkfontein and a moderate sample from Makapansgat, while to it, also, have been assigned a number of fossils from Omo, and perhaps those of Lothagam and Kanapoi. For some, a few Olduvai fossils belong in *A. africanus*, including the Olduvai hominid 7 mandible (though not the associated parietals) that formed part of the type specimen of *H. habilis*. Others incline to the view that all the hominids in Beds I and II may be accommodated in *A. boisei*, *H. habilis* and *H. erectus*. Hence, the question of whether the species *A. africanus is* represented at Olduvai at all remains an open one.

(Several workers believe that the mandible and parietals, which were given the same catalogue number, OH 7, though associated on the same living floor, represent two different individuals, one belonging to the big-brained *habilis* group and the other to *A. africanus*. These students assert that the mandible was misidentified by Leakey et al., 1964, and that it should be removed from the type of *H. habilis*. Others point to the juvenile age of both jaw and parietals, as supporting their assignment to the same individual; and they further point to dental features of the mandible which they claim to be distinguishable from those of *A. africanus*: the buccolingual narrowness of the premolars, for instance, could be matched by no *A. africanus* teeth, despite rather large samples available from Sterkfontein and Makapansgat - Tobias, 1966. The question of the position of the mandible of Olduvai hominid remains for the time being unresolved.

This hypodigm, even omitting for the moment the East African claimants and referrals, is extensive. If we exclude the Sterkfontein hominids from Member 5 (formerly 'Middle Breccia' of the 'Extension Site') and if we disregard here the possibility that the Taung specimen may on close comparative study prove to belong in the robust lineage, the total sample of South African fossils assigned to the species *A. africanus* up to 1966 numbered 436 (Table I).

TABLE I

South African fossils assigned to A. africanus (1966)

	Taung	Sterkfontein	Makapansgat	Total
Deciduous teeth	20	18	2	40
Permanent teeth	4	190	59	253
Cranial parts	5	55	28	88
Postcranial bones	0	44	11	55

Mann's careful palaeodemographic studies (1975) of the fossils from the South African sites led him to infer that the Sterkfontein 'Type Site' (Member 4) fossils represent between 25 and 40 individuals, though there is an outside chance that as many as 60 individuals may be represented. From Magapansgat we have fossils representing between 8 and 13 individuals, possibly even 14. When the Taung child is added, the *minimum* number of individuals assigned to *A.*

africanus from South Africa is 34, whilst the number is quite likely to be as
high as 54. There is a remote chance that the number of individuals may be
as high as 75. Even at the minimum value of 34, this is an excellent figure
for a palaeontological sample. The concept of a species that one can create
on a sample representing 34 to 54 individuals would most likely have a considerable degree of validity.

If we consider only those specimens to which an approximate age can be
assigned - as Mann (1968, 1975) and Tobias (1968b, 1974a) have done - we arrive at the breakdown given in Table II.

TABLE II

*Provisional age assignment of A. africanus specimens from South Africa**

	Taung	Sterkfontein Member 4	Makapansgat Members 3+4	Age Totals
Early Childhood	1	4	3	8
Later Childhood	-	5	3	8
Adolescence	-	5	1	6
Young Adulthood	-	2	3	5
Adulthood	-	30	6	36
Site Totals	1	46	16	63

* Table excludes the newest finds from Sterkfontein.

Thus, of 62 specimens of gracile australopithecines (excluding Taung), 41 or
66% are adult and 21 or 34% immature. Similar results were obtained in an
independent study by Mann (1968, 1975): on his analysis, in which actual
ages were assigned, 13 out of 55 specimens from Sterkfontein and Makapansgat
represent individuals who died at 15 years of age or less; 34 are from individuals who died at 21 years and over; while 8 are derived from individuals
who died between the ages of 16 and 20. If, for purposes of comparison with
my figures, we divide the latter category equally between the immature and
mature groups, we arrive at an estimate of 17 or 31% immature and 38 or 69%
mature, which figures are close to my own, based on 62 instead of 55 specimens,
namely 34% and 66% respectively.

If the provisional age estimations for *A. africanus* specimens just cited
reflect to any degree the age at which individuals died, it would seem that,
in the populations represented in the dolomitic cave deposits, one out of
three members of *A. africanus* died before anatomical adulthood was attained.
This contrasts with the position among the robust australopithecines from
Kromdraai and Swartkrans: there, 57% (Mann) or 58% (Tobias) of specimens
represented individuals who were immature at death (Table 3).

The explanation of this demographic difference between *A. africanus* and
A. robustus may be purely taphonomic, a reflection of possibly varying circumstances under which the several sets of bones accumulated in the respective
deposits. On the other hand the demographic differences may possibly reflect
varying cultural and ecological factors. The gracile *A. africanus* was the
earlier inhabitant of southern Africa and we have no evidence that a more

advanced tool-making hominid was contemporary with him in the same region. On
the other hand, synchronic and sympatric with the later *A. robustus* were tool-
making hominids of the genus *Homo*, whose handiwork and skeletal remains are
evident in Member 1 at Swartkrans and, as we have very recently found, in
Member 5 at Sterkfontein.* This additional competitive element in the life of
the late-surviving robust australopithecines may account for the larger proportion of young specimens belonging to individuals of pre-childbearing age
at Swartkrans and Kromdraai than at the earlier *A. africanus* deposits (Tobias,
1974a).

TABLE III

*Specimens of immature and mature early hominids
from South Africa* †

	% Immature	% Mature	*n*
A. africanus			
(from Sterkfontein	34*	66*	62
and Makapansgat)	31¹	69¹	55
A. robustus			
(from Kromdraai	57*	43*	94
and Swartkrans)	58¹	42¹	117

¹ Mann, 1968, 1975

* Tobias, 1968b, 1974a

† Excluding the newest finds from Sterkfontein.

New additions to A. africanus

To the available samples of hominids of *A. africanus*, researches of the
last decade have added an appreciable number of new specimens. From Makapansgat, where no new systematic work has been prosecuted, has come one new canine
tooth of *A. africanus*, discovered by J. Kitching in 1975. The majority of new
finds have been made at Sterkfontein where A.R. Hughes and I have completed
a decade of continuous excavation, which started on 1st December 1966, one
day after the centenary of the birth of the late Dr. Robert Broom, F.R.S.
In that time, our excavation and the studies it made possible have thrown a
flood of light on the extent, stratigraphy, dating, fauna and palaeoecology
of the deposit, which will be reported on elsewhere. Table IV reflects the
dates and anatomical identity of the new hominid fossils found during the decade under review, while Table 5 enumerates and classifies the parts of the
skeleton recovered.
Most of the new specimens are derived from Partridge's Member 4 and have
been assigned provisionally to *A. africanus* (Table VI). Thus, the hypodigm
of Sterkfontein fossils attributed to *A. africanus* has been swelled by the

---

* See Addendum to this chapter.

addition of: 8 cranial parts (including mandibular fragments); 62 teeth and
dental fragments; and 13 postcranial bones - namely a total of 83 items.
  Especially valuable parts of this list are 9 new front teeth, comprising
6 incisors and 3 canines; and 13 new post-cranial bones, comprising 6
thoraco-lumbar vertebrae, 1 humeral and 1 radial fragment, 3 metacarpals and
finger phalanges, and 2 femoral heads.

TABLE IV

Discovery of new hominid specimens from Sterkfontein
1968 - July 1976

1968	2 isolated teeth
1969	5 isolated teeth
	4 articulated lumbar vertebrae
1970	1 isolated tooth
1971	4 isolated teeth
	1 cranium with teeth *in situ*
	1 mandible with 8 teeth *in situ*
1972	1 maxillary fragment with 2 teeth
	1 maxillary fragment with 3 teeth
	3 isolated teeth
1973	6 isolated teeth
	1 maxillary fragment with 2 teeth
	1 metacarpal
	1 finger phalanx
1974	3 isolated teeth
	1 maxillary fragment with 1 tooth
	1 distal humerus
	1 metacarpal
	1 finger phalanx
	2 heads of femora
	3 miscellaneous fragments
1975	3 isolated teeth
	1 maxillary fragment with tooth fragments
	1 mandibular corpus with 5 teeth
	2 articulated thoraco-lumbar vertebral bodies
1976	6 isolated teeth
	1 maxillary fragment with 2 teeth
	1 distal radius

TABLE V

Summary of new hominid fossils from Sterkfontein
1968 − July 1976

65 permanent teeth
 1 cranium
 6 maxillary fragments
 2 partial mandibles
 6 thoraco-lumbar vertebrae
 1 humeral fragment
 1 radial fragment
 2 metacarpals
 2 finger phalanges
 2 femoral heads

Figure 1. Left norma lateralis of cranium StW 13, various parts of
which were found by S. Gasela, A.R. Hughes and P.V. Tobias
between August 1971 and April 1972.   The cranium was
found in situ in Member 4 some 6 metres from where Sts 5
('Mrs Ples') was found in 1947.

Figure 2.   *Right view of six hominid thoraco-lumbar vertebrae from*
Sterkfontein. The lower four still articulated vertebrae
were discovered to be hominid in 1969 though their pro-
venance was uncertain. The upper two vertebrae were dis-
covered by P.V. Tobias on 10 January 1975 and are derived
from Member 4 of the Sterkfontein Formation, that is, the
layer that is so rich in remains of Australopithecus
africanus. On the basis of concordance of size, shape, state
of preservation, colouration and matching areas of damage,
there is little room for doubt that the upper two and the
lower four vertebrae are consecutive parts of the same
vertebral column.

TABLE VI

Division of new hominid fossils between Members 4 and 5
of Sterkfontein formation
1968 – July 1976

Member 5 (former Middle Breccia)	2 isolated teeth 1 maxillary fragment with 2 teeth 1 metacarpal	*Homo species*
Member 4 (former Lower Breccia)	61 permanent teeth 1 cranium 5 maxillary fragments 2 partial mandibles 6 thoraco-lumbar vertebrae 1 humeral fragment 1 radial fragment 1 metacarpal 2 finger phalanges 2 femoral heads	*A. africanus*

*Figure 3. Metacarpals and phalanges from Sterkfontein. The two bones
on the left are a metacarpal and proximal phalanx, probably of
the same digital ray of the same individual from Member 4.
The bone in the middle is an incomplete proximal phalanx probably of a middle finger, also from Member 4. On the right
is the distal part of a metacarpal from Member 5, the stratum
that has yielded stone implements and the recently discovered
skull provisionally assigned to Homo. These bones were found
by A.R. Hughes in 1973 and 1974.*

All of these specimens are under study at the present time. They have
brought the grand total for Sterkfontein hominid fossils to 395 (Table VII)
and for Sterkfontein hominid specimens attributed to *A. africanus* to 18 deciduous teeth, 252 permanent teeth, 63 cranial parts, and 57 postcranial bones.
These totals exclude the items previously recovered from Member 5 (the former
'Middle Breccia' of the 'Extension Site') and regarded by myself since 1965
as belonging to *Homo* sp. Important new additions have been made to the
Member 5 fossils recently and these serve to confirm the view that the identifiable hominid fossils from Member 5 do not belong to *A. africanus* but to
Homo sp.*

The new augmented totals from Sterkfontein Member 4, coupled with those
from Makapansgat and Taung, bring the grand totals of South African fossils
assigned to *A. africanus* to the figures given in Table VIII.

The 316 permanent teeth represent an increase of about 25%, the 96 cranial
parts a 9% increase and the 68 post-cranial bones an increase of about 24%.

For those who may be inclined to doubt the reality of *A. africanus* as a
valid early hominid taxon, it is important to contemplate the very bulk of
material available, probably exceeding that referred to any other taxon of

---

* See Addendum to this chapter.

early hominid save, perhaps, for that of *A. robustus*. The stockpile of hominid fossils from Sterkfontein and Makapansgat remains a standard reference collection with which it is necessary to compare all new discoveries of early hominid fossils, before the latter may be classified.

TABLE VII

Tally of hominid fossils from Sterkfontein (July 1976)*

	Recent Discoveries	Previous Total	New Total
Deciduous Teeth	0	18	18
Permanent Teeth	65	190	255
Cranial parts	9	55	64
Post-cranial bones	14	44	58
Total	88	307	395

TABLE VIII

South African fossils assigned to A. africanus (July 1976)

	Taung	Sterkfontein	Makapansgat	Total
Deciduous Teeth	20	18	2	40
Permanent Teeth	4	252	60	316
Cranial parts	5	63	28	96
Post-cranial bones	0	57	11	68

Underlined items show an increase over 1966 totals.

The place of A. africanus in time

Researches since 1970 have at last provided reasonable estimates of the age of the *A. africanus*-bearing deposits in South Africa. Faunal comparisons with the well-dated faunae of East Africa concur in pointing to an age of about 2.5-3.0 Myr for the fossils derived from Member 4 of Sterkfontein and from especially Member 3 of Makapansgat (on Partridge's proposed stratigraphic terminology, 1975). This estimate is supported by another and independent line of new evidence just announced, namely preliminary palaeomagnetic results from Makapansgat: these results indicate a maximum age of between 2.8 and 3.3 Myr for Member 2 of the Makapansgat Formation (Brock et al., 1977).

The large time-lapse between Members 4 and 5 (Robinson's former Lower and Middle Breccias) is testified to by the presence of *A. africanus* and of archaic (Sterkfontein Faunal Span) mammals and the absence of stone tools in Member 4,

---

* Subsequent to the compilation of Tables IV to VIII, there were found between August and November 1976, the following additional Sterkfontein hominid specimens: 1 cranium, including 9 perfect teeth, and a mandibular fragment (see Addendum to chapter); 8 isolated teeth or dental fragments; 1 maxilla with 7 deciduous and permanent teeth; and a further metacarpal.

and the presence of *Homo* sp., stone tools and mammals of the later Swartkrans or Cornelia Faunal Span in Member 5 (Tobias, 1965; Vrba, 1974).

The place of A. africanus in phylogeny

Recently, claims have been made for the existence of very early members of the genus *Homo* in East Africa. This has led some workers, perhaps a little prematurely, to discount the probable role of *A. africanus* in hominid phylogeny. Thus, Oxnard (1975) states, "The genus *Homo* may, in fact, be so ancient as to parallel entirely the genus *Australopithecus* thus denying the latter a direct place in the human lineage". This inference is open to criticism on several grounds:

(1) The evidence is not at all convincing for the very early age of those East African fossils that most scholars agree should be assigned to *Homo*. The fossils originally ascribed to *H. habilis* were those of Olduvai Bed I and lower Bed II and the age of the oldest of this group of fossils is 1.8 Myr. At Omo, in southern Ethiopia, the earliest appearance in the record of fossils suggesting affinities with *H. habilis* occurs at almost the same time, ca. 1.85 Myr (Howell and Coppens, 1976). The *Homo* fossils from east of Lake Turkana were originally said to be of an age of ca. 2.9 Myr: this was based on the supposed age of the KBS tuff, originally said to be 2.6 Myr. It was solely the supposed early age of this group of large-brained fossils, assigned by some to *H. habilis*, that gave rise to the view that the genus *Homo* had been in existence for over a million years before the Olduvai *H. habilis* fossils. However, recently, Curtis et al. (1975) have revised the age of the KBS tuff to approximately one million years younger, i.e. about 1.8 to 1.6 Myr. If the earliest *Homo* fossils from east of Lake Turkana are still considered to be about 0.3 Myr older than the KBS tuff, and if the Curtis revision is correct, the earliest *Homo* fossils from east of Lake Turkana would be about 2.1-1.9 Myr, almost the same age as at Omo (1.85 Myr) and Olduvai (1.8 Myr). In South Africa, the early *Homo* of Swartkrans (formerly 'Telanthropus') would seem, on Vrba's bovid evidence, to be about 2.0-1.5 Myr old (Brain, 1976): that is, it is of the same order of age as East African early *Homo*. It seems likely, from the work of Vrba and other evidence, that early *Homo* in Member 5 of Sterkfontein is of the same order of age.

(2) The evidence that any East African fossils, clearly dated to earlier time periods, belong to *Homo* is far from satisfactory. The most obviously distinguishing feature of early *Homo* is the cranial capacity that is significantly larger than that of *A. africanus*. This crucial evidence is apparently lacking from all the very old (3-4 Myr) finds of Hadar, Ethiopia, and Laetolil, Tanzania. The dental evidence for early *Homo* (albeit the criteria are not accepted by all) has either not been demonstrated or is not available at these early sites. The evidence of those mandibles which lack teeth is not at all convincing: we have as yet no clear, demonstrated and accepted set of criteria to distinguish between the jaws of *A. africanus* and those of early *Homo* (see Tobias, 1974b). The evidence of post-cranial bones remains suspect, partly at least because of the difficulty of assigning such bones to taxa whose recognition was based largely on dental and cranial features; in any event, we have no body of agreed wisdom on the differences between the post-cranial bones of *Homo* and those of *A. africanus*. For these reasons, it is concluded that, at this stage, there is no convincing evidence to assign to *Homo* these exciting new fossil finds from Hadar and Laetolil; on the contrary, at least some of the preliminary accounts suggest strong affinity with South

African fossils of *A. africanus* (Leakey, M.D. et al., 1976).

(3) For the reasons given in paragraphs (1) and (2), it may reasonably be stated that there is at present no good evidence for the occurrence of *Homo* before 2.0-1.8 Myr.

(4) The statement that early *Homo* may parallel *Australopithecus* in time or may even antedate it fails to take cognisance of the new evidence on the early age of *A. africanus*-bearing strata at Sterkfontein and Makapansgat. Indirect faunal dating has set *A. africanus* at 3.0-2.5 Myr at these two sites and at one of them the palaeomagnetic evidence supports this. Similarly, at Omo, with its excellent suite of radiometric dates, *A. africanus* (or a closely related species) is characteristic of the 3.0-2.5 Myr time range (Howell and Coppens, 1976). From Hadar, Johanson and Taieb (1976) have reported the presence of specimens showing affinities with *A. africanus* from about 3 Myr. Thus, in both South and East Africa, we have good evidence for *A. africanus* or *A. aff. africanus* in the Upper Pliocene from 3.0-2.5 Myr and, in both areas, we have evidence for *convincing* early *Homo* appearing in the Lower Pleistocene (2.0-1.8 Myr).

Furthermore, it must not be forgotten that the yet earlier fossil mandible of Lothagam (ca. 5.5 Myr) closely resembles some of the *A. africanus* jaws. Likewise, the humeral fragment of Kanapoi (ca. 4.4 Myr) has been likened to that of *A. africanus*, although here the same caveat should be applied, as mentioned previously. It remains to be seen whether other early fossils such as those of Laetolil include specimens of *A. africanus* or *A. aff. africanus*.

Thus, even if we exclude the 4 Myr and 5 Myr fossils, the evidence of the later (3.0-2.5 Myr) hominid fossils shows that *A. africanus* (including the East African *A. aff. africanus*) is well-placed in time for some populations of this medium-toothed australopithecine species to have been ancestral to early *Homo*.

(5) It has been recognized since 1925 that, in encephalic and in facial, gnathic and dental characters, *A. africanus* shows the presence of morphological trends known to be essential components of the process of hominization. Later finds of *A. africanus* have served not only to strengthen the weight of evidence supporting this conclusion but to broaden the notion to more aspects of the total morphological complex. Oxnard (1975) has recently challenged this commonly held view on the morphology of *A. africanus*, by his multivariate statistical studies on several postcranial fragments. Of eight anatomical areas studied, he has concluded that four are unique, being like neither the pattern of modern man nor that of modern apes; whilst four show resemblances to the orang-utan. All 8 regions or bones are post-cranial. It is perhaps significant that Oxnard does not apply his approach to teeth, jaws, calvariae, endocasts and endocranial capacities - the features on which the hominid affinities of the australopithecines have been determined and which have been repeatedly shown to sort the australopithecines with the Hominidae rather than with the Pongidae. A further consideration is that, for most parts of the post-cranial skeleton of the australopithecines, Oxnard's work is on very small samples and sometimes unique specimens. This compels him to disregard the variability of each anatomical region in the australopithecines. In contrast, a great deal is known about cranial, encephalic and especially dental variability.

I have already referred to the large problem of knowing to which taxon (defined largely cranio-dentally) each post-cranial fragment is to be assigned. If, with Oxnard, we ignore for the moment the cranio-encephalo-dental evidence, what is to be inferred from Oxnard's two sets of results referred to in the

previous paragraph? As to the uniqueness of the australopithecines, confirmed
by the Oxnard studies on four parts of the post-cranial skeleton, is this finding really surprising?

Australopithecus is generically distinct, in the judgment of most specialists, and it is to be expected that in many respects it would be unique -
especially in those features which are apomorphous or derived. The fact that
in four other features the australopithecine uniqueness bears some resemblance
(if indeed it does!) to the orangoid uniqueness certainly does not suggest
that *Australopithecus* is more closely related to the orang than to any other
hominoid lineage. The resemblances may point to nothing more than examples
(mainly in the upper limb) of convergent evolution. In the light of all the
other evidence - both postcranial and cranio-dental - could this suggested
occurrence of convergent evolution between two hominoid lineages disqualify
Australopithecus - or at least *A. africanus* - from being ancestral to *Homo*?
I do not believe that it would be valid to draw such an inference from this
evidence alone; and especially when it is contradicted by the large body
of cranio-dental evidence on the affinities of the australopithecines, as well
as by the latest dating evidence just reviewed.

Another difficulty resides in the method itself. The method of multivariate
statistical analysis, as used by some palaeo-anthropologists, has come in for
much criticism of late. It is not proposed to review here the pros and cons
of the method; but it is only fair to point out that conflicting results
have been obtained by the application of multivariate morphometric approaches
to the *same* post-cranial bones of the early hominids (for example, those of
McHenry and Corruccini, 1975). This fact - that, in the hands of different
and highly competent specialists, the same bones can yield different results
when similar analytical tools are applied to them - makes Oxnard's claimed
resemblances (or differences) somewhat suspect. Certainly this methodological
problem should be resolved before sweeping claims are based upon the results
of such multivariate studies.

Thus, both the use of the method, the selection of the materials to which
it is applied and the interpretations of the results leave room for doubt.

Although an analysis or re-analysis of the morphology of the australopithecines has no place in this paper, it remains fair to claim that nothing
in the morphology of *A. africanus* would necessarily exclude some populations
of the species from having been ancestral to later hominids of the genus *Homo*.

CONCLUSIONS

We have already concluded that a critical look at the dating evidence dis-
counts the view that early *Homo* was earlier in time than, or even parallel
with, *A. africanus*. The proposition stands that *A. africanus* - as defined
from the South African fossils, reinforced by the Omo sequence and by some
at least of the Laetolil and Hadar remains - is well-placed in time for some,
probably earlier, populations of that species to have been ancestral to later
hominids of the genus *Homo*.

Both the chronological and the morphological evidence have led to the same
conclusion, that *A. africanus* - or at least some populations of it - *could
have been* ancestral to later hominids of the genus *Homo*. We can go a little
further and ask: is there any other hominid at the right level in time and
which is morphologically a stronger claimant than *A. africanus*? Despite the
claims that the new Tanzanian and Ethiopian fossils are closer to *Homo*, or
even members of the genus *Homo*, they do not seem to me to show features that

betoken greater hominisation than that shown by *A. africanus*. Closer study than these new fossils have so far received may compel me to change my mind on this point; but at the moment, I am not at all convinced that there are any morphological traits in the two new sets of ancient fossils which either point to a significantly greater degree of hominisation than that of *A. africanus*, or cannot be accommodated within the estimated population range of variability for *A. africanus*. In other words, it seems to me, on what evidence I have seen, that the new fossils probably fall into *A. africanus* (or, at the least, *A. aff. africanus*) and that they help us to expand our concept of the variability - regional, racial, ecological - of that late Pliocene species.

Hence, faced with what I consider to be the absence of a more hominised early hominid round about 3 million or more years ago, one is led to conclude not only that some populations of *A. africanus* could have been ancestral to later *Homo*, but also that among all the earlier hominids known, this species is *the most likely claimant* to have been ancestral to *Homo*.

ADDENDUM: STW 53: A NEW HOMINID SKULL FROM THE STONE-TOOL-BEARING LAYER AT STERKFONTEIN

A new fossil skull of an early hominid has just been found* in Member 5 of the Sterkfontein cave deposit. It was discovered by Mr. Alun R. Hughes who has been in charge of field-operations in the excavation at Sterkfontein since 1966. The skull is of considerable importance, as it provides the first good evidence bearing on the nature of the hominid who was probably responsible for making the stone tools (Tobias, 1976, Hughes and Tobias, 1977).

Figure 4. Part of norma facialis of the new Sterkfontein skull discovered by A.R. Hughes in Member 5 of the Sterkfontein Formation. The cranium has been provisionally assigned to an early member of the genus Homo and has some affinities with the lower Pleistocene species Homo habilis from East Africa. It also has some strong resemblances with the specimen of Homo sp. (previously called 'Telanthropus capensis') from the neighbouring site of Swartkrans.

* This first announcement was made on 26th August 1976, only 9 days after the last part of the new skull was discovered on 17th August.

The first portions of the cranium came to light on 9th August 1976, forty years to the day since the first visit was paid to Sterkfontein by the late Dr. Robert Broom F.R.S. More parts came to light over the ensuing week: all came out of a pocket of decalcified breccia, the walls of which were formed from calcified breccia of Member 5. It is from Member 5 (as it is now called, on Dr. T.C. Partridge's new classification) that stone tools had been excavated in 1956 by C.K. Brain and by J.T. Robinson. Member 5, too, contains a bovid fauna much younger than that of the underlying Member 4. The subjacent Member 4 is very rich in remains of *A. africanus*, has a fauna that resembles East African forms dated radiometrically to between 3.0 and 2.5 Myr, but totally lacks stone implements.

For 20 years, it had been uncertain whether the Member 5 stone tools were made by *A. africanus* or by another hominid. Only a few fragments of hominid skeletal material had previously been found in Member 5 and these were so incomplete that it had been impossible to determine which hominid they represented. One view was that they belonged to *A. africanus* (J.T. Robinson) and the other was that they belonged to an early member of the genus *Homo* (Tobias).

It was thus extremely important to be sure that the new skull in the decalcified pocket did, in fact, emanate from Member 5. The difficulty was that the contents of decalcified pockets, that form in the upper surface of deroofed breccia, comprise not only material derived from the breccia itself, to a volume of about 60%, but also material derived from the surface higher up. Had we been forced simply to make the assumption that the new cranium came from Member 5, there would always have remained room for doubt on its exact provenance.

This is the background to the final spectacular find made by Hughes - on 17th August, 40 years to the day after Broom had found the first adult specimen of an *A. africanus* at Sterkfontein!

On 17th August 1976, a substantial part of the vault of the same cranium was found *in situ*, in the side-wall of the pocket, where the breccia was still solid and calcified. Broken fragments of the vault from the pocket actually fitted on to the large moiety of the calvaria in the wall. This provided the critical evidence that the new cranium (Stw 53) had been derived from the layer represented in the wall of the pocket, namely Member 5.

Stw 53 is very different from the crania of *A. africanus*. In a number of respects, it suggests a hominid that is closely related to the early species of *Homo* to which was given the name *H. habilis*. Elsewhere in Africa, too, the remains of *H. habilis* are associated with signs of early tool-making, though not necessarily the earliest stone tools. Its appearance in Sterkfontein, at a level provisionally assigned a date of 2.0 to 1.5 Myr, adds one more to the African sites where there is evidence of *Homo habilis* not appearing until about 2.0 Myr ago. Its stratigraphic relationship to the lower member containing *A. africanus* neatly illustrates the way in which hominid phylogeny may well have occurred and provides indirect evidence supporting the status of *A. africanus* as older than and ancestral to *Homo*.

The new discovery is the most important fruit of our 10-year programme of excavations at Sterkfontein and certainly the most significant specimen to emerge from southern Africa since the finds of the late 'forties and early 'fifties.

REFERENCES

Amsden, T.W. (1970). *In* "The Genus: a Basic Concept in Paleontology",
    (T.W. Amsden, ed.), p. 156. Proc. N. Am. Paleont. Conv. (1969), Allen
    Press, Lawrence, Kansas.
Beck, C.B. (1970). *In* "The Genus: a Basic Concept in Paleontology", (T.W.
    Amsden, ed.), pp. 173-193. Proc. N. Am. Paleont. Conv. (1969), Allen
    Press, Lawrence, Kansas.
Blackwelder, R.E. (1964). *In* "Phenetic and Phylogenetic Classification",
    (V.H. Heywood and J. McNeill, eds), pp. 17-28, Publication No. 6, The
    Systematics Association, London.
Boardman. R.S., Cheetham, A.H. and Cook, P.L. (1970). *In* "The Genus: a Basic
    Concept in Paleontology", (T.W. Amsden, ed.), pp. 294-320, Proc. N. Am.
    Paleont. Conv. (1969), Allen Press, Lawrence, Kansas.
Brain, C.K. (1976). *S. afr. J. Sci.*, 72, 141-146.
Brock, A., McFadden, P.L. and Partridge, T.C., (1977). *Nature,Lond.*,266, 249-250.
Broom, R. and Robinson, J.T. (1949). *Nature, Lond.*, 164- 322-323.
Campbell, B.G. (1972). *Ann. Rev. Anthropol.*, 1, 27-54.
Cooper, G.A. (1970). *In* "The Genus: a Basic Concept in Paleontology", (T.W.
    Amsden, ed.), pp. 194-263, Proc. N. Am. Paleont. Conv. (1969), Allen Press,
    Lawrence, Kansas.
Curtis, G.H., Drake, R.F. Cerling, T.E. and Hampel, J.M. (1975). *Nature,
    Lond.*, 258, 395-398.
Dart, R.A. (1925). *Nature, Lond.*, 115, 195-199.
Edey, M. (1973). "The Missing Link", Time-Life International (Nederland) BV.
Groves, C.P. and Mazak, V. (1975). *Cas. Miner. Geol.*, 20, 225-247.
Hennig, W. (1965). Quoted by Goodman, M. (1974) *in* "Protein Sequence and
    Immunological Specificity in the Phylogenetic Study of Primates", pp. 1-30,
    Wenner-Gren Symposium No. 61.
Hennig, W. (1966). "Phylogenetic Systematics", (translated by D.D. Davis and
    R. Zangerl), 263 pp., University of Illinois Press, Urbana, Illinois.
Howell, F.C. (1975). *In* "Mammalian Evolution in Africa", (V.J. Maglio, ed.),
    Princeton University Press, Princeton.
Howell, F.C. and Coppens, Y. (1976). *In* "Earliest Man and Environments in the
    Lake Rudolf Basin", (Y. Coppens, F.C. Howell, G.L. Isaac and R.E.F. Leakey,
    eds), pp. 522-532, Chicago University Press, Chicago and London.
Hughes, A.R. and Tobias, P.V. (1977). *Nature, Lond.*, 265, 310-312.
International Code of Zoological Nomenclature (1961). (N.R. Still et al.,
    eds), International Trust for Zoological Nomenclature, London.
Johanson, D.C. and Taieb, M. (1976). *Nature, Lond.*, 260, 293-297.
Leakey, L.S.B., Tobias, P.V. and Napier, J.R. (1964). *Nature, Lond.*, 202, 7-9.
Leakey, M.D., Hay, R.L., Curtis, G.H., Drake, R.E., Jackes, M.K. and White,
    T.D. (1976). *Nature, Lond.*, 262, 460-466.
Leakey, R.E.F. (1971). *Nature, Lond.*, 231, 241-245.
Leakey, R.E.F. (1972). *Nature, Lond.*, 237, 264-269.
Le Gros Clark, W.E. (1955, 1964). "The Fossil Evidence for Human Evolution",
    University of Chicago Press, Chicago and London (1st and 2nd edns).
McHenry, H.M. and Corruccini, R.S. (1975). *Am. J. phys. Anthrop.*, 43, 263-270.
Mann, A.E. (1968). The paleodemography of *Australopithecus*, Ph.D. thesis,
    University of California, Berkeley.
Mann, A.E. (1975). "Paleodemographic Aspects of the South African Australo-
    pithecines", pp. 1-171, Pubs. Univ. Pa. Anthrop. 1.
Oxnard, C.E. (1975). *Nature, Lond.*, 258, 389-395.

Partridge, T.C. (1975). Communication to a conference of the Southern Africa
    Society for Quaternary Research, Cape Town, June 1975.
Pilbeam, D.R. (1972). "The Ascent of Man", Macmillan, New York.
Robinson, J.T. (1954). *Am. J. phys. Anthrop.*, 12, 181-200.
Robinson, J.T. (1965). *Nature, Lond.*, 205, 121-124.
Robinson, J.T. (1972a). "Early Hominid Posture and Locomotion", Chicago Univ.
    Press, Chicago and London.
Robinson, J.T. (1972b). *Nature, Lond.*, 240, 239-240.
Rowell, A.J. (1970). *In* "The Genus: a Basic Concept in Paleontology", (T.W.
    Amsden, ed.), pp. 264-293, Proc. N. Am. Paleont. Conv. (1969), Allen Press,
    Lawrence, Kansas.
Simons, E.L. (1972). "Primate Evolution: an Introduction to Man's Place in
    Nature", Macmillan, New York.
Sweet, W.C. and Bergström, S.M. (1970). *In* "The Genus: a Basic Concept in
    Paleontology", (T.W. Amsden, ed.), pp. 157-173, Proc. N. Am. Paleont.
    Conv. (1969), Allen Press, Lawrence, Kansas.
Tobias, P.V. (1965). *S. Afr. archaeol. Bull.*, 20, 167-192.
Tobias, P.V. (1966). *Nature, Lond.*, 129, 953-957.
Tobias, P.V. (1967). "Olduvai Gorge - Vol. II. The Cranium and Maxillary
    Dentition of *Australopithecus* (*Zinjanthropus*) *boisei*", University Press,
    Cambridge.
Tobias, P.V. (1968a). *In* "Taxonomy and Phylogeny of Old World Primates with
    References to the Origin of Man", pp. 277-318, Rosenberg & Sellier, Turin.
Tobias, P.V. (1968b). *Anthropologist* (Delhi), special volume, 23-28.
Tobias, P.V. (1973). *In* "L'Origine dell'Uomo", pp. 63-85, Accademia Nazionale
    dei Lincei, Rome.
Tobias, P.V. (1974a). *Leech, Johannesburg*, 44, 119-124.
Tobias, P.V. (1974b). *In* "Perspectives in Palaeoanthropology", (A.K. Ghosh,
    ed.), pp. 9-17, Firma K.L. Mukhopadhyay, Calcutta.
Tobias, P.V. (1976). *S. Afr. J. Sci.*, 72, 227.
Vrba, E.S. (1974). *Nature, Lond.*, 250, 19-23.
Walker, A. (1976). *In* "Earliest Man and Environments in the Lake Rudolf
    Basin", (Y. Coppens, F.C. Howell, G.L. Isaac and R.E.F. Leakey, eds),
    pp. 484-489, Chicago University Press, Chicago and London.
Wolpoff, M.H. and Lovejoy, C.O. (1975). *J. Hum. Evol.*, 4, 275-276.

SOME PROBLEMS IN MIDDLE AND UPPER PLEISTOCENE HOMINID RELATIONSHIPS

C.B. STRINGER

Sub-department of Anthropology, British Museum (Natural History), Cromwell Road, London SW7 5BD, UK.

INTRODUCTION

Some recent analyses of late Pleistocene human crania suggested that there were quite distinct forms of hominids within the group sometimes called "Neandertals" (Howells, 1974, 1975; Stringer, 1974b). Certain crania appeared to be close to a modern morphology and were easily distinguishable from the later European and near-eastern/middle-eastern Neandertals. In general, a chronologically-based sequence of European hominids made up a reasonable morphocline leading to the later Neandertals (Stringer, 1974b) but analyses indicated that the later Neandertals were not the most likely ancestors of the European Upper Palaeolithic populations.

Other workers, for example Wolpoff, have continued to develop the idea of a world-wide Neandertal stage of human evolution. In a recent paper Wolpoff (1975) reiterated the view that anterior dental reduction was the factor behind the evolution of a modern cranial form from that of the classic Neandertals, although he recognised the individuality of the Neandertal facial and mandibular morphology, which he considered to be due to climatic and functional factors. He did not consider that any of the near-eastern or middle-eastern Upper Pleistocene hominids (including Skhūl and Qafza) were modern in morphology and stated that in almost every cranial and facial feature the sample was intermediate between the European Neandertals and the earliest European, anatomically modern *H. sapiens*. Furthermore, he stated that in most of the especially diagnostic features such as tooth size, cranial height, occipital breadth and facial form, the entire sample was Neandertal rather than modern in morphology. Such differences as existed were due to a slightly later date and a lesser degree of climatic adaptation. Wolpoff (1975) stated also that in the case of the Upper Palaeolithic populations, the samples recognised as early modern men were not in fact modern. They were probably as different from living populations as they were from preceding ones.

This paper extends my previous analyses of data from Pleistocene crania (Stringer, 1972, 1974a, 1974b) by the use of cranial angles, indices and the Penrose size and shape statistic and includes new observations on some of the Qafza and Arago (Tautavel) material. In the case of Qafza 9, which is at present unpublished, data have not been presented separately although they were used in the Penrose size and shape analyses. Whilst multivariate techniques continue to be used in the study of fossil hominids, there has been a tendency in recent years for biometricians, in particular, to urge greater caution in their application (for example, Kowalski, 1972; Corruccini, 1973). Corruccini suggested that rather than use complex multivariate techniques where the data may not fulfil all the criteria required for an analysis, it would be more satisfactory to use simpler techniques where the results are also easier to interpret.

By plotting scatter diagrams from the metrical characteristics of *individual* specimens the validity of terms such as "archaic *Homo sapiens*", "Neandertal", and "anatomically modern" can be tested. In the following results particular attention is paid to the position of the European Neandertals to see whether they, as a group, are intermediate between pre-last glaciation forms and modern specimens. The selected modern population means are plotted from large samples using Howell's body of data (Howells, 1973) or my own observations. Modern populations which are extreme in particular features have been plotted as well as more typical populations.

RESULTS

Symbols used in the graphs

● *Homo erectus* (Peking)
○ *Homo erectus* (Java)
✪ Solo
■ Classic Neandertal
▲ Upper Palaeolithic
△ Mesolithic
◆ means of modern populations ♂
◇ means of modern populations ♀
★ other specimens

Abbreviations used in the graphs

Ak	Arikara	Mo	Le Moustier (cast)
Am	Amud 1 (cast)	Mok	Mokapu
An	Andamans	No	Norse
Ar	Arago 21	Om	Omo
At1	"*Atlanthropus*" 4 (cast)	Per	Peru
Au	Australian	Pet	Petralona
Be	Berg	Qu	La Quina 5
Bu	Bushman	Rh	Rhodesian (Kabwe or Broken Hill)
Bur	Buriat	Sac	Saccopastore
Co	Cohuna (cast)	Sala	Sala
D.I.	Djebel Ighoud 1	Sald	Saldanha (cast)
Do	Dogon	Sh	Shanidar 1 (cast)
D.Q.	Djebel Qafza (or Qafzeh or Kafzeh)	Si	Singa
Eg	Egyptian	Sk	Skhūl (original and/or cast)
Eh	Ehringsdorf 9 (cast)	Sp	Spy
Es	Eskimo	St	Steinheim (cast)
Fe	La Ferrassie 1	Sw	Swanscombe
Fo	Fontéchevade 5	Tab	Tabūn 1
Ga	Galilee (cast)	Tas	Tasmanian
Gi	Gibraltar 1	Te	Teita
I.E.	Iwo Eleru	To	Tolai
Kan	Kanjera 1	T.T.	Teshik-Tash (cast)
Kr	Krapina C,E	Ve	Vértesszöllös 2 (cast)

Laz   Lazaret 3 (cast)   Za   Zalavar
L.Ch. La Chapelle   Zu   Zulu
M.C.   Monte Cristo

Cranial angles and indices (Fig. 1)

    Cranial angles provide simple measures of differences in cranial form and
of relationships between various parts of the face and vault. Although they
have been criticised as "disguised ratios", they nevertheless contain much
useful morphological and shape information and I have calculated individual
cranial angles for the various fossil hominids studied wherever possible.
    The angles employed here are those defined by Howells (1973); one addi-
tional angle is used, called by me the bregma angle (BRA). This is the
angle at bregma defined by the nasion-bregma chord and the basion-bregma
height. This angle discriminated well between anatomically modern and non-
modern crania and was easily obtained when the basion-bregma and nasion-
bregma angles were known.

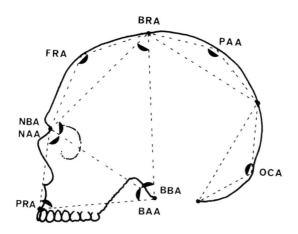

Figure 1. Diagram showing some of the cranial angles used in the analyses.

Frontal angle (FRA) v. Parietal angle (PAA) (Fig. 2)

   This figure shows a clear association between decrease in frontal angle
and decrease in parietal angle during the Pleistocene evolution of *Homo*.
Peking *H. erectus* specimens and the Solo crania (except Solo 5) are peripheral, with high parietal and frontal angle values. The Neandertals show
a wide range of values, slightly overlapping the modern range in the case of
the adolescent Le Moustier specimen and the immature Teshik-Tash cranium.
Tabūn, Ehringsdorf, Omo 1, Skhūl 5 and 9, and Qafza 6 appear rather "advanced"
(in the sense of being close to modern values), whereas La Quina, Djebel
Ighoud, Singa, Omo 2, Iwo Eleru and Cohuna appear "archaic". In the case
of Cohuna this is due solely to the remarkable dimensions of the frontal bone.

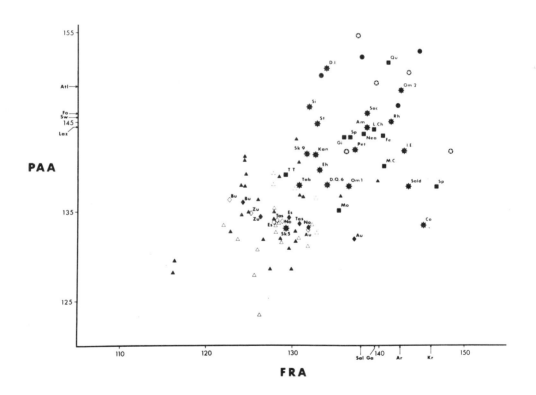

Figure 2. Plot of frontal angle (FRA) against parietal angle (PAA).

Parietal angle (PAA) v. Occipital angle (OCA) (Fig. 3)

This figure shows a fairly clear separation between "modern" and "non-modern" crania, although Teshik-Tash, Omo 1, Skhūl 5, and Qafza 6 fall within the modern range, and La Ferrassie and Swanscombe display "modern" occipital angle values. Petralona has a particularly low occipital angle, placing it close to *Homo erectus* specimens and to Broken Hill. Omo 1 and 2 are again contrasted strongly, whilst a single Upper Palaeolithic specimen (San Teodoro) and Iwo Eleru appear rather peripheral to the modern sample. The position of Swanscombe is interesting since on the basis of the occipital angle it would appear very modern, yet the parietal angle value is high as in the early Neandertal specimens. The estimated value for Vértesszöllöz (104.9) obtained from a reconstructed cast of the occipital, places it in an intermediate position as either an evolved *H. erectus* or an early *H. sapiens* fossil.

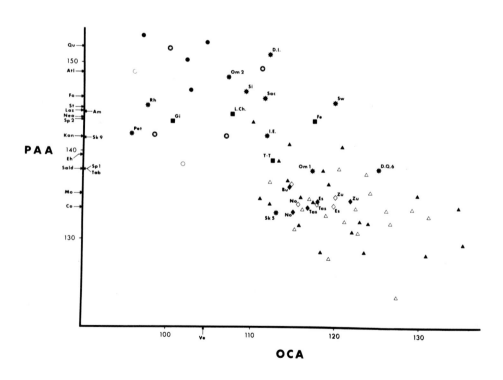

Figure 3. Plot of parietal angle (PAA) against occipital angle (OCA).

Frontal angle (FRA) v. Occipital angle (OCA) (Fig. 4)

In Fig. 4 Petralona and Broken Hill again scatter with the *H. erectus* and Solo specimens. Singa and Djebel Ighoud 1 (without the values of their rather archaic parietal arches included on the graph) now fall on the edge of the range of the modern populations as does Omo 1. Teshik-Tash, Skhūl 5 and Qafza 6 are within the "modern" cluster of values. Iwo Eleru, by contrast, now falls well away from the modern group, close to Omo 2.

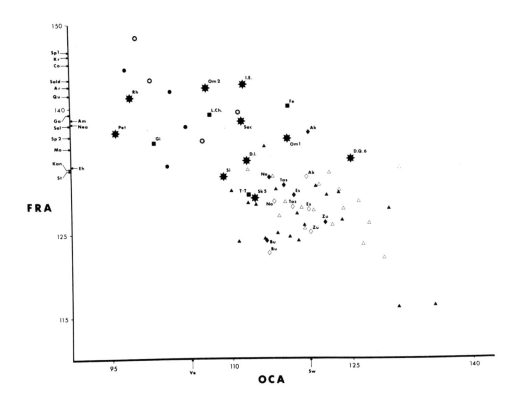

Figure 4. Plot of frontal angle (FRA) against occipital angle (OCA).

Parietal indices (parietal subtense/chord and bregma-asterion chord/
biasterionic breadth) (Fig. 5)

This distribution diagram illustrates well the morphocline in parietal arch
shape from *H. erectus* to early *H. sapiens* and anatomically modern *H. sapiens*
Most of the *H. erectus* crania from Java and Peking and the Solo crania (with
the exception of the large Solo 5 specimen) are characterised by broad short
and low parietal arches. The single *"Atlanthropus"* parietal would also fall
close to the other *erectus* fossils, if given a similar biasterionic breadth
value. Fontéchevade and Swanscombe cluster with other early *sapiens* parietals
such as Steinheim, Broken Hill and Saccopastore, as does the Lazaret fossil to
judge by the parietal subtense/chord value. The Neandertals show a wide variation and slightly overlap the modern range. Iwo Eleru, Skhūl 9 and one Mesolithic cranium (Afalou 29) appear rather archaic compared with most anatomically modern specimens, whereas Le Moustier, Spy 1, Saldanha and Ehringsdorf
appear within the range of "modern" values. Skhūl 5, Qafza 6 and Omo 1 are
completely modern in these indices of parietal form and Omo 2 is sharply
contrasted by its archaic characteristics. Cohuna, despite its "archaic"
frontal bone shape, is completely modern in parietal form and in fact falls
close to the mean for a modern Australian sample. As in the Kow Swamp crania
(Thorne and Macumber, 1972) the overall cranial form may not be as archaic
as certain areas (for example the frontal bone and face) would imply. An
explanation for this frontal bone morphology has recently been suggested
by Brothwell (1975).

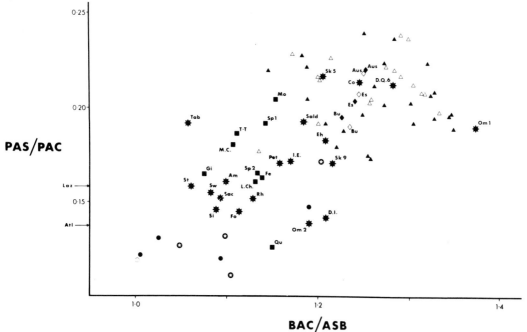

Figure 5.   Plot of parietal subtense/chord (PAS/PAC) against bregma-
asterion chord/biasterionic breadth (BAC/ASB)

Bregma angle (BRA) v. Nasion-bregma angle (NBA) (Fig. 6)

The Neandertal specimens as well as Saccopastore 1, Steinheim and Solo 6 and 11 are here characterised by exceptionally high values of the BRA and low values for NBA. One Neandertal specimen (Gibraltar 1), Petralona, Broken Hill, Singa and Qafza 6 are less extreme in their separation from modern crania, and Skhūl 5 is once again within the modern range. The low NBA values are related partly to a low cranial height, and in the case of the Neandertals the forward position of nasion is no doubt also involved in the low NBA value and high BRA value. Modern crania are characterised by low values of BRA partly because the high frontal bone and cranial height tend to increase the values of the other angles of the triangle (BRA and NBA), and in the case of Broken Hill and Singa the low BRA is probably related to a relatively short nasion-basion length.

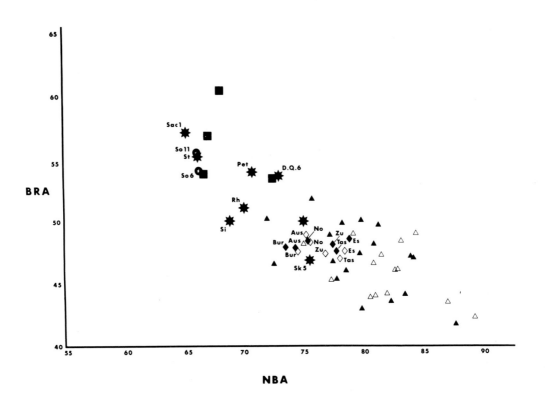

Figure 6. Plot of bregma angle (BRA) against nasion-bregma angle (NBA)

Basion angle (BAA) v. Prosthion angle (PRA) (Fig. 7)

In this figure the most archaic crania appear to be characte ised by a high BAA and low PRA. The former is related partly to a large facial height, and the latter to a projecting face (also shown here by Skhūl 5). The Neandertal specimens appear intermediate between archaic and modern *Homo sapiens* crania, but in the case of the anatomically modern specimens the high PRA value appears to be related to a near-vertical face rotated under the cranial vault, whereas in the Neandertals the high PRA value may also be due partly to the forward position of nasion and the large nasion-basion length. The Qafza 6 cranium here falls within the range of the anatomically modern specimens.

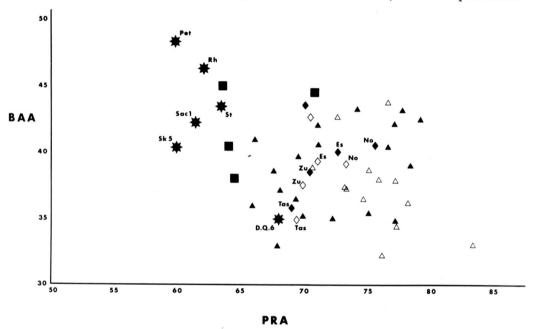

Figure 7. Plot of basion angle (BAA) against prosthion angle (PRA).

Subspinale (Zygomaxillare) angle (ZMA or SSA) v. nasiofrontal angle (NFA) (Fig. 8)

This scatter diagram of two angles of mid-facial projection relative to the cheek and frontal bone shows most clearly the distinctive classic Neandertal morphology of a medially projecting and laterally retracted face. This morphology is shared to a lesser extent by the Broken Hill and Teshik-Tash fossils. Amud (using measurements on the reconstructed face) also clusters with the classic Neandertals. Two European Middle Pleistocene fossils - Petralona and Steinheim - do not show this distinctive facial projection and fall within the range of modern specimens. The Arago 21 face (allowing for the distortion of the biorbital dimension) and the Skhūl 5 face (with an estimated NFA of 149°) are also clearly distinct from the classic Neandertals. Similarly the Solo, Ehringsdorf and Galilee specimens would not have fallen into the classic Neandertal group to judge by their NFA values, but La Quina, Spy 1, Sala, Le

Moustier, 2 Krapina specimens and some Peking crania may have done so.
Saccopastore 2 would probably have fallen into an intermediate position between classic Neandertals and crania such as Petralona and Steinheim to judge
by the ZMA of 117°.

It is interesting to note two other specimens which have estimated ZMA
values within the modern range. These are the Peking *H. erectus* face as
reconstructed by Weidenreich and Swan (NFA = 144°, ZMA = 126°) and the Broken
Hill 2 right maxilla (estimated ZMA = 139°). The value for the Peking reconstruction, taken in combination with NFA values for three Peking specimens,
suggests that these Middle Pleistocene hominids, like Petralona and Steinheim,
may have been distinct from the Neandertals in angles of facial projection.
Different views have been expressed about the relationship of the Broken Hill
maxilla to the Broken Hill cranium (Pycraft et al., 1928; Clark et al., 1950).
However the estimated ZMA value obtained for the maxilla is very much higher
than that obtained from the cranium and, taken in combination with other features, must once again raise doubt about the contemporaneity of all the Broken
Hill remains.

In this diagram, The Qafza 6 cranium shows no trace of distinctive Neandertal facial characteristics and the same is true of Djebel Ighoud 1 and all the
Upper Palaeolithic and Mesolithic specimens with the sole exception of Predmost
3 - one of the most robust of Upper Palaeolithic specimens - which has a very
low value for ZMA. This graph presents further evidence for a late Neandertal
"specialisation" of facial morphology, since both earlier and later non-
Neandertal fossil samples do not show their combination of low ZMA and NFA
values, and even Broken Hill is not as extreme as the classic Neandertals in
this respect.

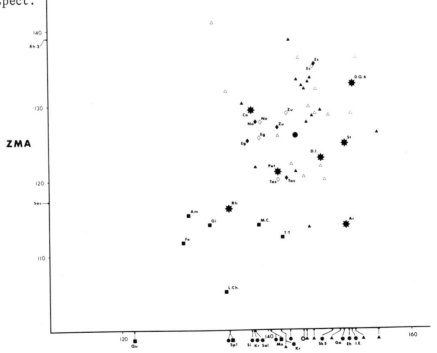

*Figure 8. Plot of subspinale (zygomaxillare) angle (ZMA) against nasio-
frontal angle (NFA).*

Plot of facial measurements (Fig. 9)

One of the most individual Neandertal characteristics suggested by the results of previous analyses (Stringer, 1974b) concerned the narrowness of the classic Neandertal upper face relative to orbital and nasal height. Crania both earlier and later than the Neandertals appeared to be characterised by faces which were proportionately broader in the upper face and shorter in nasal height. The more archaic crania were also apparently characterised by low orbits and a wider interorbital breadth. (It is interesting to note that the frontal bone recently found at the Holsteinian site of Bilzingsleben [Grimm et al., 1974] also displays a large interorbital breadth like Petralona and Arago.)

To demonstrate the distinctiveness of the Neandertal face in these features a scatter diagram of ratios calculated from four measurements of the upper and middle face is shown in Fig. 9. The means for all the modern populations studied by Howells (1973) and the overall modern mean (indicated by a diamond marked "M") are plotted, as are Upper Palaeolithic and Mesolithic crania. These show a very wide range of variation with some specimens, notably Afalou 10 and Chancelade, falling close to Neandertal values. The four classic Neandertal crania (La Chapelle, Gibraltar, Monte Circeo and La Ferrassie 1) cluster well together (mean value indicated by a square marked M), and Saccopastore 1, Shanidar 1 (cast) and Amud 1 (reconstructed cast) also fall close to this Neandertal group. A "group" of pre-last glaciation European crania comprising Saccopastore 1 and 2, Steinheim, Arago and Petralona has a distinctive mean value (star marked M). Additional individual specimens such as Djebel Ighoud 1, Skhūl 5 (with reconstructed face), Qafza 6 and Broken Hill are all quite distinct from classic Neandertal values and those of Shanidar and Amud. So there appear to have been two distinct morphoclines in Pleistocene facial evolution, one characterised by a relative narrowing of the face and "nasal dominance" and the other characterised by retention of a relatively broad upper face and smaller nasal portion of the face. Both these morphoclines, however, appear to be associated with reduction of the interorbital breadth and heightening of the orbits through time. In this respect some of the Upper Palaeolithic and Mesolithic crania and the Qafza, Skhūl, Broken Hill and Djebel Ighoud crania appear relatively archaic rather than like the classic Neandertals.

Previously, lack of material has made the morphology of the *Homo erectus* face uncertain. The Weidenreich's reconstruction of the Peking *H. erectus* reconstruction has a value for OBH/DKB of 1.27, and a value for NLH/EKB of 0.50. These values put the reconstructed face close to Broken Hill and Saccopastore 2 in Fig. 9. The new Koobi Fora *H. erectus* specimen KNM-ER 3733 is said to have a morphology comparable to that of the Peking crania and to the Weidenreich reconstruction (Leakey and Walker, 1976). Using measurements obtained from photographs, the specimen appears to fall within the range of the European archaic *Homo sapiens* specimens in Figure 9. This seems to confirm that the *H. erectus* and archaic *H. sapiens* face was distinct in shape from that of both anatomically modern man and Neandertal fossils.

Penrose size and shape analyses.

The Penrose size and shape statistic (Penrose, 1954) is a relatively simple method of analysing differences in measurements between specimens or populations in terms of size and shape. Results of Penrose size and shape analyses are comparable with those obtained by more complex multivariate techniques and

Corrucini (1973) reported that the shape distance (C^2_Z) produced an effective measure of true similarity in shape, irrespective of size difference, which was in turn accurately reflected in the size distance (C^2_q) statistic of Penrose. In the following analyses measurements were first standardised using a mean standard deviation obtained from each modern and fossil group of sufficient sample size. The measurements used (Howells, 1973; Stringer, 1974b) were as follows:

Vault analysis: maximum cranial breadth; bistephanic breadth; biasterionic breadth; supraorbital projection; frontal chord; frontal subtense; parietal chord; parietal subtense; occipital chord; occipital subtense.

Facial analysis: nasion-prosthion height; nasal height; orbital height; orbital breadth; nasal breadth; palate breadth; biorbital breadth; cheek height; supraorbital projection; frontal chord; frontal subtense.

Vault and face analysis: combination of above 18 variables.

Variables were chosen to permit analyses of some of the most important crania, and in the vault analysis, to allow the use of large samples for most measurements in the case of the Upper Palaeolithic and Neandertal groups (samples essentially as in Stringer, 1974b), and a sample of 4 crania for both the Solo and Peking *Homo erectus* groups. The facial analysis was performed using variables present in Arago 21, care being taken to allow for the effects of distortion in that specimen. The graphs were plotted using the Neandertal, Upper Palaeolithic, Qafza or Arago fossils as reference groups against which the other specimens were plotted. A positive sign next to a specimen indicates if it is of larger overall size than the reference group - otherwise it is smaller.

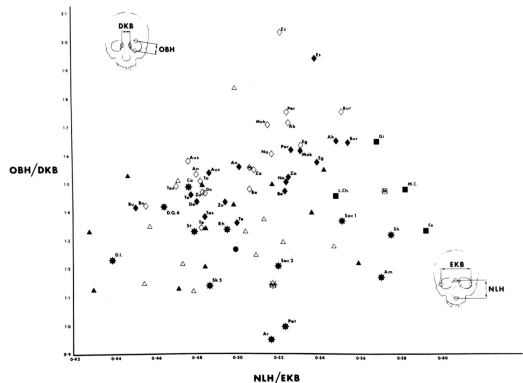

Figure 9. *Plot of orbital height/interorbital breadth (OBH/DKB) against nasal height/biorbital breadth (NLH/EKB).*

VAULT ANALYSES - 10 VARIABLES

Figure 10: Using the Neandertal sample as the reference population the closest in terms of shape is Saccopastore, then Djebel Ighoud and a group including Solo, Petralona and Amud. Steinheim is more distant and even more distinct are Omo , and 2, Skhūl 5 and Qqfza. The modern and Upper Palaeolithic groups are very distant. This graph suggests that the main cranial vault changes from the earlier European crania to the classic Neandertals were concerned with cranial expansion and this size increase is even more marked in Amud. However, the evolution of modern crania involved greater changes in shape and, ultimately, reduction in size.

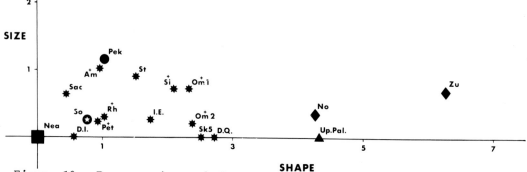

*Figure 10.   Penrose size and shape analysis on 10 vault measurements –
        Neandertal reference group.*

Figure 11: A very different pattern of relationship emerges when the Qafza crania are used as the reference group. Closest to Qafza are Skhūl 5 and the Upper Palaeolithic group followed by Norse, Omo 1, Iwo Eleru and Zulu. The next nearest, but smaller in size, is Saccopastore and then, about the same overall size, the Neandertal group. Amud is nearly as distant as Peking, Solo and Petralona, although in the former case the size difference is a positive one.

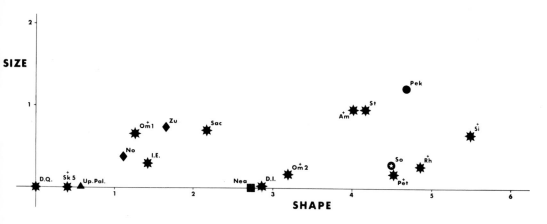

*Figure 11.   Penrose size and shape analysis on 10 vault measurements –
        Qafza 6 and 9 reference group.*

FACIAL ANALYSES - 11 VARIABLES

Figure 12: Using the Penrose size and shape statistic to assess the positio
of the Arago face it is interesting that the closest specimen in terms of shap
(but very much larger in size) is Petralona, followed by the Saccopastore earl
Neandertal fossils. However the Steinheim face, perhaps chronologically the cl
sest to Arago, is slightly more distant than the Neandertal and Broken Hill sp
cimens and is very much smaller in size than Arago. The anatomically modern
groups are very distant.

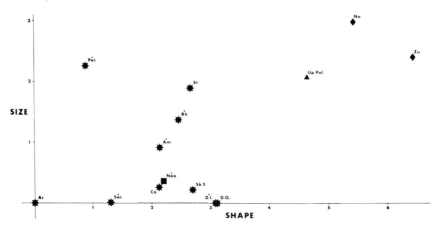

*Figure 12.   Penrose size and shape analysis on 11 face measurements –
           Arago 21 reference specimen.*

Figure 13: Using the Qafza 6 and 9 faces as a reference group, Skhūl 5 is very
close, then the Upper Palaeolithic group (smaller size) and Djebel Ighoud (slightly
larger size). Amud and the Neandertals are now amongst the next closest, in contrast
to the vault analysis.   Arago, Steinheim and Petralona are the most dictinct.

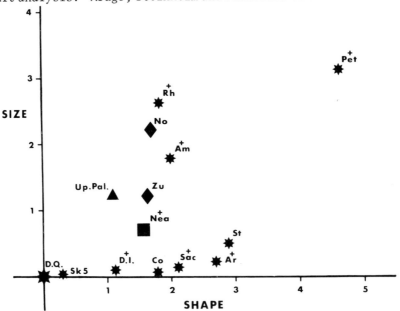

*Figure 13.   Penrose size and shape analysis on 11 face measurements –
           Qafza 6 and 9 reference group.*

Figure 14: In this figure the Norse, Zulu, Qafza and Skhūl 5 crania are closest to the Upper Palaeolithics followed by Cohuna and then a group with about equal shape distance but (in increasing magnitude) a positive size difference - Steinheim, Djebel Ighoud, Saccopastore, Neandertal, Amud and Broken Hill. The most distant, once again, are Arago and Petralona.

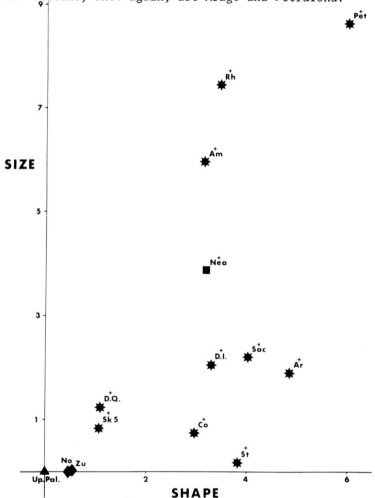

Figure 14. Penrose size and shape analysis on 11 face measurements – Upper Palaeolithic reference group.

CRANIAL ANALYSES - 18 VARIABLES

Figure 15: Closest to the Neandertal reference group are Saccopastore, Amud, Broken Hill and Djebel Ighoud. Steinheim seems fairly similar in shape but much smaller in size. Qafza, Skhūl 5 and Petralona are increasingly distant with the anatomically modern groups very distinct both in size and shape.

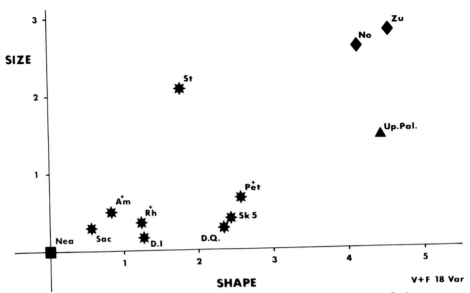

Figure 15.   *Penrose size and shape analysis on 18 vault and face*
               measurements – Neandertal reference group.

Figure 16:   With the Qafza crania as the reference group Skhūl 5 is very
close, followed by the Upper Palaeolithic and modern groups (smaller size).
Djebel Ighoud, Neandertal and Saccopastore are next closest but Amud is differentiated by both greater size and a larger shape distance.   Petralona is
very distant.

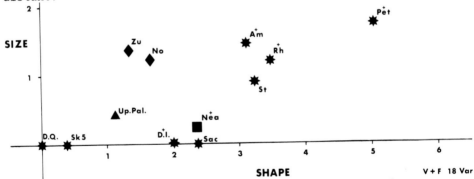

Figure 16.   *Penrose size and shape analysis on 18 vault and face*
               measurements – Qafza 6 and 9 reference groups.

Figure 17: A comparable overall pattern to that of Figure 16 is shown
using the Upper Palaeolithics as reference group instead of the Qafza crania.
The nearest groups are now the Norse and Zulu followed by Skhūl, Qafza and
Saccopastore. Very much more distinct are Steinheim, Djebel Ighoud, Neandertal, Amud, Broken Hill, with Petralona extremely distant on both size and
shape.

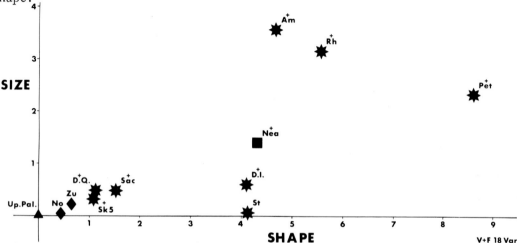

*Figure 17. Penrose size and shape analysis on 18 vault and face mea-
surements - Upper Palaeolithic reference group.*

CONCLUSIONS

Homo erectus and archaic Homo sapiens

   The *Homo erectus* crania form a meaningful group in most distribution diagrams although other crania such as Petralona, Broken Hill, Singa and, in
particular, Solo share many of the same characteristics. In fact there is now
much evidence for the classification of the Solo material as *H. erectus* rather
than archaic *H. sapiens*. In *H. erectus* crania the high values of FRA and PAA,
together with low values for OCA and the parietal indices in Fig. 5, confirm
the presence of a low frontal bone, a low short parietal arch broad at the
base, and a short broad projecting occipital bone. Certain aspects of the
facial morphology of *H. erectus* seem similar to European Middle Pleistocene
archaic *H. sapiens*, and to a lesser extent anatomically modern *H. sapiens*,
rather than similar to the late Neandertals. Other Penrose size and shape
analyses (not shown here) relate the Peking crania to the Solo fossils and to
specimens such as Petralona and Broken Hill, confirming analyses performed by
Corruccini (1974). Much of the distinctiveness of the Peking crania from
later crania, including the Neandertals, is dependent on factors of size
rather than shape (see Fig. 10). However differentiation from anatomically
modern crania is, in contrast, due to differences in shape rather than in size.
   The European pre-last glaciation hominids form a reasonable morphocline
through time, assuming that Petralona and Arago are the oldest, and the Saccopastore specimens are youngest. Petralona and Arago 21 seem comparable in
facial morphology, although the massive build of the former specimen produces
a large size distance in Penrose analyses (Fig. 12). Further (unpublished)

analyses using the Petralona skull as reference specimen link it most closely
to Peking *H. erectus*, Djebel Ighoud, Solo, Broken Hill and the Neandertals
using vault measurements; to Arago, the Upper Palaeolithics, Skhūl 5, Qafza,
Saccopastore and Broken Hill using facial measurements; and to Broken Hill,
Saccopastore and the Neandertals using combined face and vault measurements.
In facial form the distance between Petralona and many other fossils is primarily due to *size* differences, but in the case of the Neandertals the distance is due to *shape* differences.

Petralona does not seem to resemble the Steinheim skull in these same analyses, as is also the case when Steinheim is compared with the Arago 21 specimen (Fig. 12). The distinctiveness of Steinheim seems rather marked to be
explained by sexual dimorphism alone, although the contrast between mandibles
2 and 13 from Arago shows that Brace (1962) and Wolpoff (1975) are right to
emphasise the importance of sexual dimorphism in assessing Middle Pleistocene
hominids. However, even though the Steinheim skull differs in morphology
from the Petralona and Arago 21 fossils somewhat more than might have been
expected, they share a number of non-Neandertal facial characteristics (Figs
7, 8, 9). Differences between Petralona and Swanscombe seem more marked in
the occipital region than in the parietal region (Figs 3, 5), but the Vértesszöllös occipital shows an intermediate value for the occipital angle (Figs 3,
4), illustrating the wide variation present in Middle Pleistocene European
hominids.

Evidence is accumulating that the Petralona cave deposits are at least
Holsteinian in age (Kurtén, pers. comm.; Poulianos, pers. comm.) and it is
to be hoped that the hominid cranium can be related to the known stratigraphy
and fauna. There are several parallels here with the discovery of the Broken
Hill skull which is still not satisfactorily dated since it is difficult to
relate to the artefacts, fauna and other fossil human material from the site
(but see Klein, 1973). However, much of the vast Petralona cave system contains fossiliferous deposits, so it is not too late in this case for the time-
range of the cave deposits to be established beyond doubt. There are also
marked morphological similarities between the Petralona and Broken Hill
crania (Stringer, 1974a, 1974b) which are probably due to their similarity
in evolutionary grade. The very completeness of these two specimens has
made them difficult to classify since they show a mosaic of features found
in *H. erectus* and *H. sapiens*.

It has recently been suggested that the Arago material is of greater antiquity than the Swanscombe and Steinheim material (de Lumley, 1975) and it
is possible that the Swanscombe material may not even be truly "Hoxnian" in
age (Gladfelter and Singer, 1975). The possible existence of a large time-
gap between Petralona and Arago, on the one hand, and Swanscombe and Steinheim
on the other, would further the concept of "phyletic gradualism" (Pilbeam,
1975) in European Middle Pleistocene hominid evolution, but until better samples of well-dated fossil hominids are available from this period, it is impossible to say how much variation existed due to sexual dimorphism, mosaic
evolution or perhaps even the existence of more than one lineage of hominids.
Nevertheless there are common features linking the European Middle Pleistocene
hominids to one another and to early Neandertal fossils such as Saccopastore,
Ehringsdorf and Fontéchevade (e.g. see Figs 2-9 and Fig. 12).

Upper Pleistocene Homo sapiens

The Saccopastore crania and other European last interglacial fossil hominids are excellent models for the ancestors of the later "classic" Neandertals and link them to earlier European hominids. Evolution of the late Neandertals from such ancestors would have involved less change in the vault (beyond lengthening and general expansion in size) than in the face and its relationship to the vault (Stringer, 1974b). There must have been a general "opening out" of the cranial base and lengthening of the nasion-basion and nasio-occipital chords, an associated forward migration of the tooth rows relative to the maxilla and mandible, and the evolution of a voluminous and projecting nasal area and associated maxillary sinuses (Stringer, 1974b; Howells, 1975; Wolpoff, 1975).

Confusion has been caused in the past by the use of different definitions of the term "Neandertal" (Howells, 1974; Stringer, 1974b). A combination of morphological and metrical data from both cranial and post-cranial material would seem the most satisfactory way of attempting to define the term "Neandertal". Chronological or cultural criteria are likely to be less certain and more open to dispute. When correlations of glacial stages cannot even be fixed between adjacent areas of Europe, it seems pointless to introduce a term such as "end of the Riss glaciation" (Brose and Wolpoff, 1971) into a definition of Neandertals which is intended to be applied world-wide. The constituents of a Neandertal group so defined would vary according to prevailing views of the dating of the fossils. The latest limit of Brose and Wolpoff's definition of "Neandertal" concerns the appearance of anatomically modern man, and this is similarly subject to the problems of dating the earliest emergence of modern *Homo sapiens* which will be discussed later, and might exclude fossils which were morphologically Neandertal from the group (Stringer, 1974b). If methods such as amino-acid racemization can fulfil their promise then perhaps chronological criteria will eventually be incorporated into definitions of the term "Neandertal", but such data must always remain less certain than data from the morphology of the fossils themselves. An alternative definition using cultural criteria (Hrdlicka, 1930; Brace, 1964) is equally unsatisfactory, since specimens of uncertain cultural associations are difficult to place and the term "Mousterian" is of doubtful application in much of Africa and Asia. The other part of Brace's definition, the retention of a "Middle Pleistocene face", is also unsatisfactory since as already discussed the Neandertals have an individual facial morphology rather than a "Middle Pleistocene" face.

It is this facial morphology which also proves to be a major problem in linking the European or near-eastern/middle-eastern Neandertals to the Upper Palaeolithic, Skhūl and Qafza crania. It is still uncertain whether the Neandertal facial morphology was a functional development, an adaptation to climate, or a result of both these factors working in combination. However, since this facial form is reflected in the morphology of the lower dentition and mandible, it is possible to infer its absence in many *H. erectus* and *H. sapiens* fossils (Arago 2 may be an exception here), thus providing negative support for the cold-adaptation hypothesis. African fossil mandibles such as those from Ternifine, Thomas quarry, Rabat, Haua Fteah, Omo (Kibish formation) and the Makapansgat Cave of Hearths also show no signs of the typical Neandertal morphology (Tobias, 1968; Howells, 1975; Wolpoff, 1975; and my own observations). Additionally there are post-cranial features which distinguish the Neandertals from modern populations, from the Upper Palaeolithic and Skhūl

fossils (Musgrave, 1973; Trinkaus, 1976) and from the Omo I skeleton
(Trinkaus, pers. comm.). The true near-eastern/middle-eastern Neandertals
(here I am excluding Skhūl and Qafza) also show most of the European Neandertal cranial and post-cranial characteristics and this would imply a close relationship and at least intermittent genetic contact perhaps following the
model proposed in another context by Brues (1972).

The European and near-eastern/middle-eastern Neandertal crania form a
valid group in the distribution diagrams of angles and indices (Figs 2-9),
exhibiting variation comparable with that found in Upper Palaeolithic and
modern crania, and overlapping the ranges of *Homo erectus* and modern samples
on several graphs. Immature Neandertal crania (Teshik-Tash and Le Moustier)
are closer to modern populations than the fully adult specimens in several
of the analyses illustrated here. Moreover, in some analyses the Neandertals
could be said to be extreme rather than intermediate (Figs 6, 8, 9, 17) compared with other fossil material. Thus it is possible that the Neandertals
are rather "advanced" or "specialised" in a number of respects compared to
their precursors, but not necessarily "advanced" in a direction leading to
modern man. In other words, the late Neandertals show derived characteristics, some of which are not shared with anatomically modern populations. Such
characteristics should form part of any meaningful definition of the term
"Neandertal", and may differentiate the last glaciation "Neandertals" from
their precursors, the "early Neandertals".

In the case of the Skhūl and Qafza material, results of analyses conducted
here are in agreement with the conclusions of Vallois and Vandermeersch (1972),
Vandermeersch (1972) and Howells (1974, 1975), in that the crania studied so
far are metrically modern in form with few archaic characteristics (Figs 2-5,
7-11, 14-17). Those archaic characteristics they show (e.g. Fig. 6) relate
them as much to hominids such as Saccopastore, Djebel Ighoud and Omo as to
the late Neandertals. The Qafza 6 and 9 crania appear to be very similar to
the Skhūl 5 specimen (Figs 2-5, 8-9, 11, 13, 15) and could conceivably be
drawn from the same population, although it cannot be demonstrated that they
are, in fact, contemporaneous. The Skhūl and Qafza fossils are both associated
with Levalloiso-Mousterian industries, but other evidence suggests that Skhūl
may be later than the Tabūn Neandertal fossils, whereas Qafza may be earlier
(Vandermeersch, 1972; Farrand, 1972). As there was Upper Palaeolithic occupation of the interior of the Qafza cave, doubts have been raised about the
possibility of intrusive burials, or whether all the Qafza sample is equally
modern in form. From published evidence there seems no possibility of intrusive burials since the recent Qafza finds were covered by 2 metres of Levalloiso
Mousterian deposits and we know that Palaeolithic graves were usually very shallow. Furthermore, examination of the sample of adult and immature material
from Qafza does not indicate any real variation from a modern morphology (Vandermeersch, pers. comm.). If the Qafza material is contemporary with, or earlier than, fossils such as Amud, Tabūn and Shanidar, then we must surely exclude the later near-eastern/middle-eastern Neandertals from the ancestry of
the Qafza hominids and of modern man. This possibility is supported by the
size and shape analyses given here, in which the Amud skull, although showing
some resemblance in facial measurements to Qafza (Fig. 13), nevertheless
occupies an individual position (Figs 11, 14, 16 and 17) as in previous analyses (Stringer, 1974b). The scheme of human evolution leading from Neandertal to modern man advocated by Brace (1962, 1964) Wolpoff (1975) and others
(displayed in simplified form in Fig. 18) is completely contradicted by the
Qafza evidence. The Qafza specimens have a modern vault shape and facial form

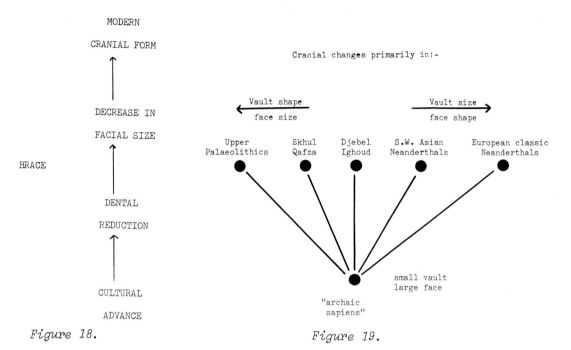

MODERN

CRANIAL FORM

↑

DECREASE IN

FACIAL SIZE

↑

BRACE

DENTAL

REDUCTION

↑

CULTURAL

ADVANCE

Figure 18.

Cranial changes primarily in:-

← Vault shape

face size

Vault size →

face shape

Upper
Palaeolithics

Skhul
Qafza

Djebel
Ighoud

S.W. Asian
Neanderthals

European classic
Neanderthals

small vault
large face

"archaic
sapiens"

Figure 19.

yet are associated with artefacts which are apparently not of the more advanced Mousterian varieties known from the near-east, middle-east and Europe. Furthermore, although the Qafza mandibles are modern in morphology, the dentitions show no sign of reduction from Neandertal dimensions (Vallois and Vandermeersch, 1972; and my own observations). Such reduction must have occurred at a later date and in some unquestionably modern populations it may not have occurred at all.

If the Qafza hominids prove to be younger than the Tabūn, Amud and Shanidar fossils, then the situation would parallel that in Europe. However, it would still be difficult to accept an ancestor-descendant relationship between the "Neandertal" and "modern" fossils unless there was a large time difference involved, or some special factors other than those proposed by Wolpoff (1975) were at work in temporarily accelerating the rate, and altering the direction, of evolutionary change.

There still remains the problem of the ancestors of the Qafza-Skhūl hominids, if we assume they did not evolve from the near-eastern and middle-eastern Neandertals. If the Omo crania are correctly dated as late Middle or early Upper Pleistocene then they (and in particular Omo 1) are possible ancestral forms (Day, 1969; Stringer, 1974b). Analyses conducted here confirm that Omo 1 is basically modern in cranial morphology (Figs 2-5) and resembles the Upper Palaeolithic and Qafza crania (Fig. 11). Neither of the Omo crania, although very robust, could meaningfully be called "Neandertal" (Fig. 10; and cf. Howells, 1974) but other recent work suggests that they are linked with material from southern Africa (Rightmire, 1976). The recently discovered Sale (Jaeger, 1975) and Ndutu (Clarke, 1976) crania are also possible ancestors for advanced *Homo sapiens* in Africa. The former specimen certainly confirms the reality of that nebulous term "mosaic evolution" since the cranial vault is of small capacity and has an angular appearance, particularly in *norma occi-pitalis*, recalling *Homo erectus* and Omo 2. However, the mastoid processes are well-developed and the occipital shows apparent resemblances to Swanscombe, Saccopastore, Amud and Omo 1. The Ndutu cranium appears from photographs to represent *H. sapiens* rather than *H. erectus* (the value for BAC/ASB, as calculated from data supplied through the courtesy of P. Andrews and R. Clarke is 1.10; this value compares well with either advanced *erectus* or true *sapiens* parietal dimensions in Fig. 5). However, further study and more complete dating evidence for both Sale and Ndutu are required before these important fossils can be incorporated into the picture of Middle-Upper Pleistocene human evolution. A re-evaluation of the Singa skull (Woodward, 1938) is also necessary now that this and other studies have shown it to represent an archaic *H. sapiens* form (Figs 2, 3, 5, 11) rather than a proto-Bushman (Tobias, 1968; Brothwell, 1974; Stringer, 1974b).

The Djebel Ighoud crania are also possible ancestors for modern man. They show an unusual mixture of archaic and advanced features. In previous multivariate analyses Djebel Ighoud 1 seemed most closely related to the early Neandertals, near-eastern/middle-eastern Neandertals and Skhūl 5. There were also some resemblances to more archaic specimens such as Broken Hill and Petralona (Stringer, 1974b). In general cranial form the Djebel Ighoud 1 fossil is similar to the Neandertals (Figs 10, 15) but displays relatively archaic parietal features like Omo 2 (Figs 2, 3, 5) and a relatively advanced frontal bone form (Figs 2, 4). In facial form the Djebel Ighoud specimen resembles the Qafza and Skhūl fossils (e.g. Figs 8, 9, 13, 16) and contrasts markedly with the Neandertals in some respects (e.g. the ratio of NLH/EKB in Fig. 9, and the "modern" values for ZMA and NFA in Fig. 8). Whether North African populations evolved from forms like Djebel Ighoud into forms like Qafza and Skhūl is unknown, but the morphological gap between these fossils is still large (at least in the case of Djebel Ighoud 1 - the second specimen may be more "advanced"). All these crania can perhaps be seen as the result of evolution from an archaic *H. sapiens* or broadly defined "early Neandertal" form, but in this case their evolution occurred *without* the influence of the climatic or functional factors which produced "classic" Neandertal populations in Eurasia. A possible model of cranial evolution based on the results of size and shape analysis is shown in Fig. 19.

Finally, the Upper Palaeolithic and Mesolithic crania analysed here conform completely to the anatomically modern pattern. They show individual variation but are far closer to the various modern populations than to more archaic crania (e.g. Figs 2-8, 14, 17). Certain specimens are extreme in certain angles or indices (e.g. Figs 2, 9) but no crania are consistently close to the Neandertals, nor are the earlier Upper Palaeolithic specimens consistently closer to the Neandertals than are later specimens. Examination of original or cast material of the early Upper Palaeolithic crania from sites such as Lautsch (Mladec), Pavlov, Predmost, Brno etc. indicates that these are extremely robust anatomically modern crania, not only metrically quite distinct from the Neandertals but quite different in overall cranial form, since their rugged angular contours contrast markedly with the smooth rounded contours of later Neandertal skulls. Their robustness can be seen as a feature also present in fossils such as Qafza, Skhūl and Omo, rather than as signifying a particularly close relationship to the "classic" Neandertals. Dating evidence is accumulating that anatomically modern man may have appeared in Australasia, southern Africa and America as early as, or even earlier than, in Europe. Some of these early dates have been obtained by the amino-acid racemization technique and need further confirmation. But other dates seem mor- securely based and indicate that the emergence of modern man was not a phenomenon related to the transition from Mousterian to Upper Palaeolithic industries, although the *spread* of modern man may have been related to that change in certain areas. I still feel that the climatic instability of the middle period of the last glaciation in Europe may have led to population movement into Europe during climatic ameliorations such as the one recorded between the late Mousterian and early Upper Palaeolithic of Cottés (Bastin et al., 1976), despite the lack of evidence for major faunal change during that period. Classic Neandertals could well have contributed to the gene pool of early anatomically modern *Homo sapiens* populations but the contributions as manifested in the skeleton of the known early modern fossils must have been small. Our picture of modern human evolution is based to a great extent on the circum-Mediterranean area and it is possible that future discoveries from the early Upper Pleistocene will reveal fossils like Omo and Qafza in other parts of the world which have developed not from "classic" Neandertal populations but from a less "specialised" form of *Homo sapiens*.

ACKNOWLEDGEMENTS

I should like to thank the staff of the many institutes who allowed me to study fossil material in their care (particularly, in this case, Dr. B. Vandermeersch and Dr. H. de Lumley) and Professor W.W. Howells for providing comparative data on modern populations. I should also like to thank Mr. R.G. Kruszynski who drew most of the figures and Dr. C.G. Adams and Dr. P. Andrews for providing critical comments on the manuscript of this paper. Some of the data used here were collected whilst I held a Medical Research Council Studentship in the Department of Anatomy, University of Bristol. Funds for travel were provided by the Medical Research Council, the Boise Fund, Oxford and the British Museum (Natural History).

REFERENCES

Bastin, B., Leveque, F. and Pradel, L. (1976). *C.r. hebd. Seanc. Acad. Sci., Paris D,* 282, 1261-1264.
Brace, C.L. (1962). *In* "Culture and the evolution of men", (M.F.A. Montagu, ed.), pp. 343-354, Oxford University Press, New York.
Brace, C.L. (1964). *Curr. Anthrop.,* 5, 3-43.
Brose, D.S. and Wolpoff, M.H. (1971). *Am. Anthrop.,* 73, 1156-1194.
Brothwell, D.R. (1974). "The Upper Pleistocene Singa skull: a problem in palaeontological interpretation", Bevolkerungsbiologie, pp. 534-545, Fischer, Stuttgart.
Brothwell, D.R. (1975). *J. archaeol. Sci.,* 2, 75-77.
Brues, A.M. (1972). *Am. J. phys. Anthrop.,* 37, 389-400.
Clark, J.D., Oakley, K.P., Wells, L.H. and McClelland, J.A.C. (1950). *J.R. Anthrop. Inst.,* 77, 7-32.
Clarke, R.J. (1976). *Nature, Lond.,* 262, 485-487.
Corruccini, R.S. (1973). *Am. J. phys. Anthrop.,* 38, 743-753.
Corruccini, R.S. (1974). *Yearb. phys. Anthrop.,* 18, 89-109.
Day, M.H. (1969). *Nature, Lond.,* 222, 1135-1138.
Farrand, W.R. (1972). *In* "The Origin of *Homo sapiens*", (F. Bordes, ed.), pp. 227-235, Unesco, Paris.
Gladfelter, B.G. and Singer, R. (1975). *In* "Quaternary studies", (R.P. Suggate and M.M. Cresswell, eds), pp. 139-145, Royal Society of New Zealand, Wellington.
Grimm, H., Mania, D. and Toepfer, V. (1974). *Z. Archäol,* 8, 175-176.

Howells, W.W. (1973). *Pap. Peabody Mus.*, 67, 1-259.
Howells, W.W. (1974). *Am. Anthrop.*, 76, 24-38.
Howells, W.W. (1975). *In* "Paleoanthropology, morphology and paleoecology", (R.H. Tuttle, ed)), pp. 389-407, Mouton, The Hague.
Hrdlicka, A. (1930). *Smithson. misc. Collns.*, 83, 1-379.
Jaeger, J. (1975). *Colloques int. Cent. natn. Rech. scient.*, 218, 897-902.
Klein, R.G. (1973). *Nature, Lond.*, 244, 311-312.
Kowalski, C.J. (1972). *Am. J. phys. Anthrop.*, 36, 119-131.
Leakey, R.E.F. and Walker, A.C. (1976). *Nature, Lond.*, 261, 572-574.
de Lumley, H. (1975). *In* "After the Australopithecines", (K.W. Butzer and G. Ll. Isaac, eds), pp. 745-808, Mouton, The Hague.
Musgrave, J.H. (1973). *In* "Human Evolution", (M.H. Day, ed.), pp. 59-85, Taylor and Francis, London.
Penrose, L.S, (1954). *Ann. Eugen.*, 18, 337-343.
Pilbeam, D.R. (1975). *In* "After the Australopithecines", (K.W. Butzer and G. Ll. Isaac, eds), pp. 809-856, Mouton, The Hague.
Pycraft, W.P., Elliot-Smith, G., Yearsley, M., Thornton Carter, J., Smith, R.A., Tindell Hopwood, A., Bate, D.M.A. and Swinton, W.E. (1928). "Rhodesian man and associated remains", British Museum (Natural History), London.
Rightmire, G.P. (1976). *Nature, Lond.*, 260, 238-240.
Stringer, C.B. (1972). *J. Anat.*, 114, 295.
Stringer, C.B. (1974a). *J. Hum. Evol.*, 3, 397-404.
Stringer, C.B. (1974b). *J. archaeol. Sci.*, 1, 317-342.
Thorne, A.G. and Macumber, P.G. (1972). *Nature, Lond.*, 238, 316-319.
Tobias, P.V. (1968). *In* "Evolution und Hominisation", (G. Kurth, ed.), pp. 176-194, Fischer, Stuttgart.
Trinkaus, E. (1976). *Z. Morph. Anthrop.*, 67, 291-319.
Vallois, H.V. and Vandermeersch, B. (1972). *Anthropologie, Paris*, 76, 71-96.
Vandermeersch, B. (1972). *In* "The Origin of *Homo sapiens*", (F. Bordes, ed.), pp. 49-54, Unesco, Paris.
Wolpoff, M.H. (1975). *In* "Determinants of mandibular form and growth", (J.A. McNamara, ed.), Centre for Human Growth and Development, Ann Arbor.
Woodward, A.S. (1938). *Antiquity*, 12, 193-195.

HOMO ERECTUS OR *HOMO SAPIENS?*

JAN JELINEK

Anthropos Institute, Moravian Museum, Brno, Czechoslovakia.

The relationship between *Homo erectus* and *Homo sapiens* (especially *Homo sapiens sapiens*) has often been discussed in the last few years. The reason for this is not only the number of new finds from Asia, Africa and Europe, but also the new knowledge of the dating of these finds, of their environment (climate, fauna, flora) and their social behaviour and cultural background.

Students of human evolution have expressed differing views as to the fate of *Homo erectus*. Many consider this species to have been a specialised side branch that became extinct and had no genetical contact with *Homo sapiens*, this latter species having existed in some regions contemporaneously with *Homo erectus* or even having preceded the latter. These new finds of the direct forefathers of *Homo sapiens* (named *Homo sp.*) in some African finds have been considered to be responsible for the extinction of the evolutionary side branch of *Homo erectus* (Leakey, 1966, 1972). Some have speculated that extra-European populations of *Homo erectus* could have been the ancestors of some later populations (e.g. Australian aboriginals) or at least that they participated in their formation. To consider *Homo erectus* as direct ancestors of Australian aboriginals is no longer a suitable idea in contemporary palaeo-anthropology, despite new finds and datings in Australia that have reawakened the question of the origin of the archaic character of early Australian finds (Macintosh, 1965; Macintosh and Larnach, 1972). Finally the opinion is still held that all populations of modern man have direct ancestors in *Homo erectus* (Brace, 1964, 1968), even though this opinion is not accepted by the majority of palaeo-anthropologists. Opposition to this theory was based upon European anteneandertalian and *H. erectus* finds that are morphologically distinct from the classical Javan Trinil finds of *Homo erectus erectus*, and from the Choukoutien finds of *Homo erectus pekinensis*. The Trinil and Choukoutien finds belong to the first known, best studied and most broadly discussed *H. erectus* remains. They form, as a background to our speculations, the picture of the "typical" *Homo erectus* and other finds are sometimes consciously or subconsciously considered to resemble them.

The East African finds, mainly the result of the Leakeys' research at Olduvai, brought to light the longest chronological sequence in one locality and with this the question of the limits and boundaries between different hominid species, in particular, the limits and species boundaries of *Homo erectus*. The chronological criteria, brain capacity, cranial morphology, cultural remains, behaviour and other features were used as a possible differential diagnosis to help with the taxonomic problem. Let us consider briefly the contemporary situation in this field with the help of some important finds.

In East Africa the new palaeomagnetic data proved that all three layers

(Bed I-III) were formed before the Matuyama reversal and therefore were Lower
Pleistocene in age. This means that the Acheulian tool assemblages and *Homo
erectus* remains are here older than any similar European find. The hominid
no. 9 is characterized by its robusticity. Holloway (1973) estimated its
cranial capacity at 1067 ccm. Its extraordinary brow ridges are unique in size
and shape when compared with other *H. erectus* finds. At site MNK II were found
hominid remains provisionally described as *Homo habilis*, but Tobias and von
Koenigswald (1964) have found some similarity with *Homo erectus* from the Djetis
beds of Java. Both of these finds were above the Lemuta member, that is in the
upper part of Bed II. From Bed IV, associated with Acheulian tools postcranial
remains of *Homo erectus* were found. At Olduvai therefore *Homo erectus* appears
about 1,500,000 years ago. It is important to stress also that the facial part
of a hominid skull found in Yayo (Chad) (Coppens, 1967) was originally ascribed
to *Australopithecus sp.* but now is recognised to be late *Australopithecus*
or early *Homo erectus*.

The Broken Hill skull with which the South African Saldanha skull was morphologically associated, even though usually considered somewhat older, was
originally designated by Coon (1962) as a very late representative of *H. erectus*
type persisting in African Upper Pleistocene times. Arambourg (1963) similarly
considers Broken Hill and Saldanha as late *H. erectus* finds and Heberer et al.
(1975) speak about "isolated evolution of *H. erectus* in this (South African)
region". The Broken Hill skull has a fairly high cranial capacity (1280-1400
cc). Systematically Heberer et al. consider these two important finds (Broken
Hill and Saldanha) as *Homo erectus* (seu *sapiens*) *rhodesiensis* (Heberer et al.,
1975). Pilbeam (1972) classifies Broken Hill as *Homo sapiens* and Kurth (1965)
again as *H. erectus*. Recently Klein's view (Klein, 1973) that both finds,
Broken Hill and Saldanha, are considerably older than was previously considered,
supported by the faunal and archaeological evidence, has gained more and more
support though J. Bada (Howells, 1973) gives for an animal bone from Broken
Hill a date of about 36,000 and considers Saldanha something over 40,000 B.C.

In the Broken Hill skull it was mainly the higher cranial vault, the large
sinuses in the large supraorbital torus, the character of the second maxilla
found in the same locality and the modern morphology of the postcranial remains
that supported the *Homo sapiens* classification. The supraorbital region, the
profile of the facial skeleton, the occipital crest and shape and situation
of *planum nuchale* led, on the other hand, some other experts to classify it
as an advanced *H. erectus*. The classification of the Broken Hill skull as an
African or "tropical" neandertaloid or tropical neandertal (von Koenigswald,
1956) is now nearly abandoned.

Quite a different picture is given by the North African finds. Geographically they belong to the circummediterranean region which was separated by the
Sahara from the rest of Africa, and the whole environment of this region was
a more Mediterranean than an African one. The finds from Rabat, Casablanca,
Ternifine and Sale are instructive. The three Ternifine mandibles and one
parietal are the oldest and several authors stress their resemblance to *Homo
erectus pekinensis* or even with Trinil finds. But unfortunately only one
parietal bone is preserved showing that the brain case was a flat one, but
bigger than in *Homo erectus pekinensis* or *Homo erectus erectus*. The thickness of
this bone is inside the range of variation of *Homo sapiens*. The mandibles are
robust. They are without chin, only mandible II has a slight chin-like protuberance. Slight cingular remains can be seen in the molar teeth. These
cranial remains are associated with a rich stone tool assemblage of Acheulian
II type. The Rabat mandible is supposed to be later but exact dating does not

exist and no tools were found with it. The tentative Choubert and Marcais
(1947) dating shows Rabat Man as an Acheulian tool maker. Arambourg (1963)
considered this mandible as belonging to *Homo erectus mauritanicus,* but
Vallois (1960) describes it as having some neandertaloid characters and being
phylogenetically between *H. erectus* and neandertal types. He designates it
as anteneandertalian.

The three other North African mandibles coming from two quarries from Casablanca are also classified as *Homo erectus mauritanicus* (Arambourg and Biberson,
1955-56; Biberson, 1963). The first two mandibles are from the Schneider Quarry
(Arambourg and Biberson, 1955, 1956), the third mandible from Thomas I quarry
(Sausse, 1975).

In 1973 Jaeger (1973) published a short report announcing the discovery of
another skull of *H. erectus* type at Sale, not far north of Rabat in Morocco,
in a quarry which was opened in fossil gravels dated provisionally to the
penultimate cold period (Riss), dated usually around 200,000 years B.P. This
means that this skull is later than the finds from Casablanca and Rabat. It
is small with a cranial capacity of 930-960 ccm. The upper dentition conserves
the cingulum and the teeth are large and robust. The position of the parietals
gives a characteristic "gabled" look in the occipital view but the greatest
breadth of the cranium, even when situated at its base on the temporals, is no
more pronounced than in older *H. erectus* finds. *H. erectus* characters of the
occipital bone are present (torus occipitalis and general morphology of squama
occipitalis), but again the sagittal curve of this bone is better vaulted than
in typical older *H. erectus* finds. This is well seen in lateral view where
the occipital part of the skull looks more modern. The Sale skull has well
developed parietal bosses and its transverse arch is considerably better vaulted when compared with the Ternifine parietal bone. There is no doubt that
this skull of late *H. erectus* type is much more advanced in physical type than
the other, older, North African finds of *Homo erectus.*

Interestingly enough Jaeger shows that the morphology of the occipital and
parietal bones is here more advanced than that of the frontal bone and the
dentition.

It is perhaps illustrative to mention also four younger North African finds
chronologically outside of our study, namely Témara, Tangier and Djebel Irgoud
(two finds). Biberson (1963) writes in his detailed ecological study: "This
human fossil (Témara) studied by Vallois, has several remarkable characteristics. Some are archaic, close to those of *Atlanthropus,* Sidi Abderrahman Man
and Rabat Man; others are more evolved. But Témara Man cannot be allied with
the Neanderthals, makers of the Mousterian in France which seem nevertheless
to have been their contemporaries in Europe." In 1961 and 1963 two skulls were
found at Djebel Irgoud again in Morocco. Some characters separate them from
European Neandertals with which superficial observations sometimes allied them.
In this way the morphology of an extremely broad and, in lateral view, rounded
occipital bone is characteristic. This view shows also a considerably higher
cranial vault, higher at the vertex than in typical Neandertals, and the facial
skeleton lacks the typical height of classical European neandertal crania. As
far as we can follow the dental dimensions they were much greater than in European Neandertals. Tangier Man, found in Mougharet-el-Aliya near Tangier, brought
difficulties when some experts tried to associate it with other European finds
of *Homo sapiens sapiens* or *Homo sapiens neanderthalensis.* Some archaic characters show the possibility that he belongs as a late descendant to the *Homo erectus mauritanicus* evolutionary line, joining in North Africa older finds from
Ternifine, Sidi Abderrahman, Rabat and Témara.

Piveteau in his study of the Acheulian parietal bone from Grotte du Lazaret (1967) states clearly that Jebel Irgoud cannot be associated with European Neandertals. The Moroccan palaeo-anthropological material is still not numerous enough to give a fully clear picture. So the precise dating of the Jebel Irgoud skulls is unknown, since Ennouchi (1968) states only that the age of this fossil was greater than the possibilities of the C14 laboratory in Nancy. By comparison with other North African localities an age round 45,000 years is suggested. Nevertheless, the relatively small Moroccan territory gives several fossil human remains of different prehistoric periods which show that most probably local evolution in this region took place for a considerable period of time and that the North African *Homo erectus* stock gave rise to, or at least participated in the formation of later populations. This is fairly important and an illustrative situation for the consideration of European Middle Pleistocene and later developments.

Another illustrative situation, with new discoveries and rich additional knowledge, comes from South Eastern Asia. Most authors of the papers dealing with new aspects of the South East Asian finds accept three separate subspecies of Javanese *Homo erectus* namely *H. e. modjokertensis, H. e. erectus,* and *H. e. soloensis.* New finds from Sangiran (Pithecanthropus 17) and from Sambungmatjan east of Sragen (S 1) are considered by Jacob (1975a, b) as *Homo erectus soloensis.* They are chronologically different from the original XII individuals from Ngandong. They belong, together with numerous finds of *Homo erectus erectus,* in the time span between 1 million and 600,000 years B.P. (Sartono, 1975) and separate the *Pithecanthropus* finds of Middle Pleistocene age (he considers find no. 17 not as *H. e. soloensis* but as *Pithecanthropus*) into two types: the small brained specimens (Nos. I, II, III, VII) from the older part of the Middle Pleistocene, and the large brained cranium (No. VIII) from the younger middle part of the Middle Pleistocene. But the small brained find No. VII and large brained one No. VIII are practically contemporary. "However, *H. erectus* VII which is different in many ways from other small brained *H. erectus,* lived almost contemporaneously with *H. erectus* VIII." (Sartono, 1975, p. 353). Not only this, the new find (1973) from Sambungmatjan found on the very base of the Kabuh formation is dated 900,000 years B.P. (Jacob, 1975a). His cranial capacity is 1035 ccm (Jacob, 1975b). This does not support the schematic division of Sartono into chronologically different small brained, earlier, and large brained, later, types. There are also other discrepancies: in his Table 5 Sartono (1975) puts the large brained *Pithecanthropus* VIII together with the small brained *Pithecanthropus* VII in higher levels of the Middle Pleistocene, whereas in Table 7 in the same paper he puts *Pithecanthropus* VII in the lower part of the Middle Pleistocene and *Pithecanthropus* VIII separately in the middle part of the Middle Pleistocene. Also the capacity of *Pithecanthropus* VIII is different in Sartono (1975) (1029 ccm) and in Jacob (1975b) (1125 ccm)! It seems evident that Sartono's idea simplifies the real situation. Jacob considers both new finds, Sangiran 17 (*Pithecanthropus* VIII) and Sambungmatjan (S 1) as *Pithecanthropus soloensis.* He finds greater morphological similarity between *Homo erectus modjokertensis* (which he designates as *Pithecanthropus modjokertensis* and *Homo erectus soloensis* (*P. soloensis*) than with *Homo erectus erectus* (*P. erectus*). In such divided and classified finds there is a chronological difference between the *H. e. modjokertensis* and other finds, whereas *H. erectus* and *H. soloensis* finds (P. VIII and Sambungmatjan I) are of the same chronological group. The morphological differences between the *H. erectus* and the *H. soloensis* finds are mainly: the large mastoid processes in the *H. modjokertensis* finds and the *H. soloensis* finds and small processes

in *H. erectus* finds; the differences in the degree and morphology of ethmoidal,
maxillary and frontal sinuses; small zygomatic bones in *H. erectus* and large
and robust bones in *H. soloensis* (P. VIII and S 1); while the muscular crests
on the nuchal plane are of different degree and shape. Similarly the morphology
of the occipital torus is variable. There are some other minor differences
which I consider to be in the frame of normal variability even in one population and certainly in the finds of one species.

Detailed study to consider if the subspecies divisions, like *Homo erectus
modjokertensis, erectus* or *soloensis* are valid, is highly desirable. The tax-
onomical value even of the morphological differences mentioned above seems,
according to our experience in human palaeontology, of problematic significance.
Finally what we lack is the knowledge of the functional meaning of such variable
characters. Practical observation shows that the morphology of nuchal attachments and nuchal plane morphology is not a specific character of *Homo erectus*,
but much more that it is a functional character appearing frequently in *Homo
erectus* but existing also in some later finds, even when not to such a degree.
Some other characters (e.g. the robusticity of the zygomatic bone) are bound
to the general robusticity of the skull. The great variability of the shape
and degree of torus occipitalis or occipital crest is well known (Jelinek,
1970).

These are some ideas on the problem of the classification of the Javanese
Homo erectus finds.

When considering the other part of our question, the relationship with *Homo
sapiens*, some Australian finds can be helpful. The time between the last *H.
erectus* (or *H. soloensis* according to Jacob, 1975a and b) finds, namely the
Ngandong finds, and the first *H. sapiens* finds (Wadjak, Niah) in South East
Asian regions is - if the dating of the Ngandong finds is reliable - fairly
small. This raised doubts about these *H. erectus* finds and of their possibly
being forefathers of later *Homo sapiens* in this region. But several studies
of recent Australian aboriginal skulls proved the presence of several archaic
morphological traits in different degree and complexity (Macintosh, 1965, 1967;
Thorne, 1971). The fossil finds in Australia show two types. One, more *H.
erectus*-like (represented by the Cohuna or Kew skull) and the other more Wadjak-
like (represented by the Keilor skull). What these differences really mean we
unfortunately do not know, but certainly the *H. erectus* evolutionary stock
heritage in Australian aboriginals is evident. Nevertheless, everybody considers today all the aboriginal population in Australia - recent or fossil -
as *Homo sapeins sapiens* . Which characters do we consider decisive for our
taxonomic considerations? Chronology? Morphological traits or a complex of
morphological characters? Culture or behaviour? Where are the criteria which
can give us the solid base to separate the *H. erectus* finds from *Homo sapiens*?
If we remember how long the Ngandong finds were frequently considered as Neandertaloids, tropical Neandertals or *Homo soloensis*; how different degrees of
variability can appear even for one population; that sexual dimorphism in
Homo erectus has not yet been profoundly studied; that developmental changes
in different populations were dependent on many factors and certainly were not
contemporary, not of the same degree, nor of the same quality and quantity, then
we see how complex and important for future palaeo-anthropological research the
profound study of these questions is.

The European Middle Pleistocene finds, Mauer mandible, Swanscombe and Steinheim skulls and Montmaurin mandible, are well known and were studied several
times. We have no important news about their morphology but some about their
dating. The palaeomagnetic method proved that all these finds are later than

the Matuyama - Brunhes reversal and this means that they are of Middle Pleistocene age.Important is the chronological relation to the Olduvai finds from Beds
I-IV which are older than 700,000 years and therefore of Lower Pleistocene age
(Clark, 1976). Certainly the European finds have not the possibility of potassium-
argon dating and their dating is therefore based mostly on relative palaeontological evidence. The oldest find (the Mauer mandible) is now dated by most workers
between 320,000 and 500,000 years B.P. Older is the small fragment of the second
upper molar from Prezletice in Czechoslovakia (Fejfar, 1969), coming just from
the Matuyama reversal period (700,000 years B.P.). The skull from Petralona in
Greece is, according to new faunal studies, dated round 500,000 years B.P.,even
though we must mention that this dating is still far from being accepted. The
human remains from Vértesszöllös are dated around 400,000 years B.P. and new
finds from Bilzingsleben near Halle in East Germany as younger. The finds from
Steinheim, Swanscombe and Arago are all younger. The Swanscombe skull is around
300,000 years (at least 272,000) (Szabo and Collins, 1975), Steinheim 200,000-
230,000 years B.P. and Arago around 200,000 years B.P. The Montmaurin jaw is
probably contemporaneous with the Steinheim-Swanscombe and Arago finds and comes
from the early Riss period. This gives a rough chronological sequence and the
relative positions of the most important European finds. Let us look now at
these individual finds.

The small fragment of enamel of a human molar tooth from Prezletice is not
such an important contribution from the morphological point of view. The importance lies in the environmental conditions of this Czechoslovak locality and in
all the geomorphological, geological, archaeological and palaeontological facts
discovered here. The dating as Cromerian is based on palaeontological and geomorphological evidence. The layers of lacustrine marls have their origin in
Cromerian times, from a river system of the neighbouring Labe and Vltava rivers
of which the local pool was a part. The relative height of these marls is approx.
80 m above the recent river and they correspond to the local gravel terraces of
Günz age. Palaeontologically, the typical middle-younger Biharian fauna was found
here (Cromer). The frequent existence of a small rodent *Mimomys* and its absence at
Mauer shows (Together with other palaeontological evidence) that this locality is
the older.*

The Petrolona skull, found without proper stratigraphic or other evidence of provenance, is only now being dated by means of palaeontological studies. Morphologically,
it is very near to the African Broken Hill cranium. The occiput is, in lateral view,
slightly less angled and its upper part is lower than in the Broken Hill cranium. The
whole brain case is also lower but the supraorbital region has not such a strong torus
as in the Broken Hill cranium. The facial profile, even though hidden under a lime-
stone crust, was in its alveolar part not so prognathous as the Broken Hill cranium.
In spite of these differences the similarity of both skulls is striking. If the
datings to the Mindel period were correct (Hemmer, 1972; Kanellis and Marinos,
1969), then this cranium would be nearly contemporary with the Mauer jaw.

The Vértesszöllös occipital would be also near in age (Mindel interstadial).
Surprisingly, its general shape shows that, even though well angled and with
strong muscular relief, it belonged to a well-vaulted, voluminous skull with
a cranial capacity of around 1400 ccm. This large cranial capacity, together
with the size and dimensions of the occipital squama, and the palaeontologically
well-documented early age, led Thoma to describe this specimen as *Homo erectus
seu sapiens* (Thoma, 1966, 1967, 1969). But to suppose that the European Middle Pleistocene finds will somehow morphologically correspond to Lower Pleistocene finds

* The newest study - not yet published - brings this find in doubt because the
enamel fragment seems to belong to Ursus sp. and not to Homo.

from Africa or South East Asia does not correspond to our contemporary knowledge. Both chronological and geographical differences play a role here. The
type of tool assemblage from Vértesszöllös is interesting here as well. They
are mostly pebble tools of small dimensions. They correspond to somewhat
later finds, discovered recently in Bilzingsleben, DDR (Mania, 1974).

Mania started in 1971 the excavation of Bilzingsleben, a Holstein interglacial living site. Heavy duty tools were made from pebbles but there were
few. The majority of implements were of small size. Sometimes antler was
used as a club-tool and small pieces of bone show traces of human activity
also, having been used as implements in different ways. The site was situated
near a spring in a travertine layer, similar to Vértesszöllös and to many other
travertine sites in this part of Europe (e.g. the Last Interglacial sites of
Ganovce, Horka, Beharovce etc., Lozek & Prosek, 1954, 1957). Human remains
found here are represented by two pieces of occipital squama and by a piece of
the glabellar part of the frontal bone. The two occipital fragments were found
separated but seem to belong to one individual. In any case they belong to a
very similar, or even identical type of occipital squama and they conserve
different but neighbouring parts of this bone. The general shape is very
similar to Vértesszöllös, only the muscular relief is slightly less pronounced
and the angle of the bone is slightly more rounded. This can be an individual
or a sexual difference. Most probably, this cranium too was voluminous with
a surprisingly high cranial capacity. The piece of frontal bone shows a strong
supraorbital wall. The glabellar part is not so strongly prominent as in
Petralona. It reminds one of the Arago skull (Riss period) but it is less
depressed. Certainly it would be important to know if the skull was male or
female. This piece of frontal bone shows that the nasion was deep and the
nasal root broad with fairly broad interorbital dimensions. The supraglabellar depression is shallow, showing a low and receding frontal part of the
skull. If we consider the conserved parts of the occipital bone as if all
these remains belonged to one calvaria, then the highest point on the vault
would have been at the bregma or on the parietal bones; judging from the dimensions of the reconstructed broad occipital squama, the whole calvaria must
have been fairly broad.

Morphologically, so far as we can judge from the remains, and chronologically,
the Bilzingsleben find is halfway between Petralona and Arago skulls.

The last find I have selected for this brief comparative study is Arago
(Tautavel). On the frontal bone we see in frontal view a well depressed glabella dividing the supraorbital torus into two parts. In each part the beginning of a division into superciliary and supraorbital parts can be observed.
The supraglabellar depression is only slight in lateral view. The forehead is
slightly vaulted, not so receding as in the Petralona skull. In frontal morphology as in many other facial characters (de Lumley, 1975), the Arago skull
is certainly more developed than other archanthropine skulls and shows already
the tendency to develop in the direction of the later *Homo sapiens neandertalensis*.

DISCUSSION

I have selected some Middle Pleistocene finds of *Homo erectus* from Africa,
South East Asia, North Africa and Europe to show their common characters and
some differences. Leakey (1972) considered the Olduvai hominid no. 9 as a
locally developed *H. erectus* type. The recognition of Ngandong skulls as *Homo
erectus* brings also the knowledge of long local evolution of *Homo erectus*

populations in South East Asia. The same is clearly demonstrated for North Africa and mainly for Morocco where the finds are happily more numerous. The same is shown already with the not so numerous European material. There is no sign of larger migrations or extinctions which we often invoked to explain development (in the larger geographical areas). Correct dating and chronological sequence is the basic condition of understanding any changes illustrating the evolutionary process. In the past our study of human evolution was obscured by many wrongly dated finds and even by wholly wrong chronological systems and suppositions. The palaeomagnetic method gives us now the possibility to compare at least one time level all over the world. With its help we see that most of the East African *H. erectus* finds and even many of the South East Asian *H. erectus* finds are older than any find in Europe. When comparing such distant populations from Europe, Africa and far South East Asia with a different chronological order and developmental tempo, and certainly with different environmental conditions, we cannot think that the evolutionary results of morphological adaptation to these different conditions will be without differences. There is no reason for us to try to explain evolutionary differences in different regions of the world with the help of theories of large over-continental human migrations or extinctions.

The geographical, regional and environmental differences, different time schedules of the process and the degree of isolation, brought enough differences in human morphology at the subspecies level.

When Howell already in 1960 (and later also other palaeo-anthropologists) stated that "in Europe there is no trace of the so-called 'pithecanthropine' peoples so characteristic of the whole of the Middle Pleistocene of Eastern Asia..." he was right in principle. We may well add today "of the Lower and Middle Pleistocene of Eastern Asia" to stress some chronological differences. Yet, in Europe we find another mode of cerebralisation. This is even more clear if we include in Europe and also in Java the later finds: Anteneandertals and Neandertals in Europe and Ngandong finds in Java. The process has some similar principles. *Some* early European finds (e.g. Vértesszöllös) have great cranial capacity and only much later (*Homo sapiens neandertalensis*) are they followed by the *majority* of the European population. In Java some early finds (Kabuh) have greater cranial capacity (Sambungmatjan, Sangiran 17) and only much later are they followed by the majority of the population (Ngandong). Is this only chance similarity misinterpreted because of an insufficient number of observations? There are some other examples of cerebral changes suddenly appearing in some individuals and later followed by general change in the majority of the population. The first examples of the brachycephalisation process and its study in East European prehistoric skeletons (Necrasov, 1959) merit more detailed study on a broader comparative basis.

And what is the meaning of the exceptional cranial capacity in the famous Koobi Fora 1470 skull?

Unfortunately, mostly only cranial and dental remains have been studied in human evolution. The postcranial remains have begun to be studied in detail only recently. We know that morphological characters can be grouped to form parts of different functional units like the digestive system, brain development, locomotion, etc. These functional units have sometimes mutual boundaries like locomotion, the visual apparatus and brain development. Up to a certain degree they can develop at a different tempo according to their functional adaptability, but often we do not understand the functional meaning of important morphological characters. Speculations on skull thickness, supraorbital morphology or speech possibilities are only some examples. Without any doubt,

outside of proper dating, the study of functional anatomy in skeletal characters
is the fundamental need in modern palaeo-anthropology. It was rightly recognised by Jaeger (1973) in his report on the Sale skull that "... il apparait
ainsi que l'occipital et les parietaux sont en avance sur le frontal, la den-
ture, et les structures basicraniennes." Certainly different functional units
can, under different conditions, result in different changes, in quantity and
even in quality.

   Yet another observation can be helpful in our study of Middle Pleistocene
hominids. Similar modes of change that we follow in *H. erectus* finds can be
found also in later finds. When considering some European finds of *Homo
sapiens neanderthalensis*, we find different degrees of change both quantitatively and qualitatively. Some archaic characters persist in some individuals far longer and are more pronounced than in others. In 1960 I studied the
frequency and persistence of the *torus occipitalis* in central European prehistoric populations from the early Neolithic period up to the 10th century
A.D. and several cases documenting the morphological persistence of some archaic characters in palaeolithic and even later individuals were reported
elsewhere (Jelinek, 1970). This is in principle similar to the mode of morphological change Jaeger (1973) observed in North Africa. We face differences
between individuals inside one population, and also when different populations
are compared.

   To come back to the European Middle Pleistocene finds, some authors (Thoma
1966, 1969; Heberer et al., 1975) described some of them as *Homo erectus seu
sapiens* or even as *Homo sapiens* (Leakey, 1972) for Vértesszöllös, Steinheim,
Fontéchevade and Swanscombe. Similarly African finds from Broken Hill and Saldanha and the South East Asian Ngandong finds were taxonomically considered as
H. sapiens or neandertaloids, neandertals or *H. erectus* representatives.

   All the Middle Pleistocene hominid remains attributed to *Homo erectus*, whether in Europe, Asia or Africa, show a similar tendency towards the increase
in cranial capacity; they have perfectly erect stature and a modern way of
locomotion, they have hands like ours and with them the possibility of the
broadest range of manual activity. Culturally they proved to have the ability
to adapt, live and reproduce successfully in very different environments, they
developed variable tool assemblages, knew fire and knew how to construct pro-
tective shelters and the first human habitations. They developed social relations, co-operation and communication inside the group which allowed them to
cope with their different environments, to hunt big mammals and to ensure
further cultural development. In different parts of the world we can follow
their changes and transition into later types which we describe as *Homo
sapiens*. If we do not hesitate to do it in Europe, if we try to call *Homo
sapiens* some of the early Middle Pleistocene finds here, why should we try to
prolong the existence of *H. erectus* species late into the Upper Pleistocene in
other parts of the world? Why, when all anthropologists unanimously consider
all living populations of man as one species, the existing morphological differences
that we can see in Middle Pleistocene hominids should be evolutionarily and
biologically important enough to be regarded as indicating a different species?
Have we any solid scientific grounds on which to consider Middle Pleistocene
European finds, with earlier morphological cranial changes, as *Homo sapiens*
and the extra-European finds evolving in the same direction but in somewhat
different degree and time sequence of adaptation into different conditions as
Homo erectus? The whole mode and results of the hominid evolutionary process
show that there are not, and in the past could not have been differences at the
species level, but only at the subspecies level, whether the cerebralisation

process - as only one part of the mosaic of evolutionary changes - started earlier or later. The logical consequence of such a situation is to lead us to consider the different African, European and Asian finds of *H. erectus* type as *Homo sapiens erectus*. With the increasing chronological knowledge, with better dating and better understanding of functional morphology the picture becomes clearer and clearer.

REFERENCES

Arambourg, C. (1963). *Arch. Inst. Paleont. hum.*, 32, 37-190.

Arambourg, C. and Biberson, P. (1955). *C.r. hebd. Acad. Sci.*, *Paris*, 240, 1661-3.

Arambourg, C. and Biberson, P. (1956). *Am. J. phys. Anthrop.*, 14, 467-490.

Biberson, P. (1963). *In* "African Ecology and Human Evolution", (F.C. Howell and F. Bourlière, eds), pp. 417-490, Viking Fund Publications.

Brace, C.L. (1964). *Curr. Anthrop.*, 5, 3-43.

Brace, C.L. (1968). *Curr. Anthrop.*, 9, 30-31.

Choubert, G. and Marcais, J. (1947). *C.r. hebd. Seanc. Acad. Sci.*, *Paris*, 224, 1645-7.

Coon, C. (1962). "The Origin of Races", New York, Knopf.

Coppens, Y. (1967). Actas del V. Congreso Panafricano de Prehistoria y de Estudio del Cuaternario, (L.D. Cuscoy, ed.), pp. 329-330, Tenerife, 1963.

Clark, J.D. (1976). *In* "Human Origins", (G.L. Isaac and E.R. McCown, eds), Meulo Park, California.

Ennouchi, E. (1968). *Annls Palaeont.*, 56, 95-107.

Fejfar, O. (1969). *Curr. Anthrop.*, 10, 170-173.

Heberer, G., Henke, W. and Rothe, H. (1975). Der Ursprung des Menschen, G. Fischer, Stuttgart.

Hammer, H. (1972). *Anthropologie, Paris*, 76, 155-162.

Holloway, R.L. (1973). *Nature, Lond.*, 243, 97-99.

Howell, F.C. (1960). *Curr. Anthrop.*, 1, 195-232.

Howells, W. (1973). "Evolution of the Genus *Homo*", Addison Wesley, Reading, Massachussetts.

Jacob, T. (1975a). *Recherche*, 62, 1027-1032.

Jacob, T. (1975b). *In* "Paleoanthropology, Morphology and Paleoecology", (R.H. Tuttle, ed.), Mouton, The Hague.

Jaeger, J.J. (1973). *Recherche*, 39, 1006-1007.

Jelinek, J. (1970). *Homo*, 72, 89-100.

Kanellis, A., and Marinos, G. (1969). Proceedings of the Fourth International Congress of Speleology in Yugoslavia, 4-5, 355-362.

Klein, R.G. (1973). *Nature, Lond.*, 244, 311-312.

Kurth, G. (1965). Die (Eu) Hominiden *in* "Menschliche Abstammungslehre", (G. Heberer, ed.), pp. 357-425, Fischer, Stuttgart.

Leakey, L.S.B. (1966). *Nature, Lond.*, 209, 1279-1281.

Leakey, L.S.B. (1972). *In* "The Origin of *Homo sapiens*", (F. Bordes, ed.), pp. 25-29, Paris, UNESCO.

de Lumley, A.M. (1975). *In* "Paleoanthropology, Morphology and Paleoecology", (R.H. Tuttle, ed.), pp. 381-387.

Macintosh, N.G.W. (1967). "Recent discoveries of Early Australian Man".

Macintosh, N.G.W. and Larnach, S.L. (1972). *Archaeol. & phys. Anthropol. Oceania*, 7, 1-7.

Mania, D.L., Grimm, H., Vlcek, E. (1976). *Z. Archaeol. Mittelalters*, 10.

Necrasov, O., Cristescu, M., Maximilian, C., and Nicolaescu-Plopsor, D. (1959).

   Probleme Antrop., IV, 21-45.
Pilbeam, D. (1972). "The Ascent of Man", Macmillan, New York.
Piveteau, J. (1967). *Annls Paleont.*, 53, 167-199.
Prosek, F., and Lozek, V. (1954). *Pamatky Archaeologicke a Mistopisne*, 45,
   35-74.
Prosek, F. and Lozek, V. (1957). *Eiszeitalter Gegenw.*, 8, 37-90.
Sartono, S. (1972). *Curr. Anthrop.*, 13, 124-125.
Sartono, S. (1975). *In* "Paleoanthropology, Morphology and Paleoecology",
   (R.H. Tuttle, ed.), Mouton, The Hague.
Sausse, F. (1975). *L'Anthropologie, Paris*, 79, 81-112.
Szabo, B.J. and Collins, D. (1975). *Nature, Lond.*, 254, 680-681.
Thoma, A. (1966). *L'Anthropologie, Paris*, 70, 495-533.
Thoma, A. (1967). *Z. Morph. Anthrop.*, 58, 152-180.
Thoma, A. (1969). *Z. Morph. Anthrop.*, 60, 229-241.
Thorne, A.G. (1971). *Mankind*, 8, 85-89.
Tobias, P.V. and von Koenigswald, G.H.R. (1964). *Nature, Lond.*, 204, 515-518.
Vallois, H.V. (1960). *Bull. Archeol. Maroc.*, 3, 87-91.
von Koenigswald, G.H.R. (1956). "Meeting Prehistoric Man", Thames and Hudson,
   London.

A PRELIMINARY DESCRIPTION OF THE SHANIDAR 5 NEANDERTAL PARTIAL SKELETON

E. TRINKAUS

Department of Anthropology, Harvard University, Cambridge, Mass. 02138, USA.

Excavations between 1951 and 1960 in Shanidar Cave, northern Iraq, unearthed the remains of nine Neandertals, seven adults (1, 2, 3, 4, 5, 6 & 8) and two infants (7 & 9). These include the hitherto undescribed Shanidar 5 adult partial skeleton. Although fragmentary, the Shanidar 5 remains are sufficiently intact to provide valuable morpholotical and metrical data.

Shanidar 5 was discovered in 1960 near the top of Layer D, the Mousterian level. It was at the same stratigraphic level as Shanidar 1 (associated radiocarbon determination of 46,900 ± 1,500 B.P. (GrN-2527)) and stratigraphically above the other Shanidar Neandertals.

Among the Shanidar 5 remains are portions of the cranium, dentition, axial skeleton, forearms and hands, and thigh and leg bones. The cranium consists of a facial-frontal piece with most of the frontal, zygomatics, nasals and maxillae, a separate left temporal, and fragments of the parietals and the occipital. The anterior maxillary alveolus is preserved with four identifiable teeth, an I^1, a $P^{\underline{3}}_{4}$, an $M^{\underline{1}}_{2}$ and an M^3. Nothing remains of the mandible or the lower dentition. The axial skeleton is represented by a fragment of a cervical vertebra, portions of eight ribs, and a crushed ilium, all of which are relatively undiagnostic. The most complete bones are from the forearm and hands. Two ulnae, an almost complete right one and a left diaphysis, and two major portions of the right radius, proximal diaphyseal and distal articular, were preserved. The right hand is represented by three carpals (hamate, pisiform and triquetral), three metacarpals (2, 4 and 5), and seven phalanges (PP1, PP5, MP2, MP4, DP1, DP2 and DP4). The lower limbs are much less complete. Both femoral diaphyses are present, but only the posterior surfaces and some of the medial and lateral surfaces are intact. A left femoral head and neck and portions of the left condyles were also recovered. The patellae are present, but only the left one is reasonably intact. The medial half of the left tibia remains; it is divided into a proximal articular and diaphyseal piece and a distal diaphyseal piece. There are also fragments of the left fibula. Portions of the skeleton were left *in situ* due to excavation difficulties, so more remains, such as foot bones, may be recovered at a future date.

The Shanidar 5 cranium is typically neandertaloid in several features (Figure 1). The supraorbital torus is prominent, but not massive, and is filled with large, multichambered frontal sinuses. The supratoral sulcus is slight and leads on to a low, wide and very heavy frontal squama. The midfacial region is highly prognathic and exhibits prominent nasal bones, a large pyriform aperture, an absence of canine fossae, and bilaterally double infraorbital foramina. The prognathism is accentuated by the retreating zygomatic profile (zygomatic root at approximately M^2) and moderate alveolar prognathism.

The temporal bone has a relatively large mastoid process, similar to those of Shanidar 1 and 2 and Amud 1. The external acoustic meatus is elongated and obliquely oriented, while the glenoid fossa is wide and deep. It is a large cranium which conforms entirely to the morphological pattern of the Neandertals.

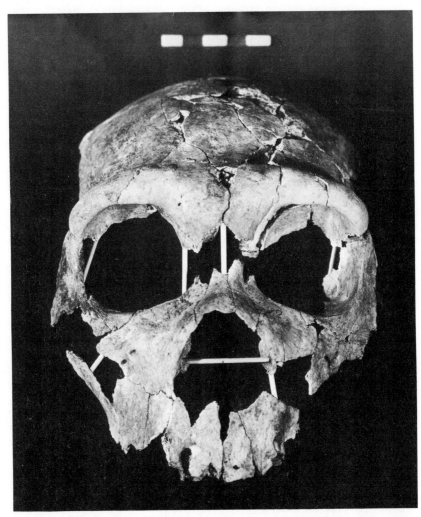

Figure 1. Shanidar 5 cranium norma frontalis, oriented approximately on the Frankfurt horizontal. Photograph by courtesy of the Iraq Museum.

Shanidar 5 suffered considerable alveolar resorption and at least three teeth, the left I^2, P^3 and P^4, were lost antemortem. All of the preserved teeth show extensive occlusal wear, especially the anterior teeth. Furthermore the right I^1 exhibits the anterior rounding previously noted on Shanidar 1 and other Neandertals. It is also present on the Shanidar 3 and 4 anterior teeth, but not on those of Shanidar 2 and 6. Low power magnification (14 linear diameters) of the occlusal surface reveals only normal polishing and microspalling of the secondary dentine. The microwear pattern does not, therefore, appear to support directly the frequent non-masticatory use of the anterior dentition.

The ulnae and radius are quite similar to those of recent humans; they are noteworthy only for their strong muscular markings. The right hand, however, exhibits several Neandertaloid features. The metacarpals and phalanges have relatively large articular surfaces, and all exhibit strong muscular and ligamentous attachment areas, especially on their proximal ends. The tuberosities of the distal phalanges are enlarged and strongly marked. The relative development of these aspects of the hand implies an extremely powerful grip. None of the preserved portions of the hand, however, implies manipulative abilities inferior to those of recent humans.

The femoral diaphyses are large and highly robust, with thick cortices wider medially than laterally. The femora are strongly curved anteroposteriorly and lack pilasters but the muscular markings are still quite strong. In this pattern they resemble the femora of Shanidar 1, 4 and 6 and of other Neandertals. The patellae, tibia and fibula present no unusual features. The medial tibial condyle is highly retroverted but within modern human ranges of variation. All three are robust but similar to those of recent humans.

The Shanidar 5 remains are clearly those of an elderly individual, as determined by the advanced state of sutural closure and especially the extreme dental wear, antemortem tooth loss and alveolar resorption. It was probably the oldest individual in the Shanidar sample. Accurate sex determination is not possible, since the diagnostic portions of the pelvis were not preserved, but on the basis of its large overall size in relation to the other Shanidar Neandertals, it was probably a male.

Morphologically the Shanidar 5 skeleton is therefore similar to the other Near Eastern and European Neandertals. This is particularly evident in the facial and temporal configurations and in the robustness of the hand bones and femora. Interestingly, Shanidar 5 is especially close to the other Shanidar adult Neandertals for which there is comparable material. Despite the considerable variation in size, chronological age and geological age, the Shanidar Neandertal sample is remarkably uniform.

Subsequent analyses of the Shanidar 5 remains, as well as those of the other Shanidar Neandertals, will help to document this important hominid fossil sample and should contribute to our understanding of their position in hominid phylogeny.

ACKNOWLEDGEMENTS

My sincere gratitude is due to Dr. Isa Salman, Director-General of Antiquities, Iraq, and Dr. Fausi Raschid, Director of the Iraq Museum, Baghdad, for permission to study the Shanidar remains. This research was supported by NSF grant BNS76-14344.

RECENT RESULTS CONCERNING THE BIOMECHANICS
OF MAN'S ACQUISITION OF BIPEDALITY

H. PREUSCHOFT

Ruhr University, Bochum, West Germany.

INTRODUCTION

Upright, bipedal posture is one of the outstanding characteristics of modern man. The reasons behind the assumption of this unique posture will not be discussed here but it raises unique functional and mechanical problems. These problems have been solved by the evolution of special morphological features that might have been obtained, theoretically, in two ways. One is by the well-known evolutionary mechanism of mutation and selection and the other is by a process which adapts, during the life of the individual, parts of the locomotor system to the mechanical requirements of their function (Amtmann, 1971, 1978; Kummer, 1959, 1972; Pauwels, 1960; Preuschoft, 1969, 1970a and b, 1971a and b, 1973a and b, 1975).

Fossils provide us with little more than the shapes of bones, or sometimes only parts of bones. Because bone shapes are so closely related to function, they offer an opportunity for an understanding of how animals that no longer exist have mastered the mechanical problems connected with the acquisition of upright posture. Recent advances in several disciplines justify a reconsideration of these problems.

A. Increasingly, attention is being paid by morphologists to functional problems. We have learned much about the way in which stresses are produced inside an animal's body and more and more parts of the skeletal system have been subjected to an analysis which throws light on the way in which they fulfil their functions. Former studies selected one or two static positions for analysis based on observations or on still photography (Kummer, 1959; Pauwels, 1935; Preuschoft, 1969, 1970a and b). Now we know more about the movement sequences and about motion by means of cinematography (e.g. the many films on primates in their habitat, such as those of Kortlandt, 1966-1974; Grand, 1976; Ishida et al., 1975). Also x-ray cinematography has given precise evidence as to joint movements and positions of bones in a moving body (Jenkins, 1972, 1974; Jenkins and Fleagle, 1975). In some cases high speed cinematography combined with the subsequent calculation of accelerations has provided data about the influence of movement on the stressing of bones and joints (Plagenhoef, 1971; Wells and Wood, 1975). Some direct measurements of the forces that act from the ground against the body by force plates are now available (Kimura et al., 1978). An important method of checking theoretical predictions about muscular activities is the EMG-technique (Ishida, et al., 1975; Tuttle and Basmajian, 1974; Tuttle et al., 1975; and Kimura et al., this volume). Finally we have some understanding of how mechanical stresses lead to the deposition and resorption of bone - the intra vitam process which makes the bone reflect the motor pattern of the *individual* animal.

External forces normally act on a long "load arm" whereas the muscles which maintain equilibrium at the joints have in most cases a much shorter "power arm". This means that the greatest stresses inside an animal are evoked by muscles. Since the arrangement of muscles is, as far as we know, controlled genetically, the variation in muscle forces and thus the individual variations of bone shape are limited.

Besides the arrangement of muscles, the ways of transmitting body weight to a substrate (e.g. a branch below or a twig above the animal), as well as the positions of the joints in various phases of movement, determine the stress upon and shape of the bones.

In the extant primates, which seem to stake out the ranges of variability of the primate order, the morphology of the muscular system and the variations in the skeletal system between *populations* are well known. This is to a large extent due to the common use of refined multivariate statistical methods (see below).

B. Concerning behaviour, again recent primates can be used to demarcate the range of variance. We now know more about the locomotion of primates under natural conditions than before. Attempts have been made to develop criteria for description and for comparisons (Carpenter, 1974; Fleagle, 1977; Morbeck, 1977; Prost, 1965; Prost and Sussmann, 1969; Kortlandt, 1966-74, in films). The environment of the wild living populations and the demands made on them for survival are no longer theoretical assumptions and local legends. Some of the earlier classifications of motor behaviour have been questioned and some need further specification. The variability of primate movements is impressive although perhaps understandable: branches of adjacent trees do not offer a substrate upon which a gait with a regular footfall sequence can be executed - as is usual in terrestrial runners. The versatility of primate movements raises an obstacle against clearcut correlations of motor behaviour with morphology. On one hand, it does not encourage a new version of locomotor categories; on the other, no comprehensive description or set of quantitative data can provide a system which can be related by means of statistical methods to the elaborate classifications that are based on morphology alone (see below).

C. The number of available fossil cranial remains has increased so rapidly that the finds have outpaced descriptions as well as interpretations. In spite of this, newly discovered fossil elements have given reasons for the consideration of points which have been neglected up to now.

According to our overall knowledge of human evolution, considerations about the assumption of bipedal posture must be related to the *early hominoidea* and to the *Plio-Pleistocene hominids* (see Andrews, Conroy, Joysey in "Primate Evolution" or Bilsborough, Tobias, Wood in "Hominid Evolution").

Early Hominoids

The earliest hominoid to which some postcranial fossils may belong (a first metatarsal, a basal phalanx and a nearly complete ulna) is *Aegyptopithecus*. According to an earlier study (Preuschoft, 1975), they are well suited to sustain the stresses evoked during quadrupedal walking in a lowered, crouched posture (on any substrate) and during climbing in postures in which body weight tends to extend the elbow rather than to flex it. This happens in climbing on perpendicular branches and in all kinds of suspensory behaviour. Arboreality is indicated by the prehensile feet and hands.

Fleagle et al. (1975) in an independent study arrived at very similar results and they pointed out that both forms of locomotion occur in the living form *Alouatta* (see also Conroy, 1976a and b).

In the Miocene, the hominoids closest to the line of human ancestry were the *Dryopithecinae*. The discussion on their locomotor habits followed, in parallel, several lines: Le Gros Clark and Leakey (1951) in their publication of the more or less damaged long bones and tarsals found indications of arboreality and some sort of bipedality. Their opinion was expressed by Napier (1963) that "Proconsul represents an important structural and functional stage in the phylogeny of hominid locomotion."

Napier and Davis (1959) noted in the hand and arm skeleton of *D. africanus* "a particular combination of quadrupedal and brachiating features". "The arboreal quadrupedal features of the fossil hand are most evident in the carpus, while those structural changes which suggest adaptation to a brachiating mode of locomotion are most evident in the metacarpus and phalanges."

Observations in the wild show that chimpanzees (Kortlandt, 1968 and in several films) and gorillas (Schaller, 1963) seldom or never brachiate. Later Tuttle (1967, 1969a, b and c, 1970) demonstrated that several features of the hand morphology of the African pongids can be explained on the basis of adaptation to knuckle walking.

Day and Wood (1969) showed that three of the East African fossil tali included in a canonical analysis were significantly closer to those of the pronograde quadrupedal apes than to those of striding bipedal man. A minor ramification of the discussion branched off here: Oxnard (1972) raised objections against Day and Wood because of the statistical methods used. These objections led to a new multivariate analysis of the tali by Wood (1973), now comparing them with "knuckle walkers" (= *Gorilla* and *Pan*), "Quadrupeds" (*Papio*, *Cercopithecus*, *Colobus*) and a "brachiator" (*Pongo*), whereby the closest affinities were shown between the fossils and the tali of *Papio* and *Cercopithecus*. Lisowski et al. (1976) used a different set of measurements and included 21 fossil tali in their analysis. They found that two *D. nyanzae* and one *D. major* talus tended towards *Papio*, *Pongo* and *Pan*. "However, ... these three specimens are not intermediate between these living forms but are uniquely separated in ways additional to those that separate the living forms. Each is, for instance, morphologically as far distant from *Papio*, *Pan* and *Pongo* as these latter genera are from each other."

Pilbeam (1969) and Simons and Pilbeam (1972) tried to integrate the differing viewpoints. They imagine *D. africanus* as "knuckle walking, arm swinging, but not identical with living hominoids". The larger *D. major* is assumed to have been less active than *D. africanus* and "at least as terrestrial as the chimpanzee, and may well have been capable of knuckle-walking behaviours, although...we can only assume, that the functional differences are as great". The authors rely largely on Walker (pers. comm.) and on Day's and Wood's (1969) study on the talus.

Stern (1975) objected to the over-estimation of terrestrial habits. "The arboreal adaptation of African apes is sometimes forgotten as a result of studies by Tuttle" (see above). Being "as terrestrial as a chimpanzee is still to be arboreal most of the time". Stern also reminds us that many arboreal features in the chimpanzee are combined with a small number of terrestrial ones. According to him neither the adaptation to one behaviour, nor the lack of adaptation to another movement has ever been demonstrated, whereas the focus of the discussion has changed.

Lewis (1971, 1972) reconstructed the wrist joint of *D. africanus* including the soft parts and concluded that the complex was adapted to forelimb suspension. He recommends "a return to the view that man and the African apes share a history of specialised arboreal locomotion". "...they are also clearly preadaptive to efficient use of the hands as manipulative organs in such activities as food gathering and transport, and tool use and tool manufacture."

Lewis's papers were opposed by Schön and Ziemer (1973) who argued also on the basis of the "close packed positions" (Barnett et al., 1961), and consider the weight-bearing phase of *Dryopithecus* to be "a palmigrade quadrupedal stance". Objections were raised also by Jenkins and Fleagle (1973) who have observed the carpus of knuckle-walking chimpanzees by means of cine-radiography and found the postulated situations not realised. The entire concept of close-packed positions has been criticised by Preuschoft (1969). A thorough investigation of the wrist joint in primates is contained also in O'Connor (1975).

Conroy and Fleagle (1971) accepted Lewis's reconstructions, but interpreted this morphology in the opposite way, namely as an adaptation not to brachiation, but to knuckle-walking instead. This interpretation was reconfirmed by Zwell and Conroy (1973) who have computed indices taken from Napier and Davis (1959) and performed a principal components analysis. Corruccini, Ciochon and McHenry (1976a) objected to this on methodological grounds. They have based a statistical study on measurements of the wrist-forearm complex, and compared hominoids cercopithecoids, *Ateles*, *Alouatta* and *Pithecinae* with *D. africanus*. The result is a clear separation of the hominoids from the densely crowded monkeys, among which *D. africanus* assumes a rather central position.

Morbeck (1972), in a very detailed and careful investigation of the whole bulk of postcranial material attributed to the various *Dryopithecus* species, was unable to agree with ready and clearcut categorising of the fossils. "In the majority of cases, the evidence derived from the study of the joints indicates basic primate quadrupedal patterns." "...it is difficult to isolate and define specific knuckle-walking characters except those pertaining to the wrist joint, metacarpal and phalanges (Tuttle, 1970)". In 1975 Morbeck demonstrated similarities to *Cercopithecus* in the movement capabilities of the *Dryopithecus* wrist and distal forearm. She contradicts Lewis's reconstruction of a meniscus. She also points out shortcomings in Zwell's and Conroy's (1973) study. "Miocene hominoidea such as *D. africanus* may have resembled modern apes in some features of the postcranial anatomy as well as in the dentition. These, however, do not suggest knuckle-walking adaptations." Morbeck accepts only one direct relationship between form and function: the unquestionable interdependence of joint surfaces and ranges of movement. "The ranges and directions of movement within the wrist and the hand in *D. africanus* are indicative of movement capabilities similar to those of contemporary, quadrupedal, palmigrade monkeys, not African pongids." And she concludes: "Interpretation of Miocene hominoids as possible ground-adapted or knuckle-walking primates is unclear.... Adaptations in the elbow and shoulder complexes to this specialised locomotor pattern have not yet been demonstrated (Morbeck, 1976).

I read this with pleasure since my own biomechanical analysis (Preuschoft, 1973a and b), based on all postcranial elements which may belong to a species of *Dryopithecinae* in Nairobi and in the British Museum of Natural History, yielded a similar result. All the elements meet the mechanical requirements of prehensile hands and feet with strong finger- and toe-flexors (= arboreality, climbing habits). All are well suited to a pronograde body posture with flexed elbows and semiplantigrade and palmigrade movements. On the basis of these considerations of course no conclusions regarding the preferred substrate of locomotion can be drawn. It may have been flat ground, cliffs, branches, tree trunks or twigs. None of the fossil elements is of the shape necessary to avoid very high stresses during brachiation or during knuckle-walking.

Corruccini, Ciochon and McHenry (1976b) found also, by means of a multivariate study, that the *Dryopithecinae* remains "all fall in a somewhat intermediate position between apes and monkeys, but were phenotypically nearer monkeys and joined them in a cluster analysis". Considerations involved in the multivariate statistical treatment and in a biomechanical analysis will be discussed later.

The last contribution to the motor behaviour of the once discussed *Oreopithecus* after Schultz (1960) was made by Kummer (1965a). The femora of the nearly complete skeleton of *Oreopithecus* show a clear valgus position of the distal joint and a narrow angle between shaft and neck. Although they are severely crushed and perhaps deformed, the traits are visible in both femora. These characteristics are advantageous for sustaining the stresses which occur in bipedal posture (Kummer, 1965a; Lovejoy and Heiple, 1970; Preuschoft, 1971a, 1970b). This contradicts the overall picture derived from Schultz of a slow quadrumanal climber in swamp forest like the present day orang-utan.

Only feeble attempts have been made to clarify the locomotor habits of *Limnopithecus* since Ferembach (1959) and of *Pliopithecus* since Zapfe (1960) (see Ankel, 1965; Ciochon and Corruccini, 1977; Preuschoft and Weinmann, 1973).

Although many workers agree about the probability or about the "beauty" of some suggestions or "theories" concerning reasons and pathways of evolutionary changes, there is still no full agreement about the motor habits from which bipedality evolved.

However, at some stage of evolution (and I would assume in the late Pliocene), the change from pronograde to permanently orthograde posture must have taken place (see below).

Plio-Pleistocene Hominids

The hominids of the South African and East African deposits are commonly considered to have been bipeds (see for instance Bilsborough, Tobias, Wood in this volume or Heberer, 1965; Le Gros Clark, 1964, 1967; Robinson, 1972). The pelvis, the proximal and distal sections of the femur, and the foot skeleton have been investigated morphologically by many authors. All used the same "principal methodological approach, namely comparing the fossil postcranials with corresponding elements of extant primates" and drew (based on the similarity) conclusions "about the locomotor habits of fossil primates" (Ankel-Simons, 1975).

The overall similarity of the fossil hominid remains with modern man left little doubt that the forms used an upright, bipedal mode of posture and locomotion. But simple morphological observation also revealed some traits which are not in complete accordance with the characters in present day man:
1. The Kromdraai talus shows a deviating neck and an unusual trochlear shape.
2. The Sterkfontein shoulder blade exhibits resemblances to pongids.
3. Some of the femora from east of Lake Turkana and the two from Swartkrans have very long necks and small heads, as well as a less marked greater trochanter.
4. South African and East African upper extremities are more massive and seem to be relatively longer than the lower extremities (Genet-Varcin, 1961; Helmuth, 1968; Howell and Wood, 1974; Robinson, 1972; McHenry et al., 1976).
5. There are differences in body size and perhaps in weight (Helmuth, 1968; Kay, 1973, 1975; McHenry, 1973a, 1974, 1976; Pilbeam and Gould, 1974; Wolpoff and Brace, 1975).

6. All fossil pelves differ from those of modern man, and they are not all
of the same shape (Robinson, 1972; Robinson et al., 1972; McHenry, 1975a and
c).
7. From Olduvai we know of two sorts of finger bones (Napier, 1962).
    One of the possible conclusions that can be drawn from these observations
is the assumption of two phylogenetically distinct hominids, which existed at
roughly the same time and moved differently. These may be called *Homo* and
Paranthropus as Robinson (1972) does, or *Homo* and *Australopithecus* (Day,
1973; Day and Leakey, 1973; Day et al., 1975; Leakey, 1971, 1973; Leakey
et al., 1971, 1972; Leakey and Wood, 1973; Leakey and Walker, 1973; Tobias,
1973a, b and c, 1975a, b and c) or *Australopithecus* and *Paranthropus* (Heberer,
1956; Le Gros Clark, 1964, 1967) or gracile and robust *Australopithecus*
(Cachel, 1975; Wolpoff, 1975; McHenry, 1975a).
    In any case, we are left with the question: what do the differences mean
functionally or, more precisely, how far did the creatures differ in body posture and locomotion, both being essential and determining factors in their
making use of the environment. To approach the question in a generalised,
comparative way has not led to much progress. In the case of the Miocene
fossils, comparisons can be made with a number of differently moving primates
which can be separated and grouped by using terms like "brachiators" or
"arboreal quadrupeds" or "knuckle-walkers", although every author associates
different meanings and different behaviours with the words. The shortcomings
and insufficiency of these motor categories have been discussed in detail by
Oxnard (1975). In the case of the hominids, however, such a rough classification can hardly yield another result than "biped" (see also Robinson, 1972),
thus suppressing the evident differences in morphology and their possible
meaning. Some authors have felt some unease and so ascribed to modern man
the epithet "striding" or "propulsive", which leads to a negative characterisa
tion of the fossil forms as being incapable of exercising a "striding gait".

APPLICATION OF NEW METHODS: MULTIVARIATE STATISTICS

    At this stage, two attempts have been made to obtain further information.
One is using a metric-statistic approach, normally based on measurements of a
single element. For instance:
    Distal phalanx of the hallux (Wood, 1974b; Day, 1967)
    First metatarsal of the hallux (Rightmire, 1972)
    Tali (Day and Wood, 1968; Lisowski et al., 1976; Oxnard, 1972; Wood,
    1974a and b)
    Proximal ends of femora (Day, 1973; McHenry and Corruccini, 1976; Walker,
    1973)
    Pelvis (McHenry, 1975a and b; McHenry and Corruccini, 1975; Zuckerman et
    al., 1973)
    First metacarpal of the thumb (Rightmire, 1972)
    Humerus (McHenry, 1973b; Patterson and Howells, 1967)
    Shoulder joint (Ciochon and Corruccini, 1976)
    Scapula and clavicle (Oxnard, 1968a and b)
    Ulna (McHenry et al., 1976)
These studies have confirmed or established the above listed statements 1-7.
    This treatment has important advantages over the classical morphological
comparisons:
1. The details are introduced in the form of measurements, therefore in a
reproducible way.

2. During processing of the measurements, no uncontrolled weighting of importance takes place.
3. Details can be included in the study, which are not yet understood from a "functional" point of view. It is possible to base an analysis exclusively on metrical characters which have no known functional meaning.
4. They allow the handling of great numbers of individuals and the taking into account of the variability.
5. Retrospectively, those traits can be identified which have contributed to establish the result.
6. The results come out in the form of relative distances. These can be interpreted as taxonomic differences, because they are highly correlated with the results of comparisons based on characters said to bear taxonomic evidence (Creel, 1968; Oxnard, 1975). If a classification is the sum of all information known about the animals (not only morphology of teeth), there must be considerable taxonomic evidence in the postcranial skeleton. This, however, must be weighed carefully against the likewise possible "functional" interpretation.
7. We should never forget that the foregoing selection of measurements will to a considerable extent anticipate the result of the analysis.
8. No matter which traits are measured, the ultimate result of such a comparison is always the statement that: fossil X is similar to living Y, but not similar to living Z.
9. The functional aspect is taken into account only by the background knowledge that living form Y performs a special behaviour, equipped with bones of a shape comparable to that of the fossil (see also Stern and Oxnard, 1973). These last points also constitute the basic shortcoming of the above mentioned discussion about the *Dryopithecus* wrist joint or talus.
    This leads to the other way of treating fossils, namely, to start with a biomechanical analysis.

NEW APPROACH: BIOMECHANICS

   There is considerable confusion in the literature about what this really means. Certainly it does not mean a pure comparison between anatomical details of animals which are classified as "brachiators", "knuckle-walkers", "bipedalists", etc. This has been made very often. As an example, I mention Marzke (1971), who fails to show how far a morphological trait, which she claims to be "adaptive" or "non-adaptive", is really related to a certain behaviour. Undue simplifications of this sort are obstacles on our way to a real understanding and must be avoided in the future.
   The essential point is a precise understanding of what "function" really means. The smaller the anatomical item under investigation, the closer the answers come to the question in terms of pure mechanics. The whole body can be looked at as a moving entity, as a part of the ecosystem. The trunk "functions" as an enclosure for the organs. From a mechanical point of view, this rather rigid part contains the centre of gravity and connects the limbs. The legs can do many things, for example "function" as a support or a hang-up of the body; they can perform locomotion and assist in gathering food etc. The talus connects foot and leg together with muscles, tendons, ligaments and other soft parts. Its "function" in the strictest sense of the word is to sustain compressive forces, while allowing movements which are easily visible. These tasks are the determinants of its shape and the most important question left is: to which stresses is the element adapted in a given form, (see Preuschoft, 1970b).

Among the variety of stresses created during motion only a few seem to determine shape, namely the greatest, those which occur frequently and those which act during long periods (Preuschoft, 1976).

A biomechanical analysis yields evidence concerning only one single element or concerning a "functional complex". The definition of this phrase depends upon the analysis and cannot be given in a general way. In rare cases functional complexes comprise several sections of a limb, so that traits in a distant area can be predicted. For instance, a non-adductable first ray, the shape of the talus, a valgus-position of the knee joint and a high lateral border of the patellar surface (Heiple and Lovejoy, 1971; Preuschoft, 1970b, 1971a), seem to be causally connected to form the hind limb of a terrestrial, bipedally walking hominoid. This correlation of traits, however, is not confirmed by the available evidence on *Oreopithecus*.

In other cases, neighbouring joints move and the positions of segments vary independently, so that predictions are not possible. This is the case in the forelimb of a quadruped. The morphological characteristics of the parts that constitute an elbow may be such as to give evidence concerning the usual or maximum stress position of arm, forearm, elbow joint in relation to the trunk (Morbeck, 1972; Preuschoft, 1973a and b, 1975). From this the position of the hand, however, cannot be predicted. The hand may be used in a palmigrade, or semi-palmigrade, or digitigrade or a knuckle-walking posture.

In the apes, the length of the fingers, the shortness of the flexor tendons and the subsequent inability to dorsiflex the extended fingers can be understood only as adaptations to climbing (Preuschoft, 1973a). About one half out of 22 living apes observed in Frankfurt and Stuttgart and in dissections of 4 gorillas and 3 chimpanzees was unable to extend the fingers while extending the wrist joint at the same time. These characters leave no other possibility for the transmission of weight through the hands to the ground in quadrupedal stance except in unguligrade or in knuckle-walking postures. The disadvantages of the former are evident. The elbow does not exert limitations in this respect, since there are no structures to connect its movements with the movements of the wrist.

Another example is provided by the ulna. The mechanical demands to be met by it are only quantitatively different in knuckle-walking, in hanging by the arms or in manipulating heavy objects (Preuschoft, 1973a). Therefore, the shape, for instance of the Omo ulna (Howell and Wood, 1974) is nearly as independent of the position and the use of the hand as is the elbow joint.

Since the results of a mechanical analysis are confined to isolated elements or to functional complexes in the above sense, these results are not invalidated by changing attributions of fossil remains to one or another species. A terminal phalanx of a big toe (Olduvai OH 10) has been shown to exhibit traits that are advantageous for walking on more or less level surfaces, but not suited to sustain the stresses which occur in gripping (Preuschoft, 1971a, 1970). This was taken as a further proof for the hypothesis that East African hominids were true bipeds. According to Wood (1974b) this terminal phalanx OH 10 does not belong to the same taxon as the Olduvai foot skeleton OH 8. If this is correct, the phalanx no longer supports the conclusion about the Olduvai Hominid 8. We now have to assume that two bipedal forms existed, because both remains are well suited to sustain those stresses which are known to be evoked by this posture. Whether the motor habits were identical or not, is not discernible at this time. The dependence of the functional complexes upon one another also lends support to the idea of "mosaic evolution" as discussed by McHenry (1975b) or, in this volume, by Bilsborough.

Up to now, not many satisfying results of biomechanical studies on fossil hominids are available.

In an extensive and useful study, Robinson (1972) reviewed the postcranial material from South Africa. "*H. africanus* evidently was, for all practical purposes, basically as well adapted to erect bipedal posture and locomotion as we are, although some slight improvements in efficiency still had to be made here and there even occasional knuckle-walking is improbable in the extreme. The one suggestion of a not very human structure is in the scapula and perhaps in a slightly long arm" (p. 245). Whereas *Paranthropus* "...clearly must have had appreciable capability as an erect biped, it was not as effectively specialised in this direction as was *H. africanus* since it still had a long ischium and therefore still used its propulsive mechanism at least partly in a power-specialised manner" (= long power arm combined with short load arm). "This suggests a compromise adaptation - perhaps not completely efficient bipedality on the ground, coupled with spending some time in the trees." These ideas are expressed also by Sigmon (1971). Objections to this last assumption were raised by McHenry (1975b) and by McHenry and Corruccini (1975).

Robinson argues strongly against a "one reason theory" for explaining bipedality, but he offers - admittedly on a higher level - a new version thereof: the tendency to shift the centre of gravity downward. Unfortunately, his arguments contain a number of basic errors in mechanics, which raise doubts about the validity of his conclusions. He speaks, for instance, about the whole body centre of gravity when discussing the equilibrium in the hip region. But only the weights of trunk plus arms must be balanced in the hip joints (see below). The perhaps elongated lower limbs do not exert any influence on the gravity centre of the trunk + head + arm system. Another example of an unsatisfactory approach is also contained in Robinson et al. (1972) and in McHenry (1975c): in principle, an elongation of the load arm indeed increases the speed of its movement, but the elongation of the hominid leg cannot directly be transformed into speed as Robinson says. This is because the leg, as a whole, does not form the "load arm", no matter whether it is extended (Robinson) or flexed (McHenry). As the "load" in the case of a walking hominoid must be considered as a resultant of a (small) horizontal force component (= acceleration), and a (large) vertical force component (= body weight), the load arm of this resultant is always much shorter than the length of the limb, even if this is flexed. In the quoted papers the vertical component is completely neglected.

The short-legged modern apes are not slower than the notoriously slow humans. To make use of the possibilities offered by a short power arm requires much force of specially developed muscles - a point not considered by the authors.

Robinson's conclusions are based largely on the Sterkfontein pelvis and vertebrae. He postulates a correlation between the curvature of the vertebral column and the pelvis, without presenting evidence, either mechanical or statistical for this.

Zihlman (1971) and Zihlman and Hunter (1972) also tried to examine the details of australopithecine locomotion, looking at the hip and thigh complex. Her approach is basically a descriptive one. "Interpretation of fossils depends upon analogies with living primates: the degree of correlation between structure and locomotor behaviour and the locomotor category into which each species is then placed." "...Given analogies with living primates, it is difficult at this time to define the structural differences which are the basis for locomotor behavioural differences...." "There is no apparent correlation between ecological niche, speciation and locomotor differences in closely related living species" (Zihlman 1971).

The basic physical facts are not considered in a satisfying way in her analyses. According to Zihlman the "structural pattern" of the australopithecines "suggests one basic locomotor potential for both species. This configuration is distinct from modern man and from the extant apes." "On the basis of existing postcranial evidence, it is neither necessary nor defensible to conclude that different locomotor adaptations existed...."

Sigmon (1969) and Sigmon and Robinson (1967) pointed out the incorrectness of Washburn's (1950, 1963) interpretation of bipedality as a consequence of the development of a gluteus maximus muscle. The pongids clearly possess a superficial gluteus, which is one of the heaviest muscles of the lower limb and which acts as a hip extensor (see also Preuschoft, 1961, 1963).

Day (1969) found in the femoral neck of robust *Australopithecus* an impression of the external obturator tendon, which is indicative of habitual extension of the hip. A well-defined ilio-psoas groove in the ilium also may serve as an indication of a habitually extended position of the hip (Le Gros Clark, 1967; Lovejoy et al., 1973).

Lovejoy et al. (1973) and Lovejoy (1973 and 1975) analyse hip and thigh mechanics following the principles worked out by Pauwels (1935). The same considerations are used also by McHenry (1975b). Lovejoy et al. make a most serious attempt to demonstrate that the australopithecines were able to walk like modern man. They follow lines of thinking similar to Robinson (1972) but dig deeper into the mechanics. Lovejoy et al. (1973) and Lovejoy (1973, 1975) showed the elliptical condyles of the femur to be adaptations to the loading of the knee in more or less extended position.

In view of the quality of these papers, some very detailed criticism seems to be justified:

In contrast to their Fig. 8 which shows bending, Lovejoy et al. say that "the femoral neck usually is loaded only in compression during gait". Later the formula for "relative femoral head area", available for sustaining the resultant force in the hip should read

$$\text{Area} = \left(\frac{\text{head diameter}}{2}\right)^2 \times \pi$$

The problems of force per unit area in the hip joint and bending of the femoral neck have been treated in more detail than is possible in this context by Pauwels (1935) and by Amtmann and Kummer (1968).

Lovejoy (1973) says "In the evolution of bipedal walking the mass of the thigh has been decreased." Since we do not know the soft parts, statements of this sort are open to doubt. In addition, my data on muscular weights of African apes (Preuschoft, 1961, 1963, 1965) do not support this opinion. The mechanical reason is sought by Lovejoy in the comparison of the moving lower limb with a pendulum, but he is wrong in assuming an influence of mass on the frequency of a pendulum. Instead,

$$\text{Duration of the period} = \frac{\text{Length}}{\text{Earth acceleration}}$$

This is the complete reason why in running, the legs of all animals are so strongly flexed: less muscular force is necessary to move a shortened pendulum rapidly.

Lovejoy and his co-workers did not try to evaluate minor morphological differences in terms of variations of motor behaviour. Human locomotion is defined as a "striding gait" and consequently the authors confine themselves to analysing those traits which are related to this movement. Thus they put emphasis on those traits which all australopithecines have in common.

McHenry's (1975c) findings also "add support to the view that there is little functional difference in gait among them" (= South African early hominids and modern *H. sapiens*). This statement, however, is qualified by McHenry and Corruccini (1975) who found the australopithecine pelves removed farther from each other than the pongid pelves are separated. The existing differences cannot yet be related to differences in motor habits; the mechanical analyses do not yet yield sufficiently precise results.

Heiple and Lovejoy (1971), Preuschoft (1971a, 1970b), Lovejoy and Heiple (1970, 1972) interpreted the valgus position of the knee joint and the elevation of the lateral lip of the patellar joint surface mechanically. According to the authors mentioned last, the necessity of placing the foot during the stance phase of striding below the centre of gravity requires a valgus position of the knee joint. The author mentioned second, however, considers the different situation in bipedal apes and emphasises the reduction of bending stresses in the human leg as the result of a valgus position. Also the narrowness of the tibia and the position of its distal joint surface can be associated with the non-prehensile foot (Preuschoft, 1971a and b). In chimpanzees as well as in humans, the bending moments in the frontal plane influence the stressing of the fibula (Preuschoft, 1970b, 1971b). The morphology of the Olduvai fibula (OH 35) is not yet understood from a mechanical point of view.

Preuschoft (1971a) has based a biomechanical analysis on the Olduvai foot skeleton OH 8, the distal phalanx OH 10, the tibia and fibula OH 35, the Kromdraai FKR3 talus and the Sterkfontein distal femora FST 10 and FST 19. He found many morphological details well adapted to habitual upright posture and locomotion. He focussed on the principal characteristics of bipedalism and therefore did not pay much attention either to the attribution of the parts analysed or to minor differences in movements.

Archibald et al. (1972) compared the relative midshaft diameters ("robusticity") of the metatarsals in OH 8 and concluded that the lateral metatarsals are relieved by the strong first ray. This is in accordance with the aforementioned analysis, which showed the inclusion of the hallux in the longitudinal bowstring construction of the foot to be a characteristic of the human foot, in contrast to that of other primates. The authors noted the relative thickness of the fifth metatarsals and connected it "with other biomechanical attributes such as body weight, stature and bicondylar angle of the femur".

In summary, the state of knowledge can still be characterised as by Day in 1973: "the gait of early hominids seems to have been different from our movements today, but we do not know how".

In the future, both "new" methods should be used, but no longer independently of each other. If at all possible, the results of biomechanical analyses should be tested by statistics: those traits which, on the grounds of biomechanics, are considered to be adaptive to a specific behaviour, must be shown really to exist in animals which perform it (see also Stern and Oxnard, 1973). The occurrence of similar or of the same traits in other forms is no objection. On the other hand, metric-statistical approaches should be based on those characters which are understood from a functional point of view and which have been demonstrated to be meaningful in a mechanical sense. If other traits are used, in order to do phylogenetic or taxonomic work, any mention of terms like "functional" should be carefully avoided. The metric-statistical analyses which aim at functional interdependences can be improved by consideration of more detailed information on movements and postural behaviour than just locomotor categories.

STRESSES IN THE TRUNK DURING TRANSITION TO BIPEDALITY

At some stage during evolution, and it seems to have been in the Middle or
Late Pliocene, the change from pronograde to permanently orthograde posture
must have taken place.

This change is accompanied by little noted but deep rooted alterations in
the mechanical situation of the trunk. By the use of considerations of Strasser
(1913), Slijper (1946) and Kummer (1959 and 1960) as a starting point, the trunk
of a quadruped can be understood as a beam which rests on two supports - forelimb (A) and hindlimb (B) in Fig. 1. The same mechanical situation holds true
in any pronograde animal, no matter whether walker, runner or climber.

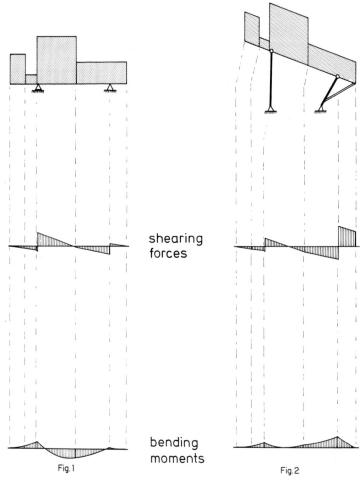

shearing
forces

bending
moments

Fig.1   Fig.2

*Figure 1. Stresses which occur in the trunk of a pongid in quadrupedal
posture.*
*The weights of the sections are indicated by varying thickness
of the beam.*

*Figure 2. Stresses which occur in the trunk of a pongid in quadrupedal
posture with hindlimbs supporting an enlarged part of the total
body weight. To equilibrate the hip joint, the hip extensors
must be activated. To make the model more similar to a pongid,
the arms are elongated, so that the trunk is inclined by 20°.*

The different weights of the trunk segments are given in Table I. The weights have been determined by cutting a deep frozen subadult male chimpanzee into sections. The weight of the trunk segments tend to bend the "beam" downward in the cranial (left) part, upward in the height of the shoulder and downward again in the thoracic and abdominal region. More exactly, shearing forces are created by body weight which cause bending moments of differing magnitude over the length of the body (Fig. 1).

TABLE I

Body Sections	Weights in kp	Weights in %	Distances from hip joint to cranial border in cm	Segments
Head	2,36	8,3	76,8	
Neck	0,56	1,8	63,9	C_1-C_7
Thorax	9,88	32,3	54,5	Th_1-Th_{10}
Abdomen	5,41	17,6	28,0	Th_{10}-L_4
Upper extremities	6,86	22,0		
Lower extremities	5,69	18,0		
			hip joint ischial tuberosity	
Total	30,76	100	11,3	

Shearing forces reach their maximal values in the thoracic and hip sections. In the former, resistance to shearing is offered by the ribs and the system of intercostal muscles. In addition, the articular processes of the vertebrae and the *m. serratus dorsalis* as well as the muscles of the shoulder girdle may take over the shearing forces. In the hip region, the mass of bone material concentrated in the pelvis also offers high resistance to shearing. A contribution from the hip and thigh muscles seems plausible.

If the joints of the anterior extremity are fixed by the intrinsic arm muscles, the trunk can be thought to be hung up by the lateral serratus muscle like a string-bridge from the shoulder blade (for further details about the shoulder-thorax connection, see Preuschoft, 1973a and Preuschoft and Fritz, 1977). The hindlimb supports the body at the hip joint. The bending moments have two maxima: a negative one in the shoulder, a positive one in the abdominal or lumbar section. Due to the fan-shaped arrangement of the *m. serratus lateralis* the transmission of body weight is distributed over a certain length, instead of being concentrated in a single point like the hip joint. This results in a cutting off of the peak of the stresses.

The bending moments cause compression at the lower margin of the beam in the anterior (= left) part and at the upper margin in the right part. Inversely, tensile stresses are evoked at the upper margin in the anterior section and at the lower margin in the posterior half of the trunk. The magnitudes of the compressive or tensile stresses are proportional to the size of the bending moments in the respective region. In a vertebrate's body, the only material available for sustaining the compression is bone and cartilage, as

present in the vertebral column and in the sternum. The materials available
for sustaining the tensile forces are collagenous or elastic connective tissue,
as in the ligaments, and muscles. This means that bone must be expected to be
removed from the dorsal and concentrated near the lower margin of the beam in
the region of the anterior support, while shifted towards the dorsal margin
in the area between the limbs. The muscular, tendinous and ligamentous structures can be expected to have the reverse distribution (cf. Gray, 1968).

This situation can be explained also in another way: a vertebrate's trunk
is a mobile system. As long as it maintains its shape, equilibrium must be
kept between each pair of neighbouring vertebrae. In the neck and thorax this
is accomplished by the dorsal muscles, which prevent the spine from bowing with
a dorsal convexity. In the abdominal part, the tension of the ventral muscles
prevents the trunk from bowing with a dorsal concavity. Equilibrium is defined
by

$$\text{Weight} \cdot \text{weight arm} = \text{force} \cdot \text{power arm},$$

wherein "weight" means weight of a trunk segment; "Weight arm" is the shortest
distance between the centre of this segment and support A or support B; "force"
is the tension of the muscular or ligamentous structure which prevents the
vertebral column from bending; "power arm" is the shortest distance between
the middle of the intervertebral joint and the equilibrating force. As an approximation, this is assumed to act halfway between the dorsal margin of the transverse processes and the tips of the spinous processes in the case of dorsal muscles.

Energy can be saved by approaching the backbone to the compression-bearing
margin of the trunk and by shifting the tension producing structures away
from the vertebral column. This principle can be observed in the bodies of
all quadrupeds, often much more clearly than in a pongid: the backbone in
the neck and upper thoracic region approaches the ventral margin, thus giving
the neck muscles long lever arms. This effect is enforced by the lengths of
the spinous processes. The sternum can take a part of the total compressive
force in the most highly compressed region. In the lumbar section, the vertebral column is near the dorsal margin, thus providing the long lever arms for
the abdominal muscles. This constructional principle works better the longer
the dorsoventral diameter of the animal. The trunk volume can be kept at a
given level by reducing the frontal diameter; the result of this is the often
mentioned keel-shaped trunk of typical quadrupeds, for instance monkeys.

The compressive forces that occur even in a horizontal position of the
trunk, can be sustained by no other structure than the vertebral column and
sternum. As Figs. 6a and b show, these compressive forces increase caudally.
If the cancellous bone of the vertebral bodies is assumed to be of about
equal strength in all segments, we have to expect the caudal increase of the
vertebral diameters tha can be observed in our pongid (Fig. 5) as in nearly
all tetrapods. The last lumbar and first sacral segments offer smaller joint
surfaces, because the ilium takes over a part of the compressive forces. If
the animal assumes an inclined position by elongation of the arms (as in the
African pongids), its feet and hands come closer together, but the stress
distribution in the trunk is not essentially changed (Fig. 2 and Stern, 1975).

A considerable change can take place by approximating the feet but not the
arms to the centre of gravity (Fig. 2). By this, a larger part of the body
weight is taken over by the hindlimbs.

The transmission of a larger part of the body weight to the hindlimb makes
the assumption of the upright posture easy. The tendency of all primates to
assume this posture not only in bipedal standing but as well in sitting has

been emphasised by Napier and Napier (1967). Thus erect posture is foreshadowed in all primates. The transition to bipedality can be observed often in
African pongids, but also in baboons (Rose, 1976). Actually, the carrying of
a large part of the body weight on the hind limb seems to be typical for primates in contrast to other ground-living mammals (Kimura et al., 1978).

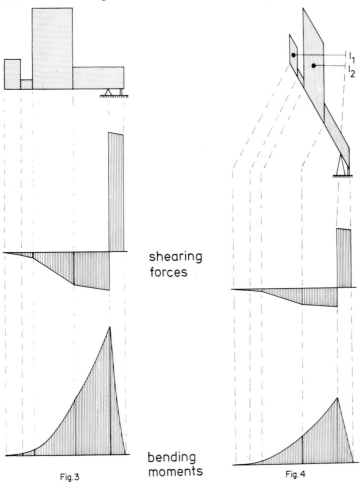

shearing
forces

bending
moments

Fig.3   Fig.4

Figure 3.   *Stresses which occur in the trunk of a pongid after lifting the
            hands from the supporting ground. The weights of the arms con-
            tribute to bending and the hip extensors have to spend more
            force than in Fig. 2.*

Figure 4.   *Stresses which occur in the trunk of a pongid in a more upright
            position than in Fig. 3. The influence of the weight component
            is increased, whereas the compression derived from bending be-
            comes smaller. The load arms (l_1 and l_3) for 2 of the 4 sec-
            tions are indicated. They are much shorter than in Fig. 3.*

Of course all stages of transition between Figs 1, 2 and 3 are possible,
depending on the exact part of the entire body weight which is supported by
the hindlimb.

The angulation of the hindlimbs on the trunk necessitates a contraction of the hip extensors. These pull the postcoxal part of the pelvis downwards. Thus, great shearing forces occur in the hip region. In the neck section, the shearing forces remain unchanged, in the lumbar section they assume higher values. In the lower part of the thorax, the direction of the shearing forces changes. The existence of fibres in the direction of the external as well as the internal intercostal and oblique abdominal muscles makes the system insensitive to changes of sign.

Bending stresses also differ markedly. Tension occurs at the upper margin of the lumbar segments and tensile forces must be provided by the dorsal musculature, while the abdominal muscles may relax. According to the formula (weight · weight arm = force · power arm) the short lever arms of the dorsal musculature require more force and thus evoke greater compression in the vertebral column than in Fig. 1. The compression in the lumbar segments of the spine reach higher values in this variant of quadrupedal posture than in that shown in Fig. 1.

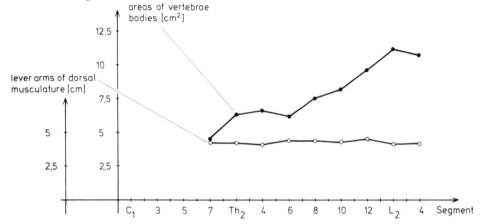

Figure 5. *Lever arms of the dorsal musculature and cross sectional areas of the vertebrae at various segments of a pongid. In Fig. 1 the ventral muscles must be activated. Their lever arms are 4-6 times longer than those of the dorsal muscles and are not shown here.*

As mentioned above, the compressive forces are sustained by the vertebral bodies. Their caudal increase parallels the caudal increase of their diameters (Fig. 5). This is not surprising: if spongy bone is considered a material of defined strength properties (Evans and King, 1961), the amount of material on which the compression is distributed must increase proportional to the stresses.

Lifting the arms from the ground changes the stress pattern further (Fig. 3). The weight of the anterior extremities now contributes to the dorsally convex bending moments. In the modern apes, and probably in the early hominids as well, the arms are heavily muscled and therefore increase the moments considerably.

The stress pattern now is that of a beam on one support. The changes in the shearing forces and in the bending moments discussed above for Figs 1 and 2 are much more pronounced. The structures that resist the further increased shearing forces are the same as those noted above. The sternum receives higher

compression than before. Actually, it is stouter and broader in the apes than
in the more strictly quadrupedal monkeys. Its existence relieves the spine so
much that a seemingly disadvantageous kyphotic curvature of the thoracic section is possible. In the abdominal part of the trunk, however, there are no
obvious adaptations to stresses of this type. The short lever arms of the
dorsal musculature necessitate an increase of the tension which must be exerted. Consequently the compression in the lumbar vertebral column is increased
by the same factor. This seems to narrow the safety margin, since the area
of the cross sections does not increase so rapidly (Figs 5 and 6). Frequent
assumption of this semi-erect posture requires particular strength in the backbone and in the dorsal muscles. Benton's (1974) data on the lumbar and sacral
vertebrae and on the pelvis can be understood as adaptations to the frequency
with which upright postures are assumed in apes and monkeys.

A limited improvement of the situation can be accomplished by the assumption of a lordosis which gives the muscles longer lever arms. Although Kummer
(1965) doubts its existence in bipedal pongids, my own observations make me
believe that a moderate lordotic curvature is quite usual in at least the
African apes. In spite of this, the absolute values of compression are, in
this inclined posture, greater than in Figs 1, 2 and 4 (see Fig. 6).

Bending stresses reach their maximum at the hip. The hip joint itself is
balanced by the hamstring muscles, the adductors and the gluteal group. At
least in the African apes, the medial gluteus muscle contributes to equili-
bration of the hip (Preuschoft, 1961, 1970; Sigmon, 1969; Tuttle et al., this
meeting, at least in critical phases). The obvious and well known frontal orientation of the ilium and its elongation enables this very powerful muscle (Preuschoft, 1963) in the apes to take over a share of the tensile force in the
most highly stressed section of the body. In the fully upright posture of
humans, this particular function of the *m. gluteus medius* is dispensable,
so that the form of the iliac blade can be changed to fulfil other functions.

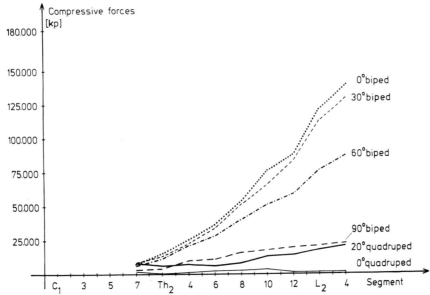

*Figure 6a. Compressive forces acting in the vertebral column of a pongid,
calculated on the basis of Figs 1-5. The values increase dra-
matically with the lifting of the arms from the ground (0°),
and decrease with erection of the trunk.*

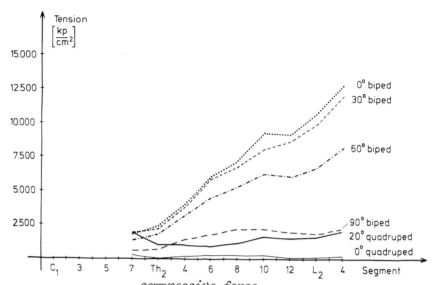

Figure 6b. Tension (= $\dfrac{compressive\ force}{cross\ sectional\ area}$) in the vertebral column of
a pongid. The same positions as those in Fig. 6a are used here.
All values remain far below the breaking tension of about
600 kp/cm² (Evans & King, 1961).

If the trunk is further elevated, the stresses decrease again (Fig. 4), because the lever arms of the body segments become shorter. Instead, the weights
of the segments exert more and more compression of the vertebral column. In
Fig. 3 a body segment of the weight G_p exerts a bending moment
$$M = G_p \cdot \ell$$
ℓ being the horizontal distance between the centre of this segment and hip
joint. If the body axis is elevated by an angle α ℓ follows the cosine of
this angle and goes down to 0 at 90°. In more detail, G_p breaks down into a
component K_c, which is parallel to the trunk axis and evokes nothing but compressive stress and a component K_B, which is perpendicular to the trunk axis
and causes bending. The angle α between body axis and the horizontal is the
same as that between K_B and the vertical.
$$K_B = cosine\ \alpha \times G_p$$
$$K_C = sine\ \alpha \times G_p$$
Therefore in horizontal posture the compressive component is $K_C = 0$ and in
vertical posture $K_C = G_p$, while the bending component K_B at the same time
changes from G_p to 0.
The compressive stresses caused by body segments never reach the magnitude
of the compression resulting from the dorsal muscles in Fig. 3, which always
have the same, short lever arms (Fig. 5).
What has been shown above makes a very good reason for a hominid to raise
its trunk completely, instead of maintaining a stooped, semi-erect posture.
The static situation can be further improved:
1. By a trunk cross section which is broad, rather than deep. A marked dorsal curvature of the proximal parts of the ribs (see Slijper, 1942) also entails an elongation of the lever arms of the iliocostal muscles.
2. By a displacement of the shoulder backward instead of the maintenance of
a hunched posture.

3. By reduced weight of the upper extremity, which is possible if it is re-lieved of locomotor tasks.

4. Most important, by a permanent lordotic curvature of the lumbar vertebral column (compare above). Elasticity in the craniocaudal direction is often as-sumed (for example Mörike, 1975) to be an important characteristic of the human backbone. But a series of cylinders which are set one on top of the other do not allow changes in length. The lordosis is not sufficiently pronounced by far for allowing elastic changes in curvature.

All of these traits are typical for modern man, more or less in contrast to the apes. All contribute to a shifting backward of the centre of gravity of the trunk and upper extremity in relation to the lower limb (compare Preuschoft, 1971a; Robinson, 1972). They form a "functional complex" in the sense used above.

Unfortunately these insights do not help us in solving our problem at this time. There are no sufficiently complete skeletal remains from the most cri-tical period. Ribs, such as are available from Hadar (Johanson and Taieb, 1976) or Sterkfontein (Robinson, 1972) could give indications about the shape of the thorax by their curvature. But ribs quite often undergo deformation during embedding. In most cases they are broken and crushed. The occurrence of a lordotic curvature in the Sterkfontein 14 skeleton must indeed be inter-preted as very strong evidence for bipedality as was done by Robinson (1972). The shape of the sternum alone yields no convincing evidence, since it is stressed principally the same way in quadrupeds and bipeds.

To make further progress, a precise biomechanical understanding of the pel-vis is essential. A quantification of forces and stresses may provide a means for obtaining knowledge about the morphological "stages" between the fully de-veloped extremes: apes and man. Just these stages seem to be available in the form of fossils.

Whether an occasional lifting of the anterior extremity from the ground can be maintained as a permanent posture, depends upon the possibility of realising the external and internal equilibrium in an economic way. Bipedal locomotion does not seem to require per se more energy than quadrupedalism (Taylor and Rowntree, 1973). Aside from a secure external equilibrium, stresses that are too high in the elements of the passive locomotor system must be avoided and the energy spent by the musculature must be kept low. This can be, and is accomplished in a number of non-primate bipedalists in differing ways (Kummer, 1965; Slijper, 1942 and 1946).

Of course the acquisition of bipedality must have a reason, a selective advantage. All the points mentioned here are related to a "constructional analysis" (Peters et al., 1974) of the primate body, which says nothing about the reasons for the changes, perhaps in terms of selective pressures exerted by the environment on the population of ancient prehominids. The most complete system of considerations about this is that proposed by Kortlandt (1972, 1974).

Lovejoy has proposed to accept encephalisation and subsequent enlargement of the newborn's head as the evolutionary reason of the morphological changes in the pelvis, namely a widening of the birth channel. In my opinion this can in no way be excluded, but it should not be the end point of the consideration. Firstly, the causal dependence of shape on the mechanical stresses allows one in other cases to make a connection between locomotor and postural behaviour - why not here also? Secondly, even if the morphological differences between the hominids cannot be derived from reasons of locomotor behaviour, the exist-ing deviations should have consequences. Whether or not the assumption of bi-pedality can be or must be a consequence of a widening of the birth channel, is still open to doubt.

REFERENCES

Amtmann, E. (1971). *Ergebn. Anat. EntwGesch.*, 44, 3, 89.
Amtmann, E. (1978. *In* "Environment, Behavior, and Morphology. Dynamic inter-
    actions in primates", (Morbeck and Preuschoft, eds), G. Fischer, Stuttgart
    and New York, in press.
Amtmann, E. and Kummer, B. (1968). *Z. Anat. EntwGesch.*, 127, 286-314.
Ankel, F. (1965). *Folia primatol.*, 3, 263-276.
Ankel-Simons, F. (1975). *In* Symp. 5th Congr. Int. Primat. Soc., Nagoya, 1974,
    (Kondo, Kawai, Ehara, Kawamura, eds), Japan Science Press, Tokyo.
Archibald, J.D., Lovejoy, C.O. and Heiple, K.G. (1972). *Am. J. phys. Anthrop.*,
    37, 93-95.
Barnett, C.H., Davies, D.V. and MacConaill, M.A. (1961). "Synovial Joints",
    Longman, Green and Co., London.
Benton, R.S. (1974). *Yearb. phys. Anthrop.*, 18, 65-88.
Cachel, S. (1975). *In* "Paleoanthropology", (Tuttle, ed.), Mouton Publishers,
    The Hague, Paris, pp. 183-201.
Carpenter, C.R. (1974). Activity characteristics of Gibbons (*Hylobates lar*) II
    Locomotion. Film. Penn. State Univ., Univ. Park, Pa.
Ciochon, R.L. and Corruccini, R.S. (1976). *S. Afr. J. Sci.*, 72, 80-82.
Ciochon, R.L. and Corruccini, R.S. (1977). *Systematic Zoology*, in press.
Conroy, G.C. (1976a). *Nature, Lond.*, 262, 684-686.
Conroy, G.C. (1976b). *Contrib. Primatol.*, 8, 1-134.
Conroy, G.C. and Fleagle, J.G. (1971). *Nature, Lond.*, 237, 103-104.
Corruccini, R.S. and Ciochon, R.L. (1976). *Am. J. phys. Anthrop.*, 45, 19-37.
Corruccini, R.S., Ciochon, R.L. and McHenry, H.M. (1976a). *Folia Primatol.*,
    24, 250-274.
Corruccini, R.S., Ciochon, R.L. and McHenry, H.M. (1976b). *Primates*, 17, 205-223.
Creel, N. (1968). Dissertation, Tübingen, 140 pp.
Day, M.H. (1967). *Nature, Lond.*, 215, 323-324.
Day, M.H. (1969). *Nature, Lond.*, 221, 230-233.
Day, M.H. (1973). *Symp. zool. Soc. Lond.*, 33, 29-51.
Day, M.H. and Leakey, R.E.F. (1973). *Am. J. phys. Anthrop.*, 39, 341-354.
Day, M.H., Leakey, R.E.F., Walker, A.C. and Wood, B.A. (1975). *Am. J. phys.
    Anthrop.*, 42, 461-476.
Day, M.H. and Wood, B.A. (1968). *Man*, 3, 440-455.
Day, M.H. and Wood, B.A. (1969). *Nature, Lond.*, 222, 591-592.
Evans, F.G. and King, A.J. (1961). *In* "Biomechanical studies of the musculo-
    skeletal system", (Evand, ed.), pp. 49-67, Thomas, Springfield, Ill.
Ferembach, D. (1959). *Annales Paleont.*, 44, 151-249.
Fleagle, J.G. (1977). *Yearb. phys. Anthrop.*, in press.
Fleagle, J.G., Simons, H.L. and Conroy, G.C. (1975). *Science, N.Y.*, 189, 135-137.
Genet-Varcin (1961). *Seances mensuelles*, 3, 106-107.
Grand, T.J. (1976). *Am. J. phys. Anthrop.*, 45, 101-108.
Gray, J. (1968). "Animal Locomotion", Widenfeld & Nicolson, London.
Heberer, G. (1956). *In* "Primatologia" I (Hofer, Schultz, Starck, eds),
    Karger-Verlag, Basel, New York.
Heberer, G. (1965). *In* "Menschliche Abstammungslehre", (Heberer, ed.),
    pp. 301-356, G. Fischer, Stuttgart.
Heiple, K.G. and Lovejoy, C.O. (1971). *Am. J. phys. Anthrop.*, 35, 75-84.
Helmuth, H. (1968). *Z. Morph. Anthrop.*, 60, 147-155.
Howell, F.C. and Wood, B.A. (1974). *Nature, Lond.*, 249, 174-176.

Ishida, H., Kimura, T., and Okada, M. (1975). *In* 5th. Symp. 5th. Congr. Int. Primat. Soc. 1974, pp. 287-301, Nagoya (Kondo, Kawai, Ehara, Kawamura, eds), Japan Science Press Co., Ltd., Tokyo.

Jenkins, F.A. Jr. (1972). *Science, N.Y.,* 178, 877-879.

Jenkins, F.A. Jr. (1974). "Primate Locomotion", Academic Press, New York.

Jenkins, F.A. and Fleagle, J.G. (1975). *In* "Primate functional morphology and evolution", (Tuttle, ed.), pp. 213-227, Mouton Publishers, The Hague, Paris.

Johanson, D.C., and Taieb, M. (1976). *Nature, Lond.,* 260, 293-297.

Kay, R.F. (1973). *Science, N.Y.,* 182, 396.

Kay, R.F. (1975). *Science, N.Y.,* 189, 62.

Kimura, T., Okada, M. and Ishida, H. (1978). *In* "Environment, behaviour and morphology. Dynamic interactions in primates", (Morbeck and Preuschoft, eds), Fischer, Stuttgart, New York, in press.

Kortlandt, A. (1966-1974). Films shown at the Congresses of the Internat. Primat. Soc. (in Frankfurt, Atlanta, Zurich, Portland, Nagoya).

Kortlandt, A. (1968). *In* "Handgebrauch und Verständigung bei Affen und Frühmenschen", (Rensch, ed.), pp. 59-100, Huber, Bern-Stuttgart.

Kortlandt, A. (1972). "New Perspectives on Ape and Human Evolution", Department of Animal Psychology and Ethology, Zoological Laboratory, University of Amsterdam.

Kortlandt, A. (1974). *Curr. Anthrop.,* 15, 427-447.

Kummer, B. (1959). Bauprinzipien des Säugerskelettes. Thieme Verlag, Stuttgart.

Kummer, B. (1960). *Z. Tierzücht.Zuchtbiol.,* 74, 159-167.

Kummer, B. (1965). *In* "Menschliche Abstammungslehre", (Heberer, ed.), pp. 225-248, Fischer Verlag, Stuttgart.

Kummer, B. (1972). *In* "Biomechanics: its foundations and objectives", (Fung, Perone, Anliker, eds), pp. 237-271, Prentice-Hall, Englewood Cliffs, N.J.

Leakey, R.E.F. (1971). *Nature, Lond.,* 231, 241-245.

Leakey, R.E.F. (1973). *Nature, Lond.,* 242, 447-450.

Leakey, R.E.F., Mungai, J.M. and Walker, A.C. (1971). *Am. J. phys. Anthrop.,* 35, 175-186.

Leakey, R.E.F. (1972). *Am. J. phys. Anthrop.,* 36, 235-252.

Leakey, R.E.F. and Walker, A.C. (1973). *Am. J. phys. Anthrop.,* 39, 205-222.

Leakey, R.E.F. and Wood, B.A. (1973). *Am. J. phys. Anthrop.,* 39, 355-368.

Le Gros Clark, W. (1964). "The fossil evidence for human evolution", 2nd ed., University of Chicago Press, Chicago.

Le Gros Clark, W. (1967). "Man-apes or ape-man", Holt, Rinehart & Winston, New York.

Le Gros Clark, W. and Leakey, L.S.B. (1951). *Fossil Mammals Afr.,* 1, 1-117.

Lewis, O.J. (1971). *Nature, Lond.,* 230, 577-578.

Lewis, O.J. (1972). *Am. J. phys. Anthrop.,* 36, 45-58.

Lisowski, F.P., Albrecht, G.H. and Oxnard, C. (1976). *Am. J. phys. Anthrop.,* 45, 5-18.

Lovejoy, C.O. (1973). *Yearb. phys. Anthrop.,* 17, 147-161.

Lovejoy, C.O. (1975). *In* "Primate Functional Morphology and Evolution", (Tuttle, ed.), pp. 291-326, Mouton Publishers, The Hague, Paris.

Lovejoy, C.O. and Heiple, K.G. (1970). *Am. J. phys. Anthrop.,* 32, 33-40.

Lovejoy, C.O. and Heiple, K.G. (1972). *Nature, Lond.,* 235, 175-176.

Lovejoy, C.O., Heiple, K.G. and Burstein, A.H. (1973). *Am. J. phys. Anthrop.,* 38, 757-779.

Marzke, M.W. (1971). *Am. J. phys. Anthrop.,* 34, 61-84.

McHenry, H.M. (1973a). *Science, N.Y.*, 182, 396.
McHenry, H.M. (1973b). *Science, N.Y.*, 180, 739-741.
McHenry, H.M. (1974). *Am. J. phys. Anthrop.*, 40, 329-340.
McHenry, H.M. (1975a). *Am. J. phys. Anthrop.*, 43, 245-261.
McHenry, H.M. (1975b). *Science, N.Y.*, 190, 425-431.
McHenry, H.M. (1975c). *Am. J. phys. Anthrop.*, 43, 39-46.
McHenry, H.M. (1976). *Am. J. phys. Anthrop.*, 45, 77-83.
McHenry, H.M. and Corruccini, R.S. (1975). *Am. J. phys. Anthrop.*, 43, 263-280.
McHenry, H.M. and Corruccini, R.S. (1976). *Nature, Lond.*, 259, 657-658.
McHenry, H.M., Corruccini, R.S. and Howell, F.C. (1976). *Am. J. phys. Anthrop.*, 44, 295-304.
Morbeck, M.E. (1972). University Microfilms Ltd., Ann Arbor, Michigan, USA.
Morbeck, M.E. (1975). *J. hum. Evol.*, 4, 39-46.
Morbeck, M.E. (1976). *J. hum. Evol.*, 5, 223-233.
Morbeck, M.E. (1977). *Primates*, 18, in press.
Mörike, K. (1975). *Orthopädic-Technik*, 2, 23-27.
Napier, J. (1962). *Nature, Lond.*, 196, 409-411.
Napier, J. (1963). *In* "Classification and Human Evolution", (Washburn, ed.), Aldine, Chicago.
Napier, J. and Davis, P.R. (1959). *Fossil Mammals Afr.*, 6, 78.
Napier, J. and Napier, P. (1967). "A handbook of living primates", Academic Press, London, New York.
O'Connor, B.L. (1975). *Am. J. phys. Anthrop.*, 43, 113-121.
Oxnard, C.E. (1968a). *Am. J. phys. Anthrop.*, 28, 213-217.
Oxnard, C.E. (1968b). *Am. J. phys. Anthrop.*, 29, 429-431.
Oxnard, C.E. (1972). *Am. J. phys. Anthrop.*, 37, 3-12.
Oxnard, C.E. (1975). *In* Symp. 5th Cong. Int. Primatol. Soc., 1974, Nagoya (Kondo, Kawai, Ehara, Kawamura, eds), pp. 2671286, Japan Science Press, Tokyo.
Patterson, B. and Howells, W.W. (1967). *Science, N.Y.*, 156, 64-66.
Pauwels, F. (1935). *Beilagheft Z. orthop. Chir.*, 63. (Also in Pauwels, 1965.)
Pauwels, F. (1960). *Z. Anat. EntwGesch.*, 121, 487-515.(Also in Pauwels, 1965).
Pauwels, F. (1965). "Gesammelte Abhandlungen zur Funktionellen Anatomie des Bewegungsapparates", (C. Pauwels, ed.), Springer Verlag, Berlin.
Peters, D.S., Franzen, J.L., Gutman, W.F. and Mollenhauer, D. (1974). *Umschau*, 74, 501-506.
Pilbeam, D. (1969). *Nature, Lond.*, 223, 648.
Pilbeam, D. and Gould, S.J. (1974). *Science, N.Y.*, 186, 892-901.
Plagenhoef, St. (1971). "Pattern of Human Motion", Prentice-Hall, Englewood Cliffs, New Jersey.
Preuschoft, H. (1961). *Morph. Jb.*, 101, 432-540.
Preuschoft, H. (1963). *Anthrop. Anz.*, 26, 308-317.
Preuschoft, H. (1965). *Morph. Jb.*, 107, 99-183.
Preuschoft, H. (1969). *Z. Anat. EntwGesch.*, 129, 285-345.
Preuschoft, H. (1970a). *Z. Anat. EntwGesch.*, 131, 156-192.
Preuschoft, H. (1970b). *In* "The Chimpanzee", (Bourne, ed.), III, pp. 221-294. Karger-Verlag, Basel and New York.
Preuschoft, H. (1971a). *Folia Primatol.*, 14, 209-240.
Preuschoft, H. (1971b). *Gegenbaurs. morph. Jb.*, 117, 211-216.
Preuschoft, H. (1973a). *In* "The Chimpanzee", (Bourne, ed.), VI, pp. 34-115. Karger-Verlag, Basel, München, Paris, London, New York, Sidney.
Preuschoft, H. (1973b). *In* "Human Evolution", (Day, ed.), pp. 13-46, Taylor & Francis, London.

Preuschoft, H. (1975). *In* Proc. Symp. 5th Congr. Internat. Primat. Soc. 1974, Nagoya, (Kondo, Kawai, Ehara, Kawamura, eds), pp. 345-359, Japan Science Press, Tokyo.

Preuschoft, H. (1976). *Aufs. Reden senckenb. naturf. Ges.*, **28**, 98-117.

Preuschoft, H. and Fritz, M. (1977). *Fortschr. Zool.*, **24**, 2-3, 75-98.

Preuschoft, H. and Weinmann, W. (1973). *Am. J. phys. Anthrop.*, **38**, 241-250.

Prost, J.H. (1965). *Am. J. phys. Anthrop.*, **23**, 215-240.

Prost, J.H. (1970). *Am. J. phys. Anthrop.*, **32**. 121.

Prost, J.H. and Sussman, R.W. (1969). *Am. J. phys. Anthrop.*, **31**, 53.

Rightmire, G.P. (1972). *Science, N.Y.*, **176**, 159-161.

Robinson, J.T. (1972). "Early Hominid Posture and Locomotion", University of Chicago Press, Chicago and London.

Robinson, J.T., Freedman, L. and Sigmon, B.A. (1972). *J. Hum. Evol.*, **1**, 361-369.

Rose, M.D. (1976). *Am. J. phys. Anthrop.*, **44**, 247-261.

Schaller, G.B. (1963). "The Mountain Gorilla", University of Chicago Press, Chicago.

Schön, M.A. and Ziemer, L.K. (1973). *Folia Primatol.*, 20, 247-261.

Schultz, A.H. (1960). *Z. Morph. Anthrop.*, **50**, 136-149.

Sigmon, B.A. (1969). Thesis, University of Wisconsin.

Sigmon, B.A. (1971). *Am. J. phys. Anthrop.*, **34**, 56-60.

Sigmon, B.A. and Robinson, J.T. (1967). *Am. J. phys. Anthrop.*, **27**, 245-246.

Simons, E.L. and Pilbeam, D.R. (1972). *In* "The Functional and Evolutionary Biology of Primates", (Tuttle, ed.), 2, 36-62.

Slijper, E.J. (1942). *Sciences*, 65, 288-295, 409-415.

Slijper, E.J. (1946). *Verh.Kon. Ned. Akad. Wetensch., Afd. Natur-kde.*, 2 Sect. D42.

Stern, J.T. (1975). *Yearb. phys. Anthrop.*, **19**, 59-68.

Stern, J.T. and Oxnard, C.E. (1973). *Primatologia*, **4**, 1-93.

Strasser, H. (1913). "Lehrbuch der Muskel- und Gelenk-Mechanik", 2 Teil, Berlin.

Taylor, C.R. and Rowntree, V.J. (1973). *Science, N.Y.*, **179**, 186-187.

Tobias, P.V. (1973a). *Problemi att. Sci. Cult.*, **182**, 63-85.

Tobias, P.V. (1973b). *Nature, Lond.*, **246**, 79-83.

Tobias, P.V. (1973c). *Ann. Rev. Anthropol.*, **2**, 311-334.

Tobias, P.V. (1975a). *In* "The Role of Natural Selection in Human Evolution", (Salzano, ed.), 6, pp. 89-118, North Holland Publishing Co.

Tobias, P.V. (1975b). *In* "Quaternary Studies", (Suggate & Cresswell, eds), pp. 289-296, Royal Soc. New Zealand, Wellington.

Tobias, P.V. (1975c). *Optima*, 25, 24-35.

Tuttle, R.H. (1967). *Am. J. phys. Anthrop.*, **26**, 171-206.

Tuttle, R.H. (1969a). *J. Morph.*, 128, 309-364.

Tuttle, R.H. (1969b). *Science, N.Y.*, 166, 953-961.

Tuttle, R.H. (1969c). *Science Journal*, 5A, 66-72.

Tuttle, R.H. (1970). *In* "The Chimpanzee" (Bourne, ed.), II, pp. 167-253, Karger-Verlag, Basel, New York.

Tuttle, R.H. and Basmajian, J.V. (1974). *In* "Primate Locomotion", (Jenkins, ed.), pp. 293-347, Academic Press, New York.

Tuttle, R.H., Basmajian, J.V. and Ishida, H. (1975). *In* "Primate Functional Morphology and Evolution", (Tuttle, ed.), pp. 253-269, Mouton Publishers, The Hague, Paris.

Walker, A. (1973). *J. hum. Evol.*, **2/6**, 545-555.

Washburn, S.L. (1950). *Cold Spring Harb. Symp. quant. Biol.*, **15**, 67-78.

Washburn, S.L. (1963). *In* "Classification and Human Evolution", (Washburn, ed.), pp. 190-203, Aldine, Chicago.

Wells, J.P. and Wood, G.A. (1975). *Am. J. phys. Anthrop.*, 43, 217-226.

Wolpoff, M.H. (1975). *In* "Palaeoanthropology", (Tuttle, ed.), pp. 245-284, Mouton Publishers, The Hague, Paris.

Wolpoff, M.H. and Brace, C.L. (1975). *Science*, *N.Y.*, 189, 61-62.

Wood, B.A. (1973). *Nature*, *Lond.*, 246, 45-46.

Wood, B.A. (1974a). *Nature*, *Lond.*, 251, 135-136.

Wood, B.A. (1974b). *J. hum. Evol.*, 3/5, 373-378.

Zapfe, H. (1960). *Schweiz. palaeont. Abh.*, 78, 1-293.

Zihlmann, A.L. (1971). Proc. 3rd Int. Congr. Primatol. Zurich, 1970 (Biegert and Leutenegger, eds), 1, 54-66, Karger-Verlag, Basel, München, Paris, London, New York, Sidney.

Zihlmann, A.L. and Hunter, W. (1972). *Folia Primatol.*, 18, 1-19.

Zuckerman, S., Ashton, E.H., Flinn, R.H., Oxnard, C.E. and Spence, T.F. (1973). *Symp. zool. Soc. Lond.*, 33, 71-165.

Zwell, M. and Conroy, G.C. (1973). *Nature*, *Lond.*, 244, 373-375.

ACTIVITIES OF HINDLIMB MUSCLES IN BIPEDAL GIBBONS

H. ISHIDA*, M. OKADA**, R.H. TUTTLE***, T. KIMURA****

*Kyoto University, Kyoto, Japan; **Tokyo University of Education,
Tokyo, Japan; ***University of Chicago, Chicago, Illinois, USA.
****Teikyo University, Tokyo, Japan

INTRODUCTION

Comparative studies on the bipedal gaits of primates are essential to provide a sound base for inferences about the evolution of hominid bipedalism.
Several authors have compared aspects of the bipedal gaits of anthropoid primates, including gibbons (Ishida et al., 1975; Yamazaki et al., in preparation; Kimura et al., this volume). The study presented here is a more detailed complement to these studies. This paper is focussed on the activities
of hindlimb muscles of gibbons during bipedal walking. We shall also present
brief comments on other kinesiological parameters of gibbon bipedalism.

SUBJECTS AND METHODS

The subjects were two juvenile gibbons: a male *Hylobates agilis* (3.5 kg)
and a female *Hylobates lar* (3.0 kg). Electromyograms (EMG) of hip, thigh and
leg muscles were recorded as the subjects walked on a level platform 4 m long,
50 cm wide and 20 cm high. Bipolar fine-wire electrodes (50 μ in diameter)
were used (Basmajian, 1974). Light anaesthesia was necessary for electrode
implantation (Tuttle et al., 1972). Recording was begun at least 3 hours after
anaesthesia.
Cadence was monitored with foot contact signals generated by micro-switches
attached to the soles of both feet. The EMGs and foot contact signals were
recorded together with behaviour of the subject on video-tape. In some sessions, data from a force plate and 16 mm cine films were taken simultaneously
with EMGs and foot contact signals. The former provided data on vertical,
sagittal and transverse components of the force exerted by the foot against
the substrate. The films provided information on joint excursions and foot
contact sequences.

RESULTS AND DISCUSSION

A. *Electromyography*

We conducted 16 experiments on the 2 subjects. Electromyograms were successfully recorded from 22 muscles of the hip, thigh and leg. Figure 1 is a schematic summary of EMG activity during a typical cycle of gibbon bipedal walking.
Five out of 22 muscles were active in the swing phase: the gracilis, adductor
longus, iliopsoas, rectus femoris and tibialis anterior. The gracilis and
adductor longus muscles acted at the beginning of the swing phase. Activities

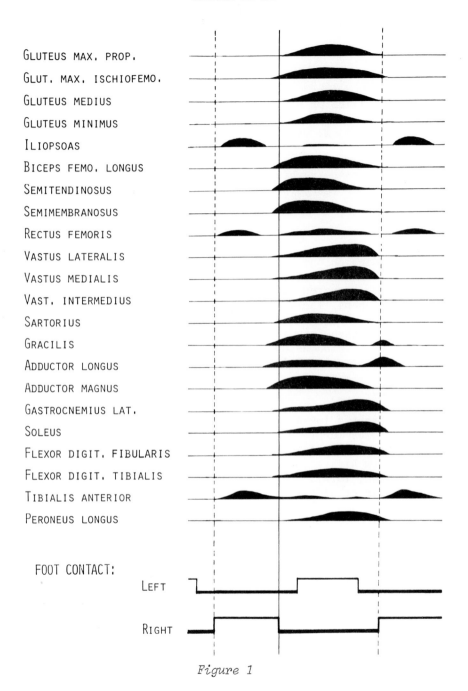

Figure 1

of the other three muscles occurred in the middle of the swing phase. The gracilis and adductor longus muscles probably initiate the swing phase by slightly raising the thigh. The iliopsoas and rectus femoris muscles flex the thigh further, thereby carrying the hindlimb forward. The rectus femoris muscle extends the knee joint and the tibialis anterior muscle dorsiflexes the ankle joint to position the foot for contact with the substrate.

All 22 muscles acted throughout the stance phase (Figure 1). This indicates that all muscles contract to prevent undue flexion of hip and knee joints and dorsiflexion of the ankle joint, as well as serving as a power source for forward motion. Further, the muscles exhibit increases and decreases in activity during the stance phase. We divided the stance phase into three parts, designated first, middle and final, to discern particularities of the activity of each muscle.

In the first part of the stance phase, the hamstring and adductor groups of thigh muscles acted most strongly. In the middle part of the stance phase, the gluteal muscles acted most strongly; and in the final part of the stance phase, the vasti and the flexor group of crural muscles acted most strongly. The high activity of the hamstring and adductor muscles of the thigh in the first part of the stance phase may function to prevent the body from bending downward at foot contact. The high activity of the gluteal muscles in the middle part of the stance phase may fix the hip joint and initiate propulsion by extending the thigh at the hip joint. The high activity of the vastus muscles and the flexor group of crural muscles probably provides the main source of forward motion.

B. *Effects of Walking Speed*

The patterns of hip and knee joint excursion, foot contact sequence, and foot force did not change with increasing walking speed. But the gastrocnemius muscle was remarkably more active in the final part of the stance phase during rapid walking.

C. *Comparisons with Other Primates*

The pattern of muscle activity in the hindlimb muscles during gibbon bipedal walking differs greatly from that of man because in the gibbon many muscles act during the stance phase. It is impossible to compare precisely the results obtained from this study with those from studies on other nonhuman primates, because there are not enough data about them (Kondo and Ishida, 1971; Ishida et al., 1974, 1975). However, the following general comparisons seem to be reasonable.

In Japanese monkeys, the hamstring and gastrocnemius muscles act simultaneously and probably provide the chief power for propelling the body, while in gibbons, chimpanzees and spider monkeys, the vasti and the flexor group of crural muscles act simultaneously to give propulsive force to the body. But the gibbon differs from chimpanzee and spider monkey in that the vertical force of its foot never wanes in the middle of the stance phase as it does in chimpanzee and spider monkey.

CONCLUSIONS

We conclude that during the bipedal walking of gibbons propulsion is characteristically provided by extension of the hip and knee joints and plantar flexion of the ankle joint, powered by relatively fast co-contractions of the gluteal, vastus and flexor group of crural muscles. The bipedal gait of gibbons is probably a specialization for rapid locomotion which complements their special adaptations for ricochetal arm-swinging and versatile climbing in arboreal environments (Tuttle, 1975).

ACKNOWLEDGEMENTS

We are indebted to Professors S. Kondo, Primate Research Institute of Kyoto University, J. Ikeda and J. Itani, Laboratory of Physical Anthropology, Kyoto University, for their constant encouragement during this study. We are especially thankful for the assistance of Mr. S. Hayakawa.

This investigation was supported by a grant from the Primate Research Institue to H. Ishida, M. Okada and T. Kimura. R. Tuttle was supported by a grant from the Japan Society for the Promotion of Science and NSF Grant Nos. GS-3209 and SOC75-02478.

REFERENCES

Basmajian, J.V. (1974). "Muscles Alive, Their Functions Revealed by Electro-myography", 3rd ed., Williams & Wilkins Co., Baltimore.

Ishida, H., Kimura, T. and Okada, M. (1975). *In* "Symposia of the Fifth Congress of the International Primatological Society", (S. Kondo, M. Kawai, A. Ehara and S. Kawamura, eds), pp. 287-301, Japan Science Press, Tokyo.

Ishida, H., Okada, M. and Kimura, T. (1974). *J. Anthrop. Soc. Nippon*, 82, 227 (abstract in Japanese).

Kondo, S. and Ishida, H. (1971). *Proc. 1st Symp. on Posture*, pp. 209-261, Shisei-Kenkyu-Sho, Tokyo (Japanese).

Tuttle, R.H. (1975). *In* "Phylogeny of the Primates, a Multidisciplinary Approach", (W.P. Luckett and F.S. Szalay, eds), pp. 447-480, Plenum, New York.

Tuttle, R.H., Basmajian, J.V., Regenos, E., and Shine, G. (1972). *Am. J. Phys. Anthrop.*, 37, 255-266.

Yamazaki, N., Ishida, H., Kimura, T. and Okada, M. (in preparation). A simulation study of dynamics of bipedal walking of primates.

ELECTROMYOGRAPHY OF PONGID GLUTEAL MUSCLES AND HOMINID EVOLUTION

R.H.TUTTLE*, J.V. BASMAJIAN**, H. ISHIDA***

*University of Chicago, Chicago, Illinois, USA, **McMaster University, Hamilton, Ontario, Canada, ***Osaka University, Osaka, Japan.

INTRODUCTION

One of the chief hallmarks of hominid status is habitual bipedal posture and locomotion. Among living primates, *Homo sapiens* is sharply distinguished by the bipedal adaptive complex. Questions of how and when hominid bipedalism evolved have commanded the attention of scientists intermittently since the beginning of the nineteenth century (McHenry, 1975). In recent years the problem of bipedalism has undergone a remarkable resurgence of interest. This was stimulated by burgeoning discoveries and new studies of Plio-Pleistocene hominid postcranial remains, largely from Africa, and the application of refined kinesiological methods and biomechanical perspectives to elucidate the mechanisms of habitual bipedalism in man and facultative bipedalism in nonhuman primates.

The imprint of bipedal positional behaviour is clearcut in the lumber vertebral column and lower limbs of *Homo sapiens*. Ideally, the transformation from antecedent positional behaviour to full-fledged bipedalism could be documented through a series of relatively complete, accurately dated hominoid fossils. Unfortunately, the fossil record falls far short of this ideal. Especially deficient are vertebral and lower limb remains from the period between 15 and 3 million years B.P., when the initial transformation and much progress toward the modern human condition probably occurred. Consequently, theories on the phylogeny of hominid bipedalism will continue to be premised heavily upon information derived from comparative studies of extant forms.

Now there is a rather wide variety of models about the positional behaviours that were proximately antecedent to hominid bipedalism (Tuttle, 1974). These ancestral models include (1) large-bodied brachiating troglodytians, i.e., arboreal arm-swinging chimpanzee-like apes (Keith, 1923, 1934; Gregory, 1927, 1934; Washburn, 1967, 1968, 1972; Tuttle, 1969; Lewis, 1971, 1972, 1974), (2) knuckle-walking troglodytians derived from arboreal apes (Washburn, 1967, 1968, 1972, 1974), (3) somewhat orangutan-like forms whose elongate forelimbs assisted terrestrial locomotion until the bipedal adaptive complex was refined (Stern, 1976), and (4) hylobatian (or small-bodied) apes which developed a high degree of arboreal bipedalism before their descendants engaged in regular, long-term bipedal ventures on the ground (Morton, 1924; Morton and Fuller, 1952; Tuttle, 1974, 1975; Tuttle and Basmajian, 1975).

Before evolutionary anthropologists can arrange these conjectural models in a probabilistic series and perhaps also devise new ones, the mechanisms of facultative bipedalism must be documented in representative anthropoid species. Here we report on the electromyographic (EMG) activities of gluteal muscles in pongid subjects as they engaged in bipedal positional behaviour. We hope that this will contribute to a base for informed surmises about the evolution of hominid bipedalism.

ANATOMY AND FUNCTIONS

According to Sigmon (1975), and as confirmed by our own dissections, in the Hominoidea there are three basic anatomical patterns of the gluteal muscles. One is unique to *Pongo pygmaeus*. It is characterized chiefly by discreteness of the lower segment of the gluteus maximus to form the ischiofemoralis muscle and discreteness of the lateral part of the gluteus minimus to form the scansorius muscle.

The second pattern is shared by the Hylobatidae and *Pan* (including *Gorilla*). Discrete ischiofemoralis and scansorius muscles do not occur in the lesser and the African apes, even though the attachments and extents of the gluteus maximus and gluteus minimus muscles are reminiscent of the condition in *Pongo*.

The third pattern, unique to *Homo*, is characterized by absence of an ischiofemoral portion of the gluteus maximus muscle. Unlike the apes, in man the gluteus maximus muscle has a proximal attachment on the ilium and distally it is confined to the proximal one-third of the thigh. In apes, the gluteus maximus complex extends distally at least midway and sometimes well into the distal one-third of the thigh. Man lacks a scansorius muscle. But, as in apes, the gluteus minimus complex has a notable proximal attachment on the ilium.

On the basis of dissections, Sigmon (1974, 1975) inferred that in apes, the upper part of the gluteus maximus was an abductor of the thigh, while the ischiofemoral portion was an abductor, lateral rotator and, especially, a powerful extensor of the thigh. The gluteus medius also was inferred to be a powerful extensor and to act additionally during abduction and medial rotation of the thigh. The gluteus minimus was inferred to be an abductor, lateral rotator and perhaps extensor of the thigh (Sigmon, 1974, 1975).

EMG studies on man demonstrated that the gluteus maximus muscle is an extensor and lateral rotator of the thigh. In addition, its upper portion is a reserve source of power during abduction of the thigh (Basmajian, 1974, p. 248). The human gluteus medius and minimus muscles are the chief abductors of the thigh (Basmajian, 1974, p. 249).

ELECTROMYOGRAPHY OF PONGID APES

Subjects and Methods

We have described our methods for recording and analysing EMG data in other papers (Tuttle and Basmajian, 1974a, 1974b; Tuttle et al., 1975). The subjects in this study were a 7 year old female Sumatran orangutan (two experiments), a $4\frac{3}{4}$ to $5\frac{1}{2}$ year old female lowland gorilla (seven experiments), and a 5 to $5\frac{3}{4}$ year old male common chimpanzee (two experiments).

Results

1. *Gluteus maximus proprius*

Generally during quiescent bipedal stance in which the knees and hips evidenced flexure, the gluteus maximus proprius exhibited low or nil EMG potentials in chimpanzee, gorilla and orangutan. When the hindlimb was eccentrically loaded during bipedal stance, higher potentials occurred in chimpanzee and gorilla. The orangutan rarely stood bipedally without manual support and with flexure of the hindlimbs. Instead she more commonly stood quietly with the hip and knee joints extended or overextended. This was accompanied by nil EMG in the gluteus maximus proprius.

High potentials occurred as the subjects stretched upward until the hip was markedly extended or overextended and laterally rotated. When they relaxed somewhat in this position, the EMG dropped to nil. The chimpanzee usually and the gorilla occasionally jumped for objects overhead. This generally was accompanied by high EMG potentials.

The chimpanzee and gorilla always evidenced flexure of the hindlimbs during bipedal walking. In the chimpanzee, the stance phases of bipedal steps were accompanied by uniphasic bursts of low or more commonly moderate and higher potentials. In gorilla, uniphasic bursts of low EMG potentials occurred during the stance phase, unless the limb was eccentrically loaded, in which case higher potentials occurred. The orangutan did not walk bipedally unassisted during these sessions. As she shuffled sideways, the gluteus maximus proprius exhibited moderate EMG as the hip was abducted in the swing phase and nil EMG during support with the hip overextended.

2. *Ischiofemoralis*

Quiescent plantigrade bipedal stances with flexure of the hindlimb were usually accompanied by low EMG potentials in the ischiofemoralis portion of the gluteus maximus muscle in chimpanzee and gorilla. It was somewhat more active than the gluteus maximus proprius in this position. High potentials occurred in the ischiofemoralis muscle of the orangutan on the few occasions when she stood unassisted and assisted by her hands with flexure of the hindlimb.

In chimpanzee and gorilla, EMG potentials (commonly high ones) accompanied the initial phase of rising to more fully extended bipedal positions. In orangutan, some rises were accompanied by silence in the ischiofemoralis muscle while others were accompanied by low or high potentials. In all subjects, the EMG ceased as full extension and overextension of the hip were achieved, for instance when they stretched upwards. Jumping was characteristically accompanied by high potentials during propulsion but EMG ceased prior to full extension of the hip joint.

In chimpanzee, gorilla and orangutan, the stance phase of bipedal plantigrade steps with flexure of the hindlimb was accompanied by uniphasic bursts of low or, more frequently, high EMG potentials in the ischiofemoralis. Nil EMG occurred in the ischiofemoralis muscle as the orangutan shuffled sideways and forwards with her hips extended and overextended.

3. *Gluteus medius*

In chimpanzee and gorilla, nil or low EMG potentials occurred in the gluteus medius muscle during quiescent bipedal stance with flexure of the hindlimb. Silence was more common in chimpanzee than gorilla. If the hindlimb was eccentrically loaded, as when the ape was standing on an incline with the ipsilateral foot downhill, high potentials occurred in the gluteus medius.

Low or high potentials occurred in the gluteus medius muscles of the three apes as they rose to bipedal positions, but the EMG dropped to nil or to very low levels when the hip joint was in extended and overextended positions. The propulsive phases of vigorous upward jumps were accompanied by high potentials

in the gluteus medius muscles of chimpanzee and gorilla. Low jumps by the
gorilla were accompanied by low EMG potentials.
   The chimpanzee consistently exhibited uniphasic bursts of high potentials
in the gluteus medius muscle during the stance phase of bipedal steps. The
stance phase of one forward hand-assisted step by the orangutan was accompanied
by high potentials. Uniphasic high or low potentials occurred in the gorilla
during the stance phase of bipedal steps, many of which were slow and brief.

4. *Gluteus minimus*

   We have EMG information about the gluteus minimus proprius in *Pongo* and the
middle of the gluteus minimus muscle in *Pan gorilla*.
   Low potentials characteristically occurred in the gluteus minimus proprius
muscle of *Pongo* during bipedal stances with hindlimb flexure, during the stance
phases of two short bipedal steps, and as she rose to bipedal postures. Nil
EMG occurred when the orangutan's hip joint was in extended and overextended
positions and as she shuffled sideways while touching the wall with her hands.
Moderate potentials occurred once when the hindlimb was eccentrically loaded
and twice as she crouched bipedally prior to reaching upward.
   When the gorilla stood bipedally with hindlimb flexure, high EMG potentials occurred nearly as often as low ones did in the gluteus minimus muscle.
High potentials always accompanied eccentric loading of the ipsilateral hindlimb
and jumping upwards. Nearly all rises to bipedal postures were accompanied by
high EMG potentials. Similarly, uniphasic bursts of high EMG potentials were
exhibited by the gluteus minimus muscle during the stance phases of nearly all
bipedal steps by the gorilla.

FUNCTIONAL INFERENCES

   We conclude that during the bipedal behaviour of pongid apes, the gluteus
maximus proprius acts predominantly as a lateral rotator and abductor of the
thigh at the hip joint, while the ischiofemoralis, gluteus medius and gluteus
minimus act chiefly as extensors of the thigh at the hip joint.
   None of the gluteal muscles in our study appears to be vital to sustain
fully extended and overextended positions of the hip joint in free-standing bipedal pongid apes. Other muscles, osseo-ligamentous mechanisms or a combination
of these factors must suffice to maintain such postures.

IMPLICATIONS FOR HOMINID EVOLUTION

   Electromyographic studies on human subjects demonstrate that, as in the
pongid apes, the gluteal muscles play a negligible role in maintaining fully
upright bipedal posture (Basmajian, 1974). During the transformation from a
hominoid which stood with hindlimb flexure to the modern human condition, loss
of the ischiofemoral portion of the gluteus maximus muscle probably occurred
gradually, as fully extended postures became habitually employed for bipedal
standing and locomotion. The ischiofemoralis need not have been a discrete
entity, as it is in *Pongo*, prior to its disappearance in the Hominidae.
   Anterior curvature of the iliac blades positioned the lesser gluteal muscles
to act as abductors in the pelvic tilt mechanism of striding *Homo*. The hominid
gluteus maximus muscle gained a notable attachment to the posterior iliac blade,
thereby enhancing its function as an extensor of the hip joint. Development
of complex biphasic and triphasic activities of the gluteal muscles (Basmajian,

1974, p. 323) probably developed pari passu with the evolution of habitual full extension of the hip and knee joints and the pelvic tilt mechanism in bipedal Hominidae. The rate of transformation to the full-fledged human condition is probably related primarily to intensity of selective forces, since even the extant apes, including hylobatids, offer suitable morphological and behavioural substrates upon which such forces could operate.

ACKNOWLEDGEMENTS

This research was supported by NSF Grant Nos. GS-3209 and SOC75-02478, the Wenner-Gren Foundation for Anthropological Research, Inc., and the Japan Society for the Promotion of Science. We thank K. Barnes, G. Bourne, G. Duncan, J. Hudson, M. Keeling, S. Kondo, S. Lee, J. Malone, R. Mathis, L. Osmundsen, J. Perry, R. Pollard, E. Regenos, J. Roberts, E. van Ormer, and M. Vitti for their assistance.

REFERENCES

Basmajian, J.V. (1974). "Muscles Alive, Their Functions Revealed by Electro-myography", 3rd ed., Williams & Wilkins Co., Baltimore.
Gregory, W.K. (1927). *Q. Rev. Biol.*, 2, 549-560.
Gregory, W.K. (1934). "Man's Place Among the Anthropoids", Oxford University Press, London.
Keith, A. (1923). *Br. Med. J.*, 1, 451-454, 499-502, 545-548, 587-590, 624-626, 669-672.
Keith, A. (1934). "The Construction of Man's Family Tree", Watts & Co., London.
Lewis, O.J. (1971). *Nature, Lond.*, 230, 577-578.
Lewis, O.J. (1972). *In* "The Functional and Evolutionary Biology of Primates", (R.H. Tuttle, ed.), pp. 207-222, Aldine, Chicago.
Lewis, O.J. (1974). *In* "Primate Locomotion", (F.A. Jenkins, Jrs., ed.), pp. 143-169, Academic Press, New York.
McHenry, H. (1975). *Science, N.Y.,* 190, 425-431.
Morton, D.J. (1924). *Am. J. Phys. Anthrop.*, 7, 1-52.
Morton, D.J. and Fuller, D.D. (1952). "Human Locomotion and Body Form", Williams & Wilkins Co., Baltimore.
Sigmon, B.A. (1974). *J. Hum. Evol.*, 3, 161-185.
Sigmon, B.A. (1975). *In* "Primate Functional Morphology and Evolution", (R.H. Tuttle, ed.), pp. 235-252, Mouton, The Hague.
Stern, J.T., Jr. (1976). *Yearb. Phys. Anthrop.*, 19, 59-68.
Tuttle, R.H. (1969). *Science, N.Y.*, 166, 953-961.
Tuttle, R.H. (1974). *Curr. Anthrop.*, 15, 389-426.
Tuttle, R.H. (1975). *In* "Phylogeny of the Primates, a Multidisciplinary Approach", (W.P. Luckett and F.S. Szalay, eds), pp. 447-480, Plenum, New York.
Tuttle, R.H. and Basmajian, J.V. (1974a). *Am. J. Phys. Anthrop.*, 41, 71-90.
Tuttle, R.H. and Basmajian, J.V. (1974b). *In* "Primate Locomotion", (F.A. Jenkins, Jr., ed.), pp. 293-347, Academic Press, New York.
Tuttle, R.H. and Basmajian, J.V. (1975a). *In* "Symposia of the Fifth Congress of the International Primatological Society", (S. Kondo, M. Kawai, A. Ehara, and S. Kawamura, eds), pp. 303-314, Japan Science Press, Tokyo.
Tuttle, R.H., Basmajian, J.V. and Ishida, H. (1975b). *In* "Primate Functional Morphology and Evolution", (R.H. Tuttle, ed.), pp. 253-269 and 1 plate, Mouton, The Hague.

Washburn, S.L. (1967). *Proc. R. Anthrop. Soc., London,* 21-27.
Washburn, S.L. (1968). "The study of human evolution", Condon Lectures, Oregon State System of Higher Education, Eugene.
Washburn, S.L. (1972). *In* "Evolutionary Biology", (T. Dobzhansky, M.K. Hecht, and W.C. Steere, eds), Vol. 6, pp. 349-361, Appleton-Century-Crofts, New York.
Washburn, S.L. (1974). *Yearb. Phys. Anthrop.,* 17, 67-70.

A MECHANICAL ANALYSIS OF BIPEDAL WALKING OF PRIMATES BY MATHEMATICAL MODEL

T. KIMURA*, M. OKADA**, N. YAMAZAKI***, H. ISHIDA****

*Teikyo University, Tokyo, **Tokyo University of Education, Tokyo,
***Keio University, Tokyo, ****Kyoto University, Kyoto.

INTRODUCTION

We have studied bipedal walking experimentally in primates from the dynamic viewpoint to learn about the acquisition of orthograde bipedal walking in man (Ishida et al., 1975; Kimura et al., in press). We have shown in the experiments that human walking resembles the bipedal walking of chimpanzee and spider monkeys, who live habitually in trees, relatively more than that of terrestrial Japanese monkeys and hamadryas baboons. In this paper we have attempted to analyse quantitatively the relationship between the form and walking motions of animals by means of simulation studies using a mathematical model.

MATERIALS AND METHODS

The mathematical model proposed here is constructed of seven rigid segments of the upper body, thighs, shanks and feet which move only in the sagittal plane. Muscle forces of eight muscle groups on a leg have been computed from force equilibria and from the condition of minimum power production. The magnitude of the electromyogram is calculated from the muscle force. The calculated values from the model in human walking agree fundamentally with the experimental data obtained by us and by other investigators on acceleration, foot force, joint force at hip, electromyography and energy consumption (Yamazaki, 1975a and b).

The physical constants of the subjects, such as the length, mass and position of the centre of gravity of segments, were measured directly by us or were estimated from data given by many other authors. Kinematic data were experimentally obtained by means of multiple exposed photographs or 16mm cinephotographs.

In the light of previous results, we investigated the walking of man, chimpanzee and Japanese monkey. We have also postulated the following four hypothetical types in order to learn more precisely the relationship between form and motion:
1. "ape-type" man who has the same segments as man but walks with the "ape-type motion;
2. "man-type" apes;
3. man with undeveloped calcaneum that ends just under the ankle joint; and
4. man with a hyperdeveloped calcaneum which protrudes backward, the distance from the ankle being the same as the forward extent of the metatarsals.

RESULTS AND DISCUSSION

Bipedal walking in the chimpanzee and in the Japanese monkey differs from that of man in that they have a forwardly bent upper body and flexed hips and knees. Calculated power generated by a leg is shown in Fig. 1. Power is

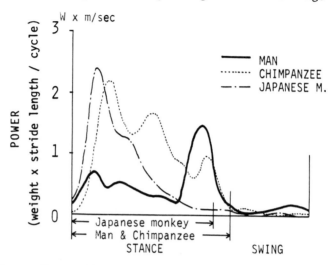

Figure 1. Power from a leg.

expressed as energy consumption in unit time. Three peaks appear in the stance phase. The meaning of the peaks can be discussed by comparing them with the computed muscle forces in Fig. 2. The hamstring muscles work at the beginning of the stance phase when the first peak appears, they extend the hip joint and pull the upper body to the leg. The nonhuman primates show a very large peak because they need much energy to support the flexed hip joint. At the middle of the stance phase, the thigh-extensors work strongly in accord with the hamstrings to prevent flexion of the hip and knee joints. Man produces lower muscle forces and power than nonhuman primates because of the extended joints. The third peak is caused by the action of the triceps surae just before the push-off. This peak is large in man and chimpanzee. Chimpanzee has some resemblance in the patterns of triceps surae force and power to man as compared with the Japanese monkey.

The joint forces are shown in Fig. 3. The large muscle forces of the Japanese monkey at the beginning of the stance phase cause large joint forces at the hip and knee, especially at the knee.

The "man-type" ape shows the same muscle force pattern as man, whereas the "ape-type" man shows the nonhuman primate pattern (Fig. 2). The mechanical characteristics of bipedal walking depend mainly on walking posture and not on physical dimensions. Bipedal walking in man is characterized by extended hip and knee joints.

The energy consumption of nonhuman primates is twice to three times larger than that of man because of their large muscle forces. Man with an undeveloped calcaneus shows about 50% increase in energy consumption compared with normal man. Man with an hyperdeveloped calcaneus shows about a 10% decrease. It seems that the moderate development of the calcaneus in recent man has an advantage in bipedal walking.

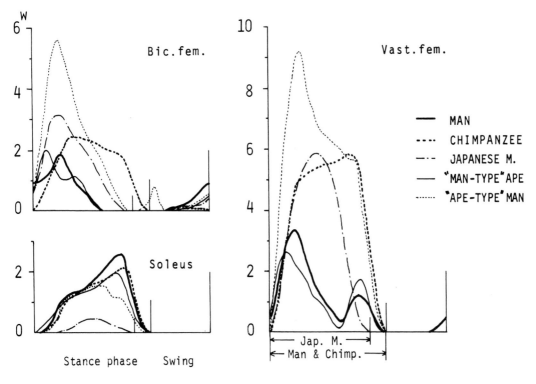

Figure 2. Muscle forces.

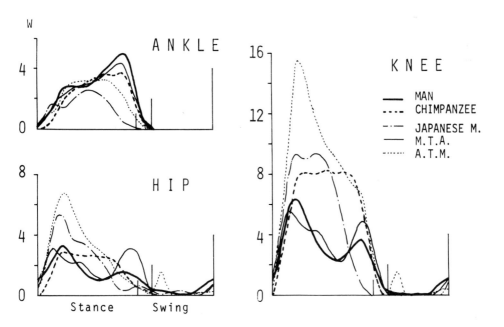

Figure 3. Joint forces.

REFERENCES

Ishida, H., Kimura, T. and Okada, M. (1975). "Symposia of the Fifth Congress
    of the International Primatological Society", (S. Kondo, M. Kawai, A.
    Ehara, S. Kawamura, eds), pp. 287-301, Jap. Sci. Press, Tokyo.
Kimura, T., Okada, M. and Ishida, H. (in press). *Primates*.
Yamazaki, N. (1975a). "Biomechanism 3", pp. 261-269, Yokyo University Press,
    Yokyo (in Japanese).
Yamazaki, N. (1975b). *Jap. J. Ergonomics*, 11, 105-110.

CERVICO-CEPHALIC ANATOMICAL SETS, HEAD CARRIAGE AND HOMINID EVOLUTION: PHYLOGENETIC INFERENCES ON *AUSTRALOPITHECUS*

M. SAKKA

31, rue Erlanger, 75016 Paris, France.

INTRODUCTION

Material

Study of the cervico-cephalic anatomical sets or complexes and research into the biological links between the developmental pattern of the anatomical structures and the pattern of the head carriage have been conducted in *Pan gorilla, Pongo pygmaeus, Pan troglodytes*, modern *Homo sapiens* and *Papio papio*. Dissection, *in vivo* and post mortem X-ray studies, films of living primates and biometry of the skeleton have been used.

In this paper only a few questions will be tackled; the relationship between form and function in the evolution of the human skull, the hypothetical reconstruction of cervico-cephalic structures in *Australopithecus* and the phylogenetic inferences for its taxonomic status.

Anatomical Sets

The comparative anatomy of musculo-skeletal structures has allowed me to speak previously of the methodological concept termed the "anatomical set" (Sakka, 1973b). The study of the same anatomical areas in differing species demonstrates evolutionary changes of a group of structures and of the reciprocal links between elements of this group. For instance, the pectoral girdle is not, functionally, ontogenetically and phylogenetically only a bony skeletal girdle. It is actually an "osteo-syndesmo-myological anatomical set". This concept has been useful to this study in three fields; morphology, function and evolution. It is verified by ontogeny and phylogeny and perhaps will be useful in taxonomy.

RESULTS

Living Primates

1. I have described three layers of muscles on the dorsal aspect of the trunk (Sakka, 1973 and 1974a, vol. II).
(a) The spino-omo-humeral muscles. The muscles of this layer belong partly to the anatomical set of the pectoral girdle.
(b) The "prefatory" muscles: serratus posterior.
(c) The dorso-spino-cephalic muscles. This layer has been completely described by Vallois (1922).

2.   Evolutionary changes of the cervico-cephalic set. When this set in modern
man is compared with that in *Pan* and *Pongo* a twofold change is observed.

a   b

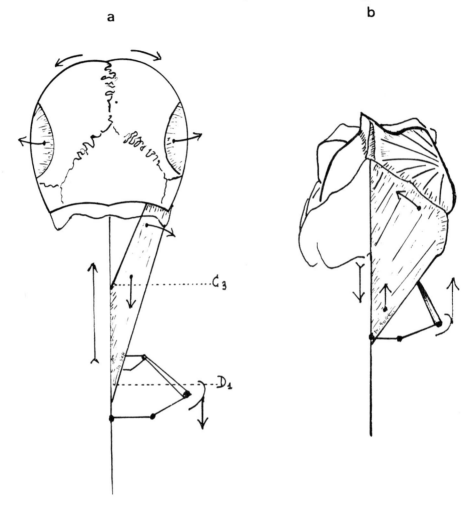

Figure 1.

(For list of abbreviations see end of chapter.)

   In man an ascent of the cervico-cephalic set is associated with a descent or caudal movement of the pectoral girdle (Sakka, 1972 and 1973a).
   In the great apes a regression of the cervico-cephalic set is accompanied
by an ascent or cephalic movement of the pectoral girdle (Sakka, 1974a).
   This divergence involves the size of the bones and muscles, the shape of
these structures and the level of the muscle insertions on the spine and
skull.

3. Man.
(a) The comparative anatomy of the skull shows an expansion of the bones of
the brain case, accompanied by a "disengagement" of the muscles which become
more gracile at the same time. This change separates one temporalis muscle

from the other as they move away from the midline. Temporalis and the nuchal muscles also move away from each other. This latter movement gives evolutionary significance to the feature that I term the *sulcus mastoideus* (Nom. nov., Sakka, 1974a, vol. I).

The nuchal muscles migrate laterally, e.g. the insertion of splenius capitis leaves the external occipital protuberance and moves toward the lateral end of the superior nuchal line. In modern man the medial fibres of splenius capitis are thinner than the lateral ones in keeping with the rule of muscle migration.

(b) The cervical spine expands superiorly while the insertion of some muscles of the superficial layer (spino-omo-humeral muscles) is extended inferiorly along the spine.

(c) Elements of the pectoral girdle's anatomical set descend. To conclude, the cervical spine grows upward like the stalk of a flower, the petals of which would be the skull bones and the sepals the muscles. This flower tends to be in full bloom throughout anthropogenesis.

4. Developmental pattern

(a) Skull changes. The expansion of the skull that is observed in man is the human phase of a process in the vertebrates that may be termed *externation*. I define externation as a trend observed in vertebrate skulls, exemplified by the change from the double-walled (dermal bone and chondral bone) skull of the Crossopterygians to the single-walled (dermal bone only) skull of Tetrapods. In the Reptilia the temporal fenestra is evidence of the externation of the masticatory sets. *Diarthrognathus* shows a characteristic phase of this process. The etiology of this trend in the skull is complex, an important factor being neuro-encephalic expansion during evolution. It would of course be important if we were to find evidence of this process in *Australopithecus*.

(b) Phylogenetic meaning of this divergence. These changes are evidence for two different types of locomotor specialisation.

i) Human locomotion involves a predominance of the inferior limb. The superior limb has become useful for informative processes (neurosensorial processes) and has become able to change the surrounding ecosystem (work and social progress).

ii) Locomotion of the great ape involves the development of the anterior limb.

III) A morphological inference may now be drawn. The difference in size and power between the nuchal muscles of man and apes is explained more by the loss of the locomotor function of the superior limb in man than by the different weight and balance of the head, especially if we take into account the relationship between them and the pectoral girdle and in particular the superficial layer of dorsal trunk musculature (Sakka, 1976b).

5. The Deep Planes

(a) Observation of the deep planes allows us to describe two cervical subsets from a morphological and functional point of view. The superior subset is a cervico-cephalic one while the inferior is a cervico-thoracic one and the boundary that divides them passes through the axis at the level of the spinous process. This boundary is a functional one. The inferior subset is a postural one while the superior is concerned more with movement. The suboccipital muscles are part of this latter subset. Superficially the two subsets are covered by long muscles which constitute a thick layer that, in biorheologics, may be compared with a "visco-plasto-elastic" shock absorber which has an important function in head carriage.

(b) The Suboccipital Muscles. I have previously described these muscles in
higher primates (Sakka, 1973a and 1974a). The comparative anatomy of these
muscles in apes and man has suggested to me some points that are of interest
with regard to the phylogeny of *Australopithecus* and which have allowed me to
propose a reconstruction for KNM-ET 406 (Leakey, et al., 1971).

Obliquus capitis superior. In the great apes this muscle is larger than
obliquus capitis inferior while in man it is smaller. This is very obvious in
Pan gorilla, where obliquus capitis superior is very large, its attachment
reaching the superior nuchal line. The orientation of obliquus capitis superior is such that it makes an obtuse angle with obliquus capitis inferior
in apes but a very acute one in modern man. As the process of externation of
the skull continues, the skull becomes more globular and the insertion of
obliquus capitis superior descends, moving away from the superior nuchal line.
At the same time the muscle becomes thinner and the angle between it and the
obliquus capitis inferior becomes graually less obtuse and finally acute. This
movement depends on the extent of externation, on the degree of cerebello-
telencephalisation and on the biomechanics of the cervico-cephalic joints.
(c) Digastricus. This of course is not a nuchal muscle but the pattern of
its insertion on the skull (situation, surface area) and its topographic relationship with the insertion of obliquus capitis superior are useful in the
diagnosis of fossil hominids. Digastricus has the same action as obliquus
capitis superior. It also becomes thinner and its cranial insertion moves
antero-inferiorly. During this migration it is at first lateral to the insertion of obliquus capitis superior (as in *Pan gorilla*), gradually moving anteriorly until it reaches the mastoid process. Its movement is linked with the
reduction of the masticatory sets but it is related also to externation and
biomechanics. The above features indicate an adaptive change.

6. Anatomical Set of the Cranial Vault
In evaluating the anatomical elements of this set I have found the study of
Crotaphytus (Sakka, 1973c and 1974a) to be very important because of the close
relationship of this muscle to the sagittal crest and because of its influence
on skull morphogenesis.
(a) Sagittal crest. The ontogeny of the gorilla's skull demonstrates the
complexity of factors involved in the building up of a strong cranial crest
which is situated on a cranial suture and lies between two contiguous muscles
which pull in opposite directions. (Two temporalis muscles are needed to
create a single sagittal crest.) It is also necessary to consider in *Pan
gorilla* metabolic and endocrine processes as the sagittal crest appears in
the adult male. It is thus a secondary trait, a feature of sexual dimorphism.
(b) The developmental pattern of skull crests permits one to distinguish
between a cranial crest and a torus. A torus may be distinguished from a
crest by:
  i)   Its shape. By definition a torus is never sharp.
  ii) Its situation. Well away from any cranial suture, a torus is situa-
       ted on a "zone of bending" of the skull, in an area of bony "conflict"
  iii) Its muscle relationships. While a torus may be near muscles it is
       never subjected to opposing muscle pulls.
The ontogeny of the skull of *Pan gorilla gorilla* illustrates this differentiation. In the early stages a line appears on the occiput, the superior nuchal
line. This is followed by the appearance of a torus and finally, when the
nuchal muscles have ascended far enough to reach the lambdoid suture and meet
temporalis, a nuchal crest begins to form.

(c) Skull morphogenesis. The effect of muscle is not the main factor contributing to the shape of the skull. In primates, especially in the lineage
of *Homo*, the cerebello-telencephalic factor is fundamental. This hypothesis
is borne out by a study of the skull of *Australopithecus*.

Hominids

1. The lumbar spine "Rubicon" and the origin of man.
The characteristics of the two types of specialisation that I have observed
in modern *Homo*, *Pan* and *Pongo* (supported by study of the comparative anatomy of
Papio papio), give evidence for markedly divergent adaptive evolution. The
two routes involve ancestral groups able in anatomical and physiological terms
to take these two paths, i.e. these two functional specialisations. From the
evidence of the modern *Pan* and *Pongo* we must try to distinguish the primitive
traits (i.e. those most distant from the present ones in terms of evolution
and function) from the specialised ones. This is difficult because the links
between elements within the anatomical sets are responsible for modifications
even in the primitive features.
    With regard to the cervical spine and head carriage I believe that the an-
cestors of the hominid lineage already had an "erect" or vertical cervical
spine. This is of great interest when one considers the morphogenesis of the
human skull.
    The most important anatomical problem in the acquisition of habitual bi-
pedalism probably concerned the lumbo-sacro-iliac anatomical set. It was
necessary to have a highly mobile and supple lumbar spine. Here is what one
may call the anatomical and functional "Rubicon"* of anthropogenesis. In this
regard the spines of the *Pan* and *Pongo* are very much more specialised than
that of modern man.

2. *Australopithecus*
(a) The acquisition of a bipedal gait by *Australopithecus* is generally accepted, though recently Oxnard (1975) expressed some doubts about the reconstruction of the Olduvai foot. In my opinion the description of Day and
Napier (1964) does not seem to require modification, but in any event a primate on the path of anthropogenesis probably retained primitive locomotor
abilities, at the same time as he acquired new ones (i.e. arborealism and
terrestrialism). Within the large *Australopithecus* group we may find some
arboreal features or even some quadrupedal ones. I do not feel that these
findings should modify the taxonomic status of *Australopithecus* if other traits
put it in the hominid lineage. In fact the relationship between form and function is very complex. There is a disparity in time between the evolution of
form and the evolution of function. This disparity in time is occupied by the
adaptive phase.
(b) The skull of *Australopithecus*. On the vault and nuchal area of KNM-ET
406 (Figs 2 and 3) there is an interesting complex which reminds me of the
mosaic concept of de Beer. Its features will only be listed here (but see
Sakka, 1974a and 1976b).

---

* The term "Rubicon" was used by Sir Arthur Keith and by H.V. Vallois to
characterise the lowest cranial capacity attributable to man in the process
of hominisation.

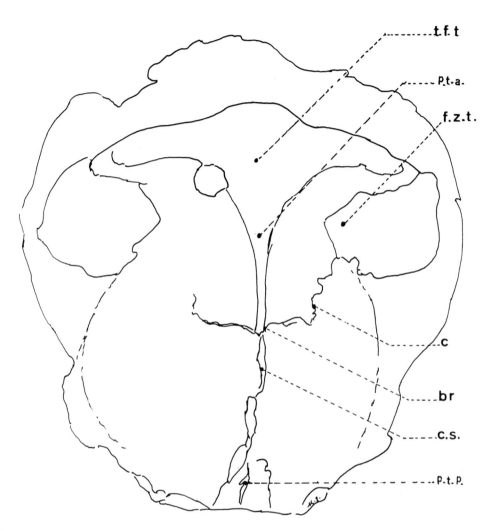

Figure 2.

  i) Primitive traits.
      1. Size of the zygomato temporal foramen (Sakka, 1973c).
      2. Sagittal crest (a feature of sexual dimorphism in the adult male).
      3. Dorsal situation of the insertion of m. obliquus capitis superior.
      4. Situation of the insertion of m. digastricus.
      5. Fronto-temporal trigone.*
 ii) Hominid traits.
      1. Parieto-occipital trigone.†
      2. Dissociation of the three points: posterior temporal point, lambda
         and inion (Fig. 4).

---

* Called *trigonum frontale* by Tobias (1967).
† It has been called the "bare area of the skull" by Dart (1948) and was described by Tobias (1967).

3. Shape of the insertion of rectus capitis minor on the nuchal area.
    (There is a slight asymmetrical depression here.)
4. The external mastoid sulcus (Fig. 5).

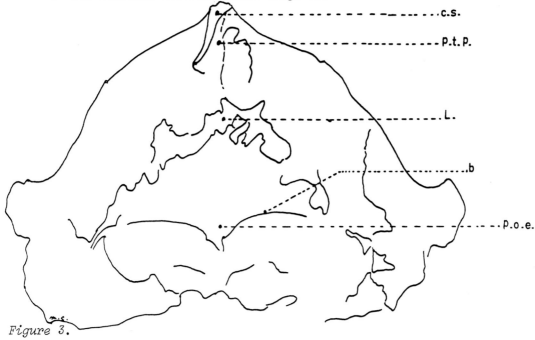

Figure 3.

The diagnostic value of these hominid traits is great as they indicate the
beginning of one of the phases of cervico-cephalic expansion and skull externation in the hominid lineage. They represent the translation of the cere-
bello-telencephalisation on to the bony exocranial cortex of the skull. I
suggest that these features corroborate Holloway's hypothesis (Holloway,
1976).

CONCLUSION

Hypothetical Reconstruction of KNM-ER 406

1. Learning to "read" the bones and gaining an understanding of the history
of the bony structures has allowed me to attempt a reconstruction. Dissection and research into the anatomical sets and the interrelationships of their
elements were necessary to this and e.g. the links between the sagittal crest
and the adipose pad. We may suppose that *Australopithecus* had a well developed muscular system with the head being very distinct from the trunk and
shoulders i.e. *Australopithecus* actually had a neck.

2. The anatomical set of the cranial vault. The scalp and galea aponeurotica
are thick and strong. There is however a contrast between the anterior and
posterior part of the cranial vault.
(a) Anterior part. Over glabella and the supra-orbital torus the skin is
thin and slides over the bone. Over the fronto-temporal trigone there is a
small layer of cellular connective tissue. Occipitofrontalis (venter frontalis)

and galea aponeurotica insert on the doral aspect of the torus.
(b) Posterior part. In the midline the sagittal crest is surmounted by a
sagittal pad which enlarges at its dorsal end to cover the parieto-occipital
trigone. Intervening between the pad and the top of the sagittal crest the
galea aponeurotica adheres to the periosteum. The skin gradually increases in
thickness with the thickest part over the nuchal region.

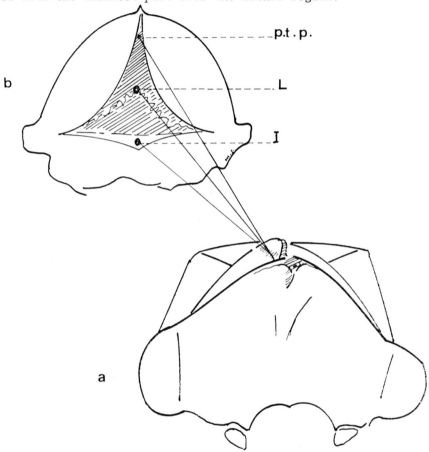

Figure 4.

(c) Lateral aspect of the skull (Fig. 6).
    i) Deep to the skin and galea aponeurotica some strong and glistening
layers of fibrous and aponeurotic tissue insert into the temporal line and
into the limits of the temporal fossa.
    ii) The fascia of temporalis is very thick especially in its anterior and
inferior part. The muscle is attached to the deep aspect of its superior part.
    iii) Crotaphytus muscle attaches to the wall of the temporal fossa. The
exocranial side of the temporal fossa is striated in the part located just
over the superior edge of the squama temporalis. I believe that this zone
corresponds to the squamosal suture. Above this zone is another one which
extends to the posterior part of the sagittal crest and which is characterised
by the presence of small crests and furrows diverging from the posterior temporal point. These correspond to the fibres of the muscle and its fascia
and this area seems to be a site of maximum traction, the direction of the

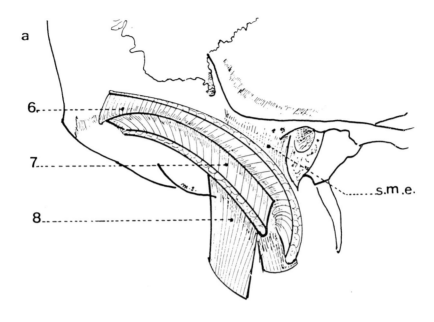

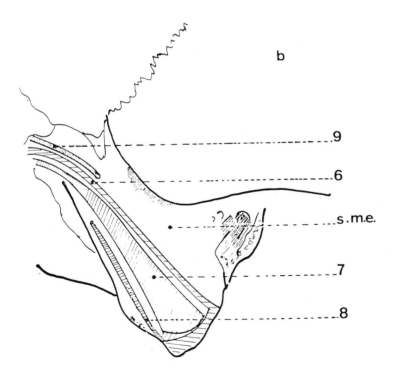

Figure 5.

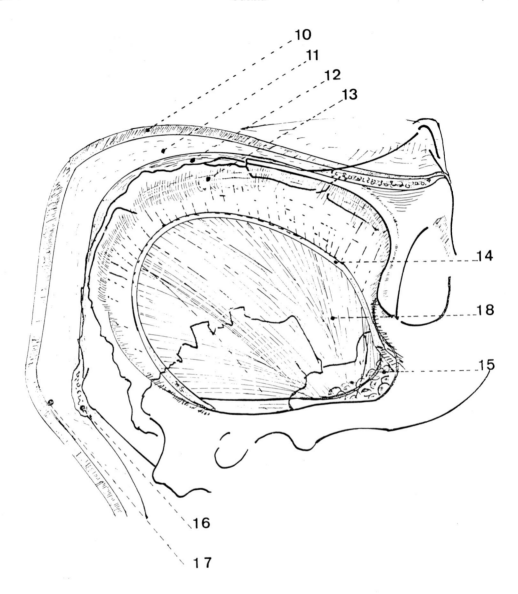

Figure 6.

furrows indicating the direction of the muscle fibres. The posterior and
superior part of the muscle is thinnest, while the middle and especially the
anterior part of temporalis are very thick. In the anterior part of the temporal fossa the post-orbital notch is very large. A boundary, not easy to
find, goes downward and backward, delineating a superior segment where the
muscle attaches (frontal and alisphenoid) and an inferior one where the muscle
slides (zygomatic). Masseter is related to the marginal process (called post-
marginal process by Tobias and R.E.F. Leakey et al.). This feature is very important because it lets us suppose that there was a strong zygomatico-mandibular
muscle. This indicates a very early "verticality" of this muscle in hominids.
   The dorsal aspect of the transverse root of the zygoma (arcus zygomaticus)
is concave and smooth where the posterior fibres of temporalis slide over it

en route to the coronoid process. The zygomatico-temporal foramen is extraordinarily wide (Sakka, 1973c and 1974a). In absolute terms, in respect to its
size, it is slightly smaller than in *Pan gorilla gorilla* adult male and larger
than in *Pan gorilla gorilla* female, *Pongo pymaeus* male and *Pan troglodytes*
male. In relation to its body or skull, it is of considerable size. It indicates that the cerebello-telencephalic development (attested by parieto-
occipital trigone, dissociation of the three points, mastoid sulcus etc.) began earlier than the regression of the masticatory anatomical set. It also
points out the link between regression of the anterior teeth and the process
of reduction of the posterior fibres of temporalis. The size of the zygomatico-temporal
foramen indicates a large muscle but this latter is anteriorly and anterolaterally covered by a layer of adipose connective tissue which separates it
from the bony wall. The coronoid process passes into the middle of the fossa
and offers attachment to the muscular and aponeurotic fibres. A large fleshy
masseter passes inferiorly just in front of the anterior border of the vertical
ramus of a massive mandible.

3. Nichal area (Fig. 7) Trapezius and sternocleidomastoideus insert on the
superior nuchal line with their superficial fibres passing into the nuchal
pad. Laterally the sternocleidomastoideus inserts on the lateral aspect of
the mastoid process, separated from temporalis by the external mastoid sulcus.
This feature, the dissociation of the three points and the parieto-occipital
trigone are some elements of the evolutionary development of bone which are
evident on the cranium, indicating the cephalic phase of the hominid process
i.e. cervico-cephalic emergence. The only place where the temporalis
and nuchal muscles are close together is the slightly crested area to either
side of the midline on the superior nuchal line.
   Semispinalis capitis. One guesses at but does not really perceive its
attachment in the nuchal area.
   Splenius capitis. It inserts on the superior nuchal line. It would be
useful to know its medial limit. The beginning of the process of cervico-
cephalic emergence allows us to suppose that its medial limit was well removed
from the external occipital protuberance but not as far as in modern man.
   Rectus capitis posterior minor. It diverges from its apex on the atlas,
lies on the nuchal area and inserts in a small, shallow, asymmetrical fossa.
It makes little impression on the bone.
   Obliquus capitis superior. This is the largest of the suboccipital muscles.
Its postero-superior area of attachment is comma shaped. The greater part of
it approaches the superior nuchal line but does not touch it. Lying on the
nuchal area it passes dorso-superiorly. Its fleshy belly is lengthened.
   Digastricus. Its posterior belly is situated lateral to the anterior part
of obliquus capitis superior.

Phylogenetic Inferences

   The pattern of bony and myological developmental history involves the early
stages of the skull's expansion and the process of cervico-cephalic emergence
in the hominids. On the cranial vault the contrast between the thinner posterior fibres of temporalis and the thicker anterior ones indicates a process
of muscle migration that I discovered through the study of splenius capitis in
man. *A migrating muscle becomes thinner on the side opposite to the general
direction of its migratory movement.* This is the case for temporalis in
Australopithecus. It becomes thinner in its superior and dorsal part, so

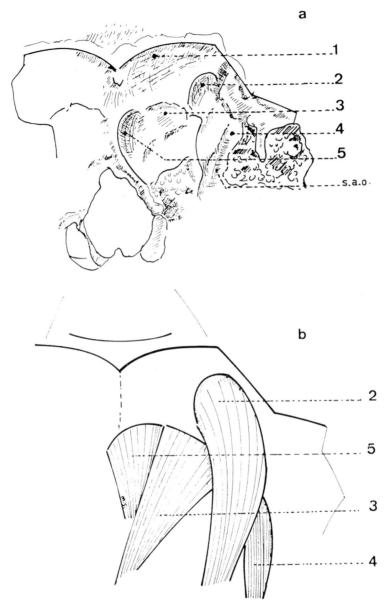

Figure 7.

preparing for an antero-lateral and inferior migration, which will leave exposed on the cranial vault first the parieto-occipital trigone and then the
top of the cranium. This process is more advanced in the specimen without a
sagittal crest, KNM-ET 732, which may be female.

   On the dorsal aspect of the skull the mosaic pattern of muscle attachments
to the nuchal area is a topographic and functional translation of this process. The muscle movement indicates that *Australopithecus* belongs in the
hominid lineage. All of these bony and muscular features give to the skull
as great a value as the post-cranial skeleton for the taxonomic diagnosis of
Australopithecus.

Functional Inferences

1. Obliquus capitis superior.
(a) The size of obliquus capitis superior and the posterior extent of its attachment on the nuchal area are "primitive" features which remind one of *Pan gorilla*'s pattern. The question now is, does this pattern relate to a style of head carriage in which the spine is not fully erect? Does it indicate an inclined cervical spine and a head tilted forward by the weight of masticatory anatomical set? I do not think so. Throughout hominid evolution the cranial attachment of obliquus capitis superior moves anteriorly and inferiorly on the nuchal area. This movement began in *Australopithecus*. This migration is linked more, in my opinion, to the skull's externation process than to head carriage.
(b) Functionally, the shape and length of this muscle are related to the movements of the atlanto-occipital joint. This muscle is a lever arm the length of which is a very important factor in the biomechanics of head movements, especially flexion and extension of the head on the occipital condyles. The more vertical the nuchal area becomes the more a long lever arm is necessary. The shape of this muscle is more related to movements of the head (within the superior cervical subset) than to head carriage as a whole. If my hypothesis concerning the cervical muscles and their bioreheological function as shock-absorbers (and as head supporters) is verified, one may say that the head carriage depends on the superficial and pericervical muscles. In this case the suboccipital muscles, involved with head movements, have maximum efficiency with a minimum of expended energy. The variability of their anatomical features depends on the biomechanics of the joints involved.

2. Head carriage.
The concept of the dissociation in time between the functional process and the morphogenetic one, a dissociation which corresponds to the adaptive phase, allows one to consider that the contingent lack of features which might be linked with an "erect" head carriage in a fossil specimen, would not be a good basis on which to conclude that this specimen did not have "erect" carriage of the head and consequently did not have bipedal gait. Only if one finds some anatomical traits incompatible with such a head posture may one think there is good reason to suppose that there was not "erect" head carriage. These elements of diagnosis are valid for the ancestors of *Australopithecus*. Concerning*Australopithecus*, morphological developmental history does not support any argument which would deny "erect" head carriage. On the contrary the pattern of the skull, especially of the external aspect of the base e.g. the situation of the occipital condyles, indicates a possible "erect" head carriage. Actually it would have been erect because if new features appear on the bone the "dissociation concept" indicates that the function had already been there probably for a long time. It takes a long time for a particular new function to inscribe its particular new traits on the skeleton.
   To conclude, I think that the acquisition of habitual erect posture as a whole is linked with the same functional acquisition concerning the anatomical set of the head and neck. The new ecological and etiological factors which involved an habitual bipedal gait also involved an "erect" head carriage. In fact the acquisition of "erect head carriage probably preceded the acquisition of erect posture as a whole (Sakka, 1976a) and as noted earlier, the anatomo-functional difficulties of habitual bipedal gait were related to the lumbo-sacral anatomical set.

3. The skull's morphogenesis and hominisation. This hypothesis bears anthropogenetic inferences. The skull's morphogenesis during hominisation is directly linked with cerebello-telencephalic development and only indirectly with upright posture. The elements of differential diagnosis in the skull (see earlier) between the modern great apes and *Australopithecus* on the one hand and between *Australopithecus* and modern man on the other, indicating the beginning of the human externation process, are good arguments in support of my hypothesis.

LIST OF ABBREVIATIONS

b	Superior nuchal line	1 and 7	Semispinalis capitis
br.	Bregma	2	Oliquus capitis superior
c	Coronal suture	3	Rectus capitis posterior major
c.s.	Sagittal crest	4	Digastricus
f.z.t.	Zygomatico-temporal foramen	5	Rectus capitis posterior minor
I	Inion	6	Sternocleidomastoideus
L	Lambda	8	Longissimus capitis
P.o.e.	External occipital pro-tuberance	9	Occipitofrontalis, venter occipitalis
P.t.a.	Anterior temporal point	10	Skin
P.t.p.	Posterior temporal point	11	Sagittal fat pad
		12	Galea aponeurotica
		13	Fascia temporalis
S.a.o.	Occipital arterial sulcus	14	Section through the fascia temporalis
S.m.e.	External mastoid sulcus	15	Cellulo-adipose tissue separa-ting temporalis from the antero-lateral wall of the temporal fossa
T.f.t.	Fronto-temporal trigone		
T.p.o.	Parieto-occipital trigone		
		16	External occipital protuber-ance
		17	Transverse occipital fat pad
		18	Temporalis

REFERENCES

Dart, R.A. (1948). *Am. J. phys. Anthrop.*, 6, 259-284.
Day, M.H. and Napier, J.R. (1964). *Nature, Lond.*, 201, 969.
Holloway, R.L. (1976). in IX Congres. U.I.S.P.P. Colloque VI (pretirage), 69-119.
Leakey, R.E.F., Mungai, J.M. and Walker, A.C. (1971). *Am. J. phys. Anthrop.*, 35, 175-186.
Oxnard, C.E. (1975). *Science, N.Y.*, 258, 389-395.
Sakka, M. (1972). These Doctorat ès Sc. Specialite. Anth. phys. Fac. Sc. Univ. Paris VII, 1-108.
Sakka, M. (1973a). *Mammalia*, 37, 126-191.
Sakka, M. (1973b). *C.r. hebd. Seanc. Acad. Sci., Paris*, 277, 865-868.
Sakka, M. (1973c). *Mammalia*, 37, 478-503.
Sakka, M. (1974a). These Doct. Etat ès Sciences, Univ. Paris VII, T I, 1-374, T II Notes et figures.
Sakka, M. (1974b). *Mammalia*, 38, 729-735.
Sakka, M. (1976a). *Zentbl. Vet Med.*, 5, 14-20.
Sakka, M. (1976b). in IX Congres U.I.S.P.P. Colloque VI (pretirage) 239-261.
Tobias, P.V. (1967). "Olduvai Gorge", Vol.2, University Press, Cambridge.
Vallois, H.V. (1922). *Arch. Morph. gen. exp.*, 1-536.

SEXUAL DIMORPHISM IN THE PYGMY CHIMPANZEE, *PAN PANISCUS*

D.L. CRAMER*, A.L. ZIHLMAN**

*Rutgers University, **University of California, Santa Cruz, USA.*

INTRODUCTION

Pygmy chimpanzees are the least known of the great apes but they may reveal the most about the anatomy and behaviour of hominid precursors. Apparently, they form an isolated population cut off from other chimpanzee groups by the sweeping curve of the Congo River (Valdebroek, 1969). They were described relatively recently by Schwarz (1929), and Coolidge (1933) first delineated their morphology. Very little is known of their ecology and behaviour, despite recent attempts to study them in the field (Nishida, 1972; Kano, in press). There are only a few animals in captivity, where their social behaviour is being investigated (Savage and Bakeman, 1976).

The two kinds of chimpanzees, *Pan paniscus* (pygmy) and *Pan troglodytes* (common) are monophyletic but distinct from each other biochemically and equidistant from *Homo sapiens* (Goodman et al., 1970; Cronin, 1975). Morphologically, pygmy and common chimpanzees overlap in body weight and, to varying degrees, in cranial and limb bone dimensions (Zihlman and Cramer, 1976), though the two species can be perfectly discriminated by mandibular length (Cramer, in press).

In this paper, we discuss sexual dimorphism among pygmy chimpanzees in dental, cranial, body weight and bony dimensions. The basis for the study is our data from the skeletal collection from the Musee Royale de l'Afrique Centrale, Tervuren, Belgium. Data on cranial capacity and cranio-facial mass measurements were compiled from a sample of about 60 adults. The postcranial material was meagre: our sample consisted of 20 adult skeletons,[1] some of them incomplete. Sexual dimorphism is one expression of anatomical variability and can be quantified by cranio-facial dimensions, limb bone lengths and robusticity, cranial and dental measurements and especially canine size.

RESULTS

In cranio-facial characteristics, no single dimension distinguishes sex; in no dimensions were *Pan paniscus* females significantly smaller than males. Of 20 variables measured, in none except mandibular height did the sex difference exceed 5% and in four variables, females were equal to or larger than males: vault length, postorbital breadth, nasion-prosthion distance and supraorbital torus (Cramer, in press).

---

[1] "Adulthood" was determined by dental eruption (appearance of M3) in the associated skull or by epiphyseal union.

Cranial capacity was also remarkably similar in both sexes.[1] The overall average was 350 cc with a female mean of 348.7 cc and a male mean of 351.6 cc (Cramer, in press). In dentition, there was no bimodality in incisor or post-canine tooth dimensions, but in canine mesiodistal diameter there was bimodality and unambiguous sexual dimorphism (Almquist, 1974; Johanson, 1974; Fenart and Deblock, 1973; Kinzey in Cramer, in press).

No sex differences were revealed in measurements of long bones, in either length or robusticity (femur, tibia, fibula, humerus, radius, ulna), or in innominate length or breadth, or clavicle length. In some measurements female means were larger than male means. There was overlap in all measurements, although if a larger sample were available, bimodality may well emerge. Robusticity, as measured by midshaft circumference of the long bones, shows overlap and no bimodality[2] (Table I).

TABLE I

*Averages and ranges of variation of P. paniscus bone measurements
(in mm) and body weight (in kg)*

Feature	Sex	Sample size	Mean	S.D.	Sample Range	
Clavicle length	female	9	103.9	5.1	94-110	n.s.
	male	11	103.3	4.7	93-110	
Humerus length	female	9	288.3	8.9	268-298	n.s.
	male	11	278.1	16.4	250-307.5	
Radius length	female	9	262.3	12.8	240.5-284	n.s.
	male	11	255.5	16.8	235-270	
Femur length	female	7	293.8	10.1	275-305	n.s.
	male	6	288.7	16.8	264.5-316	
Tibia length	female	9	240.9	8.3	225-253	n.s.
	male	11	240.2	13.7	218-271.5	
Innominate length	female	7	247.6	10.2	232-265	n.s.
	male	11	247.6	16.6	223-271	
Iliac breadth	female	9	93.0	7.0	80.5-103	n.s.
	male	10	98.3	9.4	88-108	
Body weight	female	8	31.2	5.0	25-38.5	signif p < .01
	male	8	39.8	5.4	30-48	

Body weights derived from 18 adults, including 13 wildshot from the Tervuren sample and five captive animals, showed a significant sex difference. The female mean (of 9) is 31.2 kg (range 25-38.5) and the male mean (of 9) is 39.8 kg (range 30-48), a male/female index of 78%, compared to 84-89% for common

---

[1] Interestingly, the type specimen of *Pan paniscus* was a female (no. 9338) whose cranial capacity of 420 cc was two standard deviations from the female and male means.

[2] Schultz (1953) notes that long bone thickness varies less by sex in common chimpanzees and gibbons, which have smaller body size differences by sex, than in gorillas and rhesus macaques, which have greater body size differences.

chimpanzees and 89% for humans (Schultz, 1969).

DISCUSSION

There is essentially no sexual dimorphism in cranial capacity and cranio-
facial dimensions in a sample of 60 *Pan paniscus*. No statistically significant differences could be uncovered in long bone lengths and robusticity, vertebral lengths and breadths, in clavicular length and iliac length and breadth
in a sample of 20 adult pygmy chimpanzees. In some bones, the averages for
male and female were the same (e.g. clavicular, tibial and innominate lengths).
Sex differences in body weight were marked, the means differing by 8.6 kg or
25% of female mean body weight and 22% of male mean body weight. Canine M-D
diameters were bimodal in distribution and sexually dimorphic and may reflect
in part the body size differences

The pattern of sexual dimorphism in pygmy chimpanzees appears distinct from
those of populations of *Pan troglodytes* and *Homo sapiens*. For example, in
Pan troglodytes the average cranial capacity differs between the sexes by
about 30 cc, whereas body size differences are only slight. In human populations body size differences are also slight, but cranial capacity differs between males and females more than body stature (Tobias, 1975), whereas in
pygmy chimpanzees there is no cranial or facial dimorphism despite moderate
body size dimorphism.

What implications might be drawn from this work for the study of early hominids? This work points not only to the difficulties in reconstructing body
size in *Australopithecus* (an animal similar in size range to chimpanzees), but
also to the problem of discovering the extent of australopithecine sexual dimorphism. In pygmy chimpanzees a difference of 8.6 kg in body weight is not
reflected in bone morphology. Body size estimates for *Australopithecus* based
on a single bone, or even three or four bones, could be in error by as much as
25%. By analogy with *Pan paniscus*, sexual dimorphism in *Australopithecus* may
be expressed in ways not preserved in the skull, bone lengths or robusticity.

It appears that in the australopithecines there is little sexual dimorphism
in cranial capacity (on admittedly very small samples), but there is bimodality
in canine breadth, although somewhat less than in pygmy chimpanzees. From the
fragmentary fossils of long bones, pelvic material and estimates of body weights
in the early hominids, there may be moderate dimorphism in body size within each
species of australopithecine - and in combination with the cranial and dental
evidence, a pattern of dimorphism similar to that of pygmy chimpanzees (Zihlman,
1976).

The behavioural implications and correlates, if any, of the differences in
canine breadth and body weight remain guesswork at this time, but they appear
as the only anatomical expressions of "maleness" in pygmy chimpanzees. Field
studies will, it is hoped, clarify the function of this sexual dimorphism
pattern in the overall adaptation, particularly the social behaviour, of *Pan
paniscus* which in turn may provide a further basis for the inferring of social
behaviour in the early hominids.

ACKNOWLEDGEMENTS

We thank Dr. M. Poll and Dr. D.T. van den Audenaerde for access to the
pygmy chimpanzee materials, C. Jolly, J. Lowenstein and K. Wcislo for comments
on the manuscript, and the Wenner-Gren Foundation for Anthropological Research
for financial support.

REFERENCES

Almquist, A.J. (1974). *Am. J. Phys. Anthrop.*, 40, 359-368.

Coolidge, H.J., Jr. (1933). *Am. J. Phys. Anthrop.*, 18, 1-59.

Cramer, D.L. (in press). "Cranio-facial Morphology of *Pan paniscus*: a Morphometric and Evolutionary Appraisal", Karger, Basel.

Cronin, J. (1975). "Molecular Systematics of the Order Primates", PhD. Dissertation, University of California, Berkeley.

Fenart, R. and Deblock, R. (1973). *Annls Mus. r. Afr. cent. Ser. 4to.*, IN-80, no. 204.

Goodman, M., Moore, G.W., Farres, W., and Poulik, E. (1970). *In* "The Chimpanzee", vol. 2, pp. 318-360, Karger, New York.

Johanson, D.C. (1974). *Am. J. Phys. Anthrop.*, 41, 39-48.

Kano, T. (in press). *In* "The Great Apes", (D. Hamburg, J. Goodall and E. McCown, eds), W.A. Benjamin Press, Menlo Park, California.

Kinzey, W.G. (in press). *In* "Cranio-facial Morphology of *Pan paniscus*: a Morphometric and Evolutionary Appraisal", (D.L. Cramer, ed.), Karger, Basel.

Nishida, T. (1972). *Primates*, 13, 415-425.

Savage, E.S. and Bakeman, R. (1976). Comparative observations on sexual behaviour in *Pan paniscus* and *Pan troglodytes*. Presented at the Sixth Congress of the International Primatology Society, Cambridge, England.

Schultz, A.H. (1953). *Am. J. Phys. Anthrop.*, 11, 277-311.

Schultz, A.H. (1969). "The Life of Primates", Weidenfeld and Nicolson, London.

Schwarz, E. (1929). *Rev. zool. bot. Afr.*, 16, 425-426.

Tobias, P.V. (1975). *In* "Primate Functional Morphology and Evolution", (R.H. Tuttle, ed.), pp. 353-392, Mouton, The Hague.

Vandebroek, G. (1969). "Evolution des Vertébrés", Masson et Cie, Paris.

Zihlman, A.L. and Dramer, D.L. (1976). *Am. J. Phys. Anthrop.*, 44, 216.

Zihlman, A.L. (1976). *In* "Les plus anciens hominidés", (P.V. Tobias and Y. Coppens, eds), IXe Congrès Union Int. des Sciences préhist. et protohist. Colloque VI.

DECIDUOUS AND PERMANENT TOOTH SIZE CORRELATIONS IN
MACACA NEMESTRINA AND IN *HOMO SAPIENS*: A COMPARATIVE STUDY

JOYCE E. SIRIANNI

*Department of Anthropology, State University of New York at Buffalo,
Buffalo, New York 14226, USA.*

INTRODUCTION

With the increased use of the pig-tailed macaque (*Macaca nemestrina*) in experimental dentistry and orthodontics, there is a need to determine the degree
of similarity between the dentitions of macaques and man. Of course, it is
obvious that man's molars are not bilophodont as are the more specialized
macaque molars, and that the human dentition expresses little sexual dimorphism
in comparison to that of the macaque. In spite of these shape and size differences, however, are there basic similarities between the dentitions of these
two species? Although numerous studies have investigated the crown relationships within and between the primary and secondary human dentitions (Moorrees
et al., 1957; Moorrees and Chadha, 1962; Moorrees and Reed, 1964; Bolton,
1962; Garn et al., 1965, 1968; Sofaer et al., 1971), little is known about
these dental relationships in *Macaca nemestrina*. The purpose of this investigation is to determine the degree of association within and between the deciduous and permanent dentitions of *Macaca nemestrina* and to compare these findings with those reported for human dentitions.

MATERIALS AND METHODS

The sample for this study consisted of 32 male and 33 female *Macaca nemestrina*. These animals are part of a large colony initiated at the Regional
Primate Research Center at the University of Washington in Seattle, in order
to study longitudinally the normal craniofacial growth and development of this
species. From the time of weaning and throughout the growth period, serial
and dental stone casts were made of both upper and lower dentitions. Mesio-
distal and bucco-lingual crown diameters of the deciduous and permanent teeth
were measured by myself using an electronic caliper (Chase and Swindler, 1974).
Based on repeated measurements of the crown diameters, the measurement error
was calculated to be less than 0.1 mm.

The degree of sexual dimorphism was estimated for all teeth of the deciduous
and permanent dentitions using the formula: $(1 - \frac{F}{M} \times 100)$.

Correlations matrices correlating tooth size within the deciduous and permanent dentitions as well as between the two dentitions were obtained for both
sexes separately.

TABLE I

Correlation of the mesiodistal crown diameters of deciduous and permanent teeth in Macaca nemestrina and Homo sapiens

Teeth		Macaque				Human[a] sexes combined r[b]
		Male		Female		
		n	r	n	r	
Maxillary:						
di1	I1	31	0.26	33	0.28	0.63
di1	I2	29	0.02	31	0.54**	0.31
dc	C	11	0.27	27	0.33	0.29
dm1	P3	27	0.38	28	0.70***	0.34
dm2	P4	30	0.52**	31	0.61***	0.38
dm1	M1	32	0.64***	32	0.69***	0.36
dm2	M1	31	0.65***	32	0.76***	0.51
dm2	M2	19	0.66***	16	0.68**	0.39
Mandibular:						
di1	I1	32	0.32	31	0.37*	0.43
di2	I2	32	0.42*	33	0.42*	0.44
dc	C	8	0.48	21	0.36	0.26
dm1	P3	16	0.63**	30	0.54***	0.47
dm2	P4	31	0.61***	31	0.69***	0.41
dm1	M1	32	0.64***	33	0.74***	0.45
dm2	M1	32	0.73***	33	0.80***	0.53
dm2	M2	25	0.70***	27	0.66***	0.43

a The number of individuals ranged from 121 to 153 except for permanent molars (N = 68-72).

b All the correlation coefficients were statistically significant at $p < 0.01$ level (Moorrees and Reed, 1964).

  * $p > 0.05$

 ** $p > 0.01$

*** $p > 0.001$

RESULTS AND DISCUSSION

Both the pig-tailed macaque and man express very little sexual dimorphism in the mesiodistal crown diameters of the deciduous dentition. The average difference is 1.8% for the macaque and 2.3% for humans (Moorrees et al., 1957). While the sexual dimorphism in the human dentition is estimated to be a low 3.9% (Garn et al., 1964), it is 11.9% in the pig-tailed monkey. In the macaque, the greatest degree of sexual dimorphism is seen in the maxillary canine (37.0%) and the immediately adjacent teeth. Similarly, the mandibular canine and the sectorial premolar are also highly dimorphic, having indices of 34.5% and 31.4% respectively.

Although the concordance in crown size between right and left teeth is high for the pig-tailed monkey (the correlation coefficient (r) ranging from 0.71 to 0.92 in the deciduous dentition and 0.71 to 0.93 in the permanent dentition), this degree of association is consistently lower than that reported for human dentitions in which the correlation coefficients range from 0.85 to 0.97 (Moorrees and Reed, 1964). In the macaque, the more mesial tooth in each morphological class of the deciduous and permanent dentitions is more highly correlated with its antimere than the more distal tooth of the same class. This observation agrees with the suggestion that in man the intrinsic control of tooth size decreases in a mesial to distal direction within each morphological field (Garn et al., 1965). In investigating whether or not a single size factor affects both mesiodistal and buccolingual crown diameters in humans, Garn et al. (1968) have concluded that these crown dimensions, while significantly correlated with one another (r = 0.55), are determined by relatively autonomous factors. In the macaque, the correlations between these dimensions are significant, p < 0.01, for both the deciduous (r = 0.46) and permanent (r = 0.46) dentitions. Therefore, it may be concluded also that there is not a single size factor affecting both crown diameters in the pig-tailed macaque. As shown in the table, the mesiodistal diameter of the deciduous molars is the best predictor of permanent premolar and molar size, whereas the size of the deciduous incisors and canines are generally poorer predictors of the permanent incisor and canine dimensions. In both macaque and humans the deciduous second molar and the permanent first molar are highly correlated. This would support the thesis that the first permanent molar should be classified as a deciduous tooth (Moorrees and Reed, 1964).

ACKNOWLEDGEMENTS

This comparative study was supported in part by National Institutes of Health grants DE 02918 and RR 00166.

REFERENCES

Bolton, W.A. (1962). *Am. J. Orthod.*, 48, 504-529.
Chase, C.E. and Swindler, D.R. (1974). *J. dent. Res.*, 53, 1506.
Garn, S.M., Lewis, A.B. and Kerewsky, R.S. (1964). *J. dent. Res.*, 43, 306.
Garn, S.M., Lewis, A.B. and Kerewsky, R.S. (1965). *J. dent. Res.*, 44, 350-354.
Garn, S.M., Lewis, A.B. and Kerewsky, R.S. (1968). *J. dent. Res.*, 47, 495.
Moorrees, C.F.A. and Chadha, J.M. (1962). *J. dent. Res.*, 41, 644-470.
Moorrees, C.F.A. and Reed, R.B. (1964). *Archs. oral Biol.*, 9, 685-697.
Moorrees, C.F.A., Thomsen, S.O., Jensen, E. and Yen, P.K. (1957). *J. dent. Res.*, 36, 39-47.
Sofaer, J.A., Bailit, H.L. and MacLean, C.J. (1971). *Evolution*, 25, 509-517.

OBSERVATIONS ON THE GENUS *RAMAPITHECUS*

K.N. PRASAD

Geological Survey of India

INTRODUCTION

The study of various dental elements of *Ramapithecus* recovered from the Siwalik sediments has given increasing evidence of the progressive nature of this genus in the assessment of human ancestry. Critical evaluation of the findings has considerably enlarged the scope for a better understanding of the initial differentiation of hominids from pongids. Radiometric dating of the deposits containing these valuable hominid materials has thrown new light on the diversification of these forms during the Mio-Pliocene.

PROVENANCE OF *RAMAPITHECUS*

The earliest record of *Ramapithecus* was by Lewis (1934). The dental elements of this genus, although known to Pilgrim (1915), were erroneously assigned to *Dryopithecus punjabicus*. The new taxon proposed by Lewis was *Ramapithecus brevirostris*, the holotype (Y.PM. 13700) consisting of a right premaxilla and maxilla with alveoli of incisors and canines with three premolars to the second molar. The third molar of *Ramapithecus* (G.S.I. types 18068) is the only one recovered so far by the author (1964) and correctly assigned by Simons (1964). The maxilla recovered by Lewis is equivalent to the hypodigm material D-185 collected by Rao from Haritalyangar several years earlier. This was erroneously assigned to *Dryopithecus punjabicus* by Pilgrim. Much confusion has thus resulted from the assigning of the hominid material to *Dryopithecus*. Leakey (1962) described two maxillary fragments and an associated right lower molar from the late Miocene or early Pliocene deposits from Fort Ternan, Kenya. Fort Ternan deposits have been dated with an estimated K/Ar date of about 14 million years. *Kenyapithecus wickeri*, a new taxon proposed by Leakey, is considered synonymous with *R. punjabicus* by Simons (1964). Andrews (1971) recognises species differences in the Fort Ternan material and has assigned *K. wickeri* to *Ramapithecus wickeri*. Chow (1958) referred to the type material of *Dryopithecus keiyuanensis* as closely resembling the species *Ramapithecus punjabicus*. The Keiyuan fauna is considered equivalent to the fauna of the Chinji-Nagri zones of the Siwaliks by Chow (op.cit.). Thus, there are at least three recognisable species of *Ramapithecus* of which two are discernible.

   Genus: *Ramapithecus*, Lewis, 1934.
   Genotype: *Ramapithecus brevirostris*, Lewis, 1934.
   Revised generic diagnosis: Differs from *Dryopithecus*
and other groups of apes in the following significant features. Small, short faced forms with shallow mandibles, shallow on the lingual side; anterior dentition progressive; cheek teeth simple, crowded up with elimination of

diastema. No evidence of Carabelli cusp. Cusps of lower molars splayed apart, affording a large area for occlusion and lateral grinding and chewing. Cingulum absent.

A. *Ramapithecus brevirostris*, Lewis, 1934.

Specific diagnosis: A large species of *Ramapithecus*. Differs from *R. punjabicus* in the possession of relatively deep maxilla, highly progressive teeth with low relief of cusps. Reduced premolars coupled with low relief of cusps in molars with arcuate dental disposition are suggestive of progressive species of hominid affinity. Holotype: Y.P.M. 13799, a right maxilla and premaxilla. Hypodigm: G.S.I. 18064, a maxilla with three molars. Horizon: Nagri beds, Siwalik Hills. Type locality: Haritalyangar, Himachal Pradesh, India.

B. *Ramapithecus punjabicus*, Pilgrim, 1910.

Specific diagnosis: Differs from *Ramapithecus brevirostris* in the shallower robust mandible with a more complex pattern of tooth crenulations with a reduced cingulum. A progressive short faced hominid with molars showing low cusps. Several teeth and fragments of mandible in the collections of the Geological Survey of India and Yale Peabody Museum were assigned to *Dryopithecus* by Pilgrim or to *Bramapithecus* by Lewis. However, Simons and Pilbeam (1965) critically evaluated the material and assigned them to *Ramapithecus punjabicus*. *R. punjabicus* appears to have been a short faced form smaller in size than *R. brevirostris*. Are we justified in assigning the maxilla D-185 to *Ramapithecus* since the first upper premolar shows three roots? Le Gros Clark (1955) suggests that in modern races of *Homo sapiens* a small percentage show three roots.

DISCUSSION

The hominid status of *Ramapithecus* has been the subject of critical discussion in recent years. Lewis (1937) correctly evaluated the remains of *Ramapithecus* and *Bramapithecus* but could not positively assign them to Hominidae for want of sufficient preserved materials. The crux of the problem is whether they could be included under *Homo* or *Australopithecus* based on dental, mandibular and facial morphology besides consideration as a forerunner of *Australopithecus*. The mandibles of *Ramapithecus* differ from *Dryopithecus* in their being much smaller and thicker, resembling those of *Australopithecus*. The general decrease of the antero-posterior length of molars as known by the interstitial wear is suggestive of the dentition of australopithecines. The shortening of the jaw and crowding up of molars eliminates a diastema anteriorly. The position of overlapping canines characteristic of apes contrasts with those of *Ramapithecus*. Although *Ramapithecus* possessed certain characters in the semisectorial premolar and position of incisors however, the overall progressive trend in the mandible and maxillae considered from the functional angle demonstrates that such a situation should be expected in an early hominid ancestor. Probably, grinding and chewing were first initiated at this stage as indicated by the wear facets in the dentition. Floral evidence (Prasad, 1971) shows that these early hominids may have adapted their food habits to lateral grinding and chewing. The dental complex in *"Ramapithecus-Australopithecus"* indicates the vertical emplacement of incisors, small canines, broad rounded molars appreciably of moderate size showing difference in wear depending on the age of the individual. The wear gradient on the anterior

molars (M1-M2) is more marked than on the third molar. Preservation of the
enamel in the crown of the third molar in some of the samples studied gives
a sufficient clue that these erupted late as in Pleistocene hominids.

Another significant feature in the dentition of *Ramapithecus punjabicus*
from the Chinji Beds (older than Nagri) is in the presence of a reduced cingulum, a feature which is absent in *R. brevirostris* of the Nagri Stage. Morphological features of the dentition especially the rounded cusps with wrinkles
recall those of australopithecine stock. The pulp cavity is rather low and
teeth are worn down to the same level, with cusps showing low reliefs. Canines
being inferior, these hominids must have adapted to food grinding habits. Reduction of the premaxilla was probably related to the recession of the jaws and
crowding up ofthe dentition. This resulted in the late eruption of the third
molar.

Dental remains of *Ramapithecus* recently recovered by a team from the Geological Survey of India from Ramnagar, Kashmir, indicate several progressive
characters especially of the upper molars. However, the maxillary fragment
of *Ramapithecus* (G.S.I. 18064) recovered by the author from Haritalyangar
recalls some of the later hominds by all the molars being equal with rounded
cusps. It is suggested that the Siwaliks have provided sufficient dental
material to enable us to evaluate the status of early hominids during the Mio-
Pliocene. The occurrence of an early forerunner of *Australopithecus* in the
Siwaliks should not be ruled out.

The lingual aspect of the jaw which is relatively thin and hollow probably
accommodated a mobile tongue unlike in contemporary dryopithecine fauna. Critical evaluation of the dental remains of *Ramapithecus* indicate that a free
movement of the tongue was possible which could ultimately give expression to
various modes of sound and initiate some form of speech, a specialisation probably accomplished in australopithecines and *Homo erectus*. At this stage, it
may be suggested that these South Asian Mid-Tertiary hominids had evolved a
facial and dental mechanism, both functional and evolutionary, which was a precursor for progressive development of these characteristics in later hominids,
a clue to which was culled from *Ramapithecus* remains.

REFERENCES

Andrews, P. (1971). *Nature, Lond.*, 231, 192-194.
Clark, W.E. Le Gros (1955). "The Fossil Evidence for Human Evolution", Uni-
    versity Chicago Press.
Chow, M.D. (1958). *Palaeont. Soc. Ind.* , 3, 123-130.
Lewis, G.E. (1934). *Am. J. Sci.*, 27, 161-179.
Lewis, G.E. (1937). *Am. J. Sci.*, 34, 139-147.
Leakey, L.S.B. (1962). *A. Mag. Nat. Hist.* , 4, 686-696.
Pilgrim, G.E. (1910). *Rec. geol. Surv. India,* 40 63-71.
Pilgrim, G.E. (1915). *Rec. geol. Surv. India,* 45, 1-74.
Prasad, K.N. (1964). *Palaeont. Ass. London,* 7, 123-134.
Prasad, K.N. (1971). *Nature, Lond.*, 232, 413-414.
Simons, E.L. (1964). *Proc. Natn. Acad. Sci. USA*, 3, 528-535.
Simons, E.L. and Pilbeam, D.R. (1965). *Folia Primat.*, 3, 81-152.

HUNTING BEHAVIOUR IN HOMINIDS: SOME ETHOLOGICAL ASPECTS

H.D. RIJKSEN

Zoological Laboratory, State University, Gröningen, Netherlands.

INTRODUCTION

The orang-utan is threatened with extinction, a fact which must presently
be attributed mainly to the exponentially growing destructive influence of
mankind on the forest habitat. Yet hunting pressure by man on this slow breeding ape still takes it toll, even though this form of predation is a very ancient ecological factor to which *Pongo pygmaeus* seems to have adapted with arboreality and a particular social organisation (Rijksen, in press). In order
to get an impression of the ecology of the African apes, after having studied
orang-utans in Sumatra, I paid a visit to the Gombe Stream National Park in
Tanzania in 1975. Here I had the luck to witness two successful hunting events
by a group of chimpanzees. These observations, added to the rather extensive
literature on "predatory" behaviour of chimpanzees (for a review see Teleki,
1975) and my interest in human hunting rituals generated during my Indonesian
years, led me to speculate about the original motivational basis for hunting
behaviour in hominids. In this paper I shall not deal with possible nutritional incentives, but restrict myself to the possible socio-ethological basis of
such behaviour.

THE HUNTING PRACTICE

Several traits are characteristic for both the hunting practice of the chimpanzee and the hunting, particularly head-hunting, practices of humans. Thus,
hunting is almost exclusively practised by males in both species and just as
in many forms of human hunting, hunger seems not to be the stimulant in chimpanzees. Both chimpanzee hunting and human hunting is (or afterward becomes)
an intense social happening for the "in-group", during which rank-relationships
may temporarily become suspended (Teleki, 1975). General attention becomes
focussed on those individuals who play (or played) particularly outstanding
roles in the hunt and especially on those who own major portions of the result
of the hunt. Moreover, the results of the hunt are typically shared by many
of the "in-group" members, either in the form of consumption of the meat and/
or in the form of communal rituals and feasts.
An observation by Teleki (1975) may have particular relevance in this respect: he noticed that young adult male chimpanzees show heightened interest
in initiating hunting events at the stage when they are attempting to establish their social status; there are indications that males who successfuly
do so, may reach high ranking positions later on. It is likely that success
both in hunting and in social relationships leading to a high ranking position

are based on some common trait(s), but Teleki's observation suggests that the attention attracted during hunting may contribute to advancing social status in another context. This is certainly true for the human condition; the anthropological literature gives ample evidence that hunting and head-hunting prowess contributes to the social status of the performer. In most head-hunting tribes this function was even institutionalised by special adornments for the successful hunter (for a review see Ling Roth, 1896).

MAN VERSUS APE

   The sub-fossilised remains of orang-utans found in several caves throughout South-East Asia, often in large quantities and coinciding with traces of pre-historic man's presence, indicate that this ape played a major role in the hunting practice of primitive man. It is notable that the chimpanzee also seems to have a preference for primates in the choice of its prey (Teleki, 1975). There are several references suggesting that the Malayan head-hunters sometimes considered orang-utans as human substitutes in their head-hunting practice; several cases are known in which orang-utan skulls were venerated together with human skulls (Ling Roth, 1896). Also in the cannibalistic practices of certain tribes a very similar selective use of particular body-parts in the ritual meals, from human victims and from orang-utans, suggests a human substitute role for this ape (see also Fossey, 1974, for pseudo-cannibalistic practices directed against mountain gorillas).

THE ORIGIN OF HUNTING BEHAVIOUR IN HOMINOIDS

   The primates having evolved from a predominantly insectivorous ancestor, the potential for meat-eating in them is undoubtedly an ancient trait. It is an interesting fact, however, that according to presently available field data, active hunting (as opposed to the catching of prey by chance) in primates seems to coincide with the occurrence of multi-male groups. Of particular importance in anthropoids, applying to both the chimpanzee and to primitively living hominids, may be the fact that the communities have a stable, permanent nucleus of males (Nishida, 1968), while the females may transfer between different communities. In other words, in marked contrast to the cercopithecoids, a chimpanzee male is bound to his community throughout his lifetime. Given the adaptive advantage of this multi-male group structure it is conceivable that intra-group conflicts could be best regulated in such a way that the group structure is least disrupted; the proximity of other males inevitably elicits aggression, that results in conflict.
   For a number of species, it has been shown that aggression elicited within a pair-bond becomes redirected outside that bond and that this results in the strengthening of that bond. The pattern of social relationships within a chimpanzee community obviously comprises elements that inhibit aggression between "in-group" males to a large extent, even though a considerable amount of tension is detectable within such groups (Bygott, 1974). It seems that part of the tension elicited within ape-groups finds expression in the form of the characteristic "bluff-displays". Bluff-behaviours are typically not spatially directed at a particular opponent and often have the appearance of being delayed, aggressive redirection on to some feature of the outside environment, usually some inanimate object. Such behaviours can be triggered by environmental cues (e.g. rain) or by social events (i.e. encounters). Bluff displays appear to have considerable social relevance, especially among chimpanzees, as the on-

lookers usually show an increase in contact-seeking activities with the per-
former(s) immediately after such displays. In showing off the performer's
physical prowess and vigorous perseverance, bluff displays undoubtedly play
an important role in the establishment of an individual's social status
(Bygott, 1974). It is conceivable that, when the opportunity offers, aggression elicited within the "in-group" may sometimes become redirected on to "out-
group" conspecifics (i.e. those of a neighbouring community of "strangers"),
instead of on to inanimate objects. In these cases heavy molesting or even
killing have been observed (described as "cannibalism" in chimpanzees e.g.
Bygott, 1974, and found in the institutionalised forms of "head-hunting" and
"raiding" in human societies), this suggesting that the inhibitions, normally
present between "in-group" members, do not operate to the same extent. Probably the degree of socialisation with the "out-group", as well as the numerical size of an "out-group" party, determines the degree of inhibition in such
agonistic encounters. It is likely that such molesting or killing yields the
positive feedback of the gaining of social status among the "in-group" members,
similar to bluff displays, and may even bear a higher symbolic value in this
respect.
     One could imagine that in cases where "out-group" conspecifics do not suit
the victim role, for instance because of such higher degree of socialisation,
kinship relations etc., those animal species that show strong resemblance to
one's own kind offer the most adequate alternative stimuli to release such
"outwards directed" aggression. Such species then serve, so to speak, as
"substitute conspecifics". For humans as well as for chimpanzees, the prey
that most effectively constitute "substitute conspecifics" are obviously other
primates.

CONCLUSION

     This paper expresses the view that one of the predominant functions of
hunting behaviour in anthropoids is to improve the social status of the performer by glamorous redirection of his aggression in a way that is harmless
for the "in-group". Being an acknowledged hunter may imply the possession of
courage, physical strength, technical skill, good timing, co-operative ability
etc.; all relevant incentives in the sexual selection of the species.
     Through hunting activity the group experiences a temporal enhancement of
cohesiveness, in particular the increase of contact-seeking activities among
the "in-group" members may have strong reinforcing properties facilitating
subsequent hunting actions. As suggested by Teleki (1975) and Suzuki (1975),
social rather than nutritional factors are the main causes for the occurrence
of hunting in certain hominids; within these, the key factors may well be
the multi-male condition and the particular social structure in which the male
is restricted to his community and its community-range. A consequence of this
view on hunting, reflected in the orang-utan remains in Asian caves, and backed
up by the prevalence of head-hunting and trophy veneration in many human cul-
tures, is that man apparently actively hunted his closest relatives, mainly
to strengthen his own "in-group" cohesion and to solve the problems arising
from that cohesiveness. Seen in this light, it seems miraculous that the extant apes still exist. Yet, their future is as grim as ever; they are still
mainly seen as "substitute conspecifics", by the "primitive" hunter with his
bow and arrow, as well as by the "advanced" scientist with his syringe and
scalpel.

ACKNOWLEDGEMENTS

My special thanks are due to Dr. Jane Goodall for her kind hospitality during my visit to the Gombe Stream National Park; I greatly appreciated the good fellowship and sympathy of Emilie van Zinnicq Bergmann, Anne Pusey and all others who introduced me into the world of the chimpanzee. For the many critical discussions on this, and related versions of the manuscript, I am especially indebted to Dr. J.A.R.A.M. van Hooff, Dr. I. Bossema, Prof. Dr. J.P. Kruyt, Dr. W.J. Netto and Dr. F.B.M. de Waal.

REFERENCES

Bygott, J.D. (1974). "Agonistic Behaviour and Dominance in Wild Chimpanzees", Ph.D. thesis, University of Cambridge.
Fossey, D. (1974). *Anim. Behav.*, <u>22</u>, 568-581.
Ling Roth, H. (1896). "The Natives of Sarawak and British North Borneo", 2 vols., Truslove and Hanson, London.
Nishida, T. (1968). *Primates*, <u>9</u>, 167-224.
Rijksen, H.D. (in press). Some aspects of the biology of the Sumatran Orang-utan: Ecology, Behaviour and Conservation, Meded. Landbouwhogeschool.
Suzuki, A. (1975). *In* "Socioecology and Psychology of Primates", (R.H. Tuttle, ed.), pp. 259-277, Mouton, The Hague and Paris.
Teleki, G. (1975). *J. Hum. Evol.*, <u>4</u>, 125-184.

THE ECOSYSTEMS IN WHICH THE INCIPIENT HOMININES COULD HAVE EVOLVED

ADRIAAN KORTLANDT

University of Amsterdam, The Netherlands.

For the purpose of the present paper the "incipient hominines" are defined as the evolutionary lineage that led from a dryopithecine (predominantly quadrupedal) to a hominine (predominantly bipedal) type of creature. *Australopithecus* and *Homo* are considered to constitute the subfamily *Homininae* in order to circumvent certain taxonomic problems. The following issues are submitted for consideration:

1) It is always the ecosystem as a whole which shapes the creatures that live in it. Any speculations about the hominizing processes in evolution should therefore begin with some basic knowledge of the ecosystem in which these processes took place.

2) Incipient hominines could never have branched off from the ancestral ape stock as long as these two groups enjoyed free interbreeding. This is especially true because both taxa tend (nowadays at least) to form rather unstable social units, to behave sexually rather promiscuously, and to range over long distances. Thus a geographical barrier is necessary to explain their bifurcation.

3) The most plausible barrier that could have caused the divergence of the incipient hominines from their relatives (the incipient chimpanzees and gorillas) is the African Western Rift Valley system, in combination with the Nile and Zambezi drainage systems. They probably constituted the only combination of arid and water barriers in Africa that could have effectively prevented regular crossing in Mio-Pliocene times. The essential points are that none of the *Hominoidea* can swim naturally, and that incipient men could not have carried food and water in desertic areas until they were predominantly bipedal. Thus the ecosystem in which the incipient hominines evolved was probably eastern Africa, possibly including Arabia, i.e. the floral and faunal province that was created by the rain shadow zone behind the rising Rift Mountains and the Ethiopian Plateau with its lava capping, which were due to tectonic processes during the Miocene. Subsequent tectonic rejuvenations and regional habitat diversity in eastern Africa could have produced the selection pressures that led to further hominization towards *Homo habilis* and *H. erectus*, whereas the heavier dentition of the robust forms suggests that their adaptation originated in the ecosystem of the Arabian peninsula (Kortlandt, 1972, 1974, 1975, 1976; additional data also in Flohn, 1964; Mohr, 1971). Conversely the incipient chimpanzee and the incipient gorilla probably evolved in the West African and the Central African ecosystem respectively, due to the separation caused by the Niger-Benue (and former Sanaga-Chad?) river systems, in combination with Plio-Pleistocene climatic changes. There may have been another branch of apes or man-apes south of the Congo-Kasai River system, but it probably could not have

survived those Plio-Pleistocene periods when the Kalahari sands extended northwards.

4)   The most plausible common ancestor of the hominines and our contemporary
African apes was the smallest and most "primitive" dryopithecine in Africa,
i.e. *Dryopithecus africanus* (formerly called *Proconsul africanus*). This suggestion fits in with Cope's Law, anatomical data, palaeogeographical considerations, and biochemical evidence on phyletic relationships. *D. nyanzae* and
D. major must have been blind alleys. The hominine lineage probably began
with predominantly broken-forest and gallery-forest dwellers, but eventually
adapted itself to make longer and longer excursions into the more and more
open habitats that resulted from climatic dessication and from "parklandiza-
tion" of the vegetation by the evolving "bulldozer" herbivores, particularly
the elephants (Kortlandt, 1972; compare also the anatomical data in Napier
and Davis, 1959, and other, more recent authors). Thus they could enter a
niche which, at that time, was not yet filled by the baboons. The vegetational architecture of gallery forests solves the controversy as to the "brachiating" versus "non-brachiating" origin of man, because it must have required
quite diversified patterns of arm and hand use.

5)   In open terrain, chimpanzees often walk bipedally: (a) to spot potential dangers, (b) to carry in their arms and hands large quantities of food over
short distances to a safe place, e.g. the forest edge, (c) to carry and wield
potential weapons against predators, and (d) in intimidation displays. Comparative data from other primates suggest that the order from (a) to (d) represents the correct sequence of evolution (Kortlandt & Kooij, 1963; Kortlandt,
1968). Furthermore, in some respects the foot bones of the dryopithecines seem
to have been better adapted to bipedal stance and gait than those of our contemporary African apes, at least according to Le Gros Clark and Leakey (1951).
Thus *regular* bipedal gait in the incipient hominines appears to have evolved,
in the beginning, to enable them to carry food and weapons while moving
through unsafe areas to safer places, and, in a later phase, to carry food,
weapons and drinking water (juicy fruit) while moving over greater distances
and through areas where these were not available.

6)   Prolonged drought is the major survival hazard in arid regions. The
advantageous efficiency of long-distance bipedal walking on long legs would
have enabled the more advanced forms of incipient hominines to walk to the
rainfall areas of local thunderstorms which they could have seen and heard at
distances as great as 120 kms (J.M. Savidge, pers. comm.) i.e. far beyond the
wandering ranges of their main food competitors. This would define the ecological niche of the hominines in the East African ecosystem: they would have
been able to survive when the more territorially oriented, competing primate
species died of thirst and starvation. Such an adaptation explains many peculiarities of man among the primates and among other mammals. Baboons in Ambo-
seli National Park in Kenya are known to have moved about 10 kms to areas of
local rainfall (D. Western, pers. comm.). Elephants in the driest parts of
Tsavo National Park in Kenya, on the contrary, are known to have moved over
40-60 kms within a few days, apparently in direct response to localized rainfall,
and in the course of several months their movements involved areas up to 120 kms
distant (Leuthold and Sale, 1973; D.L.W. Sheldrick, pers. comm.). It would be
interesting to gather similar data on those bipedalists par excellence, the
ostriches.[1])

---

[1] Actually the energy cost efficiency of bipedal walking is a much more controversial issue than is usually recognized among anthropologists (Taylor,

7) Almost all nonhuman primates are semi-carnivorous and predate on small vertebrates. Some species do so quite regularly, other species only incidentally. Apparent exceptions are the orang-utan, the gorilla, and possibly some specialized leaf-eaters. Chimpanzees and baboons, on the other hand, living in a zone that ranges from the Sudan to South Africa, east of approximately the 30th meridian (i.e. in the former australopithecine range), have been observed regularly to predate on fairly large-sized prey (Kortlandt and Kooij, 1963, and several later authors, with complementary data from both sides of approximately the 30th meridian). The incipient hominines in East Africa, therefore, presumably did the same and may eventually have partly specialized to survive in times of drought by killing the dying game and other vertebrates at the last waterholes. In such situations there would be no clear-cut boundary between predation and scavenging.

8) The most difficult question is how the incipient hominines could have coped with their predators and competitors (Kortlandt, 1972). They must have been too small, and their arms and hands must have been too weak, to use clubs as effectively as chimpanzees do (Kortlandt, et al., 1968). Were there enough trees at hand to flee into? Palynologists may perhaps be able to give a tentative answer. Or were the large carnivores in Africa at that time, notably the creodonts and sabretooths, more clumsy in their movements than their successors nowadays? Palaeomammalogists may consider the question. Could throwing sand and gravel into the eyes of a beast of prey have been effective? Field tests may give an answer. (Until recently, some captors of wild animals used shotguns loaded with pepper.) Or should we simply assume that the advantages of walking to rainfall areas compensated for the hazards? At any rate, the incipient hominines must have taken tremendous risks. The Olympic speed record on foot is 36 km.p.h., whereas chimpanzees without Olympic training can achieve 45 km.p.h. in rough terrain, and cheetahs can run 90, perhaps 100 km. p.h. Such small creatures as "Lucy" (AL 288) from the Awash area must rarely have been able to outrun their enemies. Only ostriches can do so, and incipient man was certainly not an ostrich. Most far-fetched of all, and yet possibly the correct solution: did bipedal gait and the use of rudimentary spears or other hand-borne weapons evolve concomitantly? Did this trend start with the carrying around of a barbed and prickly acacia branch, just as Neptune wielded his trident? Wild animals learn to avoid hedgehogs and porcupines. Present-day zoo keepers often use rough brooms effectively against, e.g., hyaenas. Were acacia branches also used as super whips? Even such small creatures as capuchin monkeys have been reported to use sticks for poking and striking agonistically at conspecifics and at a rattlesnake (Cooper and Harlow, 1961; Kortlandt and Kooij, 1963). The shoots of several species of acacia, particularly those which have both recurved thorns and straight spines, would be amongst the most deterring hand weapons one can imagine. At any rate the idea may be tested experimentally under field conditions. The result may explain why the human ape is naked: he did not need to flee under thorn bushes, as

---

et al., 1970, 1973). On the other hand, comparative physiological tests with short-legged animals in treadmills do not account for the obvious advantages of long-legged walking in grassy and low-shrub habitats. At any rate, the long distances currently walked by hunting and gathering peoples, measured in kms per kg of body weight and compared with similar figures for other ground-walking primate species, clearly indicate that their way of locomotion must be comparatively efficient.

other animals do. It all sounds rather fantastic, but we have to consider
these questions seriously. Incipient man, after all, has survived!

REFERENCES

Clark, W.E. Le Gros and Leakey, L.S.B. (1951). "The Miocene *Hominoidea* of East
    Africa", British Museum (Natural History), London.
Cooper, L.R. and Harlow, H.F. (1961). *Psychol. Rep.*, 8, 418.
Flohn, H. (1964). *Wurzb. geogr. Arb.*, 12, 21-37.
Kortlandt, A. (1968). *In* "Handgebrauch und Verständigung bei Affen und
    Frühmenschen", (B. Rensch, ed.), Huber, Bern and Stuttgart.
Kortlandt, A. (1972). "New perspectives on ape and human evolution", Stich-
    ting voor Psychobiologie, Amsterdam.
Kortlandt, A. (1974). *Curr. Anthrop.*, 15, 427-448.
Kortlandt, A. (1975). *Curr. Anthrop.*, 16, 644-651.
Kortlandt, A. (1976). *Neth. J. Zool.*, 26, 447-449.
Kortlandt, A. and Kooij, M. (1963). *Symp. Zool. Soc. Lond.*, 10, 61-88.
Kortlandt, A., Orshoven, J. van, Pfeijffers, R. and Zon, J.C.J. van. (1968).
    "Testing chimpanzees in the wild, Guinea 1966-67", (16 mm film), Stichting
    Film en Wetenschap, Utrecht.
Leuthold, W. and Sale, J.B. (1973). *E. Afr. Wildl. J.*, 11, 369-384.
Mohr, P.A. (1971). "The geology of Ethiopia" (reprinted), Haile Selassie I
    University Press, Addis Ababa.
Napier, J.R. and Davis, P.R. (1959). "The fore-limb skeleton and associated
    remains of *Proconsul africanus*", British Museum (Natural History), London.
Savidge, J.M. (pers. comm.).
Sheldrick, D.L.W. (pers. comm.).
Taylor, C.R., Schmidt-Nielsen, K., and Raab, J.L. (1970). *Am. J. Physiol.*,
    219, 1104-1107.
Taylor, C.R. and Rowntree, V.J. (1973). *Science, N.Y.*, 179, 186-187.
Teleki, G. (1973). "The predatory behavior of wild chimpanzees", Bucknell,
    Lewisburg.
Western, D. (pers. comm.).

PRIMATOLOGY, PALAEOANTHROPOLOGY AND RETICULATE EVOLUTION: CONCLUDING REMARKS

P.V. TOBIAS

University of the Witwatersrand, Johannesburg, South Africa

The place of the Hominidae within the Primates is sufficient justification, if indeed the case needs arguing, for the organizers of the Sixth Congress of the International Primatological Congress to have included a Symposium on Hominid Evolution in the programme. Indeed, the very nature of a number of the contributions in this Symposium shows how much the study of non-human Primates has to contribute to the understanding of hominid evolution.

For example, a cardinal problem, which some believe is at the core of early hominisation, is that of the upright posture and bipedal locomotion. The biomechanics and electromyography of bipedality form the subject matter of elegant analyses by H. Preuschoft, R.H. Tuttle, J.V. Basmajian, H. Ishida, T. Kimura, M. Okada and N. Yamazaki, while the possibly not directly related carriage of the head has been the object of M. Sakka's meticulous study of the comparative anatomy of the cervico-cephalic anatomical set in a variety of higher primates.

Other examples of the impact of general primate studies on our understanding of hominid evolution are furnished by J.E. Sirianni's use of *Macaca nemestrina* as a model for testing the relationship of deciduous to permanent dental diameters and by D.L. Cramer's and A.L. Zihlman's heuristic venture into the territory of sexual dimorphism through the bones and teeth of the pygmy chimpanzee.

The message is not, however, only morphological. Both the ecology and ethology of non-human primates are serving as keys to the kingdom of knowledge on the becoming of man. They find original and vivid expression in A. Kortlandt's visualizing of early hominid ecosystems and in H.D. Rijksen's inferences about hunting behaviour in hominids from observations gleaned from the social life of orang-utans.

Thus, in no fewer than nine out of sixteen papers in this Symposium, studies have been made primarily on monkeys and apes - with the object of lighting the road to human evolutionary insights. "Man's Poor Relations", as Earnest Albert Hooton was wont to call them, are thus helping palaeoanthropologists to resolve many of their most tantalizing problems. It would be wrong, though, to think of this union of primatology and palaeoanthropology as one-sided and uni-directional. Instead, it would be true to claim that these same unresolved queries about the nature of hominisation have posed numbers of the most searching enquiries and evoked some of the most seminal researches of primatologists. Many of the questions to which answers are being sought in studies on pongids, cercopithecoids, ceboids and even on prosimians today are the spin-off from human perplexity about the humanising of our remote ancestors. Problems such as bipedalism and head carriage, encephalization, sexual dimorphism, dental, manual and pedal hominisation, behavioural patterns and the milieu of the twilight of pre-man and the dawn of man - all these and more beside are providing one

508   Tobias

of the challenging incentives to primate studies. Truly, the interaction between the man-sciences and primate studies is a richly reciprocal one.
    Some of the stimulating questions set by palaeoanthropology to primatology
are those posed by the accelerating spate of early hominid discoveries from
east and south Africa, reviewed here by B.A. Wood and P.V. Tobias respectively,
and of still earlier forms from India outlined in this Symposium by K.N. Prasad.
The place of these fossils in time and space, in natural affinities and descent,
is still occupying a seemingly inordinate amount of energy and disputation.
Perhaps, though, in these early years and decades of discovery, it is well that
they should - because, until such questions as the age, the species and the lineage of the multitudes of fossils are settled, there can be little hope of advance in the more rarefied secondary areas of such research as evolutionary
rates and patterns, the relative roles of selection and random genetic drift.
    Yet, early as it may be in the tidying up of our fossil graveyards and mau-
solea, there is already enough of a pattern detectable for A. Bilsborough's in-
genious essay into mosaicism in hominid evolution and for B. Wood's confirmation, on up-to-date finds, of the proposal that our African fossil hominids
should be seen as falling into a three-stream pattern of phylogenetic relationships.
    The same kinds of problems are encountered everywhere in palaeontology -
from mammal-like reptiles to men! Nor do they vanish when true man, *Homo*,
appears on the scene. Human palaeontology has as great a need as ever of the
punctilious descriptions of even later hominid fossils, such as that offered
here by E. Trinkhaus on the Shanidar 5 Neandertal skeleton. Yet even at this
level the problem of phylogenetic systematics remains. So we find not only
A. Bilsborough, but also J. Jelinek and C.B. Stringer, grappling anew with the
complexities of *H. erectus* and *H. sapiens* and of Neandertal-Sapient relationships. They and others - including some of the newer fossil finds - are inching
their way towards an ever-closer alignment between the later *H. erectus* and
earlier *H. sapiens* remains.
    Maybe, with the raising of such problems at this Congress in England, so
soon after the death of one of its greatest biologist-sons, Julian Huxley, we
should do well to go back to Huxley's words on reticulate evolution in Man.
With his characteristically incisive analysis, he carefully distinguished between what he called the "convergent-divergent" type of reticulate evolution,
such as occurred in roses, brambles, willows and hawthorns, and the "recombi-
national" type found in man.
    "Here, a reticulate result has been achieved by quite other means. Instead
    of the initial crossing being between distinct species, and the divergent
    variability being due to segregation of whole chromosomes or genomes, the
    crossing appears to have taken place between well-marked geographical sub-
    species, and the divergent variability is thus due to ordinary gene recom-
    bination. So far as we know, no polyploidy and no formation of specially
    stable types has occurred, but the progressive increase of general varia-
    bility...
    "Man is the only organism to have exploited this method of evolution and
    variation to an extreme degree, so that a new dominant type in evolution
    has come to be represented by a single world-wide species instead of showing
    an adaptive radiation into many intersterile species. Doubtless this is due
    to his great tendency to individual, group, and mass migration of an irregu-
    lar nature, coupled with his mental adaptability which enables him to effect
    cross-mating quite readily in face of differences of colour, appearance, and
    behaviour which would act as efficient barriers in the case of more instinct-
    ive organisms."   (Huxley, 1963: pages 353-354)

If this reticulate result is evident in recent man, is it not ineluctable that, with the special hominid qualities manifest now for so many hundreds of thousands, if not a few millions, of years, reticulate evolution has been a feature of hominid evolution from the very beginning of the Quaternary? I have earlier found reason to emphasize the reticulate nature of man's evolution during the *Middle* Pleistocene. The newer finds are forcing one, more and more irresistibly, away from 3-stream and 2-stream models, to a concept of reticulate evolution of man in the *Lower* Pleistocene, and perhaps even earlier. It is of course taxonomically untidy and awkward to embrace this concept and, as Julian Huxley wrote, "There is a natural reluctance among systematists to recognize its existence and its implications, since these run counter to the generally-accepted basis of taxonomic practice" (op.cit., p. 356). Nonetheless, if the fossil facts point towards reticulate evolution, the taxonomic preconceptions and procedures should be no valid deterrent to upholding the concept!

Perhaps in this direction we may hope to find a way through the phylogenetic and systematic enigmas that have been so ably ventilated at the Cambridge Congress.

REFERENCE

Huxley, J. (1963). "Evolution: the Modern Synthesis", 2nd ed. with a new introduction, George Allen & Unwin, London.